This new series, Advances in Biochemistry in Heath and Disease, will focus on developments in biochemical research with implications for health and disease. The series will consist of original edited volumes, monographs, references, proceedings. The series will publish a minimum of three volumes a year.

More information about this series at http://www.springer.com/series/7064

Ian M.C. Dixon • Jeffrey T. Wigle
Editors

Cardiac Fibrosis and Heart Failure: Cause or Effect?

 Springer

Editors
Ian M.C. Dixon
Institute of Cardiovascular Sciences
St. Boniface Hospital Research Centre
Department of Physiology and
  Pathophysiology
University of Manitoba
Winnipeg
Canada

Jeffrey T. Wigle
Institute of Cardiovascular Sciences
St. Boniface Hospital Research Centre
Department of Biochemistry and Medical
  Genetics
University of Manitoba
Winnipeg
Canada

Advances in Biochemistry in Health and Disease
ISBN 978-3-319-17436-5   ISBN 978-3-319-17437-2 (eBook)
DOI 10.1007/978-3-319-17437-2

Library of Congress Control Number: 2015941510

Springer Cham Heidelberg New York Dordrecht London

Printed on acid-free paper

Springer is a brand of Springer International Publishing
Springer International Publishing is part of Springer Science+Business Media (www.springer.com)

Preface

Cardiac Fibrosis and Heart Failure: Cause or Effect?

Cardiac fibrosis is the abnormal expansion of the cardiac extracellular matrix (ECM) due to excessive ECM protein deposition, which occurs in most types of heart disease. It is widespread and is found in cardiac diseases including post-MI heart failure, hypertension and diabetic cardiomyopathy. Remodeling of the cardiac extracellular matrix has become a well-known modifier of cardiac performance and on-going or chronic wound healing is closely tied to heart failure. The cardiac ECM includes proteins that occupy the space between cells in the heart. Contractile myofibroblasts are those cells that not only express α-smooth muscle actin, but more importantly also exhibit formation of stress fibres, are the major players in the secretion and physical remodeling of matrix. Myofibroblasts facilitate connections from cell surface adhesions to the matrix itself, and progressive remodeling of the matrix and cellular connections contributes to heart failure. Fibroblasts (defined as fibroblastic cells without stress fibres) and myofibroblasts contribute to the normal maintenance of matrix, wound healing, and to the pathology of cardiac fibrosis. While limited wound healing is necessarily "a good thing", the situation deteriorates in runaway wound healing or fibrosis, which is marked by progressive deposition of matrix proteins long after the acute wound healing phase is completed. In the early 1990s, and despite a widespread awareness of cardiac fibrosis, our interpretation of its role in heart disease had stalled at the concept of cardiac fibrosis as a being secondary result of disease, and not as a primary contributor to the development of heart failure. Since then, major advances in our understanding have allowed us to now identify cardiac fibrosis as being a primary and causal driver of disease. This may be thought of as independent of cardiac muscle cell injury. Thus a comprehensive treatise on this topic is well warranted.

Jeffrey T. Wigle, Ian M.C. Dixon
Winnipeg, Canada

Contents

Contributors

Mark C. Blaser Institute of Biomaterials and Biomedical Engineering, University of Toronto, Toronto, ON, Canada

Cardiovascular Sciences Collaborative Program, University of Toronto, Toronto, Canada

L. Chilton College of Public Health, Medical & Veterinary Science, James Cook University, Townsville, Australia

Nuno M. Coelho Matrix Dynamics Group, University of Toronto, Toronto, ON, Canada

Ryan H. Cunnington Department of Medicine, University of Calgary, Calgary, AB, Canada

Michael P. Czubryt Department of Physiology and Pathophysiology, Institute of Cardiovascular Sciences, University of Manitoba, R4008, St. Boniface Hospital Research Centre, Winnipeg, MB, Canada

Keith Dadson Department of Biology, York University, Toronto, ON, Canada

Ashley DeCoux San Antonio Cardiovascular Proteomics Center, San Antonio, USA

Mississippi Center for Heart Research, Department of Physiology and Biophysics, University of Mississippi Medical Center, Jackson, MS, USA

Ian M. C. Dixon College of Medicine, St. Boniface Hospital Research Centre, University of Manitoba, Winnipeg, MB, Canada

Department of Physiology and Pathophysiology, University of Manitoba, Winnipeg, MB, Canada

Institute of Cardiovascular Sciences, St. Boniface Hospital Research Centre, Winnipeg, MB, Canada

Dong Fan Department of Physiology Cardiovascular Research Center, Mazankowski Alberta Heart Institute, University of Alberta, Edmonton, AB, Canada

Paul W. M. Fedak Section of Cardiac Surgery, Department of Cardiac Sciences, Foothills Medical Centre, Calgary, AB, Canada

Nikolaos G. Frangogiannis Department of Medicine, The Wilf Family Cardiovascular Research Institute, Albert Einstein College of Medicine, Bronx, NY, USA

Darren H. Freed Department of Surgery, Physiology and Biomedical Engineering, University of Alberta, Edmonton, AB, Canada

W. R. Giles Faculty of Kinesiology, University of Calgary, Calgary, AB, Canada

Ramareddy V. Guntaka Department of Microbiology, Immunology & Biochemistry, University of Tennessee Health Science Center, Memphis, TN, USA

Meghana R.K. Helder Divisions of Cardiovascular Surgery and Diseases, Mayo Clinic, Rochester, MN, USA

Boris Hinz Laboratory of Tissue Repair and Regeneration, Matrix Dynamics Group, Faculty of Dentistry, University of Toronto, Toronto, ON, Canada

Fahmida Jahan Department of Biochemistry and Medical Genetics, University of Manitoba, Winnipeg, MB, Canada

Institute of Cardiovascular Sciences, St. Boniface Hospital Research Centre, Winnipeg, MB, Canada

Hailey J. Jansen Department of Physiology and Biophysics and School of Biomedical Engineering, Faculty of Medicine, Dalhousie University, Halifax, NS, Canada

Yu-Fan Jin San Antonio Cardiovascular Proteomics Center, San Antonio, USA

Department of Electrical and Computer Engineering, The University of Texas at San Antonio, San Antonio, TX, USA

Zamaneh Kassiri Department of Physiology Cardiovascular Research Center, Mazankowski Alberta Heart Institute, University of Alberta, Edmonton, AB, Canada

Vera Kovacevic Department of Biology, York University, Toronto, ON, Canada

Natalie M. Landry Department of Physiology and Pathophysiology, St. Boniface Research Centre, Institute of Cardiovascular Sciences, University of Manitoba, Winnipeg, MB, Canada

Merry L. Lindsey San Antonio Cardiovascular Proteomics Center, San Antonio, USA

Mississippi Center for Heart Research, Department of Physiology and Biophysics, University of Mississippi Medical Center, Jackson, MS, USA

Research Service, G.V. (Sonny) Montgomery Veterans Affairs Medical Center, Jackson, MS, USA

K. A. MacCannell Worldwide Research and Development Pfizer Inc., Cambridge, MA, USA

Christopher A. McCulloch Matrix Dynamics Group, University of Toronto, Toronto, ON, Canada

R. Justin McCullough Division of Cardiovascular Diseases, University of Tennessee Health Science Center, Memphis, TN, USA

Dr. John C. McDermott Department of Biology, York University, Toronto, ON, Canada

Holly E. M. Mewhort Section of Cardiac Surgery, Department of Cardiac Sciences, Foothills Medical Centre, Calgary, AB, Canada

Thomas Moore-Morris INSERM UMR_S910, Team physiopathology of cardiac development, Medical School La Timone, Aix-Marseille University, Marseille, Cedex 05, France

Alison L. Müller Department of Physiology, University of Alberta, Edmonton, Canada

Kevin P. Newman Division of Cardiovascular Diseases, University of Tennessee Health Science Center, Memphis, TN, USA

Gavin Y. Oudit Division of Cardiology, Department of Medicine, Mazankowski Alberta Heart Institute, University of Alberta, Edmonton, AB, Canada

Christina Pagiatakis Department of Biology, York University, Toronto, ON, Canada

Nirmal Parajuli Division of Cardiology, Department of Medicine, Mazankowski Alberta Heart Institute, University of Alberta, Edmonton, AB, Canada

Vaibhav B. Patel Division of Cardiology, Department of Medicine, Mazankowski Alberta Heart Institute, University of Alberta, Edmonton, AB, Canada

Michel Pucéat INSERM UMR_S910, Team physiopathology of cardiac development, Medical School La Timone, Aix-Marseille University, Marseille, Cedex 05, France

Kodangudi B. Ramanathan Division of Cardiovascular Diseases, University of Tennessee Health Science Center, Memphis, TN, USA

Tharmarajan Ramprasath Division of Cardiology, Department of Medicine, Mazankowski Alberta Heart Institute, University of Alberta, Edmonton, AB, Canada

Sunil G. Rattan Institute of Cardiovascular Sciences, St. Boniface Hospital Research Centre, Winnipeg, MB, Canada

Patricia L. Roche Department of Physiology and Pathophysiology, Institute of Cardiovascular Sciences, University of Manitoba, R4008, St. Boniface Hospital Research Centre, Winnipeg, MB, Canada

Robert A. Rose Department of Physiology and Biophysics and School of Biomedical Engineering, Faculty of Medicine, Dalhousie University, Halifax, NS, Canada

Amit Saxena Department of Medicine, The Wilf Family Cardiovascular Research Institute, Albert Einstein College of Medicine, Bronx, NY, USA

Mengcheng Shen Department of Physiology Cardiovascular Research Center, Mazankowski Alberta Heart Institute, University of Alberta, Edmonton, AB, Canada

Robert D. Simari Executive Dean School of Medicine, University of Kansas School of Medicine, Kansas City, KS, USA

Craig A. Simmons Institute of Biomaterials and Biomedical Engineering, University of Toronto, Toronto, ON, Canada

Cardiovascular Sciences Collaborative Program, University of Toronto, Toronto, Canada

Department of Mechanical and Industrial Engineering, University of Toronto, Toronto, Canada

G. L. Smith Institute of Biomedical and Life Sciences, University of Glasgow, Glasgow, UK

Yao Sun Division of Cardiovascular Diseases, University of Tennessee Health Science Center, Memphis, TN, USA

Gary Sweeney Department of Biology, York University, Toronto, ON, Canada

Abhijit Takawale Department of Physiology Cardiovascular Research Center, Mazankowski Alberta Heart Institute, University of Alberta, Edmonton, AB, Canada

Yuan Tian San Antonio Cardiovascular Proteomics Center, San Antonio, USA

Mississippi Center for Heart Research, Department of Physiology and Biophysics, University of Mississippi Medical Center, Jackson, MS, USA

Dr. Robert. T. Tranquillo Department of Biomedical Engineering, University of Minnesota, Minneapolis, MN, USA

Karl T. Weber Division of Cardiovascular Diseases, University of Tennessee Health Science Center, Memphis, TN, USA

Jacqueline S. Wendel Department of Biomedical Engineering, University of Minnesota, Minneapolis, MN, USA

Jared A. White Mississippi Center for Heart Research, Department of Physiology and Biophysics, University of Mississippi Medical Center, Jackson, MS, USA

Jeffrey T. Wigle Institute of Cardiovascular Sciences, St. Boniface Hospital Research Centre, Winnipeg, MB, Canada

Department of Biochemistry and Medical Genetics, University of Manitoba, Winnipeg, MB, Canada

Matthew R. Zeglinski Department of Physiology and Pathophysiology, St. Boniface Research Centre, Institute of Cardiovascular Sciences, University of Manitoba, Winnipeg, MB, Canada

Pavel Zhabyeyev Division of Cardiology, Department of Medicine, Mazankowski Alberta Heart Institute, University of Alberta, Edmonton, AB, Canada

Elena Zimina Laboratory of Tissue Repair and Regeneration, Matrix Dynamics Group, Faculty of Dentistry, University of Toronto, Toronto, ON, Canada

Fouad A. Zouein San Antonio Cardiovascular Proteomics Center, San Antonio, USA

Mississippi Center for Heart Research, Department of Physiology and Biophysics, University of Mississippi Medical Center, Jackson, MS, USA

About the Editors

Dr. Jeffrey Wigle is an Associate Professor at the University of Manitoba, Winnipeg, Canada. His research focuses on the transcriptional pathways involved in regulating the growth and development of the cardiovascular system. He has characterized a number of transcription factors that are critical for the development of the lymphatic vasculature and the phenoconversion of fibroblasts to myofibroblasts. He is currently studying the transcriptional feedback loops that determine and maintain the fibroblastic phenotype.

Dr. Ian M.C. Dixon is a Professor of Physiology and Pathophysiology at the University of Manitoba, Winnipeg, Canada. His research addresses heart failure following myocardial infarction and the contributing role of dysfunctional extracellular matrix in these hearts. He is also interested in mechanisms of fibroblast phenotype regulation, cardiac fibroblast Smad protein signalling as well as other signalling proteins that may oppose the fibrogenic influence of TGF-β_1 in health and disease. He has been engaged over the past 24 years in multidisciplinary research in cardiovascular matrix biology.

Cardiac Fibrosis and Heart Failure—Cause or Effect?

Primary Contribution of Cardiac Fibrosis to Heart Failure

Ian M.C. Dixon, Ryan H. Cunnington, Sunil G. Rattan and Jeffrey T. Wigle

Abstract Cardiac fibrosis is the pathological accumulation of cardiac extracellular matrix (ECM or matrix), which occurs in most types of heart disease. Major recent advances in our understanding have allowed us to identify cardiac fibrosis as a primary disease independent of either cardiomyocyte injury or loss. New developments within this field are burgeoning, including research that points to multiple sources for cardiac myofibroblasts participating in cardiovascular disease pathogenesis, the feasibility of bioengineered matrix tissues as well as the identification of novel targets to reduce the incidence and severity of cardiac fibrosis. A summary of the state of knowledge of the regulation of the function of fibroblasts as well as a synopsis of the current state of investigation to address the biology of cardiovascular fibroblasts, valvular interstitial cells (VICs), and myofibroblasts is warranted. This book will help to adapt the information that we have gathered in order to translate it into treatments for fibrotic cardiac diseases and thus alter the course of their progression.

Keywords Cardiac fibrosis · Heart disease · Fibroblast · Myofibroblast

I. M. C. Dixon (✉)
Department of Physiology and Pathophysiology, University of Manitoba, Winnipeg, MB R2H 2A6, Canada
e-mail: idixon@sbrc.ca

I. M. C. Dixon · S. G. Rattan · J. T. Wigle
Institute of Cardiovascular Sciences, St. Boniface Hospital Research Centre, Winnipeg, MB, Canada
e-mail: idixon@sbrc.ca

R. H. Cunnington
Department of Medicine, University of Calgary, Calgary, AB, Canada

J. T. Wigle
Department of Biochemistry and Medical Genetics, University of Manitoba, Winnipeg, MB, Canada

© Springer International Publishing Switzerland 2015
I.M.C. Dixon, J. T. Wigle (eds.), *Cardiac Fibrosis and Heart Failure: Cause or Effect?*,
Advances in Biochemistry in Health and Disease 13, DOI 10.1007/978-3-319-17437-2_1

Cardiac fibrosis is the pathological accumulation of cardiac extracellular matrix (ECM or matrix). It is generally held that fibrosis occurs in most types of heart disease wherein cardiomyocytes are lost to necrotic cell death and is the end-point of a chronic or excessive wound healing process [1]. While acute wound healing is both required and necessary for the maintenance of the interconnectivity of the muscular parenchymal syncytium of cardiomyocytes, cardiac fibrosis is by definition the pathological manifestation of the abnormal expansion of the cardiac interstitium or non-parenchymal tissues with excessive matrix component protein deposition. Remodeling of the cardiac extracellular matrix has become a well-known modifier of cardiac performance and on-going or chronic wound healing is closely tied to heart failure [2]. The traditional functional description of the cardiac matrix is limited to the tethering of myocytes to effect the efficient transfer of contractile forces and to translate those forces into cardiac pump function. However the cardiac matrix itself is diverse and dynamic as it contains glycosaminoglycans, glycoproteins, matrikines, structural proteins, both active and inactive cytokines and fibrogenic growth factors including TGF-β_1 (which may be bound to docking proteins and therefore "held in reserve") and also serves as the scaffold for the most numerous cell type in the heart—cardiac fibroblasts [3–5]. The current usage of the term "fibroblast" may be a relative oversimplification insofar as it groups together these mesenchymally derived cells from heart atria and ventricles, as well as lung, kidney, skin etcetera, despite growing evidence that these cells differ topographically within organs including the heart [6] and between different organs [7]. We and others have previously shown that relatively quiescent atrial and ventricular-derived cells rapidly phenoconvert *in vitro* and in vivo to become contractile, hypersecretory myofibroblasts [8, 9]. These myofibroblast cells are the major players in the secretion and physical remodeling of the connections from cell surface adhesions to the matrix; this remodeled ECM contributes to heart failure (see Fig. 1).

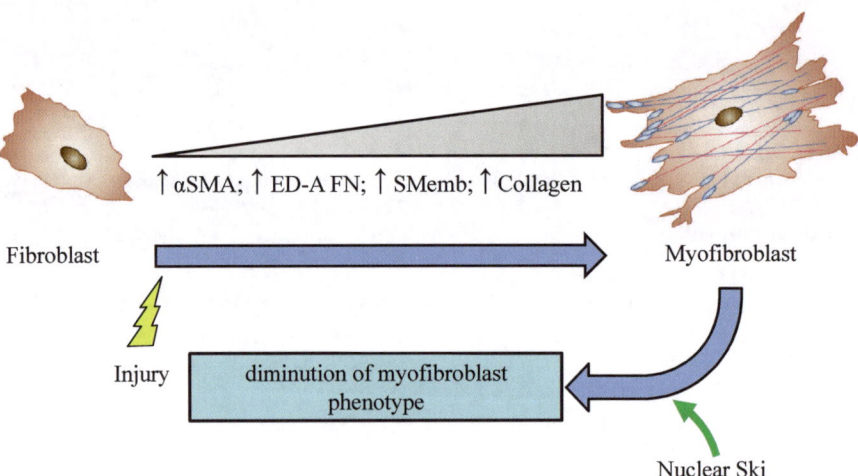

Fig. 1 Phenoconversion of fibroblasts to myofibroblasts. Activation of Ski in the nucleus of myofibroblasts is associated with the diminution of the myofibroblast phenotype

The unique biology of cardiac fibroblasts, and especially cardiac myofibroblasts [10], distinguishes them from other fibroblastic cells and this unique biology is only beginning to be appreciated in the context of their contribution to heart failure. The natural signals that stimulate and inhibit the process of cardiac fibrosis mediated by these cells are not well understood. In the early 1990s, despite a widespread awareness of cardiac fibrosis occurring in various etiologies of heart failure, our interpretation of its role in heart disease had stalled at the concept of cardiac fibrosis as being a secondary result of progressive heart failure [11, 12] Since then, major advances in our understanding have allowed us to identify cardiac fibrosis as a primary disease, which can be *independent* of either cardiomyocyte injury or loss [13]. While other comprehensive treatises on this topic have been forthcoming [1], the suggestion that cardiac fibrosis may be a primary contributor to heart failure has only recently gained ground. For these reasons, a summary of the state of knowledge of the regulation of the function of fibroblasts, including the synthesis and secretion of ECM and focal adhesion proteins, as well as a synopsis of the current state of investigation to address the biology of cardiovascular fibroblasts, valvular interstitial cells (VICs), and myofibroblasts is warranted. New developments within this field are burgeoning, including research that points to diverse sources for cardiac myofibroblasts participating in cardiovascular disease pathogenesis, the feasibility of bioengineered matrix tissues as well as the identification of novel targets to reduce the incidence and severity of cardiac fibrosis. Among the latter are recent studies to investigate endogenous inhibitors of cardiac fibrosis, including proteins such as the Ski/Sno superfamily [9, 14]. For example, we have recently discovered that Ski may strongly influence myofibroblast phenoconversion, possibly via the regulation of a novel Smad interacting protein, Zeb2 [9] and induction of apoptotic death (unpublished results), which may then influence the pathogenesis of cardiac fibrosis.

The purpose of this compendium of work is to bring together the latest findings in the investigation into matrix dysfunction in cardiovascular disease. The current book will address the molecular mechanisms that control the synthesis and secretion of the cardiac ECM. We will highlight work which sheds light on the pathogenesis of cardiovascular diseases including topics as diverse as atrial fibrillation and fibrosis, putative pools or sources for myofibroblasts, cardiac aging, endogenous inhibitors of fibrosis, autophagy and ER stress in fibrosis, and discussions to unify mechanisms of matrix remodeling in valves, atria and ventricles. As the mechanisms that underpin the stimulation of cardiac fibrosis are not fully understood, it is not surprising that no specific agents currently exist in the clinical armament to alleviate this pathology by acting specifically on cardiovascular fibroblasts and myofibroblasts. Ultimately, we hope that this book will help to adapt the information that we have gathered in order to translate it into treatments for fibrotic cardiac diseases; this will allow us to alter the course of the progression of these diseases and thus lessen their impact not only on patients with cardiovascular disease, but also on society at large.

References

1. Weber KT, Sun Y, Bhattacharya SK, Ahokas RA, Gerling IC (2013) Myofibroblast-mediated mechanisms of pathological remodelling of the heart. Nat Rev Cardiol 10(1):15–26
2. Ferreira-Martins J, Leite-Moreira AF (2010) Physiologic basis and pathophysiologic implications of the diastolic properties of the cardiac muscle. J Biomed Biotechnol 2010:807084
3. Chen W, Frangogiannis NG (2013) Fibroblasts in post-infarction inflammation and cardiac repair. Biochim Biophys Acta 1833(4):945–953
4. Hinz B (2013) It has to be the alphav: myofibroblast integrins activate latent TGF-beta1. Nat Med 19(12):1567–1568
5. Ma Y, Halade GV, Lindsey ML (2012) Extracellular matrix and fibroblast communication following myocardial infarction. J Cardiovasc Transl Res 5(6):848–857
6. Burstein B, Libby E, Calderone A, Nattel S (2008) Differential behaviors of atrial versus ventricular fibroblasts: a potential role for platelet-derived growth factor in atrial-ventricular remodeling differences. Circulation 117(13):1630–1641
7. Chang HY, Chi JT, Dudoit S, Bondre C, van de Rijn M, Botstein D, Brown PO (2002) Diversity, topographic differentiation, and positional memory in human fibroblasts. Proc Natl Acad Sci U S A 99(20):12877–12882
8. Cunnington RH, Wang B, Ghavami S, Bathe KL, Rattan SG, Dixon IM (2011) Antifibrotic properties of c-Ski and its regulation of cardiac myofibroblast phenotype and contractility. Am J Physiol Cell Physiol 300(1):C176–186
9. Cunnington RH, Northcott JM, Ghavami S, Filomeno KL, Jahan F, Kavosh MS, Davies JJ, Wigle JT, Dixon IM (2014) The Ski-Zeb2-Meox2 pathway provides a novel mechanism for regulation of the cardiac myofibroblast phenotype. J Cell Sci 127(Pt 1):40–49
10. Santiago JJ, Dangerfield AL, Rattan SG, Bathe KL, Cunnington RH, Raizman JE, Bedosky KM, Freed DH, Kardami E, Dixon IM (2010) Cardiac fibroblast to myofibroblast differentiation *in vivo* and *in vitro*: expression of focal adhesion components in neonatal and adult rat ventricular myofibroblasts. Dev Dyn 239 (6):1573–1584
11. Ju H, Dixon IM (1996) Extracellular matrix and cardiovascular diseases. Can J Cardiol 12(12):1259–1267
12. Weber KT, Sun Y, Campbell SE (1995) Structural remodelling of the heart by fibrous tissue: role of circulating hormones and locally produced peptides. Eur Heart J 16(Suppl N):12–18
13. Thum T, Gross C, Fiedler J, Fischer T, Kissler S, Bussen M, Galuppo P, Just S, Rottbauer W, Frantz S, Castoldi M, Soutschek J, Koteliansky V, Rosenwald A, Basson MA, Licht JD, Pena JT, Rouhanifard SH, Muckenthaler MU, Tuschl T, Martin GR, Bauersachs J, Engelhardt S (2008) MicroRNA-21 contributes to myocardial disease by stimulating MAP kinase signalling in fibroblasts. Nature 456(7224):980–984
14. Li J, Zhao L, He X, Yang T, Yang K (2014) MiR-21 inhibits c-Ski signaling to promote the proliferation of rat vascular smooth muscle cells. Cell Signal 26(4):724–729

Fibroblast Activation in the Infarcted Myocardium

Amit Saxena and Nikolaos G. Frangogiannis

Abstract The adult mammalian heart contains abundant fibroblasts. Cardiac fibro-blasts are versatile and dynamic cells that not only produce extracellular matrix proteins, but may also serve important functions in myocardial inflammation, angio-genesis and repair. Following injury, cardiac fibroblasts may maintain the integrity of the extracellular matrix network preserving cardiac geometry and function. Myo-cardial infarction induces dynamic alterations in fibroblast phenotype. During the early stages of infarct healing, cardiac fibroblasts may serve as sentinel cells that sense signals released by dying cardiomyocytes and activate the inflammasome, secreting cytokines and chemokines. During the inflammatory phase, fibroblasts exhibit a matrix-degrading phenotype; myofibroblast conversion may be delayed by activation of Interleukin-1 (IL-1) signaling. Suppression of pro-inflammatory signals and termination of IL-1-driven cascades during the proliferative phase of infarct healing may allow unopposed actions of Transforming Growth Factor (TGF)-β on cardiac fibroblasts, mediating myofibroblast transdifferentiation, matrix synthesis and scar contraction. Angiotensin II, the mast cell proteases chymase and tryptase, growth factors and specialized matrix proteins may co-operate to promote a synthetic and proliferative myofibroblast phenotype. The maturation phase fol-lows, as infarct myofibroblasts cross-link the surrounding matrix, become quiescent and may undergo apoptosis. However, in the non-infarcted remodeling myocar-dium, fibroblasts may remain activated in response to volume and pressure overload promoting interstitial fibrosis. This chapter discusses the role of cardiac fibroblasts in infarct healing and the mechanisms of their activation, suggesting potential thera-peutic targets aimed at attenuating adverse post-infarction remodeling.

Keywords Myofibroblast · Infarction · Proto-myofibroblast · Transdifferentiation · Ischemia

N. G. Frangogiannis (✉) · A. Saxena
Department of Medicine, The Wilf Family Cardiovascular Research Institute, Albert Einstein College of Medicine, 1300 Morris Park Avenue, Forchheimer G46B, Bronx, NY 10461, USA
e-mail: nikolaos.frangogiannis@einstein.yu.edu

© Springer International Publishing Switzerland 2015
I.M.C. Dixon, J. T. Wigle (eds.), *Cardiac Fibrosis and Heart Failure: Cause or Effect?*,
Advances in Biochemistry in Health and Disease 13, DOI 10.1007/978-3-319-17437-2_2

1   Introduction

Fibroblasts are the most abundant interstitial cells in the adult mammalian myocardium [1, 2]. There are no reliable and specific fibroblast markers, thus identification of fibroblasts in mammalian tissues is based on morphological and functional criteria. Traditionally, cardiac fibroblasts are described as elongated, spindle shaped cells that lack a basement membrane and are responsible for maintaining the integrity of the cardiac matrix network. It is increasingly recognized that fibroblasts do not only serve as matrix-producing cells, but also play important roles in regulating inflammatory, immune, reparative and angiogenic responses [3–6]. Most cardiac pathophysiologic insults activate cardiac fibroblasts. The specific fibroblast responses triggered by various injurious stimuli are major determinants of the functional and geometric consequences of the pathophysiologic condition. For example, pressure overload and volume overload may activate distinct molecular programs in cardiac fibroblasts, leading to different functional and morphologic abnormalities. The infarcted heart provides a unique opportunity to study the dynamic changes of cardiac fibroblasts and to appreciate their functional diversity in response to microenvironmental changes [7]. Because the adult mammalian myocardium has negligible endogenous regenerative capacity, loss of a large number of cardiomyocytes leads to their replacement with a collagen-based scar.

Repair of the infarcted heart is dependent on a superbly orchestrated response that can be divided into three distinct, but overlapping phases: the inflammatory, the proliferative and the maturation phase [8, 9]. In the dynamic microenvironment of the infarct, cardiac fibroblasts undergo dramatic phenotypic alterations and may serve a wide range of functions. This chapter reviews evidence on the role of cardiac fibroblasts in the phases of infarct healing and on their potential involvement in activation and suppression of the inflammatory reaction, in matrix metabolism, and in formation of a mature collagen-based scar. We will also discuss the key molecular signals directing the phenotypic transitions of fibroblasts during the phases of cardiac repair and we will attempt to identify specific therapeutic targets to prevent adverse remodeling following myocardial infarction.

2   Cardiac Fibroblasts in Normal Mammalian Hearts

The adult mammalian heart contains abundant fibroblasts (Fig. 1). Although it is often stated that cardiac fibroblasts may outnumber cardiomyocytes [10] robust documentation of this claim in human hearts is lacking. Moreover, the number of fibroblasts measured in experimental studies likely depends on the specific methodology and the markers used for fibroblast identification. Species, strain and age may also affect the relatively density of cardiac interstitial fibroblasts. Regardless of these uncertainties, it is generally accepted that fibroblasts are the predominant interstitial cells in mammalian hearts. Fibroblast studies have been hampered by the absence of a fibroblast-specific surface marker. Vimentin, a cytoskeletal protein, has been

Fig. 1 Fibroblasts harvested from mouse hearts and cultured in a collagen pad. In free floating gel pads, many cardiac fibroblasts exhibit a dendritic morphology (*arrow*), while others have a more rounded shape (*arrowhead*). Stained with Sirius red, counterstained with hematoxylin

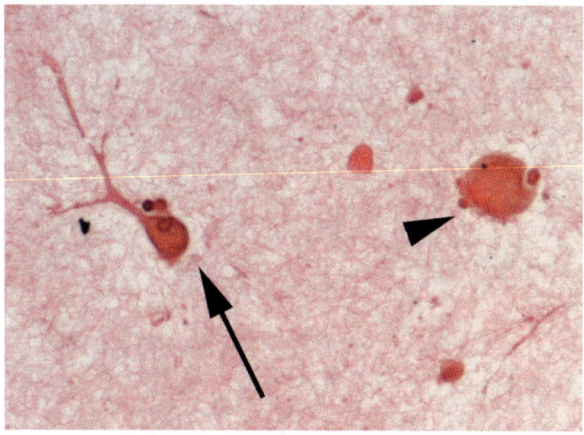

identified as an intermediate filament protein expressed by most cells of mesenchymal origin and has been extensively used to identify fibroblasts in normal and infarcted myocardium [11, 12]. Although useful in labeling fibroblasts in tissues, vimentin lacks specificity. Despite its name, fibroblast-specific protein (FSP)-1 is also known as S100A4, a calcium-binding protein which is expressed by activated macrophages, lymphocytes and endothelial cells [13, 14]. Other markers of fibroblasts (such as α-smooth muscle actin (α-SMA) and periostin) appear to label only activated cells upon acquisition of a matrix-synthetic myofibroblast phenotype. The basic helix-loop-helix (bHLH) transcription factor TCF21 has been proposed as a fibroblast-specific protein [15, 16]; however, its ability to specifically label cardiac fibroblasts in normal and injured hearts has not been systematically tested.

Little is known regarding the role of fibroblasts in cardiac homeostasis. Embryonic fibroblasts may promote cardiomyocyte proliferation through interactions involving β1 integrin signaling [17]. A growing body of evidence suggests that fibroblasts and cardiomyocytes may contribute to cardiac homeostasis by intracellular communications known as connexins, which contribute to normal electrical and mechanical function of the heart [18].

3   The Phases of Cardiac Repair

From the descriptive viewpoint, the reparative response following myocardial infarction can be divided into three overlapping but distinct phases: the inflammatory phase, the proliferative phase and the maturation phase. Each phase is characterized by recruitment/activation of specific cellular populations. Death of large numbers of cardiomyocytes in the infarcted myocardium results in the release of danger signals, triggering the activation of an inflammatory cascade. Neutrophils and inflammatory monocytes infiltrate the infarct and clear the wound from dead cells and matrix debris. Resolution of the inflammatory reaction is associated with activation and proliferation of mesenchymal cells; at this stage activated fibroblasts become the dominant cell type in the infarct and deposit large amounts of structural and matricellular matrix proteins. The maturation phase follows, as collagen is cross-linked and the cellular elements of the scar undergo apoptosis. Each phase of cardiac repair is associated with distinct phenotypic alterations of the fibroblasts.

4   Cardiac Fibroblasts During the Inflammatory Phase

4.1   Cardiac Fibroblasts as Inflammatory Cells Following an Ischemic Insult

Although robust evidence on the fate of resident cardiac fibroblasts following myocardial infarction is lacking, *in vitro* experiments suggest that fibroblasts may be more resistant to ischemic insults than cardiomyocytes [19]. Due to their relative resistance to ischemic death, their strategic location in the cardiac interstitium and potential interactions with other myocardial cells, fibroblasts may function as sentinel cells that sense injury and trigger an inflammatory reaction (Fig. 2). Because other cell types (including vascular cells, mast cells and macrophages) are also capable of releasing pro-inflammatory mediators, the relative role of fibroblasts as inflamma-

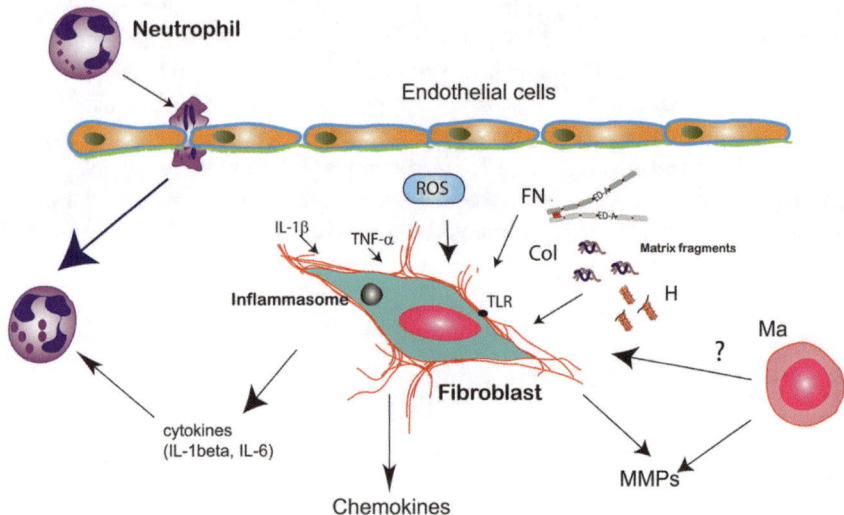

Fig. 2 The "pro-inflammatory" fibroblast. During the inflammatory phase of infarct healing, cardiac fibroblasts activate the inflammasome and may secrete matrix metalloproteinases (*MMPs*), cytokines and chemokines. Reactive oxygen species (*ROS*), activation of Toll-like receptor (*TLR*) pathways by matrix fragments (FN/fibronectin, Col/collagen, H/hyaluronan) and danger signals, and pro-inflammatory cytokines (such as Interleukin (IL)-1 and Tumor necrosis factor (*TNF*)-α may promote a pro-inflammatory and matrix-degrading fibroblast phenotype

tory cells is unclear. However, extensive experimental evidence from animal models of myocardial infarction suggests that cardiac fibroblasts exhibit activation of the inflammasome, the molecular platform responsible for caspase-mediated interleukin (IL)-1 activation [20] and downstream IL-1 signaling. Moreover, fibroblasts may serve as a source of chemokines in the infarcted myocardium.

Which molecular signals trigger pro-inflammatory activation of fibroblasts in the infarcted heart? Induction of pro-inflammatory cytokines, such as IL-1, Tumor Necrosis Factor (TNF)-α and members of the gp130 family, plays an important role in acquisition of pro-inflammatory phenotype by cardiac fibroblasts during the early stages following infarction. In vitro, IL-1, TNF-α and oncostatin-M [21–23] exert potent pro-inflammatory actions on cardiac fibroblasts. In vivo evidence suggests that IL-1β may be prominently involved in inflammatory activation of fibroblasts in the infarcted heart [24]. IL-1 signaling enhances collagenase expression by cardiac fibroblasts promoting a matrix-degrading fibroblast phenotype that may also be associated with accentuated inflammatory cytokine synthesis. In addition to its pro-inflammatory actions, IL-1 also delays myofibroblast transdifferentiation suppressing expression of α-SMA by cardiac fibroblasts [24] and may inhibit fibroblast proliferation [25] by modulating expression of cyclins and their kinases [26]. Activation of IL-1 signaling during the early inflammatory phase of infarct healing may prevent premature conversion of fibroblasts into matrix-secreting myofibroblasts at a timepoint when the infarct environment may be hostile to reparative and contractile cells. As the wound is cleared from dead cells and matrix debris, suppression and resolution of IL-1-driven inflammation may allow unopposed actions of growth factors, leading to transdifferentiation of fibroblasts into synthetic myofibroblasts. Angiotensin II and aldosterone also exert pro-inflammatory actions on cardiac fibroblasts in vitro [27, 28]; however, the in vivo significance of these actions during the inflammatory phase of cardiac repair remains unknown. Cytokines and angiotensin II may exert their pro-inflammatory actions, at least in part, by promoting generation of reactive oxygen species (ROS) [27, 28]. Matrix degradation products, generated through the early activation of proteases in the infarcted heart may also contribute to pro-inflammatory fibroblast signaling [29, 30]. In vitro studies suggest that activation of Toll-like receptor (TLR) signaling by danger associated molecular patterns (DAMPs) released from dead cells may also activate pro-inflammatory responses in fibroblasts [31]. However, the potential role of TLR actions in inflammatory activation of cardiac fibroblasts in vivo has not been investigated.

4.2 Do activated Fibroblasts Participate in Resolution of Post-Infarction Inflammation?

Effective post-infarction repair requires timely suppression of the inflammatory response; negative regulation of chemokine and cytokine signaling prevents the catastrophic consequences of uncontrolled inflammation on cardiac geometry and function [8]. Suppression and resolution of the inflammatory reaction involves the timely activation of endogenous inhibitory pathways that inhibit inflammation. Various soluble mediators, such TGF-β and IL-10, and intracellular STOP signals (such as Interleukin Receptor Associated Kinase-M) inhibit innate immune signaling [32] and have been implicated in suppression and resolution of inflammatory infiltrate. Although all cell types participating in repair following an ischemic result are likely involved in suppression and resolution of inflammation the key cellular effectors in mediating this transition have not been identified. While subsets of monocytes and lymphocytes and macrophages have been shown to negatively regulate the post-infarction inflammatory response [33–35], whether fibroblasts can acquire an anti-inflammatory phenotype, participating in resolution of inflammation in the healing infarct remains unknown.

5   Cardiac Fibroblasts During the Proliferative Phase

5.1   Activated Fibroblasts as the Dominant Reparative Cells in the Infarcted Myocardium

During the proliferative phase of infarct healing, cardiac fibroblasts undergo dramatic phenotypic changes acquiring a synthetic/proliferative myofibroblast-like phenotype. Accumulation of activated myofibroblasts in the infarct border zone has been demonstrated in experimental models of myocardial infarction (Fig. 3). The transition from an inflammatory to a synthetic myofibroblast phenotype involves the removal of pro-inflammatory signals (such as IL-1β), activation of growth factor-mediated signaling and generation of a plastic matrix network which promotes myofibroblast transdifferentiation and stimulates responses to various growth factors through matricellular interactions (Fig. 4).

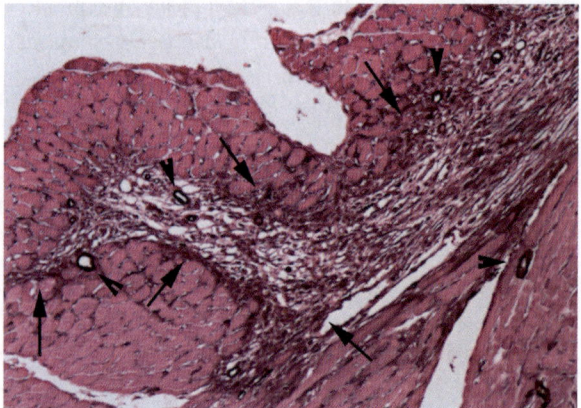

Fig. 3 During the proliferative phase, α-smooth muscle actin (*SMA*)-expressing myofibroblasts infiltrate the infarct border zone. α-SMA immunohistochemistry identifies abundant myofibroblasts in the infarcted mouse heart (1h ischemia/7 days reperfusion), as elongated cells (*arrows*) located outside the vascular media. α-SMA staining also labels vascular smooth muscle cells (*arrowheads*)

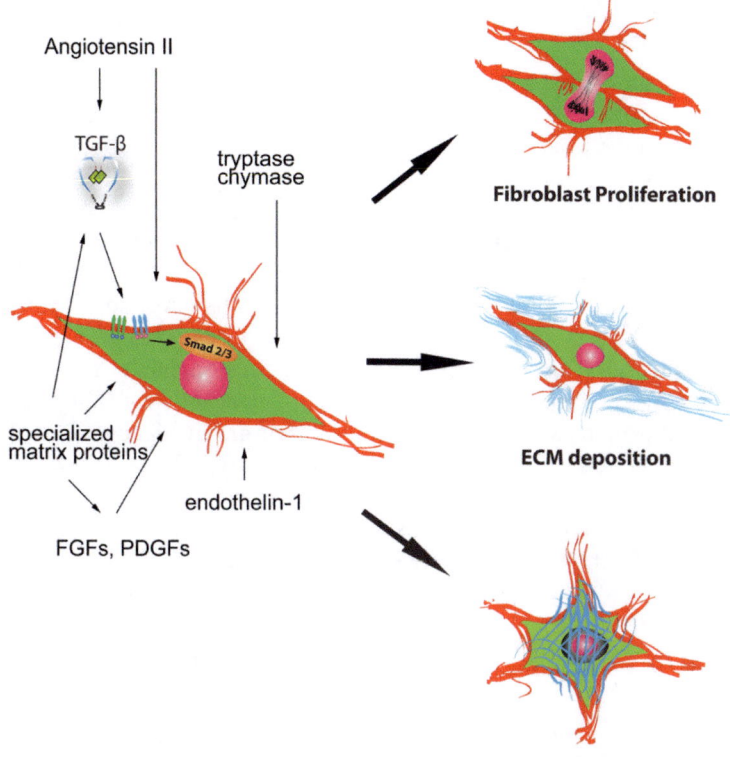

Angiotensin II

TGF-β

tryptase chymase

Fibroblast Proliferation

Smad 2/3

specialized matrix proteins

ECM deposition

endothelin-1

FGFs, PDGFs

Myofibroblast Transdifferentiation

Fig. 4 During the proliferative phase of infarct healing, fibroblasts become activated, transdifferentiate into myofibroblasts, secrete matrix proteins and proliferate. A wide range of mediators (including angiotensin II, TGF-β, growth factors, endothelin-1, tryptase, chymase and matricellular proteins) are implicated in the activation process (see text)

5.2 Characteristics and Origin of Activated Fibroblasts Following Infarction

Where do infarct myofibroblasts come from? In adult mammals, the abundant cardiac interstitial cells can be activated in response to growth factor stimulation and to alterations in their matrix environment, and may represent the most important source of myofibroblasts in the healing infarct. In vivo evidence suggests that blood-derived fibroblast progenitors and endothelial to mesenchymal transition may also contribute to the myofibroblast populations in models of cardiac fibrosis [36–38]. Additional sources of infarct fibroblasts may include epicardial cells, cardiac pericytes and vascular smooth muscle cells [39]. The relative contribution of each population remains unknown. Moreover, the infarct fibroblast population may be heterogeneous; different subsets with distinct origins and functional properties may contribute specialized roles in cardiac repair.

5.3   Myofibroblast Transdifferentiation

Conversion of fibroblasts into synthetic myofibroblasts is the hallmark of the proliferative phase of infarct healing. In the healing infarct, border zone myofibroblasts acquire some characteristics of smooth muscle cells, exhibiting formation of contractile stress fibers [12, 40], and expression of α-SMA. Although de novo synthesis of α-SMA characterizes the differentiated myofibroblast, it is important to note that it is not a necessary criterion for myofibroblast identification: "proto-myofibroblasts" form stress fibers without incorporation of α-SMA [41, 42]. Conversion of cardiac fibroblasts into myofibroblasts requires the cooperation of growth factors (such as TGF-β) and specialized matrix components (such as ED-A fibronectin) [43, 44]. Recent evidence suggests that the calcium channel Transient receptor potential canonical (TRPC)6 is critically involved in mediating the effects of TGF-β and angiotensin II on myofibroblast transdifferentiation [45, 46]. During the proliferative phase of healing, activation of TGF-β, deposition of ED-A fibronectin and matricellular proteins in the provisional matrix network [47] and removal of pro-inflammatory cytokines, such as IL-1β, create an environment that potently stimulates myofibroblast conversion.

5.4   Role of TGF-β Signaling and its Importance in Fibrosis

TGF-β is a multifunctional, versatile and highly pleiotropic growth factor that regulates a wide range of cellular responses (including cell migration, proliferation, differentiation and survival). TGF-β is found in many tissues in a latent form; generation of small amounts of bioactive TGF-β is sufficient to trigger a maximal cellular response. TGF-β expression is markedly augmented following infarction and is predominately localized in the infarct border zone; platelets, inflammatory leukocytes, fibroblasts, vascular cells and cardiomyocytes may contribute to the increased levels of TGF-β. TGF-β1, β2 and β3 isoforms exhibit distinct time courses of upregulation in myocardial infarction [48]: TGF-β1 and β2 are induced early, whereas TGF-β3 expression is upregulated at a later stage. TGF-β induction in the infarcted myocardium is associated with a marked increase in levels of phosphorylated Smad2/3 [49, 50], suggesting generation of bioactive TGF-β and downstream activation of TGF-β/Smad signaling.

TGF-β has profound effects on fibroblast phenotype and function inducing myofibroblast transdifferentiation [51], upregulating matrix protein synthesis and stimulating synthesis of protease inhibitors, such as Tissue inhibitor of metalloproteinases (TIMP)-1 and Plasminogen activator inhibitor (PAI)-1 [44, 50]. TGF-β mediated activation of Smad3 signaling plays an important role in activation of infarct fibroblasts; in the absence of Smad3, collagen deposition in the infarct, in the border zone and in the remodeling non-infarcted myocardium is significantly attenuated. The pro-fibrotic effects of Smad3 signaling are mediated, at least in part, through upregulation of matrix protein synthesis and through stimulation of α-SMA expression [52]. Our understanding of the role of non-canonical Smad-independent TGF-β signaling pathways in cardiac repair remains limited. Although in other models of tissue fibrosis, p38 mitogen-activated protein kinase (MAPK) and c-Abl tyrosine kinase have been implicated in the fibrotic response [53, 54], their significance in activation of reparative fibroblasts following infarction remains unknown. TGF-β activated kinase (TAK)-1 may also exert pro-fibrotic actions in the myocardium [55]; however, direct evidence on its role in post-infarction repair is lacking.

5.5   Signals Regulating Fibroblast Migration in the Infarcted Myocardium

Migration of fibroblasts into the healing infarct is essential for the reparative response. Formation of a fibrin/fibronectin-based provisional matrix in the infarcted area creates the environment necessary for cell migration. Despite the significance of the migratory response in cardiac repair, understanding of the signals that stimulate fibroblast migration in the healing infarct is limited. Fibroblast migration in the infarct is associated with the activation of tissue polarity genes, such as frizzled-2 [56]. Growth factor signaling and deposition of matricellular proteins likely regulate fibroblast migration; however, the specific pathways implicated in the process are poorly understood. The migratory response is tightly regulated. Activation of inhibitory anti-migratory signals, such as the anti-fibrotic chemokine CXCL10/ Interferon--inducible Protein (IP)-10 [57] reduces growth factor-induced fibroblast migration preventing excessive fibrotic remodeling of the infarcted heart. The anti-fibrotic effects of IP-10 are not mediated through activation of its main receptor CXCR3 and may involve proteoglycan-mediated interactions [58].

5.6   The Renin Angiotensin Aldosterone System (RAAS) and Growth Factors Regulate Fibroblast Function in the Infarcted Myocardium

In the healing infarct, myofibroblasts become matrix-synthetic proliferative cells and are responsible for formation of the collagen-based scar that replaces the dead cardiomyocytes. Surviving cardiomyocytes, macrophages and mast cells may contribute to activation of border zone myofibroblasts by secreting mediators that regulate their proliferative activity and matrix synthetic capacity [59, 60]. Extensive evidence suggests a crucial role for the RAAS system in activation of infarct myofibroblasts. Both pharmacologic inhibition studies and genetic loss-of-function experiments suggested that angiotensin II signaling increases the matrix synthetic capacity of infarct myofibroblasts by activating type I (AT1) receptors [61]. Aldosterone also induces expression of matrix proteins in cardiac fibroblasts [62]. The fibrogenic effects of the RAAS are mediated, at least in part, through activation of growth factors, such as Fibroblast Growth Factors (FGFs), TGF-β, and Platelet-derived growth factor (PDGF) [63, 64]. Endothelin-1 also activates cardiac fibro-blasts, by inducing collagen synthesis and by stimulating proliferation [65, 66]. Moreover, the mast cell proteases tryptase and chymase are potent activators of fibroblast proliferation and inducers of matrix protein synthesis [67–69]. In addition to its direct actions on cardiac fibroblasts, mast cell chymase may act as an impor-tant alternative mechanism for angiotensin II generation [70].

5.7   The Extracellular Matrix as a Modulator of Fibroblast Phenotype: The Matricellular Proteins

The extracellular matrix does not simply play a structural role, but also dynamically regulates cellular phenotype and function in injured and remodeling tissues. Dur-ing the proliferative phase of infarct healing, deposition of matricellular proteins in the infarct dynamically modulates fibroblast phenotype, regulating cytokine and growth factor responses. Several members of the matricellular family, including thrombospondin (TSP)-1, TSP-2, osteopontin (OPN), Secreted protein acidic and rich in cysteine (SPARC), periostin, tenascin-C and members of the CCN family are upregulated in the infarcted myocardium and modulate fibroblast function, playing an essential role in the reparative and remodeling response [47].

TSP-1, a prototypical matricellular protein is a potent angiostatic mediator with an essential role in TFG-β activation [71]. TSP-1 expression is selectively upregu-lated in the infarct border zone; TSP-1 loss results in increased dilative post-in-farction remodeling associated with prolonged and expanded inflammation [72]. TSP-1 has matrix-stabilizing effects; both direct actions of TSP-1 and its effects on TGF-β activation may drive cardiac fibroblasts towards a matrix-preserving pheno-type [73]. TSP-1 loss in a model of cardiac pressure overload was associated with increased MMP activity, impaired myofibroblast transdifferentiation, and reduced fibroblast-derived collagen synthesis [73].

Expression of Tenascin-C is also markedly upregulated during the proliferative phase of healing [74] and facilitates fibroblast migration into the infarct [75, 76]. SPARC also critically regulates fibroblast phenotype and function in the remodel-ing heart. SPARC loss in mice is associated with higher mortality following due to defective healing, associated with disorganized granulation tissue formation and deposition of immature collagen [77]. OPN upregulation is also consistently found in animal models of myocardial infarction [78]. OPN can act both as a cytokine and as a matricellular protein. Loss of OPN was associated with worse post-infarction cardiac remodeling [79]. OPN has multifunctional effects on cell survival, prolif-eration, adhesion and migration. *In vitro* experiments suggested that proliferative effects of angiotensin II in cardiac fibroblasts may be mediated by OPN [80]. Peri-ostin is also highly upregulated in the infarcted heart and plays a critical role in infarct fibroblast maturation and differentiation [81, 82].

6 Cardiac Fibroblasts During the Maturation Phase of Infarct Healing

As the scar matures, the extracellular matrix is cross-linked, while the cellular elements undergo apoptosis. Cardiac fibroblasts may contribute to formation of cross-linked collagen by secreting enzymes such as lysyl-oxidase and tissue transglutaminase. Although cardiac fibroblasts appear to undergo dramatic changes during the maturation phase, transitioning to a quiescent state and eventually disappearing from the scar, very little is known regarding their fate, and the signals that may drive suppression of their activity. In mature myocardial infarcts, myofibroblast density is decreased [83, 84]; whether this decrease is due to loss of α-SMA expression and reversal of myofibroblast phenotype, or to apoptotic death, remains unclear. Removal of matricellular proteins and growth factors from the infarct environment may be responsible for fibroblast deactivation and apoptosis in the scar, depriving the cells from key pro-survival signals [85].

7 Cardiac Fibroblasts in the Remodeling Non-Infarcted Heart

While infarct myofibroblasts may be cleared through activation of apoptotic pathways, in the remote remodeling infarcted myocardium, fibroblasts may remain activated due to pathophysiologic changes induced by pressure and volume overload. In experimental models, pressure overload induces activation of matrix protein synthesis, fibrosis and development of diastolic dysfunction. Persistent pressure overload ultimately leads to decompensation, chamber dilation and systolic dysfunction [86]. Volume overload on the other hand is predominately associated with matrix loss and cardiac dilation [87]. The phenotypic changes of interstitial fibroblasts in the remote remodeling myocardium remain poorly understood.

8 Therapeutic Opportunities: Targeting the Cardiac Fibroblast Following Myocardial Infarction

Remodeling of the infarcted heart is dependent on the mechanical properties of the scar; thus, fibroblast-mediated actions on the infarcted and remodeling myocardium have important functional implications. Because of their key role as inflammatory and reparative cells, infarct fibroblasts are important therapeutic targets. Success of several established pharmacologic strategies for patients with myocardial infarction may be dependent, at least in part, on their effects on cardiac fibroblasts. Angiotensin Converting Enzyme (ACE) inhibitors and AT1 blockers decrease mortality in patients with myocardial infarction and protect from heart failure. Although their beneficial effects are associated with attenuated cardiac fibrosis, [88], these agents have a wide range of additional actions on the myocardium. Thus, the relative role of their anti-fibrotic effects in improving clinical outcome remains unclear. Aldosterone antagonists also reduce adverse remodeling and decrease circulating levels of fibrosis-associated markers in patients with acute myocardial infarction [89]. However, because of the broad effects of aldosterone on all cell types implicated in cardiac remodeling, the relative significance of anti-fibrotic actions in mediating the observed clinical is unknown.

Investigations in animal models have identified several promising new approaches that may prevent adverse remodeling and protect from the development of heart failure following myocardial infarction by targeting fibroblast functions [6, 90]. IL-1 antagonism may attenuate dilative post-infarction remodeling by limiting fibroblast-mediated inflammatory and matrix-degrading activity [91]. On the other hand, strategies targeting fibrogenic growth factors (such as TGF-β or FGF-2) may hold promise in prevention of fibrotic cardiac remodeling [50, 92] and in attenuation of diastolic heart failure. Because all inflammatory and fibrogenic mediators affect all cell types implicated in cardiac repair, whether any beneficial effects of these interventions may be due to modulation of fibroblast function is unclear. Moreover, experience from experimental animal studies suggests caution when attempting to interfere with matrix metabolism following myocardial infarction. Overactive matrix-preserving pathways may result in excessive matrix deposition in the infarcted heart, increasing chamber stiffness and causing diastolic dysfunction. In contrast, accentuation of matrix-degrading signals (due to overactive inflammatory cascades and/or enhanced protease activation) induces ventricular dilation and causes systolic dysfunction. Thus, perturbation of the balance between matrix synthesis and degradation may have adverse consequences on myocardial geometry and function.

Implementation of anti-fibrotic strategies in patients with myocardial infarction is particularly challenging due to the complexity of the pathophysiology of cardiac remodeling in human patients. Patients surviving an acute myocardial infarction exhibit significant pathophysiologic diversity that cannot be simulated in an animal model. Gender, age, genetic background, various co-existing conditions (such as hypertension, and metabolic disease), the use of medications (including agents targeting the RAAS or β-blockers) have important effects on fibroblast responses. For example, aging is associated with impaired fibroblast responsiveness to growth factors and results in formation of a defective scar with low collagen content [93]. In contrast, genetic predispositions or metabolic diseases (such as diabetes and obesity) may accentuate fibrosis leading to development of diastolic heart failure [94]. Considering the heterogeneity of patients with myocardial infarction, biomarker-based strategies or imaging-guided approaches are needed to develop personalized therapeutic approaches for treatment of patients with myocardial infarction [90, 95, 96]. Assessment of inflammatory biomarkers (such as Monocyte Chemoattractant Protein-1) may identify patients with myocardial infarction exhibiting prominent, accentuated, or prolonged inflammatory responses who may benefit from anti-inflammatory approaches. On the other hand, individuals with overactive TGF-β responses may be identified through measurement of biomarkers that reflect matrix synthesis, or through imaging studies assessing fibrotic remodeling; these patients may benefit from inhibition of TGF-β/Smad signaling. Targeting the fibroblast may also hold promise in treatment of infarct-related arrhythmias. Experimental studies suggested that α-SMA-containing stress fibers contribute to the arrhythmogenic potential of myofibroblasts. Thus, modulation of the myofibroblast cytoskeleton may have anti-arrhythmic effects [97].

9 Conclusions

Cardiac fibroblasts are abundant and exhibit remarkable phenotypic plasticity. Following myocardial infarction cardiac fibroblasts are key effector cells in inflammation, repair and remodeling of the heart. *In vitro* studies and animal model investigations have provided us important insights into the role of fibroblasts in myocardial infarction. Unfortunately, the lack of specific and reliable fibroblast markers and challenges in targeting fibroblasts in vivo have hampered our understanding of the cell biological role of fibroblasts in healing infarction. Future studies need to focus on several important directions. First, identification of new fibroblast-specific markers and characterization of fibroblast subpopulations that may play distinct roles in cardiac repair and remodeling is crucial for understanding the role of fibroblasts in the infarcted heart. Second, dissection of the signaling pathways responsible for activation and phenotypic modulation of fibroblasts following infarction is essential in order to understand the pathophysiology of the reparative and fibrotic response and to design new therapeutic strategies. Third, understanding the fate of infarct myofibroblasts and the role of endogenous stop signals that may inhibit their activation controlling the fibrotic response may identify new strategies for the treatment of heart failure.

Acknowledgments Dr Frangogiannis' laboratory is supported by NIH grants R01 HL76246 and R01 HL85440.

References

1. Souders CA, Bowers SL, Baudino TA (2009) Cardiac fibroblast: the renaissance cell. Circ Res. 105:1164–1176.
2. Camelliti P, Borg TK, Kohl P (2005) Structural and functional characterisation of cardiac fibroblasts. Cardiovasc Res 65:40–51.
3. Bowers SL, Baudino TA (2012) Cardiac myocyte-fibroblast interactions and the coronary vasculature. J Cardiovasc Transl Res 5:783–793.
4. Smith RS, Smith TJ, Blieden TM, Phipps RP (1997) Fibroblasts as sentinel cells. Synthesis of chemokines and regulation of inflammation. Am J Pathol 151:317–322.
5. Buckley CD, Pilling D, Lord JM, Akbar AN, Scheel-Toellner D, Salmon M (2001) Fibroblasts regulate the switch from acute resolving to chronic persistent inflammation. Trends Immunol 22:199–204.

6. Shinde AV, Frangogiannis NG (2014) Fibroblasts in myocardial infarction: A role in inflammation and repair. J Mol Cell Cardiol 70 C:74–82.
7. Chen W, Frangogiannis NG (2013) Fibroblasts in post-infarction inflammation and cardiac repair. Biochim Biophys Acta 1833:945–953.
8. Frangogiannis NG (2012) Regulation of the inflammatory response in cardiac repair. Circ Res 110:159–173.
9. Frangogiannis NG (2013) The immune system and the remodeling infarcted heart: cell biological insights and therapeutic opportunities. J Cardiovasc Pharmacol
10. Nag AC (1980) Study of non-muscle cells of the adult mammalian heart: a fine structural analysis and distribution. Cytobios 28:41–61.
11. Speiser B, Weihrauch D, Riess CF, Schaper J (1992) The extracellular matrix in human cardiac tissue. Part II: Vimentin, laminin, and fibronectin. Cardioscience 3:41–49.
12. Frangogiannis NG, Michael LH, Entman ML (2000) Myofibroblasts in reperfused myocardial infarcts express the embryonic form of smooth muscle myosin heavy chain (SMemb). Cardiovasc Res 48:89–100.
13. Kong P, Christia P, Saxena A, Su Y, Frangogiannis NG (2013) Lack of specificity of fibroblast-specific protein 1 in cardiac remodeling and fibrosis. Am J Physiol Heart Circ Physiol 305:H1363–1372.
14. Li ZH, Dulyaninova NG, House RP, Almo SC, Bresnick AR (2010) S100A4 regulates macrophage chemotaxis. Mol Biol Cell 21:2598–2610.
15. Acharya A, Baek ST, Huang G, Eskiocak B, Goetsch S, Sung CY, Banfi S, Sauer MF, Olsen GS, Duffield JS, Olson EN, Tallquist MD (2012) The bHLH transcription factor Tcf21 is required for lineage-specific EMT of cardiac fibroblast progenitors. Development 139:2139–2149.
16. Braitsch CM, Kanisicak O, van Berlo JH, Molkentin JD, Yutzey KE (2013) Differential expression of embryonic epicardial progenitor markers and localization of cardiac fibrosis in adult ischemic injury and hypertensive heart disease. J Mol Cell Cardiol 65:108–119.
17. Ieda M, Tsuchihashi T, Ivey KN, Ross RS, Hong TT, Shaw RM, Srivastava D (2009) Cardiac fibroblasts regulate myocardial proliferation through beta1 integrin signaling. Dev Cell 16:233–244.
18. Camelliti P, Green CR, LeGrice I, Kohl P (2004) Fibroblast network in rabbit sinoatrial node: structural and functional identification of homogeneous and heterogeneous cell coupling. Circ Res 94:828–835.
19. Zhang X, Azhar G, Nagano K, Wei JY (2001) Differential vulnerability to oxidative stress in rat cardiac myocytes versus fibroblasts. J Am Coll Cardiol 38:2055–2062.
20. Kawaguchi M, Takahashi M, Hata T, Kashima Y, Usui F, Morimoto H, Izawa A, Takahashi Y, Masumoto J, Koyama J, Hongo M, Noda T, Nakayama J, Sagara J, Taniguchi S, Ikeda U (2011) Inflammasome activation of cardiac fibroblasts is essential for myocardial ischemia/ reperfusion injury. Circulation 123:594–604.
21. Turner NA, Das A, Warburton P, O'Regan DJ, Ball SG, Porter KE (2009) Interleukin-1alpha stimulates proinflammatory cytokine expression in human cardiac myofibroblasts. Am J Physiol Heart Circ Physiol 297:H1117–1127.
22. Lafontant PJ, Burns AR, Donnachie E, Haudek SB, Smith CW, Entman ML (2006) Oncostatin M differentially regulates CXC chemokines in mouse cardiac fibroblasts. Am J Physiol Cell Physiol 291:C18–26.
23. Zymek P, Nah DY, Bujak M, Ren G, Koerting A, Leucker T, Huebener P, Taffet G, Entman M, Frangogiannis NG (2007) Interleukin-10 is not a critical regulator of infarct healing and left ventricular remodeling. Cardiovasc Res 74:313–322.
24. Saxena A, Chen W, Su Y, Rai V, Uche OU, Li N, Frangogiannis NG (2013) IL-1 Induces Proinflammatory Leukocyte Infiltration and Regulates Fibroblast Phenotype in the Infarcted Myocardium. J Immunol 191:4838–4848.
25. Palmer JN, Hartogensis WE, Patten M, Fortuin FD, Long CS (1995) Interleukin-1 beta induces cardiac myocyte growth but inhibits cardiac fibroblast proliferation in culture. J Clin Invest 95:2555–2564.

26. Koudssi F, Lopez JE, Villegas S, Long CS (1998) Cardiac fibroblasts arrest at the G1/S restriction point in response to interleukin (IL)-1beta. Evidence for IL-1beta-induced hypophosphorylation of the retinoblastoma protein. J Biol Chem 273:25796–25803.
27. Sano M, Fukuda K, Sato T, Kawaguchi H, Suematsu M, Matsuda S, Koyasu S, Matsui H, Yamauchi-Takihara K, Harada M, Saito Y, Ogawa S (2001) ERK and p38 MAPK, but not NF-kappaB, are critically involved in reactive oxygen species-mediated induction of IL-6 by angiotensin II in cardiac fibroblasts. Circ Res 89:661–669.
28. Rude MK, Duhaney TA, Kuster GM, Judge S, Heo J, Colucci WS, Siwik DA, Sam F (2005) Aldosterone stimulates matrix metalloproteinases and reactive oxygen species in adult rat ventricular cardiomyocytes. Hypertension 46:555–561.
29. Huebener P, Abou-Khamis T, Zymek P, Bujak M, Ying X, Chatila K, Haudek S, Thakker G, Frangogiannis NG (2008) CD44 Is Critically Involved in Infarct Healing by Regulating the Inflammatory and Fibrotic Response. J Immunol 180:2625–2633.
30. Cleutjens JP, Kandala JC, Guarda E, Guntaka RV, Weber KT (1995) Regulation of collagen degradation in the rat myocardium after infarction. J Mol Cell Cardiol 27:1281–1292.
31. Pierer M, Rethage J, Seibl R, Lauener R, Brentano F, Wagner U, Hantzschel H, Michel BA, Gay RE, Gay S, Kyburz D (2004) Chemokine secretion of rheumatoid arthritis synovial fibroblasts stimulated by Toll-like receptor 2 ligands. J Immunol 172:1256–1265.
32. Chen W, Saxena A, Li N, Sun J, Gupta A, Lee DW, Tian Q, Dobaczewski M, Frangogiannis NG (2012) Endogenous IRAK-M attenuates postinfarction remodeling through effects on macrophages and fibroblasts. Arterioscler Thromb Vasc Biol 32:2598–2608.
33. Dobaczewski M, Xia Y, Bujak M, Gonzalez-Quesada C, Frangogiannis NG (2010) CCR5 signaling suppresses inflammation and reduces adverse remodeling of the infarcted heart, mediating recruitment of regulatory T cells. Am J Pathol 176:2177–2187.
34. Weirather J, Hofmann U, Beyersdorf N, Ramos GC, Vogel B, Frey A, Ertl G, Kerkau T, Frantz S.(2014) Foxp3 + CD4 + T Cells Improve Healing after Myocardial Infarction by Modulating Monocyte/Macrophage Differentiation. Circ Res
35. Nahrendorf M, Swirski FK, Aikawa E, Stangenberg L, Wurdinger T, Figueiredo JL, Libby P, Weissleder R, Pittet MJ (2007) The healing myocardium sequentially mobilizes two monocyte subsets with divergent and complementary functions. J Exp Med 204:3037–3047.
36. Haudek SB, Xia Y, Huebener P, Lee JM, Carlson S, Crawford JR, Pilling D, Gomer RH, Trial J, Frangogiannis NG, Entman ML (2006) Bone marrow-derived fibroblast precursors mediate ischemic cardiomyopathy in mice. Proc Natl Acad Sci U S A 103:18284–18289.
37. Zeisberg EM, Tarnavski O, Zeisberg M, Dorfman AL, McMullen JR, Gustafsson E, Chandraker A, Yuan X, Pu WT, Roberts AB, Neilson EG, Sayegh MH, Izumo S, Kalluri R (2007) Endothelial-to-mesenchymal transition contributes to cardiac fibrosis. Nat Med 13:952–961.
38. Mollmann H, Nef HM, Kostin S, von Kalle C, Pilz I, Weber M, Schaper J, Hamm CW, Elsasser A (2006) Bone marrow-derived cells contribute to infarct remodelling. Cardiovasc Res 71:661–671.
39. Gabbiani G (2003) The myofibroblast in wound healing and fibrocontractive diseases. J Pathol. 200:500–503.
40. Willems IE, Havenith MG, De Mey JG, Daemen MJ (1994) The alpha-smooth muscle actin-positive cells in healing human myocardial scars. Am J Pathol 145:868–875.
41. Hinz B (2007) Formation and function of the myofibroblast during tissue repair. J Invest Dermatol 127:526–537.
42. Hinz B (2010) The myofibroblast: paradigm for a mechanically active cell. J Biomech 43:146–155.
43. Serini G, Bochaton-Piallat ML, Ropraz P, Geinoz A, Borsi L, Zardi L, Gabbiani G (1998) The fibronectin domain ED-A is crucial for myofibroblastic phenotype induction by transforming growth factor-beta1. J Cell Biol 142:873–881.
44. Biernacka A, Dobaczewski M, Frangogiannis NG (2011) TGF-beta signaling in fibrosis. Growth Factors 29:196–202.
45. Davis J, Burr AR, Davis GF, Birnbaumer L, Molkentin JD (2012) A TRPC6-dependent pathway for myofibroblast transdifferentiation and wound healing in vivo. Dev Cell 23:705–715.

46. Davis J, Molkentin JD (2013) Myofibroblasts: Trust your heart and let fate decide. J Mol Cell Cardiol
47. Frangogiannis NG (2012) Matricellular proteins in cardiac adaptation and disease. Physiol Rev 92:635–688.
48. Dewald O, Ren G, Duerr GD, Zoerlein M, Klemm C, Gersch C, Tincey S, Michael LH, Entman ML, Frangogiannis NG (2004) Of mice and dogs: species-specific differences in the inflammatory response following myocardial infarction. Am J Pathol 164:665–677.
49. Hao J, Ju H, Zhao S, Junaid A, Scammell-La Fleur T, Dixon IM (1999) Elevation of expression of Smads 2, 3, and 4, decorin and TGF-beta in the chronic phase of myocardial infarct scar healing. J Mol Cell Cardiol 31:667–678.
50. Bujak M, Ren G, Kweon HJ, Dobaczewski M, Reddy A, Taffet G, Wang XF, Frangogiannis NG (2007) Essential role of Smad3 in infarct healing and in the pathogenesis of cardiac remodeling. Circulation 116:2127–2138.
51. Leask A, Abraham DJ (2004) TGF-beta signaling and the fibrotic response. FASEB J 18:816–827.
52. Dobaczewski M, Bujak M, Li N, Gonzalez-Quesada C, Mendoza LH, Wang XF, Frangogiannis NG (2010) Smad3 signaling critically regulates fibroblast phenotype and function in healing myocardial infarction. Circ Res 107:418–428.
53. Ma FY, Sachchithananthan M, Flanc RS, Nikolic-Paterson DJ (2009) Mitogen activated protein kinases in renal fibrosis. Front Biosci (Schol Ed) 1:171–187.
54. Wang S, Wilkes MC, Leof EB, Hirschberg R (2010) Noncanonical TGF-beta pathways, mTORC1 and Abl, in renal interstitial fibrogenesis. Am J Physiol Renal Physiol 298:F142–149.
55. Zhang D, Gaussin V, Taffet GE, Belaguli NS, Yamada M, Schwartz RJ, Michael LH, Overbeek PA, Schneider MD (2000) TAK1 is activated in the myocardium after pressure overload and is sufficient to provoke heart failure in transgenic mice. Nat Med 6:556–563.
56. Blankesteijn WM, Essers-Janssen YP, Verluyten MJ, Daemen MJ, Smits JF (1997) A homologue of Drosophila tissue polarity gene frizzled is expressed in migrating myofibroblasts in the infarcted rat heart. Nat Med 3:541–544.
57. Bujak M, Dobaczewski M, Gonzalez-Quesada C, Xia Y, Leucker T, Zymek P, Veeranna V, Tager AM, Luster AD, Frangogiannis NG (2009) Induction of the CXC chemokine interferon-gamma-inducible protein 10 regulates the reparative response following myocardial infarction. Circ Res 105:973–983.
58. Saxena A, Bujak M, Frunza O, Dobaczewski M, Gonzalez-Quesada C, Lu B, Gerard C, Frangogiannis NG (2014) CXCR3-independent actions of the CXC chemokine CXCL10 in the infarcted myocardium and in isolated cardiac fibroblasts are mediated through proteoglycans. Cardiovasc Res
59. Frangogiannis NG, Perrard JL, Mendoza LH, Burns AR, Lindsey ML, Ballantyne CM, Michael LH, Smith CW, Entman ML (1998) Stem cell factor induction is associated with mast cell accumulation after canine myocardial ischemia and reperfusion. Circulation 98:687–698.
60. Frangogiannis NG, Dewald O, Xia Y, Ren G, Haudek S, Leucker T, Kraemer D, Taffet G, Rollins BJ, Entman ML (2007) Critical role of monocyte chemoattractant protein-1/CC chemokine ligand 2 in the pathogenesis of ischemic cardiomyopathy. Circulation 115:584–592.
61. Sun Y, Weber KT (2000) Infarct scar: a dynamic tissue. Cardiovasc Res 46:250–256.
62. Weber KT, Sun Y, Tyagi SC, Cleutjens JP (1994) Collagen network of the myocardium: function, structural remodeling and regulatory mechanisms. J Mol Cell Cardiol 26:279–292.
63. Detillieux KA, Sheikh F, Kardami E, Cattini PA (2003) Biological activities of fibroblast growth factor-2 in the adult myocardium. Cardiovasc Res 57:8–19.
64. Zymek P, Bujak M, Chatila K, Cieslak A, Thakker G, Entman ML, Frangogiannis NG (2006) The role of platelet-derived growth factor signaling in healing myocardial infarcts. J Am Coll Cardiol 48:2315–2323.
65. Hafizi S, Wharton J, Chester AH, Yacoub MH (2004) Profibrotic effects of endothelin-1 via the ETA receptor in cultured human cardiac fibroblasts. Cell Physiol Biochem 14:285–292.

66. Piacentini L, Gray M, Honbo NY, Chentoufi J, Bergman M, Karliner JS (2000) Endothelin-1 stimulates cardiac fibroblast proliferation through activation of protein kinase C. J Mol Cell Cardiol 32:565–576.
67. Cairns JA, Walls AF (1997) Mast cell tryptase stimulates the synthesis of type I collagen in human lung fibroblasts. J Clin Invest 99:1313–1321.
68. Ruoss SJ, Hartmann T, Caughey GH (1991) Mast cell tryptase is a mitogen for cultured fibroblasts. J Clin Invest 88:493–499.
69. Zhao XY, Zhao LY, Zheng QS, Su JL, Guan H, Shang FJ, Niu XL, He YP, Lu XL (2008) Chymase induces profibrotic response via transforming growth factor-beta 1/Smad activation in rat cardiac fibroblasts. Mol Cell Biochem 310:159–166.
70. Dell'Italia LJ, Husain A (2002) Dissecting the role of chymase in angiotensin II formation and heart and blood vessel diseases. Curr Opin Cardiol 17:374–379.
71. Adams JC, Lawler J (2004) The thrombospondins. Int J Biochem Cell Biol 36:961–968.
72. Frangogiannis NG, Ren G, Dewald O, Zymek P, Haudek S, Koerting A, Winkelmann K, Michael LH, Lawler J, Entman ML (2005) The critical role of endogenous Thrombospondin (TSP)-1 in preventing expansion of healing myocardial infarcts. Circulation 111:2935–2942.
73. Xia Y, Dobaczewski M, Gonzalez-Quesada C, Chen W, Biernacka A, Li N, Lee DW, Frangogiannis NG (2011) Endogenous thrombospondin 1 protects the pressure-overloaded myocardium by modulating fibroblast phenotype and matrix metabolism. Hypertension 58:902–911.
74. Willems IE, Arends JW, Daemen MJ (1996) Tenascin and fibronectin expression in healing human myocardial scars. J Pathol 179:321–325.
75. Tamaoki M, Imanaka-Yoshida K, Yokoyama K, Nishioka T, Inada H, Hiroe M, Sakakura T, Yoshida T (2005) Tenascin-C regulates recruitment of myofibroblasts during tissue repair after myocardial injury. Am J Pathol 167:71–80.
76. Imanaka-Yoshida K (2012) Tenascin-C in cardiovascular tissue remodeling: from development to inflammation and repair. Circ J 76:2513–2520.
77. Schellings MW, Vanhoutte D, Swinnen M, Cleutjens JP, Debets J, van Leeuwen RE, d'Hooge J, Van de Werf F, Carmeliet P, Pinto YM, Sage EH, Heymans S (2009) Absence of SPARC results in increased cardiac rupture and dysfunction after acute myocardial infarction. J Exp Med 206:113–123.
78. Dobaczewski M, de Haan JJ, Frangogiannis NG (2012) The extracellular matrix modulates fibroblast phenotype and function in the infarcted myocardium. J Cardiovasc Transl Res 5:837–847.
79. Trueblood NA, Xie Z, Communal C, Sam F, Ngoy S, Liaw L, Jenkins AW, Wang J, Sawyer DB, Bing OH, Apstein CS, Colucci WS, Singh K (2001) Exaggerated left ventricular dilation and reduced collagen deposition after myocardial infarction in mice lacking osteopontin. Circ Res 88:1080–1087.
80. Ashizawa N, Graf K, Do YS, Nunohiro T, Giachelli CM, Meehan WP, Tuan TL, Hsueh WA (1996) Osteopontin is produced by rat cardiac fibroblasts and mediates A(II)-induced DNA synthesis and collagen gel contraction. J Clin Invest 98:2218–2227.
81. Oka T, Xu J, Kaiser RA, Melendez J, Hambleton M, Sargent MA, Lorts A, Brunskill EW, Dorn GW, 2nd, Conway SJ, Aronow BJ, Robbins J, Molkentin JD (2007) Genetic manipulation of periostin expression reveals a role in cardiac hypertrophy and ventricular remodeling. Circ Res 101:313–321.
82. Shimazaki M, Nakamura K, Kii I, Kashima T, Amizuka N, Li M, Saito M, Fukuda K, Nishiyama T, Kitajima S, Saga Y, Fukayama M, Sata M, Kudo A (2008) Periostin is essential for cardiac healing after acute myocardial infarction. J Exp Med 205:295–303.
83. Christia P, Bujak M, Gonzalez-Quesada C, Chen W, Dobaczewski M, Reddy A, Frangogiannis NG (2013) Systematic characterization of myocardial inflammation, repair, and remodeling in a mouse model of reperfused myocardial infarction. J Histochem Cytochem 61:555–570.
84. Ren G, Michael LH, Entman ML, Frangogiannis NG (2002) Morphological characteristics of the microvasculature in healing myocardial infarcts. J Histochem Cytochem 50:71–79.

85. Pallero MA, Elzie CA, Chen J, Mosher DF, Murphy-Ullrich JE (2008) Thrombospondin 1 binding to calreticulin-LRP1 signals resistance to anoikis. Faseb J 22:3968–3979.
86. Xia Y, Lee K, Li N, Corbett D, Mendoza L, Frangogiannis NG (2009) Characterization of the inflammatory and fibrotic response in a mouse model of cardiac pressure overload. Histochem Cell Biol 131:471–481.
87. Zheng J, Chen Y, Pat B, Dell'italia LA, Tillson M, Dillon AR, Powell PC, Shi K, Shah N, Denney T, Husain A, Dell'Italia LJ (2009) Microarray identifies extensive downregulation of noncollagen extracellular matrix and profibrotic growth factor genes in chronic isolated mitral regurgitation in the dog. Circulation 119:2086–2095.
88. Brilla CG, Funck RC, Rupp H (2000) Lisinopril-mediated regression of myocardial fibrosis in patients with hypertensive heart disease. Circulation 102:1388–1393.
89. Hayashi M, Tsutamoto T, Wada A, Tsutsui T, Ishii C, Ohno K, Fujii M, Taniguchi A, Hamatani T, Nozato Y, Kataoka K, Morigami N, Ohnishi M, Kinoshita M, Horie M (2003) Immediate administration of mineralocorticoid receptor antagonist spironolactone prevents post-infarct left ventricular remodeling associated with suppression of a marker of myocardial collagen synthesis in patients with first anterior acute myocardial infarction. Circulation 107:2559–2565.
90. Frangogiannis NG (2014) The inflammatory response in myocardial injury, repair, and remodelling. Nat Rev Cardiol 11:255–265.
91. Abbate A, Kontos MC, Grizzard JD, Biondi-Zoccai GG, Van Tassell BW, Robati R, Roach LM, Arena RA, Roberts CS, Varma A, Gelwix CC, Salloum FN, Hastillo A, Dinarello CA, Vetrovec GW (2010) Interleukin-1 blockade with anakinra to prevent adverse cardiac remodeling after acute myocardial infarction (Virginia Commonwealth University Anakinra Remodeling Trial [VCU-ART] Pilot study). Am J Cardiol 105:1371–1377 e1371.
92. Bujak M, Frangogiannis NG (2007) The role of TGF-beta signaling in myocardial infarction and cardiac remodeling. Cardiovasc Res 74:184–195.
93. Bujak M, Kweon HJ, Chatila K, Li N, Taffet G, Frangogiannis NG. (2008) Aging-related defects are associated with adverse cardiac remodeling in a mouse model of reperfused myocardial infarction. J Am Coll Cardiol 51:1384–1392.
94. Cavalera M, Wang J, Frangogiannis NG (2014) Obesity, metabolic dysfunction, and cardiac fibrosis: pathophysiological pathways, molecular mechanisms, and therapeutic opportunities. Transl Res
95. Frangogiannis NG (2012) Biomarkers: hopes and challenges in the path from discovery to clinical practice. Transl Res 159:197–204.
96. Christia P, Frangogiannis NG (2013) Targeting inflammatory pathways in myocardial infarction. Eur J Clin Invest 43:986–995.
97. Rosker C, Salvarani N, Schmutz S, Grand T, Rohr S (2011) Abolishing myofibroblast arrhythmogeneicity by pharmacological ablation of alpha-smooth muscle actin containing stress fibers. Circ Res 109:1120–1131.

Mechanical and Matrix Regulation of Valvular Fibrosis

Mark C. Blaser and Craig A. Simmons

Abstract The aortic valve lies in, arguably, one of the more complex local mechanobiological environments in the body. The inherent intricacy of this microenvironment results in multiple homeostatic mechanisms, but also a wide variety of putative disease pathways by which valve function can be compromised. Aortic valve disease (AVD) is a cell-mediated pathology whose initial stages are characterized by unchecked matrix dysregulation, leaflet thickening, and widespread fibrosis. The valve itself is composed of multiple cell populations, including endothelial cells that are sensitive to blood flow-induced shear stresses and multipotent mesenchymal progenitors which are influenced by both the mechanical properties and composition of the surrounding extracellular matrix. Dynamic mechanical loading and shear stresses over the cardiac cycle, an irregular three-dimensional shape, and a non-uniform matrix composition further influence these cellular responses. There is also abundant biochemical signaling in the aortic root, with molecular factors either produced by valve cells or transported to the root via blood flow. When these mechanical/biochemical processes become deregulated as a result of insults to their constituent components, resident valvular cells are driven to undergo myofibroblastic differentiation, a program of valvular fibrosis sets in, and valve function is compromised. Valve dysfunction affects the cardiac environment as well, as impaired opening and reductions in orifice area alter myocardial mechanics and often result in hypertrophy and/or fibrosis of the left ventricle. In this chapter, we use the aortic valve as a model tissue to discuss causative mechanisms of cardiovascular fibrosis, including the contributions of mechanotransduction, matrix dysregulation, and biochemical signaling.

C. A. Simmons (✉) · M. C. Blaser
Institute of Biomaterials and Biomedical Engineering, University of Toronto, 5 King's College Road, Rm. 221, Toronto, ON M5S 3G8, Canada
e-mail: c.simmons@utoronto.ca

M. C. Blaser · C. A. Simmons
Cardiovascular Sciences Collaborative Program, University of Toronto, Toronto, Canada

C. A. Simmons
Department of Mechanical and Industrial Engineering, University of Toronto, Toronto, Canada

© Springer International Publishing Switzerland 2015   23
I.M.C. Dixon, J. T. Wigle (eds.), *Cardiac Fibrosis and Heart Failure: Cause or Effect?*,
Advances in Biochemistry in Health and Disease 13, DOI 10.1007/978-3-319-17437-2_3

Keywords Aortic valve disease · Substrate stiffness · Cyclic stretch · Shear stress · Mechanotransduction · Extracellular matrix · Biomechanics · Myofibroblast · Fibrosis

List of Abbreviations

5-HT	5-hydroxytryptamine (serotonin)
5-HT$_{2A}$	5-hydroxytryptamine receptor 2A
α-SMA	α-smooth muscle actin
ACEi	Angiotensin converting enzyme inhibitor
ARB	Angiotensin receptor blocker
AVD	Aortic valve disease
BMP	Bone morphogenic protein
CAVD	Calcific aortic valve disease
CNP	C-type natriuretic peptide
EC	Endothelial cell (vascular)
ECM	Extracellular matrix
EndMT	Endothelial-to-mesenchymal transition
FA	Focal adhesion
FAK	Focal adhesion kinase
FGF-2	Basic fibroblast growth factor-2
GAP	GTPase-activating protein
GEF	Guanine nucleotide exchange factor
LAP	Latency associated peptide
LC	Left coronary (cusp)
LLC	Large latent complex
LTBP-1	Latent TGF-β1 binding protein-1
mDia	Mammalian diaphanous-related formin
MyHC	Heavy chain smooth muscle myosin
MLC	Myosin light chain
MMP	Matrix metalloproteinase
MRTF-A	Myocardin-related transcription factor-A
NC	Non-coronary (cusp)
NPR-B	Natriuretic peptide receptor-B
PKG	Protein kinase-G
RC	Right coronary (cusp)
ROCK	Rho-associated protein kinase
SRF	Serum response factor
TGF-β1	Transforming growth factor-β1
TGFβRI/II	TGFβ receptor I/II
TIMPs	Tissue inhibitors of metalloproteinases
TNFα	Tumor necrosis factor-α
VEC	Valvular endothelial cell
VIC	Valvular interstitial cell

1 Introduction

Mechanical forces play a broad role in a variety of physiological processes. The physical stresses in the external environment of the cell, the nature of the cell's surrounding extracellular matrix (ECM), and the properties of adjacent cells all contribute to define the local biophysical environment. Mechanotransduction is the means by which cells translate physical and mechanical stimuli into biochemical signals, and it underpins cellular sensing of and responses to mechanical forces. In development, responses to changes in tissue stiffness guide lineage specification of mesenchymal stem cells [1], and are essential for proper maturation of the embryo [2]. Mechanotransduction also has highly-conserved homeostatic roles: for example, stretch-activated ion channels are expressed in bacteria, plants, and animals [3], as are surface receptors (e.g., integrins) which transmit external forces across the cell membrane [4]. Lastly, mechanotransduction is a key causative factor in a multitude of diseases—blood flow-induced shear stresses in vascular diseases [5], impaired cellular force transmission in muscular dystrophies [6], and diminished bone elaboration in osteoporosis [7], to name a few. Indeed, a number of diseases and pathologies that were previously thought to simply be "degenerative" or due to age-induced "wear and tear" are now recognized as the result of active, mechanically-mediated cellular differentiation and/or dysfunction. In this chapter, we detail how mechanotransduction, coupled with biochemical signaling, drives matrix dysregulation and fibrosis in the complex microenvironment of the aortic valve. The concept of valvular fibrosis as an active disease is still young, and so we also draw on studies performed in other cardiovascular and connective tissues to provide further depth.

1.1 Aortic Valve Biology, Physiology, and Function

The aortic valve is one of four one-way valves in the vertebrate heart, and sits between the left ventricle and the ascending aorta. Like the pulmonary valve (which is located between the right ventricle and pulmonary artery, and is the aortic valve's analogue on the right side of the heart), the aortic valve is made up of three semilunar leaflets (or cusps) (Fig. 1). The curvature of these leaflets creates three aortic sinuses on the aortic ("top") side of the valve: the left and right coronary arteries originate in the sinuses of the left coronary (LC) cusp and right coronary (RC) cusp respectively, while no artery arises from the posterior aortic sinus formed by the non-coronary (NC) cusp. Adult human valve leaflets are approximately 1 mm thick [8], with size and structure generally well-conserved between mammalian species of similar cardiac sizes, such as the pig [9]. This well-conserved structure is rather complex, but uniquely suited to withstand the dynamic mechanical forces found in the valvular microenvironment. Each leaflet is composed of three stacked layers, each of which has unique mechanical properties that contribute to proper mechanical function of the whole cusp. Layer-specific tissue mechanics are the result of different mixes and organization of extracellular ECM components. The fibrosa layer rests on the side of the valve closest to the ascending aorta and contains dense bundles of circumferentially-oriented type I and type III collagen fibres [10, 11], whose high

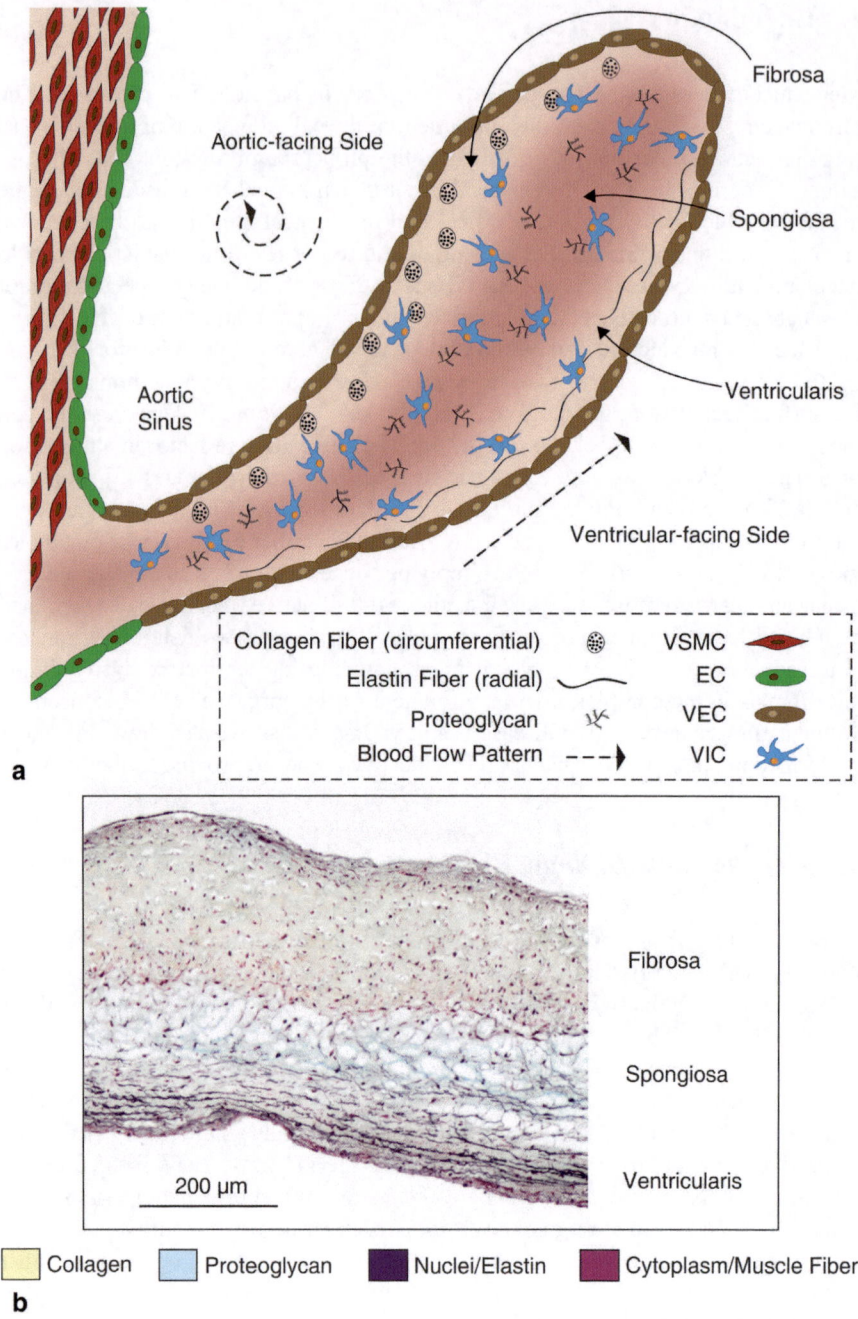

Fig. 1 Aortic valve cusp anatomy. **a** Illustration of an aortic valve cusp in cross-section, demonstrating the tri-layer morphology and composition. Valvular interstitial cells (*VICs*) are found inside the cusps, while valvular endothelial cells (*VECs*) line the surfaces. The fibrosa (*aortic side*) is rich in collagen fibers which run in a circumferential direction, while strands of radial elastin are contained in the ventricularis (*ventricular side*). The proteoglycan-rich spongiosa acts as a

tensile strength provides the majority of the valve's strength and resistance against deformation while closed [12]. On the ventricular side of the leaflet, the ventricularis is rich in collagen, but also has radially-oriented elastin which serves to control radial strains as the valve opens and closes [13]. Between these two layers, the spongiosa acts as a shock-absorber or linking buffer connecting the fibrosa and ventricularis. It is rich in highly-hydrated glycosaminoglycans and proteoglycans, which serve to resist compression and lubricate the differentially-directed shears (circumferential vs. radial) of the outer two layers [14]. Recent work has discovered finely interweaved elastin fibers in the spongiosa as well, which are oriented differently in the attachment and leaflet tip regions (rectilinear) vs. the belly or middle of the cusps (radial stripes) [15]. While the specific role of these elastin webs remains to be determined, they are hypothesized to preload and prime the outer layers for recoil during valve closure [15]. Differences in matrix composition between layers results in layer-specific matrix stiffness. Zhao and colleagues measured microscale, layer-specific mechanical properties of normal porcine aortic valve cusps at a number of points across the leaflet surface via micropipette aspiration [16]. They identified significant spatial heterogeneity in local matrix stiffness across both the fibrosa and ventricularis layers. However the fibrosa was, on average, significantly stiffer than the ventricularis. In addition, there were a number of focal regions in the fibrosa which were always stiffer than any part of the ventricularis [16]. A similar study using atomic force microscopy identified comparable trends in relative leaflet layer stiffness, with the spongiosa being the softest layer, followed by the ventricularis, while the fibrosa was roughly two times stiffer than even the ventricularis [17].

The aortic valve leaflets contain two distinct types of cell populations, both of which are responsible for cusp homeostasis and maintenance of valve health and function. Valvular interstitial cells (VICs) inhabit all leaflet layers, and are plastic and highly heterogenous. VICs from healthy valves are primarily quiescent fibroblasts, along with a small amount of activated myofibroblasts ($\sim 5\%$) [18]. In a variety of species, VICs also contain a large percentage of multipotent mesenchymal progenitors (up to 48% in porcine leaflets [19]) with myofibrogenic, osteogenic, adipogenic, and chondrogenic potential [19–21]. This cell population likely continuously synthesizes and degrades valvular ECM in a homeostatic manner, allowing healthy matrix turnover and enabling the valve to respond to microenvironmental changes [22]. Interestingly, new evidence suggests that there are even distinct cellular phenotypes and differential responses to mechanical and biochemical stimuli between VICs from different layers of the valve [23], along with clear differences in expression of hundreds of genes between male and female VIC populations [24].

While VICs reside internally, the external (blood-facing) surfaces of the valve leaflets are lined by a single layer of valvular endothelial cells (VECs), which share many similarities with endothelial cells found elsewhere throughout the body's vasculature, but are phenotypically distinct in several ways. Though vascular endothelial cells (vascular ECs) align parallel to blood flow, contact guidance from underly-

---

linking buffer between the other two layers. **b** Cross-sectional histochemical staining (Movat's pentachrome) of porcine aortic valve demonstrates layer-specific extracellular matrix composition. Image in panel **b** kindly provided by Dr. Krista L. Sider, University of Toronto

ing collagen fibers in the fibrosa or ventricularis causes VECs to orient circumferentially on the leaflet surface [25]. Thus, VECs *in vivo* are frequently aligned and extended perpendicular to the direction of blood shear in several areas of the valve. Perpendicular alignment of VECs to flow without substrate guidance *in vitro* has been reported in one study [26] but not others [27], so the true cause or mechanism of altered endothelial alignment on the valve surface remains unclear. In culture, VECs are much more proliferative than vascular endothelial cells [28], half of all active genes in cultures of VECs and vascular ECs are unique to one cell type or the other [28], gene expression profiles differ in response to shear [29], and VECs preferentially adhere on different compositions of ECM than do vascular ECs [30]. As is the case with layer-specific differences in VIC phenotype, so to do differences exist between VECs on the aortic (fibrosa) and ventricular (ventricularis) sides of the aortic valve. In healthy porcine aortic valves, over 580 genes are differentially expressed between sides, a number of which are clearly linked to susceptibility/resistance to valvular disease development [31]. Others have shown multiple regulatory microRNAs (miRNAs) are up/downregulated side-to-side in human valves [27]. Together, these findings imply that important phenotypic heterogeneity exists not only between vascular and valvular ECs, but also within the VEC cell type itself.

Movement of the aortic valve over the cardiac cycle is passive, and occurs in response to relative changes in pressure between the left ventricle and ascending aorta. The valve opens as the left ventricle contracts to expel blood during systole, and left ventricular pressure rises above that of the ascending aorta. The reverse occurs at the beginning of diastole—as the left ventricle relaxes, its pressure drops below that of the aorta, the leaflets are "sucked" back, they co-apt and close off the aortic orifice [10]. In the majority of individuals, the aortic valve functions normally for their entire lifetime—typically opening and closing an average of 3.5 billion times [10]. This resistance to cyclic failure is particularly remarkable in light of the mechanical forces to which the valve is subjected throughout the cardiac cycle. During diastole, the 1 mm thick valve leaflets must withstand pressure gradients of 80 mmHg (and well above 100 mmHg in those with hypertension) [32]. This pressure gradient exerts significant radial strain ($\sim 20\%$) on the valve leaflets, while strain in the circumferential direction is roughly 10% [33]. As they open and close, leaflets are subjected to elevated flexural stress: total diastolic stresses in a single leaflet are estimated at 250 kPa for a 15% strain magnitude [10]. Blood flowing through the valve also exerts significant hemodynamic force on the leaflets. Indeed, blood is expelled through the valve during systole at a velocity of ~ 1 m/s in healthy humans, but can reach above 4 m/s in those with advanced valve disease [34]. During systole the ventricularis is exposed to laminar blood flow from the aortic jet and a half-sinusoidal hemodynamic shear stress waveform forms which peaks at 7 Pa, with a brief reversal to retrograde shears of up to -50 dynes/cm^2 during the final ~ 20 ms of systole; diastolic shears are negligible [35]. In stark contrast, the fibrosa experiences substantial oscillatory and disturbed flow along with formation of eddies and recirculation zones [36]. Shear stress magnitudes are substantially lower as well, peaking at 15 dynes/cm^2 during systole, but generally oscillating between -2 and 5 dynes/cm^2 [37].

1.2 Aortic Valve Fibrosis

The microenvironment of the aortic valve is, then, incredibly complex: multiple cell types with progenitor potential, several layers of tissue with varied matrix composition and mechanical properties, high flexural stresses and pressure loads, and extreme regional variations in blood flow shear stress regimes. Such an intricate system is highly perturbable, and therefore prone in a number of ways to malfunction leading to disease development.

In 2011, the NIH National Heart, Lung, and Blood Institute Working Group on Calcific Aortic Stenosis defined the term calcific aortic valve disease (CAVD) to cover a wide range of valvular pathologies, from mild sclerosis (thickening of the leaflets) to advanced and severe stenosis (compromised leaflet opening resulting in an impairment of the valve function) [38]. Incidence rates are high: 25% of North Americans over age 65 years have aortic valve sclerosis, and 2–4% of those are afflicted with valvular stenosis [39]. Besides age, male sex is the most significant risk factor for CAVD—it is associated with a 100% increase in incidence rates of valve disease vs. that of females [40]. While sclerosis does not directly impact cardiac function until it has progressed to stenosis, its presence is associated with a 50% increased risk for other cardiovascular events [41]. Indeed, aortic valve dysfunction can be directly causative of cardiac hypertrophy, myocardial fibrosis, and eventual heart failure [42]. Up to 38% of patients with moderate or severe valvular stenosis develop midwall myocardial fibrosis and a more advanced hypertrophic response [43].

Compounding the severity of this disease is the complete lack of any approved pharmacological treatments. Drug classes effective against hypertension, heart failure, and cardiomyopathy such as beta blockers, angiotensin receptor blockers (ARBs), and angiotensin-converting enzyme (ACE) inhibitors have all been ineffective against valve disease pathogenesis to date [44, 45]. Retrospective trials of anti-atherosclerotic lipid-lowering statins showed promise as putative treatments for aortic stenosis [45, 46], but follow-up prospective clinical trials found no benefit [47, 48]. At present, surgical replacement of a valve is thus required once it begins to impair cardiac function—without replacement, 50% of patients will not survive two years after the onset of symptoms [49]. These surgeries have become prevalent as CAVD cases increase (an estimated 275,000–370,000 replacements worldwide per year [50]), but unfortunately they remain highly invasive, with significant side effects and variable lifespans of replacement valves [51, 52].

Aortic valve disease is, then, a deadly and growing phenomenon. For many years, onset and progression of this disease was believed to be simple age-related degeneration of valvular tissues. In the past two decades however, a multitude of *in vitro* and *in vivo* studies have clearly overturned this idea in favor of an active, cell-mediated pathogenesis [38]. Valvular sclerosis is reflective of a thickened, stiffened, and highly fibrotic disease state [8], with associated development of focal (nodular) calcification (Fig. 2). These thickened, fibrotic, and calcified valve leaflets fail to open and close properly, leading to stenosis and obstruction of blood flow [8, 39, 53]. Fibrosis in the aortic valve occurs when a wide variety of pathogenic

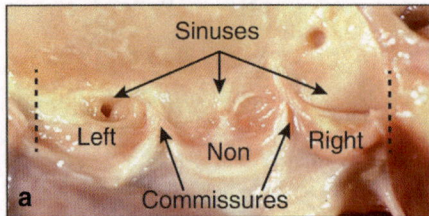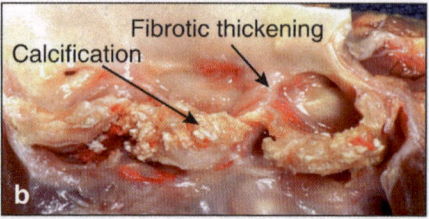

Fig. 2 Excised aortic valve specimens, viewed from parallel to the level of the valve with the aortic root cut open. **a** Healthy human aortic valve whose leaflets are translucent and flexible. *Dashed lines* denote the two edges of the single vertical cut used to open up the root. **b** Severely diseased and stenotic human aortic valve which is thickened, fibrotic, and highly calcified, with resultant impairment of cardiac function. Both panels reproduced with permission: Department of Pathology, University of Alabama at Birmingham, PEIR Digital Library (peir.patj.uab.edu/library)

insults cause previously quiescent and fibroblastic VICs to become activated and express α-smooth muscle actin (α-SMA) [54]. Incorporation of α-SMA in intracellular stress fibers is indicative of myofibroblast differentiation [55] and associated increased contractility and synthetic activity [56]. Myofibrogenesis is a key hallmark of valvular fibrosis, as myofibroblast content increases more than six-fold in diseased valves with nearly one in three VICs displaying α-SMA-positive stress fibers [57, 58]. In addition, VECs possess the potential for endothelial-to-mesenchymal transition (EndMT), where they lose expression of endothelial markers and gain a contractile, αSMA-positive apparatus consistent with that of myofibroblasts (reviewed in [59]). In response to inflammatory cytokines, VECs undergo EndMT, and endothelium-derived mesenchymal valve cells are found in close proximity to advanced valvular lesions in humans [60]. Together, VIC/VEC myofibrogenesis results in dysregulation of the valvular matrix in two key manners. First, valvular myofibroblasts synthesize large amounts of disorganized and deranged collagen and elastin [61], which dramatically thicken (by two- to three-fold) and stiffen the valve leaflets [8]. Second, they alter expression levels and activities of valvular ECM remodeling enzymes. While both matrix metalloproteinases (MMPs) and their inhibitors (tissue inhibitors of metalloproteinases, TIMPs) are upregulated in human CAVD, higher elevation of TIMP expression tilts this ratio towards MMP inhibition and resultant ECM buildup [62]. However, MMP-9 is upregulated more than its inhibitory TIMP [63, 64], thereby also resulting in fragmentation of both pre-existing and newly synthesized fibrotic collagen [63]. Fibrosis and matrix dysregulation occur along with the appearance of chronic inflammatory infiltrates [8], oxidative stress [65], cholesterol build-up [66], cartilage formation [67, 68], and calcification [8] (of which~80% is dystrophic mineralization, and the remainder is active ectopic osteogenic elaboration of bone matrix [68]). The onset of disease is side-dependent: VICs in the collagen-rich fibrosa, in close proximity to VECs experiencing low and oscillatory shear stresses, are much more prone to myofibrogenesis and calcific lesion development than those in the ventricularis [8, 53]. This side-dependent susceptibility to disease development may hold clues as to pathobiological mechanisms (reviewed in [69–71]).

2   Fibrosis, Biomechanics and the Myofibroblast

Fibrosis may be described as an out of control wound healing response. Normally, the ECM of healthy tissues is subjected to continuous remodeling and turnover, which results in a steady-state balance of matrix synthesis and degradation [72]. After tissue injury, dead or damaged cells are replaced by cells of the same type (the regenerative phase), then connective tissue replaces the damaged organ parenchyma (fibroplasia phase) [73], resulting in restoration of the organ function. However, if fibroplasia continues unrestrained and connective tissue homeostasis is no longer tightly regulated, then overexpression of disorganized collagenous ECM arises and fibrotic scar tissue develops [73, 74]. Fibrotic scars are typically permanent, dramatically modify tissue mechanical properties, and contribute to lethality in diseases of the heart, lung, liver, kidney, and skin [75]. Organ-level fibrosis is intimately regulated by biomechanics at multiple size scales (tissue to molecular, reviewed in [76]). Fibrotic scars are stiffer than their surrounding tissue, and these mechanical properties induce pathological differentiation of local mechanosensitive cells which go on to further stiffen, thicken, and contract the ECM in a system of pathological positive-feedback [77]. Furthermore, fibrotic ECM fails to shield cells from external mechanical forces, provoking further differentiation [55, 78].

Regardless of the tissue in which fibrosis occurs, there is a single cell type which plays a central role in the initiation and progression of disease: the myofibroblast. This cell was first identified over 40 years ago by Gabbiani and colleagues, who discovered fibroblasts in dermal wounds with smooth muscle cell-like packed fibrillar bundles, irregular nuclei (consistent with those found during cellular contracture of smooth muscle and myocardial fibers), and peripheral attachment sites [79]. In the intervening years, development of traction force and synthesis of ECM by myofibroblasts have been identified as key players in both physiological and pathological connective tissue remodeling (reviewed in [74]). Myofibroblast activation is transient during normal tissue repair, as the majority of these cells undergo apoptosis after their contraction contributes to rapid wound closure and mechanically stable scar tissue [80]. Apoptosis in myofibroblasts has been shown to be triggered by several factors: release from mechanical stress [81], loss of cell-cell OB-cadherin-type adherens junctions after wound closure [82], and nitric oxide signaling [83]. Other than cell death by apoptosis, the fate of the remaining myofibroblasts in a normally-healing wound is typically de-differentiation (quiescence) to a non-contractile, non-synthetic phenotype. Such quiescence has been established experimentally through loss of transforming growth factor-$\beta 1$ (TGF-$\beta 1$) signaling [84, 85], culture on soft/compliant substrates [86], and removal of underlying etiological agents (such as CCl_4 in modeled liver fibrosis), which is associated with upregulation of anti-apoptotic Hspa1a/b expression [87].

The real trouble occurs when myofibroblastic activity is not properly terminated by apoptosis or de-differentiation—unchecked fibrotic ECM production and tension generation activate feedback loops leading to fibrosis and scarring, and eventually impact proper organ function. Myofibroblasts explanted from fibrotic (sclerodermic) skin are resistant to Fas-induced apoptosis, likely due to Akt activation [88] and/or autocrine TGF-β1-induced phosphorylation of focal adhesion kinase (FAK) [89], whose downstream signaling regulates α-SMA expression in response to mechanical force application [90]. The persistence of apoptosis-resistant myofibroblasts provides a basis for the pathogenesis of fibrosis in all of the body's major organs.

The cellular origins of myofibroblasts are well-defined and represent an extremely diverse spectrum. They develop primarily from local (interstitial) fibroblasts, but pericytes, epithelial cells, endothelial cells, chondrocytes, osteoblasts, smooth muscle cells, hepatic stellate cells, and bone-marrow derived circulating fibrocytes have all demonstrated myofibroblastic potential (reviewed in [91]). Myofibroblast differentiation follows a common two-stage process in all tissues. First, elevated matrix stiffness and mechanical tension during injury repair causes precursor cells to express non-muscle myosin IIa and IIb [92] and transition to actively synthetic "proto-myofibroblasts," which are negative for α-SMA but develop contractile actin stress fibers and produce abundant collagen and the ED-A splice variant of fibronectin [79, 93, 94]. Proto-myofibroblasts also alter how they anchor to the underlying ECM: quiescent fibroblasts without stress fibers connect their cytoskeleton to the external matrix through integrin receptor binding at small (~2 μm) "focal complexes" [95], whereas proto-myofibroblasts begin to produce tension through remodeled and larger (~6 μm) "classical" focal adhesions (FAs) which connect to extracellular ED-A fibronectin [96].

During the second stage, tension force mediated by proto-myofibroblast actin stress fibers releases TGF-β1 from the surrounding ECM [97], while ED-A fibronectin activates MAPK-Erk1/2 and downstream focal adhesion kinase (FAK) signaling [98]. Together, these stimuli act to drive α-SMA synthesis/incorporation into stress fibers and to develop "supermature FAs," which are up to 30 μm in size [96], rich in FAK [99] and exert a four-fold higher traction stress (~12 kPa) than proto-myofibroblasts with classical FAs [96, 100]. Differentiated myofibroblasts also gain direct intercellular cytoplasmic connections with neighbouring cells via gap junctions, producing putative multicellular contractile units [101, 102]. This widespread system of cell-cell and cell-matrix connections underlie the perception of substrate stiffness and stress by myofibroblasts (reviewed in [103]), allowing dynamic responses to the biomechanics and biochemistry of the tissue microenvironment.

While α-SMA-positive stress fibers are therefore a specific marker of differentiated myofibroblasts, they are not a unique one. Smooth muscle cells (SMCs) also express α-SMA, along with gap junctions [104]. This necessitates the use of a relatively complex expression profile to definitively discriminate between fibroblasts, myofibroblasts, and smooth muscle cells. Though only α-SMA is most often used in practice, myofibroblasts are negative for desmin [104, 105], heavy chain smooth muscle myosin (MyHC) [104–106], smoothelin [104, 107], and N-caldesmon [108], which are all expressed by SMCs. Unlike myofibroblasts, quiescent fibroblasts are negative for α-SMA [105], SMemb [54], and MMP-13 [54, 58]. No single unique myofibroblast marker has been identified to date.

3 TGF-β-Mediated Mechanotransduction of Valvular Myofibrogenesis

Control of myofibroblastic differentiation is a complex and multifaceted affair, highlighted by the involvement of several classical and interconnected signaling cascades. However, the prototypical myofibrogenic signaling pathway is perhaps that which originates from profibrotic TGF-β1, and which is intimately associated with mechanotransduction. TGF-β1 has key homeostatic and pathological roles in a number of tissues and cellular processes, and regulates a wide variety of responses including wound repair, proliferation, inflammation, and matrix synthesis (reviewed in [109]). This molecule binds the TGFβ receptor type II (TGFβRII), which in turn phosphorylates the TGFβ receptor type I (TGFβRI). These two complexed serine/threonine kinases then induce myofibrogenesis (upregulated α-SMA and collagen synthesis) through canonical Smad signaling [110] (phosphorylation of Smad2/3, nuclear translocation of a Smad2/3/4 complex, and action as transcription factors to drive induction of profibrotic gene expression [111]).

Canonical TGF-β1 signaling is a key mechano-sensitive pathway in the stiffened environment of fibrotic scars (reviewed in [55]). TGF-β1 is secreted extracellularly in a non-bioactive form as part of a large latent complex (LLC), which consists of inactivated TGF-β1 bound to latency associated peptide (LAP) and latent TGF-β1 binding protein-1 (LTBP-1) [112]. Both LAP and LTBP-1 contain integrin binding domains [113], while LTBP-1 also binds components of the ECM such as fibrillins and fibronectins [114]. Together, this results in a pool of latent extracellular TGF-β1 that is bound mechanically to both the ECM and the contractile cytoskeleton (via integrins). This connection underpins force-sensitive TGF-β1 signaling: while the release of biologically active TGF-β1 from the LLC can occur through a variety of soluble factor/receptor interactions or through proteolytic cleavage [114], it has been conclusively demonstrated that cellular traction forces are transmitted directly to the LLC via integrins [114]. Indeed, in epithelial cells latent TGF-β1 can be freed from the LLC without any proteolytic action whatsoever, but this requires the cytoplasmic tail of the integrin to be intact and the actin cytoskeleton to be polymerized [115]; integrin/LLC interactions alone are insufficient to activate the latent complex [112]. In a classical demonstration of TGF-β1 mechanotransduction in myofibroblasts, Wipff and colleagues showed that external stretching and intracellular tension directly activate TGF-β1 in this cell type through the integrins $\alpha_v\beta_3$ and $\alpha_v\beta_5$ [97] (Fig. 3). Importantly, LAP-integrin binding, myofibroblast contracture, α-SMA-positive stress fibers, and a minimum substrate stiffness of ≥5 kPa were all required for activation of latent TGF-β1, and the magnitude of this activation grew with increases in substrate stiffness [97]. It is clear then, that a conformational change in the LLC and subsequent activation of TGF-β1 requires the ECM to provide a threshold of resistance against the "pulling" of a contractile myofibroblast. Interestingly, this 5 kPa stiffness is lower than the ~15 kPa required for incorporation of α-SMA into stress fibers [99]. This may therefore explain how proto-myofibroblasts can drive mechanical activation of TGF-β1 signaling and ECM remod-

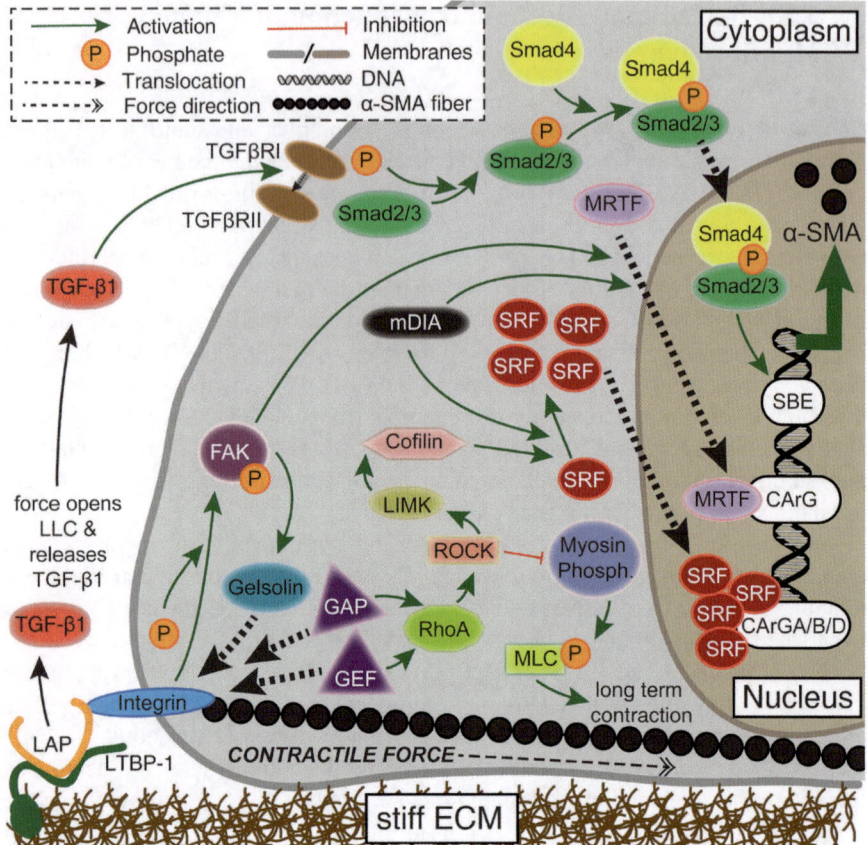

Fig. 3 Mechanotransduction leading to myofibroblast differentiation: integration of mechanics and biochemical signaling. Myofibroblasts secrete latent TGF-β1 bound to the large latent complex (LLC). When cells sit on a stiff ECM, mechanical forces from myofibroblast contracture are transferred via integrins from the cytoskeleton to the LLC. This contractile force releases biologically-active TGF-β1 which binds to the TGFβRI & II receptors, activating canonical Smad signaling. Integrin pulling also leads to activation of FAK, gelsolin, RhoA, and SRF/MRTF signaling. Together, these pathways work to upregulate expression of α-SMA

eling/stiffening, which eventually results in a matrix stiff enough to support the presence of fully differentiated α-SMA-positive myofibroblasts (reviewed in [75]).

Since TGF-β1 is positioned centrally in force-mediated myofibrogenesis, it is then of little surprise that this molecule has been highly studied in the context of valvular fibrosis. Stenotic human aortic valve leaflets are rich in TGF-β1 and contain high levels of LAP, which together implicate force-induced myofibrogenesis in the pathogenesis of CAVD [116]. The same early study found that TGF-β1 addition to cultures of VICs drives aggregation, apoptosis, and calcification, all of which are rescued by the addition of actin depolymerizing agents and/or pharmacological inhibitors of apoptosis [116]. As in many other cell types, TGF-β1 promotes a dose-dependent synthesis of α-SMA-positive stress fibers in cultured VICs [117]. Canonical TGF-β1 signaling via the Smad pathway is certainly a contributor to this myofibrogenesis: in cultured VICs, TGF-β1 drives Smad phosphorylation and nuclear translocation of Smad3 [118]. When nuclear translocation of the Smads is blocked by addition of basic fibroblast growth factor (FGF-2) to VICs, TGF-β1 is no longer capable of inducing myofibroblastic differentiation, cellular contraction, or formation of VIC aggregates [118]. Smad2/3 phosphorylation has also been implicated as a mediator of VIC motility—TGF-β1/Smad signaling regulates VIC activation, migration, and monolayer wound repair by driving α-SMA synthesis [119].

The biomechanics, composition, and stiffness of the valvular matrix are intimately involved in TGF-β1-mediated VIC myofibrogenesis (Fig. 4). Strikingly,

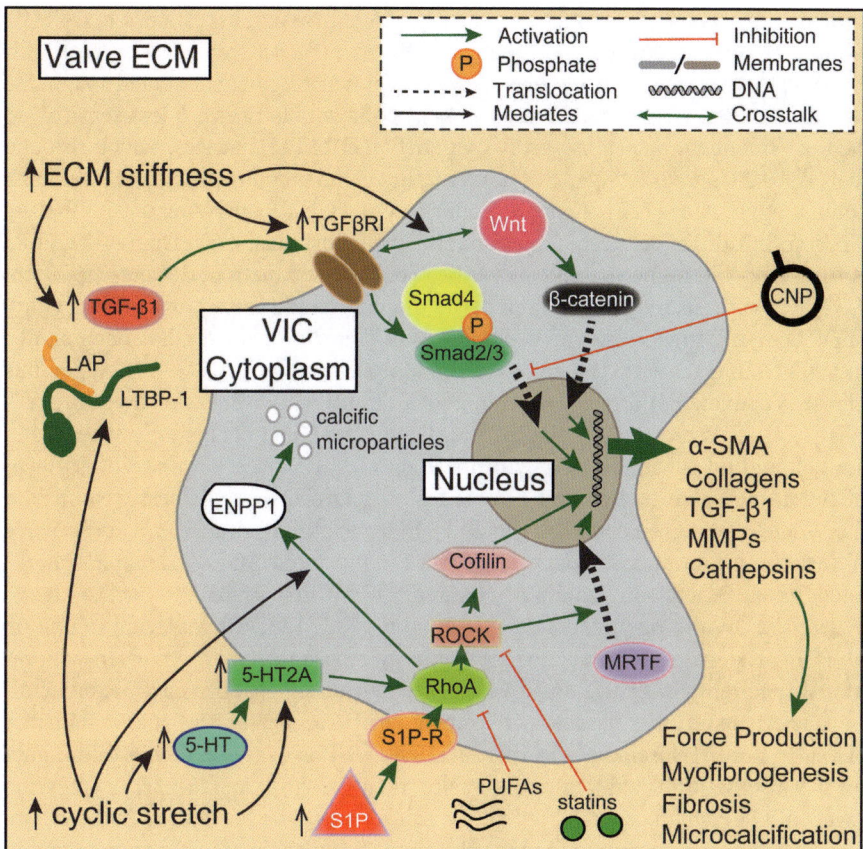

Fig. 4 Drivers of valvular interstitial cell fibrosis. A number of classical myofibroblastic pathways are at play in the aortic valve. TGF-β1 is highly present in diseased cusps, and the elevated leaflet substrate stiffnesses that occur during disease pathogenesis appear to upregulate expression of its downstream signaling pathway components, as well as potentiate cross-talk between TGF-β1 and Wnt/β-catenin. Mechanically-mediated release of TGF-β1 from the valvular ECM is believed to be due in part to this increase in cyclic stretch. Pathological levels of cyclic stretch also induce expression of MMPs, elastolytic cathepsins, and osteogenic BMPs. Elevated cyclic stretch even upregulates serotonin expression, which in turn will activate RhoA signaling, collagen synthesis, and ENPP1-dependent mineralization in VICs. Lastly, the RhoA/ROCK effector cofilin is abundant in sclerotic leaflets, as is MRTF nuclear translocation and concomitant myofibrogenesis

TGF-β1 and mechanical force act synergistically in VICs to drive stress fiber-mediated increases in cellular contractility [117]. At physiological levels of strain (15 %), exogenous TGF-β1 and cyclic stretch drive synergistic expression of α-SMA in cultured leaflets [120], possibly in part via mechanical release of bioactive TGF-β1 from the LLC [121]. Matrix composition is clearly involved in promoting this phenomenon: when cultured on fibronection or heparin substrates known to bind TGF-β1, VICs produce higher α-SMA levels than on collagen-coated control substrates [122]. Interestingly, activated myofibroblastic VICs drive alignment and reorganization of ECM fibronection fibrils by exerting force through FAs [117]. The stiffness of the surrounding matrix is of equal (or greater) importance to its composition. VIC myofibrogenesis and α-SMA expression in response to TGF-β1 administration is dependent on matrix stiffness [18], as studied with both stiff ECM substrates and with exogenous force applied to collagen-coated magnetite beads. VICs grown on compliant (~25 kPa) collagen matrices are less sensitive to TGF-β1 treatment than those grown on stiff (~110 kPa) matrices, which respond to TGF-β1 by upregulating α-SMA, increasing cellular contractility, and producing cellular aggregates [123]. Though production of TGF-β1 mRNA and TGFβRII by VICs do not differ between compliant and stiff substrates, expression of TGFβRI is over five-fold higher on stiff substrates [123]. Thus, matrix-stiffness-dependent TGFβRI synthesis may partially underlie the differential impact of matrix rigidity on VIC myofibrogenesis. The canonical Wnt signaling pathway has been shown in VICs to interact with TGF-β/Smad signaling at the level of TGFβRI. When this kinase is activated, it drives β-catenin nuclear translocation and VIC myofibrogenesis [124]. This crosstalk is also stiffness-dependent, and occurs only when VICs are cultured on substrates whose stiffness mimics the disease-prone fibrosa layer of the aortic valve leaflet, and not on softer matrices with stiffnesses similar to that of the disease-protected ventricularis. Fibrotic regions of diseased aortic valve fibrosa show colocalized expression of Wnt3A, β-catenin, TGF-β1, and phosphorylated Smad2/3, and cotreatment of cultured VICs with TGF-β1 and Wnt3A causes synergistic myofibrogenesis and α-SMA expression [124]. Downstream inhibition of TGF-β1 signaling may also be regulated by other molecules expressed with regional specificity in the aortic valve. C-type natriuretic peptide (CNP) exerts anti-fibrotic effects in other tissues such as the lungs [125], heart [126], and vasculature [127]. In normal porcine aortic valve leaflets, CNP is expressed three-fold higher on the disease-protected ventricularis side than the fibrosa side [31]. Expression levels of CNP and its specific receptor natriuretic peptide receptor-B (NPR-B) are all reduced in stenotic human aortic valves [128]. CNP-treated VICs are resistant to TGF-β1- and stiffness-mediated myofibrogenesis [129, 130], likely through a direct interaction between the activated forms of protein kinase G (PKG) and Smad2/3 which inhibits p-Smad2/3 nuclear translocation [131].

Though much work remains to be done, there are some important indications that valve biomechanics are also capable of inducing endothelial-to-mesenchymal transition in VECs. When VECs are cultured on 2D gels and subjected to 1 Hz cyclic strain, the rate of α-SMA-positive EndMT is elevated two- to three-fold compared with that of unstretched controls [132]. Remarkably, different signaling pathways drive EndMT between low-strain (10%) and high-strain (20%) environments—canonical TGF-β1 signaling via phosphorylated Smads is upregulated at 10% strain, while Wnt/β-catenin signaling dominates at 20% strain [132]. Others have demonstrated that substrate stiffness can also exert control over mesenchymal transition. In both canine kidney cells and murine mammary gland epithelial cells, TGF-β1 treatment on soft substrates (<1 kPa) induces apoptosis, but instead drives EMT on stiffer substrates to produce α-SMA-positive cells via elevated PI3K/Akt signaling (>5 kPa) [133]. VEC EndMT is also influenced by the biochemical milieu of the valve microenvironment. TGF-β1 has been found to drive EndMT in aortic VECs through the TGFβRII receptor, resulting in α-SMA-positive mesenchymal cells negative for the smooth muscle cell marker MyHC [134]. EndMT is also promoted by Notch1 signaling in both embryonic and post-natal VECs [135], as does treatment with the inflammatory cytokine tumor necrosis factor-α (TNFα) [136]. Regardless of the means through which these cells are transformed, they display upregulated MMP, Notch1, and TGF-β1 synthesis consistent with myofibrogenesis [136].

4 Mechanical Control of Valvular Myofibroblast Differentiation Independent of TGF-β

While TGF-β1 signaling plays a critical role in myofibroblastic differentiation, mechano-sensitive control of α-SMA promoter activity and resultant α-SMA synthesis is regulated in many other ways. Indeed, it appears that the very incorporation of cytosolic α-SMA into the actin cytoskeleton requires mechanical stress—if differentiated myofibroblasts are cultured on soft substrates, α-SMA is dislodged from β-actin filaments [99]. The implication here is that α-SMA is not a structural component of the myofibroblast cytoskeleton [76]. Instead, tension likely exposes cryptic cysteines on stress fibers, enabling α-SMA binding to sites that are only accessible when the cytoskeleton is mechanically loaded [137].

Another major pathway which mediates mechanical control of myofibrogenesis signals through the small GTPase RhoA and its downstream effector Rho-associated protein kinase (ROCK). RhoA activity is regulated by a collection of guanine nucleotide exchange factors (GEFs) and GTPase-activating proteins (GAPs) that lie in close proximity to the cytoskeleton and which are regulated by force transduction (reviewed in [138]). Long-term contraction and tension generation in myofibroblasts is achieved when RhoA maintains phosphorylation of myosin light chain (MLC) by actively inhibiting myosin phosphatase [139]. Mechanosensitive RhoA/ROCK signaling is stimulated by force transfer via integrins [140], and acts to phosphorylate LIM kinase and cofilin. Activated cofilin mediates an increase in the stability and formation rate of actin filaments [141]; as the ratio of actin filaments to monomeric G-actin rises, so too does serum response factor (SRF) activity [142]. The translocation of SRF and its transcriptional co-activator myocardin-related transcription factor-A (MRTF-A) is driven by integrin pulling [140], and the resultant binding of SRF to a CArG-B element in the α-SMA promoter contributes substantially to myofibrogenic α-SMA expression [143]. Partial redundancy is enabled in this system, since force applied to myofibroblast integrins also activates focal adhesion kinase (FAK) [90]. Phosphorylated FAK promotes MRTF-A nuclear translocation and recruits gelsolin, which (along with RhoA) drives ROCK/LIM kinase/cofilin activity [90]. Lastly, the actin-binding protein mammalian Diaphenous-related formin (mDia) likewise exerts control over SRF expression, MRTF-A nuclear translocation and α-SMA promoter activity [144]. Together, these components combine to initiate stress fiber assembly, and offer further force-sensitive control of myofibrogenesis (Fig. 3).

There is substantial evidence of a role for RhoA/ROCK mechanotransduction in mediating aortic valve fibrosis (Fig. 4). RhoA and ROCK activity is significantly elevated in myofibroblastic aggregates of cultured VICs, and pharmacological induction of RhoA activity drives further myofibrogenesis and α-SMA expression [145]. As would be expected, treatment with ROCK inhibitors impairs aggregate formation, α-SMA-positive stress fiber production, and gene expression related to myofibroblast activity (TGF-β1, MMP-1/13, etc) [145]. The RhoA effector protein cofilin is almost completely not expressed in normal porcine aortic valves, but is dramatically upregulated as these leaflets become sclerotic where it co-localizes with α-SMA in fibrotic focal clusters [18]. In cultured VICs, cofilin is highly upregulated during myofibrogenesis, is required for α-SMA incorporation into stress fibers, and contributes significantly to VIC myofibroblast contractile force generation [18]. Addition of the bioactive lipid sphingosine-1-phosphate (S1P, involved in a variety of cardiovascular pathophysiologies) to VICs in culture promotes RhoA activation and myofibrogenic aggregate development, which is reversed with pharmacological inhibition of S1P receptors, RhoA, and/or ROCK [146]. RhoA/ROCK signaling has also been implicated in mediating VIC calcification in response to 15% mechanical strain by facilitating export of ectonucleotide pyrophosphatase/phosphodiesterase (ENPP1, a promoter of apoptotic VIC mineralization highly expressed in stenotic aortic valves [147]) to the plasma membrane, where it mediates accumulation of spheroid mineralized microparticles [148]. While the HMG-CoA reductase inhibitor simvastatin protects against atherosclerosis by impairing cholesterol biosynthesis, it appears to have pleiotropic effects in the aortic valve and instead suppresses ROCK activation, thereby preventing formation of apoptotic and calcified VIC aggregates in culture [149]. Most recently, a clear role for the RhoA/G-actin/SRF/MRTF axis in promoting VIC myofibrogenesis has been described [150]. When RhoA activity is reduced and the VIC G/F-actin ratio is increased through administration of the polyunsaturated fatty acids (PUFAs) docosahexaenoic acid and arachidonic acid, so too is nuclear translocation of MRTF [150]. PUFA treatment suppresses VIC contractility and ameliorates α-SMA expression in explanted aortic valves—implying that therapeutic reversal of VIC myofibrogenesis may be feasible via interference with RhoA/MRTF signaling.

While cyclic stretch has clear impacts on valvular TGF-β1 signaling (see above), it also drives valve fibrosis through other means. When excised porcine aortic valve leaflets are subjected to ex vivo circumferential stretch with near-physiological

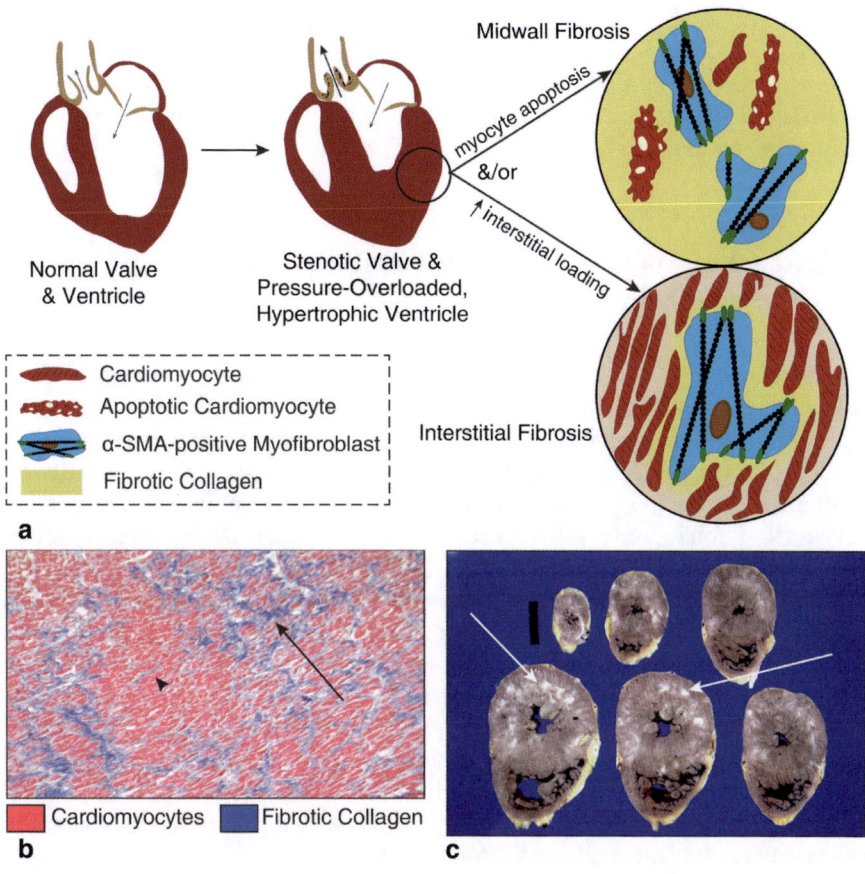

Fig. 5 Cardiac fibrosis and aortic valve disease. **a** Aortic valve stenosis increases pressure after-load and ventricular wall stress, stimulating left ventricular hypertrophy. Resultant myocyte injury/apoptosis and ventricular stiffening lead to myofibroblast infiltration and myofibrogenic differen-tiation of interstitial cardiac fibroblasts. Diffuse myocardial interstitial fibrosis, along with larger regions of midwall fibrosis follow. **b** Left ventricular histochemical staining (Masson's trichrome) of aged female patient with advanced calcific aortic stenosis demonstrates myocyte hypertrophy (*arrowhead*) and diffuse myocardial interstitial fibrosis (*arrow*). **c** Large and extensive midwall fibrotic lesions (*white tissue, arrows*) pervade these gross horizontal sections of the hypertrophic left ventricle from a patient with calcific aortic stenosis. Panels b and c reproduced with permis-sion: Department of Pathology, University of Alabama at Birmingham, PEIR Digital Library (peir.patj.uab.edu/library)

loading curves at 15 % strain, they develop upregulated levels of collagen, α-SMA, MMP-1/2/9, and the elastolytic proteases cathepsin S and K (which are also impli-cated in development of valve disease [151]) [64, 152]. Further work has demon-strated that cyclic stretch promotes expression of bone morphogenic protein-2/4 (BMP-2/4) [153], molecules with pro-osteogenic activity in stenotic aortic valves [154]. Finally, cyclic stretch upregulates both serotonin (5-HT) [155] and its recep-tor 5-HT$_{2A}$ in VECs [156]. Together, they are responsible for cyclic-stress-induced collagen synthesis, VIC proliferation, and increased leaflet stiffness, which may in turn drive further VIC myofibrogenesis [155]. Interestingly, serotonin has been shown to activate RhoA in vascular tissues [157] and to drive expression of TGF-β1 by VICs [158], placing it at a nexus between two key mechanosensitive regulators of valvular myofibrogenesis.

5   Impact of Aortic Valve Disease and Biomechanics on Cardiac Fibrosis

As valvular fibrosis develops and worsens, its impact spreads beyond just the microenvironment of the aortic root. A substantial portion of morbidity and mortality associated with aortic valve disease is due to CAVD-induced changes to myocardial health and, thus, cardiac function (Fig. 5). Some of the first studies to identify alterations in the composition and function of the myocardium in patients who had developed chronic aortic valve stenosis described hypertrophic cardiomyocytes surrounded by fibrous tissue [159, 160]. In later decades, a number of studies in humans identified cardiac muscle fiber enlargement, reduced myofibril volume fraction, reduced ejection fraction, and prolonged relaxation time constant in patients with aortic stenosis [161], and that some of these pathological changes to the left ventricle (fibrous content, muscle mass, cellular hypertrophy) were partially reversible after replacement of the aortic valve [162]. However, ventricular structure remained somewhat fibrotic and function continued to be significantly impaired even 6–7 years after valve replacement [162].

Hypertrophic myocyte enlargement and wall thickening occur in response to the increased pressure afterload and ventricular wall stress created by deficits in aortic valve function [163, 164]. Though it initially preserves ventricular function, much of this hypertrophy is increasingly believed to be maladaptive: when controlling for valve narrowing, mortality was increased in those subjects with the highest increases in left ventricular mass [42]. These patients were most prone to development of heart failure, due to myocyte apoptosis and unchecked ventricular fibrosis [165]. There are two distinct forms of cardiac fibrosis which transpire in response to stenosis-induced hypertrophy: discrete midwall lesions [43] and diffuse interstitial fibrosis [166]. Midwall fibrosis appears to be driven by hypertrophy-induced myocyte apoptosis [167], while interstitial fibrosis is thought to be the result of altered mechanical loading of cardiac fibroblasts in the pressure-overloaded heart (reviewed in [168]). In either case, myocytes/interstitial cardiac fibroblasts differentiate to myofibroblasts, initiating the classical fibrotic cycle of dysregulated matrix synthesis and resultant stiffening.

Mechanotransduction as it relates specifically to cardiac fibrosis is a relatively recent and growing field. Early indications of this mechanosensitivity were apparent when it was discovered that cardiac myofibroblasts in post-MI scars align in the direction of mechanical stresses caused by cyclic contraction of the surrounding viable myocardium [169]. When cultured cardiac fibroblasts are subjected to cyclic strain, they increase fibrotic collagen III mRNA expression after as little as 12 h [170], and elevate procollagen $a_1(I)$ synthesis after 48 hours [171]. Cell spreading and MMP-2/-9 expression are increased in both cardiomyocytes and cardiac fibroblasts when these cells are cultured on substrates of increasing stiffnesses from ~ 46 to ~ 1000 kPa [172], as are attachment, proliferation, and expression of thicker, denser, α-SMA-positive stress fibers on substrates from 140 to 590 kPa [173].

The mechanosensitivity of cardiac fibroblasts (reviewed in [174]) and their profibrotic response to mechanical forces are mediated by molecular pathways similar to those in the valve and other cardiovascular tissues. Over/under-expression of TGF-β1 respectively drives and attenuates cardiac fibrosis in aged mouse hearts [175, 176], while competitive antibody-mediated inhibition of TGF-β1 signaling in pressure-overloaded rat hearts reduces collagen mRNA production, prevents interstitial myocardial fibrosis, and rescues ventricular function [177]. Most recently, Sarrazy and colleagues showed that the integrins $\alpha_v\beta_5$ and $\alpha_v\beta_3$ are expressed in the heart, are upregulated in myofibroblast-rich fibrotic cardiac lesions, and drive cardiac fibroblast myofibrogenesis/α-SMA expression by contributing to mechanical stress-induced activation of latent TGF-β1 [178]. Other prototypical mediators of myofibroblast differentiation have also been implicated in cardiac fibrosis—for example, canonical Smad signaling is elevated in hypertensive heart failure and fibrosis [179], TGF-β1 activates RhoA/ROCK signaling and induces MRTF-A nuclear translocation in cardiac fibroblasts [180], and MRTF-A knockout mice are resistant to fibrosis post-MI [180]. Others have demonstrated that applied mechanical force stimulates RhoA and drives MRTF-A nuclear localization and α-SMA expression in cultured cardiac fibroblasts [140].

Thus, the mechanisms that regulate myofibrogenesis are highly conserved between the valves and heart. Interestingly, many genetically-induced murine models of aortic valve sclerosis and stenosis also develop cardiac hypertrophy, midwall/interstitial fibrosis, and deficits in systolic function, [181–185]. To date however, there has been little effort to specifically separate and dissect whether there are genetic pre-dispositions in humans that are causative of aortic valve disease *and* cardiac hypertrophy/fibrosis, or solely causative of the former, with the latter occurring naturally as a secondary pathology. Conditional tissue-specific knockouts are difficult to achieve in valvular tissue due to the heterogeneity of the interstitial population [19], but would be one means of answering this question. The notion of valve and cardiac fibrosis being linked by underlying genetic susceptibilities is further supported by the knowledge that after aortic stenosis onset, myocardial function is more likely to be preserved in females than in males [186]—this appears to be related to the elevated prevalence of initially restorative left ventricular hypertrophy in women [187]. LV diameter is reversed more frequently in females after valve replacement, a phenomenon speculated to be the result of differential collagen synthesis patterns in response to the sex hormone estradiol between male and female cardiac fibroblasts [187]. Regardless, careful study of shared vs. tissue-specific pathological mechanisms in the valve and heart is needed to enable efficacious treatment of both diseases.

6 Conclusions

Myofibroblastic differentiation of previously quiescent valvular interstitial cells, ECM dysregulation, and resultant unchecked synthesis of stiffened collagenous matrix are cornerstones of valvular fibrosis. VIC myofibrogenesis is driven by a number of causes, including valve stiffness increases, pathological hemodynamic shear, elevated cyclic stress, biochemical signaling, and mechanistic crosstalk and feedback therein. Valve fibrosis and associated calcification hinder leaflet motion and impair proper valve function. Compensatory remodeling of the left ventricle as a result of increased pressure afterload from the diseased valve can result in hypertrophy, fibrosis, and failure of the myocardium.

A number of gaps still exist in our knowledge of valvular fibrosis mechanobiology. First and foremost, the early pathogenesis of aortic valve disease is poorly understood and dramatically understudied. Little is known about the instigating factors that first induce myofibroblastic differentiation of VICs. A number of putative pathogenic insults have been identified *in vitro* and *in vivo*, but a consensus molecular mechanism of disease initiation has yet to be widely accepted. Characterization of valvular fibrosis biomechanics in humans and animal models of valvular sclerosis and stenosis is also largely absent [188], though a number of methods have recently been developed that enable accurate *in vivo* [189], *ex vivo* [16, 17], and *in vitro* [190] quantification of tissue or cellular mechanical properties. Importantly, fibrosis and calcification do not arise consecutively in the valve, but rather appear to develop in parallel. Even as prominent myofibrogenesis occurs during the initial stages of fibrosis, noticeable mineralization is also detectable [8]. It has therefore been difficult to separately study the relative contributions of these two phenotypes in regards to leaflet stiffening, functional impairment, and further fibrosis/calcification.

Surgical valve replacement is the only therapeutic option at present, primarily because valve disease is typically only diagnosed once functional impairments are apparent on ultrasound scans [191]. As functional impairment typically occurs only after a substantial calcific burden has developed, diagnosis and intervention are limited to the advanced stenotic forms of this disease at present [192]. Were it not for the stalled nature of valvular diagnostics, the fibrotic stage of aortic valve disease would appear ripe for pharmaceutical intervention (reviewed in [193]). Studies in mice have demonstrated that aggressive lipid-lowering can reverse pre-calcific forms of valve disease, including myofibroblast activation and fibrosis [194]. However, this lipid-lowering is unable to rescue function once the valve undergoes any significant calcification [195]. There is some evidence of the same occurring in humans: early administration of statins to asymptomatic patients with valve fibrosis/sclerosis (identified by rigorous, time-consuming, and challenging echocardiographic quantification of leaflet thickness) slowed progression to aortic stenosis by ~40 % [196]. Other anti-fibrotic therapeutic strategies may target myofibroblastic VICs themselves, through specific disruption of TGF-β1/Smad, RhoA/ROCK, or integrin-dependent signaling. Regardless, accurate earlier diagnosis of valvular fibrosis will require both an improved understanding of early valvular mechanics-mediated pathobiology, new biomarkers of early disease, and a leap forward in non-invasive imaging or detection technologies [197, 198].

Experimental systems designed to integrate a wide variety of putative pathological insults in a combinatorial and high-throughput manner will be necessary in order to fully characterize how individual mechanical forces and biological signals integrate together in the valvular microenvironment during disease development [32]. Organ-on-a-chip microsystems may enable these sorts of high-throughput studies *in vitro* [199], and could reveal novel opportunities/pathways for diagnosis and therapeutic intervention. These microfabricated devices can incorporate multiple cell types, ECM compositions, biochemical factors, and more recently, complex mechanical forces [200, 201], although none yet models a heart valve. Ultimately, multi-organ microdevices currently under development [202] will allow simultaneous study of the valve and myocardium and their interactions *in vitro,* providing new insights into the intricate and coupled mechanobiological bases of valvular and cardiac fibrosis.

Acknowledgements This work was supported by the Canadian Institutes of Health Research, the National Science and Engineering Research Council of Canada, and the Heart and Stroke Foundation of Ontario. We gratefully acknowledge Dr. Krista Sider for providing images of porcine aortic valves.

References

1. Engler AJ, Sen S, Sweeney HL, Discher DE (2006) Matrix elasticity directs stem cell lineage specification. Cell 126(4):677–689
2. Moore SW, Keller RE, Koehl MA (1995) The dorsal involuting marginal zone stiffens anisotropically during its convergent extension in the gastrula of *Xenopus laevis*. Development 121(10):3131–3140
3. Kullberg R (1987) Stretch-activated ion channels in bacteria and animal cell membranes. Trends Neurosci 10(10):387–388
4. van der Flier A, Sonnenberg A (2001) Function and interactions of integrins. Cell Tissue Res 305(3):285–298
5. Li YS, Haga JH, Chien S (2005) Molecular basis of the effects of shear stress on vascular endothelial cells. J Biomech 38(10):1949–1971
6. Heydemann A, McNally EM (2007) Consequences of disrupting the dystrophin-sarcoglycan complex in cardiac and skeletal myopathy. Trends Cardiovasc Med 17(2):55–59
7. Klein-Nulend J, Bacabac RG, Veldhuijzen JP, Van Loon JJ (2003) Microgravity and bone cell mechanosensitivity. Adv Space Res 32(8):1551–1559
8. Otto CM, Kuusisto J, Reichenbach DD, Gown AM, O'Brien KD (1994) Characterization of the early lesion of 'degenerative' valvular aortic stenosis. Histological and immunohistochemical studies. Circulation 90(2):844–853
9. Sim EK, Muskawad S, Lim CS, Yeo JH, Lim KH, Grignani RT, Durrani A, Lau G, Duran C (2003) Comparison of human and porcine aortic valves. Clin Anat 16(3):193–196
10. Thubrikar M (1990) The aortic valve. CRC Press, Boca Raton
11. Schoen FJ (1997) Aortic valve structure-function correlations: role of elastic fibers no longer a stretch of the imagination. J Heart Valve Dis 6(1):1–6

12. Schoen FJ (2012) Mechanisms of function and disease of natural and replacement heart valves. Annu Rev Pathol 7:161–183
13. Schoen FJ (2008) Evolving concepts of cardiac valve dynamics: the continuum of development, functional structure, pathobiology, and tissue engineering. Circulation 118(18):1864–1880
14. Sacks MS, David Merryman W, Schmidt DE (2009) On the biomechanics of heart valve function. J Biomech 42(12):1804–1824
15. Tseng H, Grande-Allen KJ (2011) Elastic fibers in the aortic valve spongiosa: a fresh perspective on its structure and role in overall tissue function. Acta Biomater 7(5):2101–2108
16. Zhao R, Sider KL, Simmons CA (2011) Measurement of layer-specific mechanical properties in multilayered biomaterials by micropipette aspiration. Acta Biomater 7(3):1220–1227
17. Sewell-Loftin MK, Brown CB, Baldwin HS, Merryman WD (2012) A novel technique for quantifying mouse heart valve leaflet stiffness with atomic force microscopy. J Heart Valve Dis 21(4):513–520
18. Pho M, Lee W, Watt DR, Laschinger C, Simmons CA, McCulloch CA (2008) Cofilin is a marker of myofibroblast differentiation in cells from porcine aortic cardiac valves. Am J Physiol Heart Circ Physiol 294(4):H1767–H1778
19. Chen JH, Yip CY, Sone ED, Simmons CA (2009) Identification and characterization of aortic valve mesenchymal progenitor cells with robust osteogenic calcification potential. Am J Pathol 174(3):1109–1119
20. Wang H, Sridhar B, Leinwand LA, Anseth KS (2013) Characterization of cell subpopulations expressing progenitor cell markers in porcine cardiac valves. PloS ONE 8(7):e69667
21. Nomura A, Seya K, Yu Z, Daitoku K, Motomura S, Murakami M, Fukuda I, Furukawa K (2013) CD34-negative mesenchymal stem-like cells may act as the cellular origin of human aortic valve calcification. Biochem Biophys Res Commun 440(4):780–785
22. Taylor PM, Batten P, Brand NJ, Thomas PS, Yacoub MH (2003) The cardiac valve interstitial cell. Int J Biochem Cell Biol 35(2):113–118
23. Moraes C, Likhitpanichkul M, Lam CJ, Beca BM, Sun Y, Simmons CA (2013) Microdevice array-based identification of distinct mechanobiological response profiles in layer-specific valve interstitial cells. Integr Biol (Camb) 5(4):673–680
24. McCoy CM, Nicholas DQ, Masters KS (2012) Sex-related differences in gene expression by porcine aortic valvular interstitial cells. PloS ONE 7(7):e39980
25. Deck JD (1986) Endothelial cell orientation on aortic valve leaflets. Cardiovasc Res 20(10):760–767
26. Butcher JT, Penrod AM, Garcia AJ, Nerem RM (2004) Unique morphology and focal adhesion development of valvular endothelial cells in static and fluid flow environments. Arterioscler Thromb Vasc Biol 24(8):1429–1434
27. Holliday CJ, Ankeny RF, Jo H, Nerem RM (2011) Discovery of shear- and side-specific mRNAs and miRNAs in human aortic valvular endothelial cells. Am J Physiol Heart Circ Physiol 301(3):H856–H867
28. Farivar RS, Cohn LH, Soltesz EG, Mihaljevic T, Rawn JD, Byrne JG (2003) Transcriptional profiling and growth kinetics of endothelium reveals differences between cells derived from porcine aorta versus aortic valve. Eur J Cardiothorac Surg 24(4):527–534
29. Butcher JT, Tressel S, Johnson T, Turner D, Sorescu G, Jo H, Nerem RM (2006) Transcriptional profiles of valvular and vascular endothelial cells reveal phenotypic differences: influence of shear stress. Arterioscler Thromb Vasc Biol 26(1):69–77
30. Young EW, Wheeler AR, Simmons CA (2007) Matrix-dependent adhesion of vascular and valvular endothelial cells in microfluidic channels. Lab Chip 7(12):1759–1766
31. Simmons CA, Grant GR, Manduchi E, Davies PF (2005) Spatial heterogeneity of endothelial phenotypes correlates with side-specific vulnerability to calcification in normal porcine aortic valves. Circ Res 96(7):792–799
32. Butcher JT, Simmons CA, Warnock JN (2008) Mechanobiology of the aortic heart valve. J Heart Valve Dis 17(1):62–73

33. Lo D, Vesely I (1995) Biaxial strain analysis of the porcine aortic valve. Ann Thorac Surg 60(2 Suppl):S374–S378
34. Gerdts E, Rossebo AB, Pedersen TR, Boman K, Brudi P, Chambers JB, Egstrup K, Gohlke-Barwolf C, Holme I, Kesaniemi YA, Malbecq W, Nienaber C, Ray S, Skjaerpe T, Wachtell K, Willenheimer R (2010) Impact of baseline severity of aortic valve stenosis on effect of intensive lipid lowering therapy (from the SEAS Study). Am J Cardiol 106(11):1634–1639
35. Yap CH, Saikrishnan N, Yoganathan AP (2012) Experimental measurement of dynamic fluid shear stress on the ventricular surface of the aortic valve leaflet. Biomech Model Mechanobiol 11(1/2):231–244
36. Balachandran K, Sucosky P, Yoganathan AP (2011) Hemodynamics and mechanobiology of aortic valve inflammation and calcification. Int J Inflamm 2011:263870
37. Yap CH, Saikrishnan N, Tamilselvan G, Yoganathan AP (2012) Experimental measurement of dynamic fluid shear stress on the aortic surface of the aortic valve leaflet. Biomech Model Mechanobiol 11(1/2):171–182
38. Rajamannan NM, Evans FJ, Aikawa E, Grande-Allen KJ, Demer LL, Heistad DD, Simmons CA, Masters KS, Mathieu P, O'Brien KD, Schoen FJ, Towler DA, Yoganathan AP, Otto CM (2011) Calcific aortic valve disease: not simply a degenerative process: a review and agenda for research from the National Heart and Lung and Blood Institute Aortic Stenosis Working Group. Executive summary: calcific aortic valve disease—2011 update. Circulation 124(16):1783–1791
39. Otto CM, Lind BK, Kitzman DW, Gersh BJ, Siscovick DS (1999) Association of aortic-valve sclerosis with cardiovascular mortality and morbidity in the elderly. N Engl J Med 341(3):142–147
40. Stewart BF, Siscovick D, Lind BK, Gardin JM, Gottdiener JS, Smith VE, Kitzman DW, Otto CM (1997) Clinical factors associated with calcific aortic valve disease. Cardiovascular Health Study. J Am Coll Cardiol 29(3):630–634
41. Barasch E, Gottdiener JS, Marino Larsen EK, Chaves PH, Newman AB (2006) Cardiovascular morbidity and mortality in community-dwelling elderly individuals with calcification of the fibrous skeleton of the base of the heart and aortosclerosis (the cardiovascular health study). Am J Cardiol 97(9):1281–1286
42. Cioffi G, Faggiano P, Vizzardi E, Tarantini L, Cramariuc D, Gerdts E, de Simone G (2011) Prognostic effect of inappropriately high left ventricular mass in asymptomatic severe aortic stenosis. Heart 97(4):301–307
43. Dweck MR, Joshi S, Murigu T, Alpendurada F, Jabbour A, Melina G, Banya W, Gulati A, Roussin I, Raza S, Prasad NA, Wage R, Quarto C, Angeloni E, Refice S, Sheppard M, Cook SA, Kilner PJ, Pennell DJ, Newby DE, Mohiaddin RH, Pepper J, Prasad SK (2011) Midwall fibrosis is an independent predictor of mortality in patients with aortic stenosis. J Am Coll Cardiol 58(12):1271–1279
44. Armstrong ZB, Boughner DR, Carruthers CP, Drangova M, Rogers KA (2014) Effects of an angiotensin II type 1 receptor blocker on aortic valve sclerosis in a pre-clinical model. Can J Cardiol 30(9):1096–1103
45. Rosenhek R, Rader F, Loho N, Gabriel H, Heger M, Klaar U, Schemper M, Binder T, Maurer G, Baumgartner H (2004) Statins but not angiotensin-converting enzyme inhibitors delay progression of aortic stenosis. Circulation 110(10):1291–1295
46. Shavelle DM, Takasu J, Budoff MJ, Mao S, Zhao XQ, O'Brien KD (2002) HMG CoA reductase inhibitor (statin) and aortic valve calcium. Lancet 359(9312):1125–1126
47. Rossebo AB, Pedersen TR, Boman K, Brudi P, Chambers JB, Egstrup K, Gerdts E, Gohlke-Barwolf C, Holme I, Kesaniemi YA, Malbecq W, Nienaber CA, Ray S, Skjaerpe T, Wachtell K, Willenheimer R, Investigators S (2008) Intensive lipid lowering with simvastatin and ezetimibe in aortic stenosis. N Engl J Med 359(13):1343–1356
48. Chan KL, Teo K, Dumesnil JG, Ni A, Tam J (2010) Effect of lipid lowering with rosuvastatin on progression of aortic stenosis: results of the aortic stenosis progression observation: measuring effects of rosuvastatin (ASTRONOMER) trial. Circulation 121(2):306–314
49. Otto CM (2000) Timing of aortic valve surgery. Heart 84(2):211–218

50. Zilla P, Brink J, Human P, Bezuidenhout D (2008) Prosthetic heart valves: catering for the few. Biomaterials 29(4):385–406
51. Mendelson K, Schoen FJ (2006) Heart valve tissue engineering: concepts, approaches, progress, and challenges. Ann Biomed Eng 34(12):1799–1819
52. Bonow RO, Carabello BA, Chatterjee K, de Leon AC, Jr., Faxon DP, Freed MD, Gaasch WH, Lytle BW, Nishimura RA, O'Gara PT, O'Rourke RA, Otto CM, Shah PM, Shanewise JS, American College of Cardiology/American Heart Association Task Force on Practice Guidelines (2008) 2008 focused update incorporated into the ACC/AHA 2006 guidelines for the management of patients with valvular heart disease: a report of the American College of Cardiology/American Heart Association Task Force on Practice Guidelines. Circulation 118 (15):e523–e661
53. O'Brien KD, Reichenbach DD, Marcovina SM, Kuusisto J, Alpers CE, Otto CM (1996) Apolipoproteins B, (a), and E accumulate in the morphologically early lesion of 'degenerative' valvular aortic stenosis. Arterioscler Thromb Vasc Biol 16(4):523–532
54. Rabkin-Aikawa E, Farber M, Aikawa M, Schoen FJ (2004) Dynamic and reversible changes of interstitial cell phenotype during remodeling of cardiac valves. J Heart Valve Dis 13(5):841–847
55. Tomasek JJ, Gabbiani G, Hinz B, Chaponnier C, Brown RA (2002) Myofibroblasts and mechano-regulation of connective tissue remodelling. Nat Rev Mol Cell Biol 3(5):349–363
56. Merryman WD, Youn I, Lukoff HD, Krueger PM, Guilak F, Hopkins RA, Sacks MS (2006) Correlation between heart valve interstitial cell stiffness and transvalvular pressure: implications for collagen biosynthesis. Am J Physiol Heart Circ Physiol 290(1):H224–H231
57. Taylor PM, Allen SP, Yacoub MH (2000) Phenotypic and functional characterization of interstitial cells from human heart valves, pericardium and skin. J Heart Valve Dis 9(1):150–158
58. Rabkin E, Aikawa M, Stone JR, Fukumoto Y, Libby P, Schoen FJ (2001) Activated interstitial myofibroblasts express catabolic enzymes and mediate matrix remodeling in myxomatous heart valves. Circulation 104(21):2525–2532
59. Bischoff J, Aikawa E (2011) Progenitor cells confer plasticity to cardiac valve endothelium. J Cardiovasc Transl Res 4(6):710–719
60. Mahler GJ, Farrar EJ, Butcher JT (2013) Inflammatory cytokines promote mesenchymal transformation in embryonic and adult valve endothelial cells. Arterioscler Thromb Vasc Biol 33(1):121–130
61. Eriksen HA, Satta J, Risteli J, Veijola M, Vare P, Soini Y (2006) Type I and type III collagen synthesis and composition in the valve matrix in aortic valve stenosis. Atherosclerosis 189(1):91–98
62. Fondard O, Detaint D, Iung B, Choqueux C, Adle-Biassette H, Jarraya M, Hvass U, Couetil JP, Henin D, Michel JB, Vahanian A, Jacob MP (2005) Extracellular matrix remodelling in human aortic valve disease: the role of matrix metalloproteinases and their tissue inhibitors. Eur Heart J 26(13):1333–1341
63. Satta J, Oiva J, Salo T, Eriksen H, Ohtonen P, Biancari F, Juvonen TS, Soini Y (2003) Evidence for an altered balance between matrix metalloproteinase-9 and its inhibitors in calcific aortic stenosis. Ann Thorac Surg 76(3):681–688. Discussion 688
64. Balachandran K, Sucosky P, Jo H, Yoganathan AP (2009) Elevated cyclic stretch alters matrix remodeling in aortic valve cusps: implications for degenerative aortic valve disease. Am J Physiol Heart Circ Physiol 296(3):H756–H764
65. Miller JD, Chu Y, Brooks RM, Richenbacher WE, Pena-Silva R, Heistad DD (2008) Dysregulation of antioxidant mechanisms contributes to increased oxidative stress in calcific aortic valvular stenosis in humans. J Am Coll Cardiol 52(10):843–850
66. Warren BA, Yong JL (1997) Calcification of the aortic valve: its progression and grading. Pathology 29(4):360–368
67. Caira FC, Stock SR, Gleason TG, McGee EC, Huang J, Bonow RO, Spelsberg TC, McCarthy PM, Rahimtoola SH, Rajamannan NM (2006) Human degenerative valve disease is associated with up-regulation of low-density lipoprotein receptor-related protein 5 receptor-mediated bone formation. J Am Coll Cardiol 47(8):1707–1712

68. Mohler ER 3rd, Gannon F, Reynolds C, Zimmerman R, Keane MG, Kaplan FS (2001) Bone formation and inflammation in cardiac valves. Circulation 103(11):1522–1528
69. Yip CY, Simmons CA (2011) The aortic valve microenvironment and its role in calcific aortic valve disease. Cardiovasc Pathol 20(3):177–182
70. Chen JH, Simmons CA (2011) Cell-matrix interactions in the pathobiology of calcific aortic valve disease: critical roles for matricellular, matricrine, and matrix mechanics cues. Circ Res 108(12):1510–1524
71. Gould ST, Srigunapalan S, Simmons CA, Anseth KS (2013) Hemodynamic and cellular response feedback in calcific aortic valve disease. Circ Res 113(2):186–197
72. Everts V, Beertsen W (1996) Phagocytosis and intracellular digestion of collagen, its role in turnover and remodelling. Histochem J 28:229–245
73. Wynn TA (2007) Common and unique mechanisms regulate fibrosis in various fibroproliferative diseases. J Clin Investig 117(3):524–529
74. Hinz B, Phan SH, Thannickal VJ, Prunotto M, Desmouliere A, Varga J, De Wever O, Mareel M, Gabbiani G (2012) Recent developments in myofibroblast biology: paradigms for connective tissue remodeling. Am J Pathol 180(4):1340–1355
75. Hinz B, Gabbiani G (2010) Fibrosis: recent advances in myofibroblast biology and new therapeutic perspectives. F1000 Biol Rep 2:78
76. Hinz B (2010) The myofibroblast: paradigm for a mechanically active cell. J Biomech 43(1):146–155
77. Hinz B (2009) Tissue stiffness, latent TGF-beta1 activation, and mechanical signal transduction: implications for the pathogenesis and treatment of fibrosis. Curr Rheumatol Rep 11(2):120–126
78. Eastwood M, Mudera VC, McGrouther DA, Brown RA (1998) Effect of precise mechanical loading on fibroblast populated collagen lattices: morphological changes. Cell Motil Cytoskeleton 40(1):13–21
79. Gabbiani G, Ryan GB, Majne G (1971) Presence of modified fibroblasts in granulation tissue and their possible role in wound contraction. Experientia 27(5):549–550
80. Desmouliere A, Redard M, Darby I, Gabbiani G (1995) Apoptosis mediates the decrease in cellularity during the transition between granulation tissue and scar. Am J Pathol 146(1):56–66
81. Carlson MA, Longaker MT, Thompson JS (2003) Wound splinting regulates granulation tissue survival. J Surg Res 110(1):304–309
82. Hinz B, Pittet P, Smith-Clerc J, Chaponnier C, Meister JJ (2004) Myofibroblast development is characterized by specific cell-cell adherens junctions. Mol Biol Cell 15(9):4310–4320
83. Zhang HY, Phan SH (1999) Inhibition of myofibroblast apoptosis by transforming growth factor beta(1). Am J Respir Cell Mol Biol 21(6):658–665
84. Hinz B (2007) Formation and function of the myofibroblast during tissue repair. J Investig Dermatol 127(3):526–537
85. Petridou S, Maltseva O, Spanakis S, Masur SK (2000) TGF-beta receptor expression and Smad2 localization are cell density dependent in fibroblasts. Investig Ophthalmol Vis Sci 41(1):89–95
86. Balestrini JL, Chaudhry S, Sarrazy V, Koehler A, Hinz B (2012) The mechanical memory of lung myofibroblasts. Integr Biol (Camb) 4(4):410–421
87. Kisseleva T, Cong M, Paik Y, Scholten D, Jiang C, Benner C, Iwaisako K, Moore-Morris T, Scott B, Tsukamoto H, Evans SM, Dillmann W, Glass CK, Brenner DA (2012) Myofibroblasts revert to an inactive phenotype during regression of liver fibrosis. Proc Natl Acad Sci U S A 109(24):9448–9453
88. Jun JB, Kuechle M, Min J, Shim SC, Kim G, Montenegro V, Korn JH, Elkon KB (2005) Scleroderma fibroblasts demonstrate enhanced activation of Akt (protein kinase B) in situ. J Investig Dermatol 124(2):298–303
89. Mimura Y, Ihn H, Jinnin M, Asano Y, Yamane K, Tamaki K (2005) Constitutive phosphorylation of focal adhesion kinase is involved in the myofibroblast differentiation of scleroderma fibroblasts. J Investig Dermatol 124(5):886–892

90. Chan MW, Arora PD, Bozavikov P, McCulloch CA (2009) FAK, PIP5KIgamma and gel-solin cooperatively mediate force-induced expression of alpha-smooth muscle actin. J Cell Sci 122(15):2769–2781
91. Hinz B, Phan SH, Thannickal VJ, Galli A, Bochaton-Piallat ML, Gabbiani G (2007) The myofibroblast: one function, multiple origins. Am J Pathol 170(6):1807–1816
92. Bond JE, Ho TQ, Selim MA, Hunter CL, Bowers EV, Levinson H (2011) Temporal spatial expression and function of non-muscle myosin II isoforms IIA and IIB in scar remodeling. Lab Invest 91(4):499–508
93. Hinz B, Mastrangelo D, Iselin CE, Chaponnier C, Gabbiani G (2001) Mechanical tension controls granulation tissue contractile activity and myofibroblast differentiation. Am J Pathol 159(3):1009–1020
94. Serini G, Bochaton-Piallat ML, Ropraz P, Geinoz A, Borsi L, Zardi L, Gabbiani G (1998) The fibronectin domain ED-A is crucial for myofibroblastic phenotype induction by transforming growth factor-beta1. J Cell Biol 142(3):873–881
95. Geiger B, Bershadsky A, Pankov R, Yamada KM (2001) Transmembrane crosstalk between the extracellular matrix–cytoskeleton crosstalk. Nat Rev Mol Cell Biol 2(11):793–805
96. Dugina V, Fontao L, Chaponnier C, Vasiliev J, Gabbiani G (2001) Focal adhesion features during myofibroblastic differentiation are controlled by intracellular and extracellular factors. J Cell Sci 114(18):3285–3296
97. Wipff PJ, Rifkin DB, Meister JJ, Hinz B (2007) Myofibroblast contraction activates latent TGF-beta1 from the extracellular matrix. J Cell Biol 179(6):1311–1323
98. Kohan M, Muro AF, White ES, Berkman N (2010) EDA-containing cellular fibronectin induces fibroblast differentiation through binding to alpha4beta7 integrin receptor and MAPK/Erk 1/2-dependent signaling. FASEB J 24(11):4503–4512
99. Goffin JM, Pittet P, Csucs G, Lussi JW, Meister JJ, Hinz B (2006) Focal adhesion size controls tension-dependent recruitment of alpha-smooth muscle actin to stress fibers. J Cell Biol 172(2):259–268
100. Hinz B, Dugina V, Ballestrem C, Wehrle-Haller B, Chaponnier C (2003) Alpha-smooth muscle actin is crucial for focal adhesion maturation in myofibroblasts. Mol Biol Cell 14(6):2508–2519
101. Gabbiani G, Chaponnier C, Huttner I (1978) Cytoplasmic filaments and gap junctions in epithelial cells and myofibroblasts during wound healing. J Cell Biol 76(3):561–568
102. Spanakis SG, Petridou S, Masur SK (1998) Functional gap junctions in corneal fibroblasts and myofibroblasts. Investig Ophthalmol Vis Sci 39(8):1320–1328
103. Katsumi A, Orr AW, Tzima E, Schwartz MA (2004) Integrins in mechanotransduction. J Biol Chem 279(13):12001–12004
104. Schurch W, Seemayer TA, Hinz B, Gabbiani G (2007) Myofibroblast. In: Mills SE (ed) Histology for pathologists, 3rd edn. Lippincott-Williams & Wilkins, Philadelphia, pp 123–164
105. Gabbiani G (1992) The biology of the myofibroblast. Kidney Int 41(3):530–532
106. Aikawa M, Sivam PN, Kuro-o M, Kimura K, Nakahara K, Takewaki S, Ueda M, Yamaguchi H, Yazaki Y, Periasamy M et al (1993) Human smooth muscle myosin heavy chain isoforms as molecular markers for vascular development and atherosclerosis. Circ Res 73(6):1000–1012
107. van der Loop FT, Schaart G, Timmer ED, Ramaekers FC, van Eys GJ (1996) Smoothelin, a novel cytoskeletal protein specific for smooth muscle cells. J Cell Biol 134(2):401–411
108. Eyden B (2008) The myofibroblast: phenotypic characterization as a prerequisite to understanding its functions in translational medicine. J Cell Mol Med 12(1):22–37
109. Massague J (2012) TGF-beta signalling in context. Nat Rev Mol Cell Biol 13(10):616–630
110. Schnabl B, Kweon YO, Frederick JP, Wang XF, Rippe RA, Brenner DA (2001) The role of Smad3 in mediating mouse hepatic stellate cell activation. Hepatology 34(1):89–100
111. Heldin CH, Miyazono K, ten Dijke P (1997) TGF-beta signalling from cell membrane to nucleus through Smad proteins. Nature 390(6659):465–471

112. Annes JP, Chen Y, Munger JS, Rifkin DB (2004) Integrin alphaVbeta6-mediated activation of latent TGF-beta requires the latent TGF-beta binding protein-1. J Cell Biol 165(5):723–734

113. Keski-Oja J, Koli K, von Melchner H (2004) TGF-beta activation by traction? Trends Cell Biol 14(12):657–659

114. Jenkins G (2008) The role of proteases in transforming growth factor-beta activation. Int J Biochem Cell Biol 40(6/7):1068–1078

115. Sheppard D (2005) Integrin-mediated activation of latent transforming growth factor beta. Cancer Metastasis Rev 24(3):395–402

116. Jian B, Narula N, Li QY, Mohler ER 3rd, Levy RJ (2003) Progression of aortic valve stenosis: TGF-beta1 is present in calcified aortic valve cusps and promotes aortic valve interstitial cell calcification via apoptosis. Ann Thorac Surg 75(2):457–465. Discussion 465-456

117. Walker GA, Masters KS, Shah DN, Anseth KS, Leinwand LA (2004) Valvular myofibroblast activation by transforming growth factor-beta: implications for pathological extracellular matrix remodeling in heart valve disease. Circ Res 95(3):253–260

118. Cushing MC, Mariner PD, Liao JT, Sims EA, Anseth KS (2008) Fibroblast growth factor represses Smad-mediated myofibroblast activation in aortic valvular interstitial cells. FASEB J 22(6):1769–1777

119. Liu AC, Gotlieb AI (2008) Transforming growth factor-beta regulates *in vitro* heart valve repair by activated valve interstitial cells. Am J Pathol 173(5):1275–1285

120. Merryman WD, Lukoff HD, Long RA, Engelmayr GC Jr, Hopkins RA, Sacks MS (2007) Synergistic effects of cyclic tension and transforming growth factor-beta1 on the aortic valve myofibroblast. Cardiovasc Pathol 16(5):268–276

121. Merryman WD (2008) Insights into (the interstitium of) degenerative aortic valve disease. J Am Coll Cardiol 51(14):1415 (author reply 1416)

122. Cushing MC, Liao JT, Anseth KS (2005) Activation of valvular interstitial cells is mediated by transforming growth factor-beta1 interactions with matrix molecules. Matrix Biol 24(6):428–437

123. Yip CY, Chen JH, Zhao R, Simmons CA (2009) Calcification by valve interstitial cells is regulated by the stiffness of the extracellular matrix. Arterioscler Thromb Vasc Biol 29(6):936–942

124. Chen JH, Chen WL, Sider KL, Yip CY, Simmons CA (2011) Beta-catenin mediates mechanically regulated, transforming growth factor-beta1-induced myofibroblast differentiation of aortic valve interstitial cells. Arterioscler Thromb Vasc Biol 31(3):590–597

125. Murakami S, Nagaya N, Itoh T, Fujii T, Iwase T, Hamada K, Kimura H, Kangawa K (2004) C-type natriuretic peptide attenuates bleomycin-induced pulmonary fibrosis in mice. Am J Physiol Lung Cell Mol Physiol 287(6):L1172–L1177

126. Izumiya Y, Araki S, Usuku H, Rokutanda T, Hanatani S, Ogawa H (2012) Chronic C-type natriuretic peptide infusion attenuates angiotensin ii-induced myocardial superoxide production and cardiac remodeling. Int J Vasc Med 2012:246058

127. Kuhnl A, Pelisek J, Tian W, Kuhlmann M, Rolland PH, Mekkaoui C, Fuchs A, Nikol S (2005) C-type natriuretic peptide inhibits constrictive remodeling without compromising re-endothelialization in balloon-dilated renal arteries. J Endovasc Ther 12(2):171–182

128. Peltonen TO, Taskinen P, Soini Y, Rysa J, Ronkainen J, Ohtonen P, Satta J, Juvonen T, Ruskoaho H, Leskinen H (2007) Distinct downregulation of C-type natriuretic peptide system in human aortic valve stenosis. Circulation 116(11):1283–1289

129. Yip CY, Blaser MC, Mirzaei Z, Zhong X, Simmons CA (2011) Inhibition of pathological differentiation of valvular interstitial cells by C-type natriuretic peptide. Arterioscler Thromb Vasc Biol 31(8):1881–1889

130. Wyss K, Yip CY, Mirzaei Z, Jin X, Chen JH, Simmons CA (2012) The elastic properties of valve interstitial cells undergoing pathological differentiation. J Biomech 45(5):882–887

131. Li P, Wang D, Lucas J, Oparil S, Xing D, Cao X, Novak L, Renfrow MB, Chen YF (2008) Atrial natriuretic peptide inhibits transforming growth factor beta-induced Smad signaling and myofibroblast transformation in mouse cardiac fibroblasts. Circ Res 102(2):185–192

132. Balachandran K, Alford PW, Wylie-Sears J, Goss JA, Grosberg A, Bischoff J, Aikawa E, Levine RA, Kit Parker K (2011) Cyclic strain induces dual-mode endothelial-mesenchymal transformation of the cardiac valve. Proc Natl Acad Sci U S A 108(50):19943–19948

133. Leight JL, Wozniak MA, Chen S, Lynch ML, Chen CS (2012) Matrix rigidity regulates a switch between TGF-beta1-induced apoptosis and epithelial-mesenchymal transition. Mol Biol Cell 23(5):781–791

134. Paranya G, Vineberg S, Dvorin E, Kaushal S, Roth SJ, Rabkin E, Schoen FJ, Bischoff J (2001) Aortic valve endothelial cells undergo transforming growth factor-beta-mediated and non-transforming growth factor-beta-mediated transdifferentiation in vitro. Am J Pathol 159(4):1335–1343

135. Yang JH, Wylie-Sears J, Bischoff J (2008) Opposing actions of Notch1 and VEGF in postnatal cardiac valve endothelial cells. Biochem Biophys Res Commun 374(3):512–516

136. Farrar EJ, Butcher JT (2014) Heterogeneous susceptibility of valve endothelial cells to mesenchymal transformation in response to tnf alpha. Ann Biomed Eng 42(1):149–161

137. Johnson CP, Tang HY, Carag C, Speicher DW, Discher DE (2007) Forced unfolding of proteins within cells. Science 317(5838):663–666

138. Lessey EC, Guilluy C, Burridge K (2012) From mechanical force to RhoA activation. Bio-Chemistry 51(38):7420–7432

139. Katoh K, Kano Y, Amano M, Onishi H, Kaibuchi K, Fujiwara K (2001) Rho-kinase-mediated contraction of isolated stress fibers. J Cell Biol 153(3):569–584

140. Zhao XH, Laschinger C, Arora P, Szaszi K, Kapus A, McCulloch CA (2007) Force activates smooth muscle alpha-actin promoter activity through the Rho signaling pathway. J Cell Sci 120(10):1801–1809

141. Bamburg JR, McGough A, Ono S (1999) Putting a new twist on actin: ADF/cofilins modulate actin dynamics. Trends Cell Biol 9(9):364–370

142. Miralles F, Posern G, Zaromytidou AI, Treisman R (2003) Actin dynamics control SRF activity by regulation of its coactivator MAL. Cell 113(3):329–342

143. Wang J, Su M, Fan J, Seth A, McCulloch CA (2002) Transcriptional regulation of a contractile gene by mechanical forces applied through integrins in osteoblasts. The J Biol Chem 277(25):22889–22895

144. Chan MW, Chaudary F, Lee W, Copeland JW, McCulloch CA (2010) Force-induced myofibroblast differentiation through collagen receptors is dependent on mammalian diaphanous (mDia). J Biol Chem 285(12):9273–9281

145. Gu X, Masters KS (2011) Role of the Rho pathway in regulating valvular interstitial cell phenotype and nodule formation. Am J Physiol Heart Circ Physiol 300(2):H448–H458

146. Witt W, Jannasch A, Burkhard D, Christ T, Ravens U, Brunssen C, Leuner A, Morawietz H, Matschke K, Waldow T (2012) Sphingosine-1-phosphate induces contraction of valvular interstitial cells from porcine aortic valves. Cardiovasc Res 93(3):490–497

147. Cote N, Husseini D E, Pepin A, Guauque-Olarte S, Ducharme V, Bouchard-Cannon P, Audet A, Fournier D, Gaudreault N, Derbali H, McKee MD, Simard C, Despres JP, Pibarot P, Bosse Y, Mathieu P (2012) ATP acts as a survival signal and prevents the mineralization of aortic valve. J Mol Cell Cardiol 52(5):1191–1202

148. Bouchareb R, Boulanger MC, Fournier D, Pibarot P, Messaddeq Y, Mathieu P (2013) Mechanical strain induces the production of spheroid mineralized micro particles in the aortic valve through a RhoA/ROCK-dependent mechanism. J Mol Cell Cardiol 67:49–59

149. Monzack EL, Gu X, Masters KS (2009) Efficacy of simvastatin treatment of valvular interstitial cells varies with the extracellular environment. Arterioscler Thromb Vasc Biol 29(2):246–253

150. Witt W, Buttner P, Jannasch A, Matschke K, Waldow T (2014) Reversal of myofibroblastic activation by polyunsaturated fatty acids in valvular interstitial cells from aortic valves. Role of RhoA/G-actin/MRTF signalling. J Mol Cell Cardiol 74:127–138

151. Polyakova V, Hein S, Kostin S, Ziegelhoeffer T, Schaper J (2004) Matrix metalloproteinases and their tissue inhibitors in pressure-overloaded human myocardium during heart failure progression. J Am Coll Cardiol 44(8):1609–1618

152. Balachandran K, Konduri S, Sucosky P, Jo H, Yoganathan AP (2006) An ex vivo study of the biological properties of porcine aortic valves in response to circumferential cyclic stretch. Ann Biomed Eng 34(11):1655–1665
153. Balachandran K, Sucosky P, Jo H, Yoganathan AP (2010) Elevated cyclic stretch induces aortic valve calcification in a bone morphogenic protein-dependent manner. Am J Pathol 177(1):49–57
154. Yang X, Meng X, Su X, Mauchley DC, Ao L, Cleveland JC Jr, Fullerton DA (2009) Bone morphogenic protein 2 induces Runx2 and osteopontin expression in human aortic valve interstitial cells: role of Smad1 and extracellular signal-regulated kinase 1/2. J Thorac Cardiovasc Surg 138(4):1008–1015
155. Balachandran K, Hussain S, Yap CH, Padala M, Chester AH, Yoganathan AP (2012) Elevated cyclic stretch and serotonin result in altered aortic valve remodeling via a mechanosensitive 5-HT(2A) receptor-dependent pathway. Cardiovasc Pathol 21(3):206–213
156. Balachandran K, Bakay MA, Connolly JM, Zhang X, Yoganathan AP, Levy RJ (2011) Aortic valve cyclic stretch causes increased remodeling activity and enhanced serotonin receptor responsiveness. Ann Thorac Surg 92(1):147–153
157. Guilluy C, Rolli-Derkinderen M, Tharaux PL, Melino G, Pacaud P, Loirand G (2007) Transglutaminase-dependent RhoA activation and depletion by serotonin in vascular smooth muscle cells. J Biol Chem 282(5):2918–2928
158. Jian B, Xu J, Connolly J, Savani RC, Narula N, Liang B, Levy RJ (2002) Serotonin mechanisms in heart valve disease I: serotonin-induced up-regulation of transforming growth factor-beta1 via G-protein signal transduction in aortic valve interstitial cells. Am J Pathol 161(6):2111–2121
159. Wigle ED (1957) Myocardial fibrosis and calcareous emboli in valvular heart disease. Br Heart J 19(4):539–549
160. Maron BJ, Ferrans VJ, Roberts WC (1975) Myocardial ultrastructure in patients with chronic aortic valve disease. Am J Cardiol 35(5):725–739
161. Villari B, Campbell SE, Hess OM, Mall G, Vassalli G, Weber KT, Krayenbuehl HP (1993) Influence of collagen network on left ventricular systolic and diastolic function in aortic valve disease. J Am Coll Cardiol 22(5):1477–1484
162. Krayenbuehl HP, Hess OM, Monrad ES, Schneider J, Mall G, Turina M (1989) Left ventricular myocardial structure in aortic valve disease before, intermediate, and late after aortic valve replacement. Circulation 79(4):744–755
163. Grossman W, Jones D, McLaurin LP (1975) Wall stress and patterns of hypertrophy in the human left ventricle. J Clin Investig 56(1):56–64
164. Carabello BA (1995) The relationship of left ventricular geometry and hypertrophy to left ventricular function in valvular heart disease. J Heart Valve Dis 4(Suppl 2):S132–S138. Discussion S138–139
165. Hein S, Arnon E, Kostin S, Schonburg M, Elsasser A, Polyakova V, Bauer EP, Klovekorn WP, Schaper J (2003) Progression from compensated hypertrophy to failure in the pressure-overloaded human heart: structural deterioration and compensatory mechanisms. Circulation 107(7):984–991
166. Flett AS, Hayward MP, Ashworth MT, Hansen MS, Taylor AM, Elliott PM, McGregor C, Moon JC (2010) Equilibrium contrast cardiovascular magnetic resonance for the measurement of diffuse myocardial fibrosis: preliminary validation in humans. Circulation 122(2):138–144
167. Bing OH, Ngo HQ, Humphries DE, Robinson KG, Lucey EC, Carver W, Brooks WW, Conrad CH, Hayes JA, Goldstein RH (1997) Localization of alpha1(I) collagen mRNA in myocardium from the spontaneously hypertensive rat during the transition from compensated hypertrophy to failure. J Mol Cell Cardiol 29(9):2335–2344
168. Creemers EE, Pinto YM (2011) Molecular mechanisms that control interstitial fibrosis in the pressure-overloaded heart. Cardiovasc Res 89(2):265–272
169. Willems IE, Havenith MG, De Mey JG, Daemen MJ (1994) The alpha-smooth muscle actin-positive cells in healing human myocardial scars. Am J Pathol 145(4):868–875

170. Carver W, Nagpal ML, Nachtigal M, Borg TK, Terracio L (1991) Collagen expression in mechanically stimulated cardiac fibroblasts. Circ Res 69(1):116–122

171. Butt RP, Bishop JE (1997) Mechanical load enhances the stimulatory effect of serum growth factors on cardiac fibroblast procollagen synthesis. J Mol Cell Cardiol 29(4):1141–1151

172. Xie J, Zhang Q, Zhu T, Zhang Y, Liu B, Xu J, Zhao H (2014) Substrate stiffness-regulated matrix metalloproteinase output in myocardial cells and cardiac fibroblasts: Implications for myocardial fibrosis. Acta Biomater 10(6):2463–2472

173. Kharaziha M, Nikkhah M, Shin SR, Annabi N, Masoumi N, Gaharwar AK, Camci-Unal G, Khademhosseini A (2013) PGS:Gelatin nanofibrous scaffolds with tunable mechanical and structural properties for engineering cardiac tissues. Biomaterials 34(27):6355–6366

174. MacKenna D, Summerour SR, Villarreal FJ (2000) Role of mechanical factors in modulating cardiac fibroblast function and extracellular matrix synthesis. Cardiovasc Res 46(2):257–263

175. Rosenkranz S, Flesch M, Amann K, Haeuseler C, Kilter H, Seeland U, Schluter KD, Bohm M (2002) Alterations of beta-adrenergic signaling and cardiac hypertrophy in transgenic mice overexpressing TGF-beta(1). Am J Physiol Heart Circ Physiol 283(3):H1253–H1262

176. Brooks WW, Conrad CH (2000) Myocardial fibrosis in transforming growth factor beta(1) heterozygous mice. J Mol Cell Cardiol 32(2):187–195

177. Kuwahara F, Kai H, Tokuda K, Kai M, Takeshita A, Egashira K, Imaizumi T (2002) Transforming growth factor-beta function blocking prevents myocardial fibrosis and diastolic dysfunction in pressure-overloaded rats. Circulation 106(1):130–135

178. Sarrazy V, Koehler A, Chow M, Zimina E, Li CX, Kato H, Caldarone CA, Hinz B (2014) Integrins alphavbeta5 and alphavbeta3 promote latent TGF-beta1 activation by human cardiac fibroblast contraction. Cardiovasc Res 102(3):407–417

179. de Boer RA, Pokharel S, Flesch M, van Kampen DA, Suurmeijer AJ, Boomsma F, van Gilst WH, van Veldhuisen DJ, Pinto YM (2004) Extracellular signal regulated kinase and Smad signaling both mediate the angiotensin II driven progression towards overt heart failure in homozygous TGR(mRen2)27. J Mol Med 82(10):678–687

180. Small EM, Thatcher JE, Sutherland LB, Kinoshita H, Gerard RD, Richardson JA, Dimaio JM, Sadek H, Kuwahara K, Olson EN (2010) Myocardin-related transcription factor-a controls myofibroblast activation and fibrosis in response to myocardial infarction. Circ Res 107(2):294–304

181. Tkatchenko TV, Moreno-Rodriguez RA, Conway SJ, Molkentin JD, Markwald RR, Tkatchenko AV (2009) Lack of periostin leads to suppression of Notch1 signaling and calcific aortic valve disease. Physiol Genomics 39(3):160–168

182. Nus M, Macgrogan D, Martinez-Poveda B, Benito Y, Casanova JC, Fernandez-Aviles F, Bermejo J, de la Pompa JL (2011) Diet-induced aortic valve disease in mice haplo-insufficient for the Notch pathway effector RBPJK/CSL. Arterioscler Thromb Vasc Biol 31(7):1580–1588

183. Laforest B, Andelfinger G, Nemer M (2011) Loss of Gata5 in mice leads to bicuspid aortic valve. J Clin Investig 121(7):2876–2887

184. Wilson CL, Gough PJ, Chang CA, Chan CK, Frey JM, Liu Y, Braun KR, Chin MT, Wight TN, Raines EW (2013) Endothelial deletion of ADAM17 in mice results in defective remodeling of the semilunar valves and cardiac dysfunction in adults. Mech Dev 130(4/5):272–289

185. Kokubo H, Miyagawa-Tomita S, Nakashima Y, Kume T, Yoshizumi M, Nakanishi T, Saga Y (2013) Hesr2 knockout mice develop aortic valve disease with advancing age. Arterioscler Thromb Vasc Biol 33(3):e84–e92

186. Carroll JD, Carroll EP, Feldman T, Ward DM, Lang RM, McGaughey D, Karp RB (1992) Sex-associated differences in left ventricular function in aortic stenosis of the elderly. Circulation 86(4):1099–1107

187. Petrov G, Regitz-Zagrosek V, Lehmkuhl E, Krabatsch T, Dunkel A, Dandel M, Dworatzek E, Mahmoodzadeh S, Schubert C, Becher E, Hampl H, Hetzer R (2010) Regression of myo-

cardial hypertrophy after aortic valve replacement: faster in women? Circulation 122(11 Suppl):S23–S28

188. Sider KL, Blaser MC, Simmons CA (2011) Animal models of calcific aortic valve disease. Int J Inflamm 2011:364310

189. Nadkarni SK, Bouma BE, Helg T, Chan R, Halpern E, Chau A, Minsky MS, Motz JT, Houser SL, Tearney GJ (2005) Characterization of atherosclerotic plaques by laser speckle imaging. Circulation 112(6):885–892

190. Liu H, Sun Y, Simmons CA (2013) Determination of local and global elastic moduli of valve interstitial cells cultured on soft substrates. J Biomech 46(11):1967–1971

191. Bonow RO, Carabello BA, Chatterjee K, de Leon AC Jr, Faxon DP, Freed MD, Gaasch WH, Lytle BW, Nishimura RA, O'Gara PT, O'Rourke RA, Otto CM, Shah PM, Shanewise JS (2008) 2008 Focused update incorporated into the ACC/AHA 2006 guidelines for the management of patients with valvular heart disease: a report of the American College of Cardiology/American Heart Association Task Force on Practice Guidelines (Writing Committee to Revise the 1998 guidelines for the management of patients with valvular heart disease): endorsed by the Society of Cardiovascular Anesthesiologists, Society for Cardiovascular Angiography and Interventions, and Society of Thoracic Surgeons. Circulation 118(15):e523–e661

192. Freeman RV, Otto CM (2005) Spectrum of calcific aortic valve disease: pathogenesis, disease progression, and treatment strategies. Circulation 111(24):3316–3326

193. Elmariah S, Mohler ER 3rd (2010) The Pathogenesis and treatment of the valvulopathy of aortic stenosis: beyond the SEAS. Curr Cardiol Rep 12(2):125–132

194. Miller JD, Weiss RM, Serrano KM, Brooks RM 2nd, Berry CJ, Zimmerman K, Young SG, Heistad DD (2009) Lowering plasma cholesterol levels halts progression of aortic valve disease in mice. Circulation 119(20):2693–2701

195. Miller JD, Weiss RM, Serrano KM, Castaneda LE, Brooks RM, Zimmerman K, Heistad DD (2010) Evidence for active regulation of pro-osteogenic signaling in advanced aortic valve disease. Arterioscler Thromb Vasc Biol 30(12):2482–2486

196. Antonini-Canterin F, Hirsu M, Popescu BA, Leiballi E, Piazza R, Pavan D, Ginghina C, Nicolosi GL (2008) Stage-related effect of statin treatment on the progression of aortic valve sclerosis and stenosis. Am J Cardiol 102(6):738–742

197. Aikawa E, Nahrendorf M, Sosnovik D, Lok VM, Jaffer FA, Aikawa M, Weissleder R (2007) Multimodality molecular imaging identifies proteolytic and osteogenic activities in early aortic valve disease. Circulation 115(3):377–386

198. Butcher JT, Mahler GJ, Hockaday LA (2011) Aortic valve disease and treatment: The need for naturally engineered solutions. Adv Drug Deliv Rev 63(4/5):242–268

199. Agarwal A, Goss JA, Cho A, McCain ML, Parker KK (2013) Microfluidic heart on a chip for higher throughput pharmacological studies. Lab Chip 13(18):3599–3608

200. Srigunapalan S, Lam C, Wheeler AR, Simmons CA (2011) A microfluidic membrane device to mimic critical components of the vascular microenvironment. Biomicrofluidics 5(1):13409

201. Moraes C, Chen JH, Sun Y, Simmons CA (2010) Microfabricated arrays for high-throughput screening of cellular response to cyclic substrate deformation. Lab Chip 10(2):227–234

202. Esch MB, Smith AS, Prot JM, Oleaga C, Hickman JJ, Shuler ML (2014) How multi-organ microdevices can help foster drug development. Adv Drug Deliv Rev 69–70:158–169

Bone Marrow-Derived Progenitor Cells, micro-RNA, and Fibrosis

Alison L. Müller and Darren H. Freed

Abstract Excessive extracellular matrix protein deposition, termed fibrosis, is a multi-faceted process that can exacerbate numerous cardiovascular pathologies and lead to heart failure. Classically thought to be the result of myofibroblasts activated from interstitial fibroblasts endogenously present within the heart, recent research has found that there are numerous cell sources contributing to fibrosis, including various stem cell lineages found in the heart and in the bone marrow. Mesenchymal stem cells recruited to the heart in response to inflammation in a variety of cardiovascular diseases have been directly implicated in extracellular matrix protein deposition. Circulating fibrocytes, which, in addition to expressing both hematopoietic and mesenchymal lineage markers, also express fibroblast markers, and have been shown to aggravate fibrosis. In many cardiovascular disorders that lead to fibrosis, these cells differentiate rapidly to the myofibroblast phenotype in response to injury, a process that is both facilitated and guided by microRNA. The putative role of microRNA has been implied in both the differentiation of bone marrow-derived stem cells and interstitial fibroblasts to myofibroblasts. The significant potential for a link between miRNA and the differentiation of various progenitor cell-types and their contribution to cardiac fibrosis will be explored in this book chapter. This topic invites further consideration for therapeutic potential to combat pathological cardiovascular remodeling.

Keywords Bone marrow-derived cells · Mesenchymal progenitor cells · Fibrocytes · Fibroblasts · Myofibroblasts · microRNA · Fibrosis · Cardiovascular disease · Immunomodulation

D. H. Freed (✉)
Departments of Surgery, Physiology and Biomedical Engineering, University of Alberta, T6G 2P5 Edmonton, AB, Canada
e-mail: dhfreed@ualberta.ca

A. L. Müller
Department of Physiology, University of Alberta, Edmonton, Canada

© Springer International Publishing Switzerland 2015   55
I.M.C. Dixon, J. T. Wigle (eds.), *Cardiac Fibrosis and Heart Failure: Cause or Effect?,*
Advances in Biochemistry in Health and Disease 13, DOI 10.1007/978-3-319-17437-2_4

1   Introduction

The international burden of cardiovascular disease is exacerbated by the excessive extracellular matrix (ECM) protein deposition phenomenon known as fibrosis. This process is multi-faceted and can drive numerous cardiovascular pathologies towards heart failure, the fastest growing subclass of cardiovascular diseases [1, 2]. A prominent trigger of cardiovascular fibrosis is acute cell death, which triggers an inflammatory reaction similar to what occurs during a myocardial infarction (MI) [3]. Pathological fibrosis can also occur in the context of systemic hypertension, aortic stenosis, hypertrophic and dilated cardiomyopathies, and heart transplantation, which impairs survival [2–7]. In addition, aging results in increased levels of interstitial ECM protein production which can stiffen the myocardium and impair overall cardiac function [8].

Traditionally, fibrosis was presumed to be solely the result of myofibroblasts derived from interstitial resident fibroblasts within the heart; however, recent research has found that there are numerous cell sources contributing to fibrosis, including various cell lineages found in the bone marrow [6, 9–13]. Cells of bone-marrow origin include both mesenchymal and/or hematopoietic lineages and contribute to pathological ECM protein deposition in a variety of cardiovascular diseases in both animal models and patients [12, 14–18]. A cell type of particular interest has recently been named fibrocytes, which are derived from bone marrow mesenchymal progenitor cells and express a distinct combination of markers found separately on leukocytes, hematopoietic progenitor cells, and fibroblasts [19–21]. These cells appear to be ubiquitously involved in both physiological and pathological fibrosis within the human body, including the cardiovascular system [19–22]. Other cell sources include endothelial cells undergoing endothelial-to-mesenchymal (Endo-MT) transition [12, 23, 24] and epithelial cells undergoing epithelial-to-mesenchymal transition (EMT) [25, 26].

Although a variety of scientific tools have been used to discover the culprits responsible for cardiovascular fibrosis, it is crucial to be able to understand various aspects of bone marrow derived cells in an *in vitro* setting to better interpret their physiology. This allows easier manipulation and targeted study of these cells to better elucidate the mechanisms behind beneficial fibrosis required for normal wound healing, and how it differs from pathological fibrosis that impairs cardiac functionality. *In vitro* analysis comparing primary human atrial fibroblasts (hAF) with primary human bone marrow-derived progenitor cells (hBMPCs) within the same patient was recently published [27]. It was found that hBMPCs rapidly acquire a myofibroblastic phenotype in culture, the primary cell responsible for cardiac fibrosis, evidenced by expression of α-smooth muscle actin, non-muscle myosin isoforms and EDA-fibronectin. These cells also functionally behave like myofibroblasts, evidenced by contraction of collagen gels at baseline, as well as in response to TGF-β stimulation [28]. Additionally, other research groups can induce these cells to acquire a myofibroblast-like phenotype in culture, particularly through administration of TGF-β1 [29–32]. Understanding how bone marrow-derived progenitor cell differentiate to myofibroblast-like cells could reveal therapeutic targets to prevent overzealous differentiation in a cardiovascular disease environment.

One mechanism that could be influential in the differentiation of various cells to pro-fibrotic phenotypes is microRNA (miRNA). These 20–25 nucleotide-long sequences prevent targeted messenger-RNA (mRNA) from forming proteins and have been shown to influence cell differentiation [33, 34]. Because they can simultaneously interact with multiple mRNA targets, miRNA molecules can act as a master switch to regulate multiple cellular processes, including differentiation. Many different miRNA molecules have been implicated in cell differentiation; however, miR-21, miR-24, miR-29, miR-133, miR-145, and miR-208 have been found to be the most influential in regulating cardiac fibrosis either through controlling myofibroblast differentiation or altering ECM protein deposition [35–40]. Due to the potential role of miRNA in regulating numerous protein targets, for example ECM proteins, miRNA represents an exciting potential avenue for therapeutic targets in preventing, possibly even reversing, pathological fibrosis in the heart.

2 Cell Sources of Fibrosis

The classic cellular source of fibrosis within the cardiovascular system is the myofibroblast, which is derived from resident fibroblasts present within tissue that respond to an inflammatory, pro-fibrotic signal [32]. However, there is a significant portion of ECM protein deposition that is a result of a variety of cells derived from the bone marrow. Numerous studies have been done with various cell-tracing techniques and bone marrow transplant models to understand the relative contribution of myofibroblasts from the bone marrow compartment [9–12, 22, 41]. Using enhanced green fluorescent protein (eGFP)-transgenic mice, two independent groups have shown the contribution of bone marrow-derived cells to cardiac fibrosis [41, 42]. Both studies found that approximately 20–24 % of the myofibroblast cells within the scar expressed eGFP 7 days post-infarct, indicative of their bone-marrow origin. Another comprehensive study used gender mismatched bone marrow transplant mice and evaluated the number of Y chromosome-positive cells in aortic banded and sham female hearts [12]. This study revealed that the banded hearts had 13.4 and 21.1 % of the cells being Y-chromosome FSP-1 and/or α-SMA double positive, respectively, indicating that these cells expressing fibroblast and myofibroblast markers originated in the bone marrow. Interestingly, in sham hearts, there were no FSP-1 positive cells with a Y chromosome; however, about 3.4 % of cells with a Y chromosome were also α-SMA positive which indicates that, even in normal hearts, bone marrow cells contribute to the fibroblast population within the heart. A similar study was performed in heart transplant patients with gender mismatched hearts, who suffered a myocardial infarction post-transplant, allowing tracking of cells into the transplanted heart [15]. As early as 2–4 weeks after infarction, almost 6 % of all cells were non-inflammatory cells. These could either be fibroblasts and/or mesenchymal progenitor cells out of which 24.1 % of the cells were derived from the organ host compared to only 5.3 % of cells in patients without an infarction [15]. These numbers are lower than those found in animal studies, which might indicate that it may take longer for cells to be recruited from the bone marrow to contribute to scar formation in transplant patients as a result of additional complications and/or medications that may interfere with the wound healing process. It is not enough to show that these bone marrow-derived cells are present within the heart post-injury, but also that they directly contribute to ECM protein deposition. To evaluate this, van Amerongen MG et al. [42] utilized a bone marrow transplant model with mice expressing pro-Col1A2 (coding for the α2 chain of pro-collagen I) promoter fused to both luciferase and β-galactosidase. Analysis of *in vivo* luciferase and *in vitro* β-galactosidase activity showed that cells derived from the transplanted pro-Col1A2 transgenic mice bone marrow were present in the post-MI heart and actively expressed collagen.

The studies previously discussed have found that the proportion of these cells significantly increases solely as a result of injury. As injury is attended by inflammation and fibrosis is intimately associated with inflammation [7], it is important to consider how the heart recruits these bone marrow-derived cells. There are distinct profiles of chemokines, cytokines, growth factors, and ECM structure, in heart tissue that is healthy, injured, inflamed, or fibrotic. These profiles are unique and can actually create a specific myocardial signaling milieu that can cause varied differentiation of administered multi-lineage progenitor cells at different stages of myocarditis [11, 43]. Analysis of ischemic heart disease models indicates that bone marrow-derived recruited or administered progenitor cells differentiate into hematopoietic cell types [44–46]. One lab has investigated the importance of the chemokine receptor CCR2 [27], which is found in to be expressed in > 80 % of bone marrow-derived fibroblast precursors. They showed that when CCR2-KO mice are given an angiotensin-II (AngII) infusion, there is a decrease in collagen accumulation and Col1 and fibronectin mRNA expression in the heart in comparison to wild-type controls. They also showed that bone marrow derived fibroblast precursors identified as Col1+ and CD34+/CD45+ were increased in AngII-treated wild-type mice, but these numbers were significantly lower than CCR2 KO mice. Interestingly, there was no difference in cardiac hypertrophy or hypertension between these two groups indicating a unique mechanism for recruitment of bone marrow-derived fibroblasts [27]. Cells expressing CD133 have also been observed infiltrating the myocardium [47], indicating bone marrow-derived hematopoietic lineage origins. These cells also express SDF-1α [48], a common chemokine for hematopoietic progenitor cells, which was found to be significantly up regulated in the myocardium of AngII-exposed animals as early as day one of treatment [17]. Finally, the CXCR4-CXCL12 axis is instrumental in the homing of bone marrow-derived progenitor cells [49], as CXCR4 is a crucial chemokine receptor in stem cell trafficking, and the differential expression of CXCL12 within various tissues allows for trafficking CXCR4+ cells in a gradient-dependent manner to ensure tissue specificity [49, 50]. These studies illustrate the importance of the immune system in cell recruitment, as immunodeficient mice showed no bone marrow-derived cell contribution to fibrosis [51]. This is because the response of the immune system has been shown to be imperative in not only facilitating the migration of bone marrow-derived progenitor cells to fibrogenic areas, but also encouraging their differentiation into pro-fibrotic cells [52].

2.1 Mesenchymal Progenitor Cells (MPCs)

There are two distinct lineages of progenitor cells that reside within the bone marrow that can differentiate into myofibroblasts *in vitro*: hematopoietic stem cells (HSCs) and mesenchymal progenitor cells (MPCs) [53]. MPCs have been shown to share many features with myofibroblasts, which is why their contribution to fibrosis has been studied extensively [27, 29, 30, 54–56]. They are a rare cell population that make up a mere 0.001–0.01 % of bone marrow [57] that can contribute to fibrosis, not only in response to injury or acute inflammation, but also as a result of aging [58]. MPCs are also known to be precursors for cells that can actively form ECM proteins, namely fibroblasts and smooth muscle cells [53]. When comparing MPCs to fibroblasts, it is important to note that initially, MSCs were identified as colony-forming unit fibroblast-like cells [9] and that known fibroblast cell lines, specifically HS68 and NHDF, were screened using a phenotype panel of 22 CD surface markers against MPCs and were found to be identical [24]. A study co-culturing human MPCs with human dermal fibroblasts found that there was MPC-induced fibrosis as a result of the differentiation of the MPCs into myofibroblasts with well-organized α-SMA filaments [59]. MPCs have even been found to possess basal levels of contraction similar to cardiac fibroblasts and respond in a similar way to TGFβ-1 treated cells by strengthening their ability to contract [60] through co-expression of αSMA [61]. During culturing on plastic dishes, MPCs were found to produce collagen, although this level of production did not increase in response to TGF-β1, unlike that observed with human myofibroblasts obtained from atrial tissue [60, 62, 63]. TGF-β1 is an important ligand to consider as it induces the differentiation of fibroblasts to collagen-forming myofibroblasts [61]. Human MPCs have also been found to differentiate into fibroblasts by cyclic mechanical stimulation, in addition to many other growth factors such as connective tissue growth factor and fibroblast growth factor-2 [64–67]. A unique feature of MPCs is their ability to have an intercellular connection with resident cardiomyocytes via connexin 43, which is also shown to occur with cardiac fibroblasts [68–70]. This allows for electrical signal transduction that further emphasizes the similarities between these cell types, as well as allowing for the potential for endogenous differentiation of MPCs to myofibroblasts [71, 72].

2.2 Fibrocytes

Recently a unique cell type contributing to cardiac fibrosis was identified, sharing the cell markers of leukocytes, hematopoietic progenitor cells, and fibroblasts (collagen+/CD13+/CD34+/CD45+) [21, 73]. Within the literature, these cells have been given a variety of names including telocytes, CD34+ stromal cells, and fibrocytes [74]. For the purposes of this chapter we shall refer to these cells as fibrocytes, which originate from immature mesenchymal cells within the peripheral blood but express the hematopoietic lineage marker CD34+ [74]. They are ubiquitously present throughout the body and have been found to be involved in synthesizing substrates, immunomodulation, providing scaffolding support for other cells, phagocytosis, parenchymal regulation, as well as synthesizing and remodeling ECM [74]. Although classified as a subset of fibroblasts, they have also been found to be able to differentiate into adipogenic, osteoblastic, and chondrogenic lineages [75]. They have been discovered in animal models of atherosclerosis [76], within the fibrous cap of human carotid artery plaques [77], in patients with atrial fibrosis in the context of chronic atrial fibrillation [78], and in patients with hypertrophic cardiomyopathy [79]. Repetitive episodes of ischemia-reperfusion inducing fibrotic cardiomyopathy resulted in an increase in fibrocytes, as indicated by expression of collagen-1, αSMA, CD34, and CD45 [9]. The presence of these cells in various cardiovascular pathologies, accompanied by fibrotic remodeling, indicates that they are potentially a key target in ameliorating detrimental ECM remodeling. They have shown to be robustly active in producing ECM proteins including collagen I, collagen III, and vimentin, in addition to secreting matrix metalloproteinases (MMPS), key regulators of cardiac ECM remodeling [80, 81]. As previously discussed in this chapter, immunomodulation is an important factor in promoting fibrosis and fibrocytes uniquely contribute by being a source of inflammatory cytokines, chemokines, and growth factors. During wound healing, fibrocytes express IL-1β, IL-10, TNF-α, MIP1α, MIP1β, MIP-2, PDFG-A, and TGFβ1 and actually localize to granuloma formation and connective matrix deposition [80]. Studies have shown that fibrocytes are released from the bone marrow into the peripheral blood to hone into zones of inflammation via a CCR2-mediated pathway [27]. Fibrocytes also have the ability to undergo phenotypic changes by being proliferative, synthetic, and/or contractile and, by expressing markers such as endosialin and integrin receptors, they can bind to fibronectin and fibrin to situate themselves in close proximity with resident fibroblasts [82, 83].

2.3   Other Differentiating Cell Sources

Although bone marrow-derived progenitor cells contribute significantly to fibrosis, they are not the only cell-type capable of changing their identity to become pro-fibrotic. Endothelial-to-mesenchymal transition (EndMT) has been shown to contribute to fibrosis by forming fibroblasts [12, 84]. In a cardiac context, the endothelial cells making up the endocardium in development form heart valve progenitor cells. During development, some of these cells undergo EndMT in order to form valve interstitial cells that then further mature into valve mesenchymal cells and valve leaflets [85]. An important factor in triggering this cell transformation in an embryonic context is TGF-β [86], which is also found extensively in fibrotic areas of the heart and is responsible for fibroblast activation [87]. In a lineage tracing analysis utilizing LacZ irreversibly labelled endothelial cells, Ziesber et al. [12] found that in mice exposed to pressure-overload for 5 days, nearly 33 % of their cardiac fibroblasts originated from the endothelial layer. Not only do EndMT cells further complicate fibrosis by forming fibroblasts, but there is also a net loss of endothelial cells that needs to be considered. In a study investigating the role of endothelial cells in chronic kidney injury, it was found that reduced bioavailability of nitric oxide synthesized by endothelial cells causes further endothelial dysfunction where there was up-regulation of collagen, increased TGF-β and rarefaction of capillaries [88].

Another key cellular transformation process that occurs during embryonic heart development is epithelial-to-mesenchymal transition. This mechanism has also been implicated in pathological ECM remodeling of the heart by up-regulating a number of genes encoding growth factors including VEGF, FGF2, TGF-β2, and MCP1 which promote angiogenesis and might be beneficial in reducing heart injury post-MI [89]. In addition, epithelial cells of the epicardium also have the ability to differentiate into fibroblasts [90] and, although this has been established to be a key source of fibroblasts during fibrosis in kidney [91], liver [92], and lungs [93], it is yet to be determined if it contributes to pathological fibrosis in the heart.

In addition to progenitor cells found within the bone marrow, a distinct population of cardiac stem cells has been found residing in the heart [94]. They were found to be negative for various hematopoietic lineage markers (such as CD34, CD45, CD8) but are positive for c-kit, a stem cell marker. Although these cells are being evaluated for their potential to become functional cardiomyocytes for therapeutic purposes [94, 95], there is currently no information available about the potential of these cells to differentiate into myofibroblasts or any other cell type that could contribute to cardiac fibrosis.

3 Influence of miRNA on Cell Differentiation and Fibrosis

When injury occurs in the body, cells have to respond in a promptly in order to initiate healing and prevent further damage. We have established that there are numerous cells within the body that undergo a phenotype change in order participate in fibrosis and that inflammation is responsible for guiding these cells to the damaged area to promote ECM remodeling. In many circumstances, this change has to occur rapidly where the cell essentially "switches" phenotype. As it would take time for a cell to undergo complete protein translation based off of extracellular signaling causing the activation of transcription factors, there has recently been discovered a molecule that could rapidly facilitate this "switch", known as miRNA. This molecule is made up of 20–25 nucleotides derived from mostly non-coding regions of the genome that target mRNA to fine-tune protein translation [33, 96]. The first miRNA was discovered during the adult stage of *Caenorhabditis elegans* where loss of the miRNA, let-7, caused reiteration to larval cell fates indicating that let-7 is necessary for the differentiation of adult cells [33]. Furthermore, let-7 has also been shown to be important in the development of adult fibroblasts in adult MPCs [97]. In a study evaluating the effect of Dicer, which processes miRNA permitting them to function, Dicer KO MPCs retained their mesenchymal identity. This shows that miRNAs do not necessarily govern cell identity but that miRNA processing is required for active differentiation, suggesting its presence critical for cell state transition [97]. This data supports additional findings that demonstrate the importance of miRNAs in regulating cellular transitions and physiological robustness in various model systems [98, 99]. MiRNAs have also shown to be central in the dedifferentiation of human fibroblasts. As Yamanaka demonstrated, fibroblasts can be reprogrammed back to an embryonic-like state [100], which prompted several scientists to look at various aspects of this phenomenon. The two that were found to promote the reprogramming of human fibroblasts to inducible pluripotent stem cells were miR-302b and miR-372 [101].

A miRNA has also been identified in cardiac myofibroblast differentiation, namely miR-145 [40]. This particular miRNA was initially identified as a regulator of smooth muscle cell differentiation. Several aspects of how miR-145 influences myofibroblast differentiation were investigated, with a focus on α-SMA filaments. When treated with miR-145, α-SMA positive cells readily formed stress filaments and were organized in parallel actin-filament bundles, permitting cellular contractility that was comparable to fibroblasts treated with TGFβ1 [40]. In addition, fibroblasts treated with miR-145 migrated to a comparable degree as fibroblasts treated with TGFβ. Collagen expression was also assessed, with miR-145 treated cells showing comparable decreases in pro-collagen 1A1 and pro-collagen 1A2 but significant comparable increases in mature collagen 1A1 and collagen 1A2 production. As miRNA have the potential to have numerous downstream targets, which is the nature of their behavior and natural design, miR-145 was found to target KLF5. KLF5 is instrumental in cardiovascular remodeling, particularly in smooth muscle cells, as it activates platelet-derived growth factor A/B, Egr-1, plasminogen activator inhibitor-1, inducible nitric oxide synthase, and vascular endothelial growth factor [102]. Whether miR-145 is the sole miRNA to facilitate this differentiation from fibroblast to myofibroblast is yet to be determined.

The expression of several miRNAs, including several let-7s, miR-1, miR-133a, miR-133b, miR-19a, miR-19b, miR-150, miR-195, miR-199, miR-221, miR-23a, miR-23b, miR-29a, miR-29b, the miR-30 family and miR-320, either increased or decreased during heart failure [103]. The highest expression miRNA in the heart is miR-1, which accounts for 40 % of all cardiac miRNAs [104]. It plays a role in the regulation of the cardiac conduction system, in part by controlling the expression of connexin 43 [105]. It has been shown to have differential expression based on the whether or not the heart is succumbing to short-term injury such as ischemic injury, where it is up-regulated, or long-term injury such as hypertrophy and heart failure, where it is down-regulated [103]. Its potential role in fibrosis has yet to be elucidated although, due to its high expression in the heart, it is unlikely that it is a bystander during pathological cardiac ECM remodeling. MiR-133a may be more directly implicated in fibrosis as high expression of miR-133a prevented both fibrosis and apoptosis in an animal model of trans-aortic constriction (TAC) [36]. It did not affect hypertrophy in either the TAC model or in isoproterenol-induced hypertrophy.

Another miRNA of interest is miR-208a because, when it is knocked out in TAC, there is no hypertrophy of cardiomyocytes or fibrosis detected [37]. In a different study evaluating miR-208a, it was found that in areas associated with interstitial fibrosis, there was miR-208a expression which occurred concurrently with a decrease in connexin 40, indicating that there might be a link between miR-208a, cardiac conduction, and fibrosis [38].

Another few miRNA regulators of cardiac fibrosis are miR-24, miR-21, and miR-29 [39, 106–109]. MiR-24 is found in fibroblasts and its expression has been found to be associated with the degree of fibrosis in hypertrophic hearts [110]. Post-MI, miR-24 is down-regulated only after an initial week of elevated expression where its expression correlated with the levels of collagen-1, fibronectin, and TGF-β1 [39]. When delivered *in vivo,* miR-24 reduces cardiac fibrosis and decreases collagen-1 protein expression. Not only is miR-24 expressed in fibrocytes, it is also found to decrease the expression of α-SMA, a myofibroblast marker [39]. One of the downstream targets of miR-24 is furin, which is a regulator of the TGF-β pathway by proteolytically maturing TGF-β, which correlates with the observation that TGF-β1 treatment up-regulates miR-24 expression in a time- and dose-dependent manner [39]. Both miR-21 and miR-29 have been identified in cardiac fibroblasts, where their expression is greater than that found in cardiomyocytes [106, 107]. Over-expression of miR21 in the heart results in reduced infarct size after ischemia reperfusion injury with preserved ejection fraction and decreased collagen deposition [111], in addition to promoting fibroblast survival [108, 109]. Conversely, knock-down of miR-21 attenuated interstitial fibrosis and cardiac remodeling after aortic banding [109]. In the myocardial infarct border zone, miR-21 was upregulated, inhibiting PTEN expression, which resulted in up-regulation of MMP-2 expression [108].

Finally, a very popular miRNA of interest in the fibrosis literature is miR-29, which regulates a subset of fibrosis-related gene expression including many collagens, fibrillins, laminins, integrins, and elastin [107]. When down-regulated, there is an increase of these ECM proteins found which is then attenuated with over-expression of miR-29 [107]. Although inhibiting miR-29 would be an ideal therapeutic target to prevent pathological ECM remodeling, it also regulates several anti-apoptotic genes including *Tcl-1*, *Mcl-1*, *p53*, and *CDC42* [112–114]. The continuing investigation of how miRNAs are involved in both physiological and pathological ECM remodeling is an additional therapeutic target that could be utilized in combatting fibrosis, not only within the heart, but in other fibrotic diseases as well.

4   Conclusions

It is evident that more than endogenously present interstitial fibroblasts within the myocardium contribute to fibrosis, both physiologically and pathologically. This multi-faceted process drives numerous cardiovascular pathologies toward heart

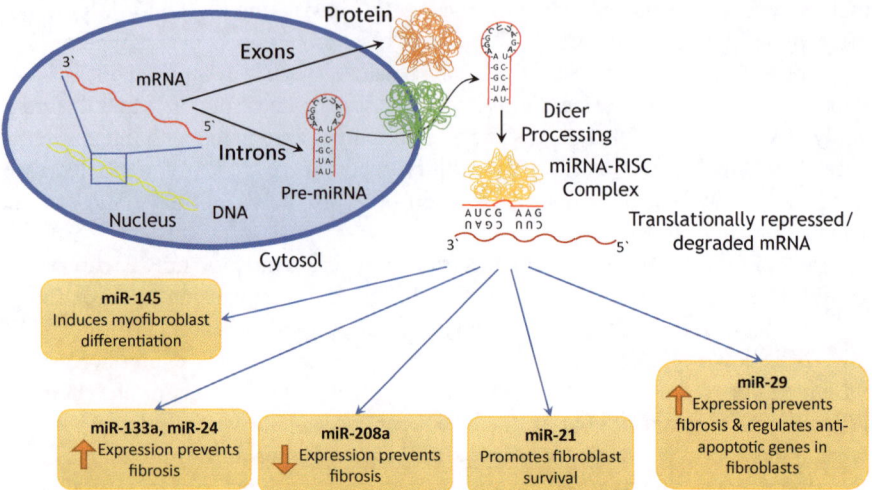

Fig. 1 Schematic diagram of microRNA formation mechanism and the roles of specific microR-NAs involved in cardiac fibrosis (*mRNA* messenger RNA, *Pre-miRNA* pre-microRNA, *miRNA-RISC* RNA-induced silencing complex)

failure and occurs via a number of different cell sources, including various cell lineages found in the bone marrow [9–12, 16]. These include fibrocytes and MPCs that contribute directly to ECM remodeling and protein deposition, and appear ubiquitously involved in both physiological and pathological fibrosis [19, 22]. We have also outlined how similar fibroblasts and MPCs act in culture and under stimulation of pro-fibrotic TGF-β. Additional sources of pro-fibrotic cells are endothelial cells undergoing endothelial-to-mesenchymal (EndoMT) transition [12, 23, 24] and epithelial cells undergoing epithelial-to-mesenchymal transition (EMT) [25, 28].

A thorough understanding of how these various cell types become pro-fibrotic could reveal therapeutic targets for preventing overzealous ECM deposition and remodeling. We explored the possibility of miRNA in inducing cell differentiation to a pro-fibrotic phenotype and/or contributing directly to fibrosis including miR-21, miR-24, miR-29, miR-133, miR-145, and miR-208a [35, 36, 38–40, 107] (Fig. 1). miRNA provides a potential therapeutic target to either promote physiological fibrosis for optimal wound healing, or to inhibit pathological cardiac ECM remodeling.

References

1. Elzenaar I, Pinto YM, van Oort RJ (2013) MicroRNAs in heart failure: New targets in disease management. Clin Pharmacol Ther 94:480–489
2. Kong P, Christia P, Frangogiannis NG (2014) The pathogenesis of cardiac fibrosis. Cell Mol Life Sci 71:549–574
3. Frangogiannis NG (2012) Regulation of the inflammatory response in cardiac repair. Circ Res 100:159–173

4. Berk BC, Fujiwara K, Lehoux S (2007) ECM remodeling in hypertensive heart disease. J Clin Invest 117:568–575
5. Ashrafian H, McKenna WJ, Watkins H (2011) Disease pathways and novel therapeutic targets in hypertrophic cardiomyopathy. Circ Res 109:86–96
6. Kania G, Blyszczuk P, Eriksson U (2009) Mechanisms of cardiac fibrosis in inflammatory heart disease. Trends Cardiovasc Med 19:247–252
7. Pickering JG, Boughner DR (1990) Fibrosis in the transplanted heart and its relation to donor ischemic time. Assessment with polarized light microscopy and digital image analysis. Circulation 81:949–958
8. Cieslik KA, Trial J, Crawford JR et al (2014) Adverse fibrosis in the aging heart depends on signaling between myeloid and mesenchymal cells; role of inflammatory fibroblasts. J Mol Cell Cardiol 70:56–63
9. Haudek SB, Xia Y, Huebener P et al (2006) Bone marrow-derived fibroblast precursors mediate ischemic cardiomyopathy in mice. Proc Natl Acad Sci U S A 103:18284–18289
10. Blyszczuk P, Kania G, Dieterle T et al (2009) Myeloid differentiation factor-88/interleukin-1 signaling controls cardiac fibrosis and heart failure progression in inflammatory dilated cardiomyopathy. Circ Res 105:912–920
11. Kania G, Blyszczuk P, Stein S et al (2009) Heart-infiltrating prominin-1+/CD133+ progenitor cells represent the cellular source of transforming growth factor beta-mediated cardiac fibrosis in experimental autoimmune myocarditis. Circ Res 105:462–470
12. Zeisberg EM, Tarnavski O, Zeisberg M et al (2007) Endothelial-to-mesenchymal transition contributes to cardiac fibrosis. Nat Med 13:952–961
13. Lei PP, Qu YQ, Tao SM et al (2013) Fibrocytes are associated with the fibrosis of coronary heart disease. Pathol Res Pract 209: 36–43
14. Korhonen J, Polvi A, Partanen J et al (1994) The mouse tie receptor tyrosine kinase gene: Expression during embryonic angiogenesis. Oncogene 9:395–403
15. Hocht-Zeisberg E, Kahnert H, Guan K et al (2004) Cellular repopulation of myocardial infarction in patients with sex-mismatched heart transplantation. Eur Heart J 25:749–758
16. Villarreal FJ, Dillman WH (1992) Cardiac hypertrophy-induced changes in mRNA levels for TGF-beta 1, fibronectin, and collagen. Am J Physiol Heart Circ Physiol 262:H1861–H1866
17. Sopel MJ, Rosin NL, Lee TD et al (2011) Myocardial fibrosis in response to angiotensin II is preceded by the recruitment of mesenchymal progenitor cells. Lab Invest 91:565–578
18. Quaini F, Urbanek K, Beltrami AP et al (2002) Chimerism of the transplanted heart. N Engl J Med 346:5–15
19. Keeley EC, Mehrad B, Strieter RM (2009) The role of circulating mesenchymal progenitor cells (fibrocytes) in the pathogenesis of fibrotic disorders. Thromb Haemost 101:613–618
20. Bucala R, Spiegel LA, Chesney J et al (1994) Circulating fibrocytes define a new leukocyte subpopulation that mediates tissue repair. Mol Med 1:71–81
21. Ebihara Y, Masuya M, Larue AC et al (2006) Hematopoietic origins of fibroblasts: II. *In vitro* studies of fibroblasts, CFU-F, and fibrocytes. Exp Hematol 34:219–229
22. Keeley EC, Mehrad B, Strieter RM (2010) Fibrocytes: Bringing new insights into mechanisms of inflammation and fibrosis. Int J Biochem Cell Biol 42:535–542
23. Goumans MJ, van Zonneveld AJ, ten Dijke P (2008) Transforming growth factor beta-induced endothelial-to-mesenchymal transition: A switch to cardiac fibrosis? Trends Cardiovasc Med 18:293–298
24. Wynn TA (2008) Cellular and molecular mechanisms of fibrosis. J Pathol 2142:199–210.
25. Russell JL, Goetsch SC, Gaiano NR et al (2011) A dynamic notch injury response activates epicardium and contributes to fibrosis repair. Circ Res 108:51–59
26. Duan J, Gherghe C, Liu D et al (2011) Wnt1/betacatenin injury response activates the epicardium and cardiac fibroblasts to promote cardiac repair. EMBO J 31:429–442
27. Ngo M, Müller AL, Li Y et al (2014) Human mesenchymal stem cells express a myofibroblastic phenotype *in vitro*: comparison to human cardiac myofibroblasts. Mol Cell Biochem 392:187–204

28. Serini G, Gabbiani G (1999) Mechanisms of myofibroblast activity and phenotype modulation. Exp Cell Res 250:273–283
29. Friedenstein AJ, Gorskaja JF, Kulagina NN (1976) Fibroblast precursors in normal and irradiated mouse hematopoietic organs. Exp Hematol 4:267–274
30. Covas DT, Panepucci RA, Fontes AM et al (2008) Multipotent mesenchymal stromal cells obtained from diverse human tissues share functional properties and gene-expression profile with CD146+ perivascular cells and fibroblasts. Exp Hematol 36:642–654
31. Roberts AB, Sporn MB, Assoian RK et al (1986) Transforming growth factor type beta: rapid induction of fibrosis and angiogenesis *in vivo* and stimulation of collagen formation *in vitro*. Proc Natl Acad Scis U S A 83:4167–4171
32. Kinner B, Zaleskas JM, Spector M (2002) Regulation of smooth muscle actin expression and contraction in adult human mesenchymal stem cells. Exp Cell Res 278:4167–4171
33. Reinhart BJ, Slack FJ, Basson M et al (2000) The 21-nucleotide *let-7* RNA regulated developmental timing in *Caenorhabditis elegans*. Nature 203:901–906
34. Choi E, Choi E, Hwang KC (2013) MicroRNAs as novel regulators of stem cell fate. World J Stem Cells 5:172–187
35. Tomé M, López-Romero P, Albo C et al (2011) miR-355 orchestrates cell proliferation, migration and differentiation in human mesenchymal stem cells. Cell Death Differ 18:985–995
36. Matkovich SJ, Wang W, Tu Y et al (2010) MicroRNA-133a protects against myocardial fibrosis and modulates electrical repolarization without affecting hypertrophy in pressure-overloaded adult hearts. Circ Res 106:166–175
37. van Rooij E, Quiat D, Johnson BA et al (2007) Control of stress-dependent cardiac growth and gene expression by a microRNA. Science 316:575–579
38. Callis TE, Pandya K, Seok HY et al (2009) MicroRNA-208a is a regulator of cardiac hypertrophy and conduction in mice. J Clin Invest 119:2772–2786
39. Wang J, Huang W, Xu R et al (2012) MicroRNA-24 regulates cardiac fibrosis after myocardial infarction. J Cell Mol Med 16:2150–2160
40. Wang YS, Li SH, Guo J et al (2014) Role of miR-145 in cardiac myofibroblast differentiation. J Mol Cell Cardiol 66:94–105
41. Möllman H, Nef HM, Kostin S et al (2006) Bone marrow-derived cells contribute to infarct remodeling. Cardiovasc Res 71:661–671
42. van Amerongen MG, Bou-Gharios G, Popa E et al (2008) Bone marrow-derived myofibroblasts contribute functionally to scar formation after myocardial infarction. J Pathol 214:377–386
43. Kania G, Blyszczuk P, Valaperti A et al (2008) Prominin-1+/CD133+ bone marrow-derived heart-resident cells suppress experimental autoimmune myocarditis. Cardiovasc Res 80:236–245
44. Balsam LP, Wagers AJ, Christensen JL et al (2004) Haematopoietic stem cells adopt mature haematopoietic fates in ischaemic myocardium. Nature 428:668–673
45. Endo J, Sano M, Fujita J et al (2007) Bone marrow derived cells are involved in the pathogenesis of cardiac hypertrophy in response to pressure overload. Circulation 116:1176–1184
46. Murry CE, Soonpaa MH, Reinecke H et al (2004) Haematopoietic stem cells do not transdifferentiate into cardiac myocytes in myocardial infarcts. Nature 428:664–668
47. Handgretinger R, Gordon PR, Leimig T et al (2003) Biology and plasticity of CD133+ hematopoietic stem cells. Ann N Y Acad Sci 996: 141–151
48. Mohle R, Bautz F, Rafii S et al (1998) The chemokine receptor CXCR-4 is expressed on CD34+ hematopoietic progenitors and leukemic cells and mediates transendothelial migration induced by stromal cell-derived factor-1. Blood 91:4523–4530
49. Murdoch C (2000) CXCR: Chemokine receptor extraordinaire. Immunol Rev 177:175–184
50. Phillips RJ, Burdick MD, Hong K et al (2004) Circulating fibrocytes traffic to the lungs in response to CXCL12 and mediate fibrosis. J Clin Invest 114:438–446
51. Yano T, Miura T, Ikeda Y et al (2005) Intracardiac fibroblasts, but not bone marrow derived cells, are the origin of myofibroblasts in myocardial infarct repair. Cardiovasc Pathol 14:241–246

52. Abe R, Donnelly SC, Peng T et al (2001) Peripheral blood fibrocytes: Differentiation pathway and migration to wound sites. J Immunol 166:7556–7562
53. Shah RV, Mitchell RN (2005) The role of stem cells in the response to myocardial and vascular wall injury. Cardiovasc Pathol 14:225–231
54. Bianco P, Robey PG, Simmons PJ (2008) Mesenchymal stem cells: Revisiting history, concepts, and assays. Cell Stem Cell 2:313–319
55. Wagner W, Ho AD (2007) Mesenchymal stem cell preparations—comparing apples and oranges. Stem Cell Rev 3:239–248
56. Ball SG, Shuttleworth AC, Kielty CM (2004) Direct cell contact influences bone marrow mesenchymal stem cell fate. Int J Biochem Cell Biol 36:714–727
57. Novotny NM, Ray R, Markel TA et al (2008) Stem cell therapy in myocardial repair and remodeling. J Am Coll Surg 207:423–434
58. Crawford JR, Haudek SB, Cieslik KA et al (2012) Origin of developmental precursors dictates the pathophysiologic role of cardiac fibroblasts. J Cardiovasc Transl Res 5:749–759
59. Dawn B, Stein AB, Urbanek K et al (2005) Cardiac stem cells delivered intravascularly traverse the vessel barrier, regenerate infarcted myocardium, and improve cardiac function. Proc Natl Acad Sci U S A 102:3766–3771
60. Paul D, Samuel SM, Maulik N (2009) Mesenchymal stem cell: present challenges and prospective cellular cardiomyoplasty approaches for myocardial regeneration. Antioxid Redox Signal 11:1841–1855
61. Ishikawa F, Shimazu H, Shultz LD et al (2006) Purified human hematopoietic stem cells contribute to the generation of cardiomyocytes through cell fusion. FASEB J 20:950–952
62. Docheva D, Popov C, Mutschler W et al (2007) Human mesenchymal stem cells in contact with their environment surface characteristics and the integrin system. J Cell Mol Med 11:21–38
63. Gronthos S, Simmons PJ, Graves SE et al (2001) Integrin-mediated interactions between human bone marrow stromal precursor cells and the extracellular matrix. Bone 28:174–181
64. Gnecchi M, He H, Liang OD et al (2005) Paracrine action accounts for marked protection of ischemic heart by Akt-modified mesenchymal stem cells. Nat Med 11:367–368
65. Gurtner GC, Chang E (2008) "Priming" endothelial progenitor cells: a new strategy to improve cell based therapeutics. Arterioscler Thromb Vasc Biol 28:1034–1035
66. Hahn JY, Cho HJ, Kang HJ et al (2008) Pre-treatment of mesenchymal stem cells with a combination of growth factors enhances gap junction formation, cytoprotective effect on cardiomyocytes, and therapeutic efficacy for myocardial infarction. J Am Coll Cardiol 51:933–943
67. Haider H, Jiang S, Idris NM et al (2008) IGF-1-overexpressing mesenchymal stem cells accelerate bone marrow stem cell mobilization via paracrine activation of SDF-1alpha/CXCR4 signaling to promote myocardial repair. Circ Res 103:1300–1308
68. Valiunas V, Doronin S, Valiuniene L et al (2004) Human mesenchymal stem cells make cardiac connexins and form function gap junctions. J Physiol 555:617–626
69. Kohl P (2003) Heterogeneous cell coupling in the heart: An electrophysiological role for fibroblasts. Circ Res 93:381–383
70. Chilton L, Giles WR, Smith GL (2007) Evidence of intercellular coupling between co-cultured adult rabbit ventricular myocytes and myofibroblasts. J Physiol 583:225–236
71. Gaudesius G, Miragoli M, Thomas SP et al (2003) Coupling of cardiac electrical activity over extended distances by fibroblasts of cardiac origin. Circ Res 93:421–428
72. Miragoli M, Gaudesius G, Rohr S (2006) Electronic modulation of cardiac impulse conduction by myofibroblasts. Circ Res 98:801–810
73. Bucala R, Spiegel LA, Chesney J et al (1994) Circulating fibrocytes define a new leukocyte subpopulation that mediates tissue repair. Mol Med 1:71–81
74. Díaz-Flores L, Gutiérrez R, García MP et al (2014) CD34+ stromal cells/fibroblasts/fibrocytes/telocytes as a tissue reserve and a principle source of mesenchymal cells. Location, morphology, function, and role in pathology. Histol Histopathol 29:831–870

75. Haniffa MA, Wang XN, Hotlick U et al (2007) Adult human fibroblasts are potent immunoregulatory cells and functionally equivalent to mesenchymal stem cells. J Immunol 179:1595–1604
76. Zulli A, Buxton BF, Black MJ et al (2005. CD34 Class III positive cells are present in atherosclerotic plaques of the rabbit model of atherosclerosis. Histochem Cell Biol 124:517–522
77. Medbury HJ, Tarran SL, Guiffre AK et al (2008) Monocytes contribute to the atherosclerotic cap by transformation into fibrocytes. Int Angiol 27:114–123
78. Xie X, Liu Y, Gao S et al (2014) Possible involvement of fibrocytes in atrial fibrosis in patients with chronic atrial fibrillation. Circ J 78:338–344
79. Fang L, Beale A, Ellims AH et al (2013) Associations between fibrocytes and postcontrast myocardial T1 times in hypertrophic cardiomyopathy. J Am Heart Assoc 2:e000270
80. Chesney J, Metz C, Stavisky AB et al (1998) Regulated production of type I collagen and inflammatory cytokines by peripheral blood fibrocytes. J Immunol 160:419–425
81. D'Armiento J (2002) Matrix metalloproteinase disruption of the extracellular matrix and cardiac dysfunction. Trends Cardiovasc Med 12:97–101
82. McClain SA, Simon M, Jones E et al (1996) Mesenchymal cell activation is the rate-limiting step of granulation tissue induction. Am J Pathol 149:1257–1270
83. Xu J, Clark RA (1997) A three-dimensional collagen lattice induces protein kinase C-zeta activity: role in alpha-2 integrin and collagenase mRNA expression. J Cell Biol 136:473–483
84. Aisagonhi O, Rai M, Ryzhov S et al (2011) Experimental myocardial infarction triggers canonical Wnt signaling and endothelial-to-mesenchymal transition. Dis Model Mech 4:469–483
85. von Gise A, Pu WT (2012) Endocardial and epicardial epithelial to mesenchymal transitions in heart development and disease. Circ Res 110:1628–1645
86. Goumans MJ, Mummery C (2000) Functional analysis of the TGFb receptor/Smad pathway through gene ablation in mice. Int J Dev Biol 443:253–265
87. Khan R, Sheppard R (2006) Fibrosis in heart disease: understanding the role of transforming growth factor-beta in cardiomyopathy, valvular disease, and arrhythmia. Immunology 1181:10–24
88. O'Riordan E, Mendelev N, Patschan S et al (2007) Chronic NOS inhibition actuates endothelial-mesencymal transformation. Am J Physiol Heart Circ Physiol 292:H285–H294
89. Zhou B, Honor LB, He H et al (2011) Adult mouse epicardium modulates myocardial injury by secreting paracrine factors. J Clin Invest 121:1894–1904
90. Smith CL, Baek ST, Sung CY et al (2011) Epicardial-derived cell epithelial-to-mesenchymal transition and fate specification require PDGF receptor signaling. Circ Res 295:507–522
91. Iwano M, Plieth D, Danoff TM et al (2002) Evidence that fibroblasts derive from epithelium during tissue fibrosis. J Clin Invest 110:341–350
92. Rygiel KA, Robertson H, Marshall HL (2008) Epithelial-mesenchymal transition contributes to portal tract fibrogenesis during human chronic liver disease. Lab Invest 88:112–123
93. Willis BC, Liebler JM, Luby-Phelps K (2005) Induction of epithelial-mesenchymal transition in alveolar epithelial cells by transforming growth factor-beta1: potential role in idiopathic pulmonary fibrosis. Am J Pathol 166:1321–1332
94. Beltrami AP, Barlucchi L, Torella D et al (2003) Adult cardiac stem cells are multipotent and support myocardial regeneration. Cell 114:763–776
95. Oh H, Bradfute SB, Gallardo TD (2003) Cardiac progenitor cells from adult myocardium: Homing, differentiation, and fusion after infarction. Proc Natl Acad Sci U S A 100:12313–12318
96. Bartel DP, Chen CA (2004) Micromanagers of gene expression: the potentially widespread influence of metazoan microRNAs. Nat Rev Genet 5:396–400
97. Gurtan AM, Ravi A, Rahl PB et al (2013) Let-7 represses Nr6a1 and a mid-gestation developmental program in adult fibroblasts. Genes Dev 27:941–954
98. Herranz H, Cohen SM (2010) MicroRNAs and gene regulatory networks: managing the impact of noise in biological systems. Genes Dev 24:1339–1344
99. Ebert MS, Sharp PA (2012) Roles for microRNAs in conferring robustness to biological processes. Cell 149:515–524

100. Takahashi K, Yamanaka S (2006) Induction of pluripotent stem cells from mouse embryonic and adult fibroblast cultures by defined factors. Cell 126:663–676
101. Subramanyam D, Lamouille S, Judson RL et al (2011) Multiple targets of miR-302 and miR-372 promote reprogramming of human fibroblasts to induced pluripotent stem cells. Nat Biotechnol 29:443–448
102. Nagai R, Suzuki T, Aizawa K et al (2005) Significance of the transcription factor KLF5 in cardiovascular remodeling. J Thromb Haemost 3:1569–1576
103. Port JD, Sucharov C (2010) Role of microRNAs in cardiovascular disease: therapeutic challenges and potentials. J Cardiovasc Pharmacol 56:444–453
104. Rao PK, Toyama Y, Chiang R et al (2009) Loss of cardiac miRNA-mediated regulation leads to dilated cardiomyopathy and heart failure. Circ Res 105:585–594
105. Yang B, Lin H, Xiao J et al (2007) The muscle-specific microRNA miR-1 regulates cardiac arrhythmogenic potential by targeting GJA1 and KCNJ2. Nat Med 13:468–491
106. Thum T, Gross C, Fiedler J et al (2008) MicroRNA-21 contributes to myocardial disease by stimulating MAP kinase signaling in fibroblasts. Nature 456:980–984
107. van Rooij E, Sutherland LB, Thatcher JE et al (2008) Dysregulation of miroRNAs after myocardial infarction reveals a role of miR-29 in cardiac fibrosis. Proc Natl Acad Sci U S A 105:13027–13032
108. Roy S, Khanna S, Hussain SR et al (2009) MicroRNA expression in response to murine myocardial infarction: miR-21 regulates fibroblast metalloprotease-2 via phosphatase and tensin homologue. Cardiovasc Res 82:21–29
106. Thum T, Gross C, Fiedler J et al (2008) MicroRNA-21 contributes to myocardial disease by stimulating MAP kinase signaling in fibroblasts. Nature 456:980–984
110. Wang J, Xu R, Lin F et al (2009) MicroRNA: novel regulators involved in the remodeling and reverse remodeling of the heart. Cardiology 113:81–88
111. Sayed D, He M, Hong C et al (2010) MicroRNA-21 is a downstream effector of AKT that mediates its antiapoptotic effects via suppression of Fas ligand. J Biol Chem 285:20281–20290
112. Pekarsky Y, Santanum U, Cimmino A et al (2006) Tcl1 expression in chronic lymphocytic leukemia is regulated by miR-29 and miR-181. Cancer Res 66:11590–11593
113. Mott JL, Kobayashi S, Bronk SF et al (2007) miR-29 regulates Mcl-1 protein expression and apoptosis. Oncogene 26:6133–6140
114. Park SY, Lee JH, Ha M et al (2009) miR-29 miRNAs activate p53 by targeting p85a and CDC42. Nat Struct Mol Biol 16:23–29

The Stressful Life of Cardiac Myofibroblasts

Elena Zimina and Boris Hinz

Abstract The ability of cardiac fibroblasts to sense and control the mechanical properties of the extracellular matrix is essential to adapt the heart tissue to mechanical load, such as in conditions of hypertension and to repair injuries after myocardial infarct. Aberrant mechanosensing and/or persistent stress results in the chronic activation of cardiac fibroblasts and other progenitors into myofibroblasts. Myofibroblasts drive the development of fibrosis by excessive collagen secretion and contraction of the neo-matrix into scar tissue. Stiff fibrotic tissue impairs heart distensibility, pumping and valve function, contributes to diastolic and systolic dysfunction, and affects myocardial electrical transmission, leading to arrhythmia and ultimately heart failure. To explore novel therapeutic strategies that specifically target the myofibroblasts in heart fibrosis, we here elaborate on the common factors that control myofibroblast activation from different precursor cells in the heart. At least two factors are pivotal for myofibroblast activation and function: mechanical stress, manifested in disease as a stiff extracellular matrix, and active TGF-β1. Because of uncontrollable side effects, global TGF-β1 inhibition has failed in clinical trials to treat fibrosis but preventing TGF-β1 activation in a myofibroblast-specific manner has promising perspectives.

Keywords Fibroblast · Fibrosis · Scar · Integrin · Extracellular matrix · Mechanical stress · Stiffness · TGF-β1 · α-smooth muscle actin · Extracellular matrix

1 Introduction

Heart failure is the leading cause of death worldwide. A variety of factors contributes to the development of heart failure, including coronary heart disease (frequently leading to myocardial infarction), chronic high blood pressure, diabetes, and cardiomyopathy. The outcomes of these conditions are structural changes in the

B. Hinz (✉) · E. Zimina
Laboratory of Tissue Repair and Regeneration, Matrix Dynamics Group, Faculty of Dentistry, University of Toronto, Fitzgerald Building, Room 234, 150 College Street, Toronto, ON M5S 3E2, Canada
e-mail: boris.hinz@utoronto.ca

© Springer International Publishing Switzerland 2015   71
I.M.C. Dixon, J. T. Wigle (eds.), *Cardiac Fibrosis and Heart Failure: Cause or Effect?*,
Advances in Biochemistry in Health and Disease 13, DOI 10.1007/978-3-319-17437-2_5

heart architecture leading to impaired left ventricular function in filling and eject-
ing blood. Pathological remodelling of the ventricle, in particular development of
fibrosis, is a key event in the progression of heart failure and there is no effective
therapy available. Fibrosis is the accumulation of excessive collagenous extracellular matrix (ECM) that can replace functional heart muscle [1–5].

Instrumental in all fibrotic conditions are myofibroblasts, which are mainly activated from local fibroblastic cells, commonly summarized under the term 'cardiac
fibroblasts' [5, 6–9]. Cardiac fibroblasts are reported to account for the majority (in
number) of all cells in the healthy adult heart where they maintain the ECM architecture [10]. In the healthy myocardium, cardiac fibroblasts form an interconnected
network of cells that are embedded within a collagen network surrounding groups
of myocytes [11] (Fig. 1). Physiological increase of the heart load (e.g., during
exercise) stimulates increased ECM production by the resident fibroblasts as an
adaptive response of the heart in addition to increasing numbers and sizes of fibroblasts and cardiomyocytes [12, 13]. However, in response to chronically increased
mechanical load (e.g. caused by hypertension leading to cardiac hypertrophy) and
to myocardial injury (e.g. after infarct), cardiac fibroblasts are activated into myofibroblasts [2, 14] that are not present in the normal heart [15] (Fig. 1).

Myofibroblasts produce excessive ECM that is contracted by the action of neo-
formed contractile actin-myosin stress fibres *in vivo* [16, 17]. The ultimate aim of
ECM secretion and remodelling is restoring the mechanical integrity of the injured
tissue even at the cost of losing tissue function [18]. Frequently, persistence of myofibroblasts and collagen contraction leads to irreversible remodelling of the heart
ECM into a stiff fibrotic scar that matures over time. The scar reduces heart pumping and distensibility, contributes to diastolic and systolic dysfunction, impairs heart
valve function, impairs oxygen availability to cardiomyocytes as perivascular fibrosis around intracoronary arterioles, and affects myocardial electrical transmission,
leading to arrhythmia [9, 10, 19–24]. Importantly, the scar environment perpetuates fibrosis by converting healthy progenitors into fibrogenic myofibroblasts and
by providing the chemical and mechanical conditions for myofibroblasts to resist
clearance by apoptosis. Hence, myocardial fibrosis can be both the cause *and* consequence of rheumatic heart disease, inflammation, pathological hypertrophy, cardiomyopathy, and post-myocardial infarct remodelling [3, 19]. It is likely, yet not
documented, that cardiac myofibroblasts can become deactivated, as described for
an animal model of liver fibrosis [25]. Alternatively, suicide may be their only way
out which is supported by the fact that myofibroblasts can persist for months and
even years in an infarct scar by escaping programmed cell death [9].

To provide novel therapeutic strategies that target the myofibroblast in heart fibrosis, it is important to understand how mechanical and chemical factors cooperatively control the myofibroblast phenotype. At least two factors are pivotal for
myofibroblast activation and function: mechanical stress, manifested in disease as
a stiff ECM, and active TGF-β1 [7, 26]. TGF-β1 is the most potent pro-fibrotic cytokine known; it causes excessive ECM production, induces its own secretion and
drives myofibroblast activation [17, 27, 28]. Although general TGF-β1 inhibition
has failed to treat organ fibrosis in clinical trials [29–35] because of uncontrol-

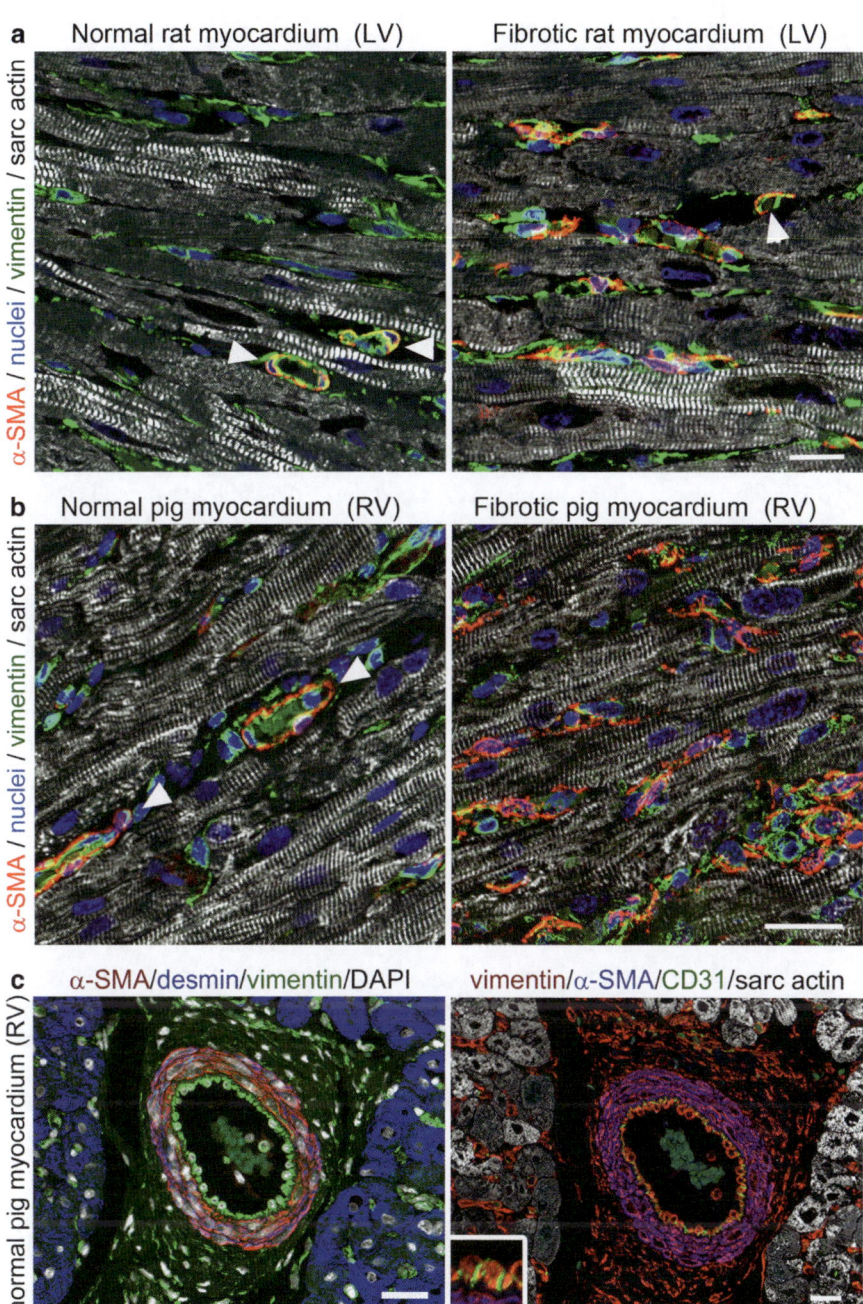

Fig. 1 Fibroblasts and myofibroblasts in the healthy and fibrotic myocardium. (**a**) A rat model of aortic coarctation was used to induce left ventricular (*LV*) cardiac hypertrophy and fibrosis. (**b** & **c**) A Yorkshire pig model of pulmonary vein stenosis for 7 weeks was used to induce right ventricular cardiac hypertrophy and fibrosis; a perivascular region has been selected for (**c**). Transmural blocks

lable side effects [36–39], preventing TGF-β1 activation in a myofibroblast-specific manner has emerged as one possible strategy to counteract fibrosis. In this chapter, we will discuss the mechanical and chemical preconditions of the ECM for latent TGF-β1 activation and how myofibroblast contraction and cell ECM receptors contribute to a mechanical feed-forward loop of latent TGF-β1 activation and myofibroblast differentiation.

2   A Myofibroblast is a Myofibroblast, Of Course, Of Course?

2.1   Myofibroblast Features

Myofibroblasts were originally discovered and defined as fibroblastic cells that simultaneously exhibit prominent endoplasmic reticulum and contractile actin microfilament bundles *in vivo* [40]. According to this most basic definition, neo-formation of contractile actin/myosin stress fibers by cardiac fibroblasts *in vivo* suffices the criterion of a myofibroblast; however, ultrastructural analysis is required to define these cell features [40, 41]. *In vitro*, fibroblastic cells spontaneously form stress fibers when cultured on standard culture dishes [7, 42] (Fig. 2). Hence, it has become common practice to use neo-expression of α-smooth-muscle actin (α-SMA) in stress fibres as distinguishing feature of myofibroblasts [43] (Fig. 2). Expression of α-SMA is the molecular basis for the higher contractile activity of myofibroblasts compared to their precursor cells [44]. The contractile apparatus of myofibroblasts is designed to remodel collagen, which is distinct from the periodic beating activity of cardiomyocytes [45].

In principal, myofibroblast-specific proteins such as α-SMA provide attractive therapeutic targets for the action and/or delivery of anti-myofibroblast/fibrotic drugs in addition to helping diagnosis and rating of the severity of heart fibrosis. However, it is important to note that no unique single marker exists to discriminate myofibroblasts from other cells in the fibrotic heart or other organs and it is questionable whether such a marker exists at all [6]. Typically, myofibroblasts are characterized by a specific set of cytoskeletal proteins, including α-SMA. Vascular smooth muscle cells and pericytes are also α-SMA-positive and additional markers are required to make the distinction [46]. In normal adult tissue, smooth muscle cells express late smooth muscle

Fig. 1 (continued) of the normal and fibrotic ventricular myocardium were sectioned and immunostained for α-SMA, CD31, sarcomeric actin (sarc actin), vimentin, desmin, and nuclei (DAPI) in the indicated combinations and colors. Fibroblastic cells are α-SMA-/vimentin+/desmin-/α-sarc-/CD31-, myofibroblasts are α-SMA+/vimentin+/desmin-/CD31-/α-sarc-/, endothelial cells are α-SMA-/vimentin+/desmin-/α-sarc-/CD31+, smooth muscle cells are α-SMA+/vimentin+/desmin+/α-sarc-/CD31-, and cardiomyocytes are α-SMA-/vimentin-/desmin+/α-sarc+/CD31-. Inset in (a) shown higher magnification of the vascular endothelium. Scale bar: 50 μm. Modified and reprinted with permission from [150]

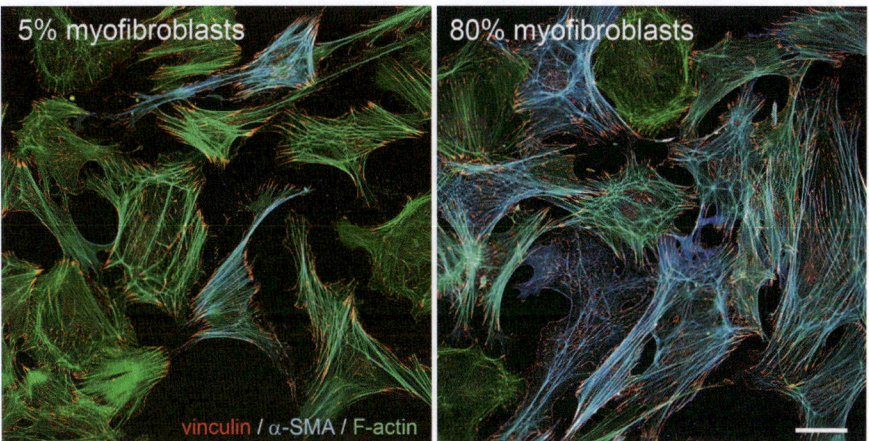

Fig. 2 Fibroblasts and myofibroblasts *in vitro*. Primary rat cardiac fibroblasts were immunostained after 4 days culture either on silicone culture substrates with a Young's modulus of 5 kPa (*left panel*) or on conventional stiff (*GPa*) culture plastic dishes (*right panel*). Cells were stained for F-actin-rich stress fibers (*Phalloidin-green*), α-SMA (*blue, turquoise in overlay*), and focal adhesions (*vinculin-red*). All cells form stress fibers. However, culture on soft substrates reduces the percentage of cells spontaneously acquiring α-SMA-positive stress fibers to ~5 % compared with ~80 % α-SMA-positive myofibroblasts on stiff culture substrates. Scale Bar: 25 μm

differentiation markers that are not expressed by myofibroblasts, including desmin, smooth muscle myosin heavy chain, h-caldesmon, and smoothelin [47]. However, in conditions of organ injury, fibrosis, and in cell culture, SMCs lose these late differentiation markers and attain a myofibroblastic and collagen synthesizing phenotype [48]. Furthermore, both smooth muscle and endothelial cells express vimentin that is also common to (myo)fibroblastic cells. Given that small vessels populate areas between cardiomyocytes, immunostaining for vimentin and α-SMA alone can be ambiguous to identify myofibroblasts. Co-staining for desmin (smooth muscle cells, cardiomyocytes) and endothelial cell specific proteins such as CD31 or VE-cadherin will discriminate between these main cell types populating the myocardium (Fig. 1).

In addition to using molecular markers in immunohistological applications, a number of different transgenic mouse models have been developed to either follow the fate or to specifically eliminate myofibroblasts in conditions of organ fibrosis. Transgenic mice that co-express fluorescent proteins under control of the α-SMA and collagen type I promoters have been used to identify myofibroblasts in lineage tracing studies in the liver and kidney [25] and should in principal be applicable to track myofibroblasts in the heart. Conditional knockouts and expression of fluorescent markers under promoter control of the platelet-derived growth factor receptor β(PDGFRβ) has been successful to eliminate myofibroblast activation from pericytes in liver (hepatic stellate cells), lung and kidney fibrosis [49] but has not been tested yet for the heart. Whereas PDGFRβ-positive pericytes emerge as major contributors to the myofibroblast population in fibrotic liver [50] and kidney (there being identified as descendants of FOXD1 lineage) [51], their role is much less defined in the heart. Another potential approach to trace myofibroblasts in fibrotic conditions is using PDGFRα promoter driven reporter constructs. PDGFRα is expressed in smooth muscle-related cell progenitors and upregulated in conditions of fibrosis and repair in the heart [52–54] and other organs [55]. Periostin promoter-driven expression constructs were employed to identify and target the fibroblastic population specifically in the heart; periostin-driven expression of β galactosidase was confirmed for fibroblastic cells and absent from cardiomyocytes and endothelial cells [56]. Other groups have used the transcription factor Tcf21 (epicardin) to specifically trace the fibroblastic population in the developing and fibrotic heart [57–59].

2.2   Cardiac Myofibroblast Precursors

In the quest to search for an anti-fibrotic cellular target, the problem of finding a specific molecular marker is intimately linked with the question of the origin of myofibroblasts. A number of different myofibroblast precursors have been reported in the heart [11], including fibroblasts [2, 60], smooth muscle cells and pericytes [61], epithelial cells [62], endothelial cells [63], resident mesenchymal progenitor cells [54, 64], and bone-marrow-derived circulating fibrocytes and mesenchymal stem cells [65–68] to list the most prominent candidates. The percentage contribution of each of these potential precursors to the myofibroblast population is a matter of ongoing and sometimes heated debate. Part of the difficulty to identify 'the' myofibroblast precursor is the lack of a unique marker and the choice of different lineage tracers in various studies as discussed above. Another confounding factor is the loose definition of the precursor population itself, such as using 'fibroblast' to classify a rather heterogeneous group of cells with no markers on their own [11, 69]. Finally, it is conceivable that the myofibroblast represents an activation state attained by multiple cell types and the precursor likely depends on the nature of the insult and the available cell populations [6]. Hence, different animal models and clinical conditions of cardiac fibrosis are likely characterized by myofibroblast populations of different origin but with similar function. Although we are far from understanding myofibroblast heterogeneity, there is a potential to exploit the composition of myofibroblast populations as indicators of disease origin and level of progression as our knowledge expands.

3   More than Just Material for Myofibroblasts: The ECM of the Heart

3.1   Composition and Function of the ECM in the Heart

Independent of their origin, all myofibroblasts are specialized to produce and re-model ECM in response to an insult. Although the most obvious function of the cardiac ECM is to provide a framework for myofibrils and to mechanically protect cardiomyocytes against overstretch, the ECM plays more than just structural roles [10, 70, 71]. In the normal heart, the ECM organizes the different cellular compartments, transmits mechanical stimuli, affects electrical transmission in the heart, stores growth factors, and thereby mediates critical cell functions such as proliferation, growth and differentiation [72, 73]. The composition of the cardiac ECM is a complex network of structural proteins such as collagens, elastin, laminins, and fibronectin [73, 74], macromolecules like proteoglycans and glycoproteins that contribute to the overall ECM architecture and bind the growth factors essential for cell migration and tissue remodelling [75], and matricellular proteins and growth factors with signalling function [70]. The fine-balanced interplay between all these elements is massively disturbed in conditions of heart fibrosis and pathological remodelling.

The principal ECM component in the heart is fibrillar collagen type I [72], which provides tensile strength by virtue of its extraordinary mechanical property with an extension modulus of several GPa [76, 77]. Other fibrillar collagens in the myocardium are types III and V. The fibrillar collagens are produced by cardiac fibroblasts and kept in homeostasis by balancing synthesis with degradation mediated by matrix metalloproteinases (MMPs) [78]. Cross-linking enzymes such as lysyl oxidases and transglutaminases provide additional resilience and structure to the collagen/ECM network [72]. Whereas fibrillar collagens provide strength, non-fibrillar collagens, such as types IV and VI mainly integrate myofibrils and cardiac ECM by being major components of the basal lamina of cardiomyocytes [79]. Collagen type IV forms a sheet-like scaffold with laminin, entactin and perlecan, and collagen type VI interacts with collagen type IV and collagen type I to create an anchoring bridge between the basal lamina and interstitial ECM and plays a role in guiding the fibroblast phenotype [79, 80]. Collagens types IV and VI mediate essential cell function through interaction with cell surface receptors such as integrins and discoidin domain receptors (DDRs) [81, 82]. It is amply clear that changes in the composition and organization of the ECM during repair after injury or in fibrosis dramatically impact cell functions. The direct instructive role of the ECM becomes evident in experiments, where de-cellularized ECM from fibrotic organs is sufficient to drive fibrogenesis of healthy cells *in situ* [83, 84].

Cellular fibronectin is a paradigm component of the cardiac ECM that is gaining particular importance during heart development and in the injured myocardium [85–87]. Alternative splicing of fibronectin occurs during wound healing and fibrosis [88] with the extradomain A (ED-A) fibronectin splice variant being highly expressed and required during the myofibroblast activation process [13, 89]. Fibronectin is a master regulator of cell signalling by providing a plethora of binding sites for cell receptors and a large number of different growth factors [90–92], including but not restricted to the latent TGF-β complex [93], vascular endothelial growth factor (VEGF) [94], bone morphogenetic protein 1 (BMP1) [95], hepatocyte growth factor (HGF) [96] and fibroblast growth factor (FGF-2) [96]. Integration of cell signalling with the ECM is also the main function of different matricellular proteins, such as thrombospondins, tenascin C, osteopontin, SPARC, periostin, and the CCN protein family members [74]. All these signalling ECM proteins are differentially regulated during myocardial remodelling and important regulators of myofibroblast functions [97, 98]. Thrombospondins and SPARC have been shown to essentially control collagen accumulation in the heart [99, 100]. The diverse functions of tenascin C include weakening of cell adhesion, up-regulating the expression and activity of MMPs, modulating inflammatory responses, promoting recruitment of myofibroblasts, and enhancing fibrosis [101]. CCN2 (CTGF) is also upregulated in various conditions of cardiac fibrosis and appears to collaborate with TGF-β1 in regulating collagen and myofibroblast induction [1].

3.2   ECM Mechanics Matters for Myofibroblast Activation

Transient activation of myofibroblasts *per se* is beneficial to preserve the structural integrity of the myocardium in response to overload or damage; this normal wound healing response is referred to 'adaptive' or 'reparative' fibrosis and considered reversible. In contrast, persistent excessive accumulation and contraction of collagenous ECM is detrimental to heart function at various levels. First, because myofibroblasts can repair but not regenerate, non-functional scar tissue replaces the damaged heart muscle ('replacement fibrosis'). Second, the stiff scar obliterates proper heart functioning by representing a sheer mechanical obstacle within the softer heart muscle ('you cannot make a scar squeeze'). Third, the electrical conduction properties of the fibrotic ECM cause arrhythmia [24, 102–105]. Fourth, scar stiffness fosters arrhythmia by directly influencing cardiomyocyte beating; embryonic cardiomyocytes beat periodically when cultured on heart-soft substrates but not on fibrotic-stiff material [106, 107]. Incompressibility and poor electrical conduction of scar tissue both contribute to diastolic and systolic dysfunction and to left ventricular hypertrophy [20, 21]. Finally, the mechanical and chemical ECM microenvironment created by myofibroblasts stimulates their own activation from normal precursor cells and perpetuates fibrosis.

One characteristic of the fibrotic scar is its high stiffness compared to the compliant texture of healthy myocardium. Tissue stiffness is measured as Young's Modulus (in Pa) and represents the force per area (stress) that is required to strain a material [108]. Normal heart muscle has a Young's modulus of ~ 10 kPa as measured by atomic force microscopy whereas fibrotic tissue is typically 2–10-times stiffer (20–100 kPa) [106, 109–111] (Fig. 3). One important condition for turning physiological remodelling (typically not involving myofibroblast activation) into pathological remodelling is the presence of an inflammatory response [112, 113]. In addition to providing pro-fibrotic cytokines, inflammatory cells produce ECM cross-linking enzymes, e.g. lysyl oxidases [114] that can initiate ECM stiffening preceding myofibroblast activation as shown in an animal model of liver fibrosis [115]. Scar stiffening is not only due to increased amounts of collagen. Increase in the Young's modulus of the scar occurs through strain-stiffening of the collagenous ECM by cell pulling forces [116, 117] which is macroscopically evident by the maturation of post-myocardial tissue into thin and highly dense scars.

Multiple studies demonstrated that ECM-transmitted stress activates myofibroblasts. Mechanical stimulation of cultured cardiac fibroblasts by twisting ECM

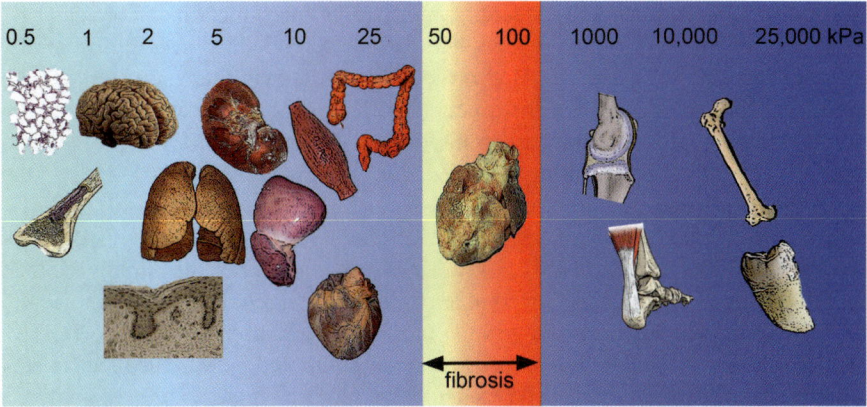

0.5   1   2   5   10   25   50   100   1000   10,000   25,000 kPa

fibrosis

Fig. 3 Stiffness of organs and fibrotic scar. The stiffness range of different organs has been assessed at the cellular perception level using atomic force microscopy indentation and expressed as Young's elastic modulus in Pa [179, 180]. Very soft organs are bone marrow [181], brain (0.1–0.5 kPa) [182] and fat (1–3 kPa) [183]. Soft organs are liver (1–2 kPa) [115], lung parenchyma (2–4 kPa) [184], and skin [185], whereas muscular tissues including the myocardium are medium stiff (10–15 kPa) [111, 186, 187]. Bone and teeth provide the stiffest structures in our body [188]. Note that fibrotic tissue (here stylized for a fibrotic heart) is always stiffer than normal tissue. Modified and reprinted with permission from [189]

protein-coated magnetite beads induces α-SMA expression [118], as does cyclic stretch applied to cultured aortic valve fibroblasts [119]. Expression of α-SMA is augmented in cardiac myofibroblasts cultured on stiff two-dimensional (2D) culture substrates but suppressed on soft polymer substrates [120] (Fig. 2). Myofibroblasts express α-SMA in attached and stressed 3D-collagen gel cultures but not in free-floating relaxed gels [121–123]. In healing rat skin wounds [44, 116] and skin scar tissue *in situ* [124], expression of α-SMA is accelerated by (re-)straining the tissue. Comparable controlled experiments with animal and human heart tissue have yet to be performed. Myofibroblast mechanoregulation occurs at different levels. (1) Mechanical load determines the intracellular stress fibre localization of α-SMA [109]; (2) Stress directly modulates α-SMA promoter activity and protein expression [118, 125]. (3) Substrate compliance and mechanical stimulation regulates the contractile activity of myofibroblasts by regulating cytosolic Ca^{2+} signalling [45, 120–126]; (4) Stress modulates the bioactivity of TGF-β1, the major cytokine inducing myofibroblast differentiation in a process that involves integrins and cell contraction.

4 TGF-β1 at the Cross-Roads of ECM and Growth Factor Signalling

TGF-β1 is a master regulator of fibrosis in all organs [127–129] including the heart where its protein and mRNA expression levels correlate with the degree of fibrosis in a variety of clinical settings and animal models [85, 87, 130–136]. Overexpression of TGF-β1 in mice causes cardiac hypertrophy and interstitial fibrosis [137, 138]; inhibition of TGF-β1 prevents late cardiac remodelling in a mouse model of heart fibrosis [139]. TGF-β1 is also the most potent cytokine known to activate myofibroblasts irrespective of the precursor [7]. Despite the clear pro-fibrotic action of TGF-β1, global therapeutic inhibition of this growth factor is problematic due to its pleiotrophic character [34, 129], e.g., knock-out of TGF-β1 leads to the development of multifocal inflammatory disease in mouse models [140]. TGF-β1 also regulates homeostasis of the vasculature and dysregulation of TGF-β1 levels can lead to tumor formation [141, 142]. More recent strategies are aiming to block TGF-β1 activation from its latent complex rather than targeting the active molecule; this approach bears the advantage of a more targeted strategy to block TGF-β1 pro-fibrotic signalling. Activation of TGF-β1 is promoted by various mechanisms which differ according to the cell type and physiological context [127, 143–147].

Myofibroblasts and other cells produce TGF-β1 together with its latency-associated pro-peptide (LAP); LAP and TGF-β1 remain non-covalently bound and are secreted as a large latent complex covalently linked to the latent TGF-β1 binding protein LTBP-1 [93, 143, 148] (Fig. 4). In cultured lung and cardiac myofibroblasts, TGF-β1 activation occurs mainly via transmission of cell traction forces at sites of integrins to an RGD binding site in the LAP moiety of the large latent complex

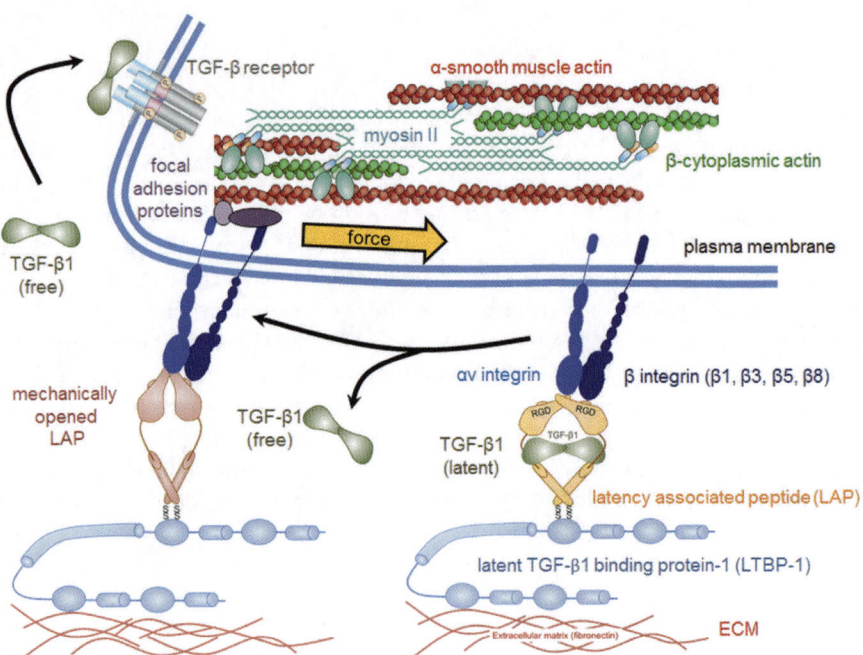

Fig. 4 Mechanical activation of latent TGF-β1. Latent TGF-β1 (TGF-β1 with it's associated pro-peptide LAP) is stored in the ECM by together with the latent TGF-β1 binding protein LTBP-1. Upon actin/myosin promoted myofibroblast contraction, interaction of integrins with RDG binding sites in the LAP activates TGF-β1 by inducing a putative conformation change in LAP

[145, 149, 150]. Binding of the latent TGF-β1 storage protein LTBP-1 to other ECM proteins, including fibrillins and fibronectin makes latent TGF-β1 an integral component of the ECM [35, 144, 151–153]. The normal myocardium is rich in latent TGF-β1 that only seems to wait for activation in conditions of heart overload and injury [98]. Binding of LAP to the ECM through the LTBP-1 is also the structural pre-condition for mechanical activation by integrins [143, 149, 154]. The LTBP-1 binding site of LAP directly opposes the RGD site in LAP for integrin attachment; integrin-mediated force transmission induces a conformational change in LAP that liberates active TGF-β1 [154, 155] (Fig. 4). Recent studies support that the mechanical state of the ECM directly contributes to the bioavailability of TGF-β1 for contraction activation by providing resistance to cell pulling on the latent complex. This effect has been demonstrated by measuring reduced levels of active (not total) TGF-β1 released by cells grown and contract on compliant versus stiff elastic culture substrates [145, 156]. In contrast to elastic cell culture polymers, the ECM of normal and fibrotic connective tissues is subject to strain-stiffening [157, 158]. Results from our lab suggest that cell remodelling strain-stiffens LTBP-1-containing ECM and thereby 'primes' latent TGF-β1 for subsequent activation, analogous to the loading of a mechanical spring [116]. This relationship between the bioavailability of TGF-β1 and the organization state of the ECM provides a mechanical threshold to generate and/or sustain myofibroblasts. In the poorly organized but latent TGF-β1-rich provisional ECM established after organ injury and overload, TGF-β1 activation by cell traction will be inefficient and α-SMA-positive myofibroblasts will not develop. In a sufficiently pre-strained ECM, even the low contractile forces exerted by migrating fibroblastic cells will promote latent TGF-β1 activation [116]. Both, the remodelling required to reaching a mechanical 'tipping point' in the ECM and active cell pulling in the acute latent TGF-β1 activation step are mediated by integrins.

5   ECM Receptors in Cardiac Myofibroblast Differentiation

Cardiac fibroblasts and myofibroblasts express a variety of different integrins, including collagen receptors integrin α1β1, α2β1, α11β1, and α1β3 and the fibronectin binding integrins α5β1, α8β1, αvβ1, αvβ3, and αvβ5 [159, 160]. The regulation of integrin expression and ligand binding is tightly linked with the ECM alterations taking place during fibrosis. For example, the α3 integrin subunit in cardiac fibroblasts interacts with type VI collagen and promotes myofibroblast differentiation post-myocardial infarction [161]. The absolute and relative expression of integrins α1, α2, and α5, all pairing with β1 integrin is differentially regulated in rat models of treadmill exercise or hypertension induced by coarctation of the abdominal aorta [162]. Integrin α8β1 is abundantly expressed in rat cardiac fibroblasts and positively modulated by angiotensin-II and TGF-β1 during cardiac myofibroblast activation [163]; overexpression of α8β1 integrin correlates with elevated production of fibronectin in the heart [164]. Similarly, the expression of integrin αvβ5 is upregulated in rat cardiac fibroblasts after treatment with TGF-β1 and angiotensin II [165]. Others have reported a critical role of β3 integrin for collagen I and fibro-nectin accumulation in conditions of pressure overload, using β3 integrin knock-out mice [166]. Pathological myofibroblast activation in diabetic cardiomyopathy is associated with highly upregulated levels of the collagen receptor α11β1 integrin and pathologically glycated collagen in conditions of diabetic cardiomyopathy en-hances both α11 integrin and α-SMA expression [167].

Whereas the role of integrins as ECM protein receptors is relatively well estab-lished, their function as growth factor receptors and activators is only beginning to be understood and exploited. We will here focus on the role of integrins on latent TGF-β1 activation. TGF-β1-activating integrins play a fundamental role in the on-set and progression of a variety of fibrotic diseases of which heart fibrosis is among the least well studied [34, 145, 168]. Whereas αvβ8 integrin mediated TGF-β1 ac-tivation depends on proteases [34, 144], integrins αvβ6, αvβ5, and αvβ3 act inde-pendently of proteolysis by transmitting cell contraction forces to the ECM-bound latent TGF-β1 complex as discussed above [149, 156, 169]. The epithelium-specific integrin αvβ6 is best studied in the context of TGF-β1 activation and fibrosis in lung and kidney; deletion or blocking of αvβ6 integrin in mice abolishes experimen-tally induced lung and kidney fibrosis [170–174]. However, αvβ6 integrin knock-out mice are not protected against carbon tetrachloride-induced liver fibrosis [174], indicating that αvβ6 does not contribute to TGF-β1 activation in certain types of organ fibrosis, in particular the heart which does not contain αvβ6 integrin express-ing cells. Our lab has recently shown that integrins αvβ5 and αvβ3 are implicated in cardiac fibroblast-to-myofibroblast activation by activating TGF-β1. Both integrins are upregulated during cardiac myofibroblast activation in a porcine model of car-diac fibrosis and correlate with the levels of α-SMA expression and myofibroblast differentiation in cultured cardiac fibroblasts [150]. Blocking both integrins sup-presses the development of contractile myofibroblasts and potentially intercepts the vicious cycle of developing fibrosis. A major finding of this study was that αvβ5 and αvβ3 integrin can compensate for each other's function in activating TGF-β1 and promoting myofibroblast differentiation. Compensation possibly explains the absence of a wound healing phenotype in integrin αvβ5 and αvβ3 knockout animals [175]. Indeed, a recent study using mouse models of lung, liver, and kidney fibrosis suggests that inhibition of all αv integrins, ideally in a myofibroblast-specific man-ner, will be most effective to suppress TGF-β1 activation and fibrosis; surprisingly administration of pan αv integrin blockers did not seem to cause adverse reactions in these models [176].

Another class of ECM receptors expressed by cardiac fibroblasts are the DDRs, receptor tyrosine kinases that are activated upon binding to both fibrillar and non-fibrillar collagens. The DDR binding sites on fibrillar collagens are different from integrin binding sites so simultaneous binding and signalling from both DDRs and integrins is possible [177]. Co-regulation of the expression levels of α-SMA with DDR1 or DDR2 in neonatal versus adult fibroblasts cultured in 3D collagen gels indicates a role in myofibroblast activation [71, 82]. The cytoplasmic tail of DDRs bears tyrosine kinase activity and function as classical receptor tyrosine kinases that recruit proteins with Src homology-2 and phosphotyrosine-binding domains after receptor autophosphorylation [178]. Cardiac fibroblasts express both DDR1 and DDR2, being exclusively expressed in fibroblasts, can serve as a molecular marker [81, 82]. Despite their high expression, little is known about the function of DRRs in the normal and diseased heart.

6 Conclusions

Not only in this myofibroblast-centric chapter, these pro-fibrotic contractile and ECM-producing cells are at the heart of fibrosis in virtually all organs. Over the past years, the view on myofibroblasts has slightly shifted - rather than considering the myofibroblast as a cell type, it is now increasingly regarded as an activation state that various different cell types can attain in response to tissue injury and excessive mechanical load. 'Cell type' or 'phenotype' is more than semantics; whereas cell type implies a final differentiation state, a phenotype is possibly reversible. Reversibility is indeed one of the great challenges in fibrosis of the heart and other organs. Even if we will be able in future to halt the progression of fibrosis, restoration of the organ's function is not automatically achieved. An increasing number of studies show that the chemical and mechanical conditions of the fibrotic ECM, even in the absence of myofibroblasts, is sufficient to instruct a pro-fibrotic behaviour of normal resident and circulating cells. In this chapter, we reviewed how targeting ECM components or ECM cell receptors provide possible strategies to persistently alter the activation of myofibroblasts independently from their origin.

Acknowledgements The research leading to this work was supported by the Canadian Institutes of Health Research (CIHR) grants #210820 and #286920, the Collaborative Health Research Programme of CIHR and the Natural Sciences and Engineering Research Council of Canada (NSERC) grants #1004005 and #413783, the Canada Foundation for Innovation and Ontario Research Fund (CFI/ORF) grant #26653, and the Heart and Stroke Foundation Ontario (grant #NA7086), all to BH. EZ was supported by post-doctoral fellowships provided by CIHR grant #246193) and the Deutsche Forschungsgesellschaft (DGF) grant #Zi 1217/2-1).

References

1. Leask A (2010) Potential therapeutic targets for cardiac fibrosis: TGFbeta, angiotensin, endothelin, CCN2, and PDGF, partners in fibroblast activation. Circ Res 106(11):1675–1680
2. Porter KE, Turner NA (2009) Cardiac fibroblasts: at the heart of myocardial remodeling. Pharmacol Ther 123(2):255–278. doi:10.1016/j.pharmthera.2009.05.002
3. Kahan A, Coghlan G, McLaughlin V (2009) Cardiac complications of systemic sclerosis. Rheumatology (Oxford) 48(Suppl 3):iii45–48. doi:10.1093/rheumatology/kep110
4. Harvey PA, Leinwand LA (2011) The cell biology of disease: cellular mechanisms of cardiomyopathy. J Cell Biol 194(3):355–365. doi:10.1083/jcb.201101100

5. Creemers EE, Pinto YM (2011) Molecular mechanisms that control interstitial fibrosis in the pressure-overloaded heart. Cardiovasc Res 89(2):265–272.
6. Hinz B, Phan SH, Thannickal VJ, Prunotto M, Desmouliere A, Varga J, De Wever O, Mareel M, Gabbiani G (2012) Recent developments in myofibroblast biology: paradigms for connective tissue remodeling. Am J Pathol 180(4):1340–1355. doi:10.1016/j.ajpath.2012.02.004
7. Hinz B (2010) The myofibroblast: paradigm for a mechanically active cell. J Biomech 43(1):146–155. doi:10.1016/j.jbiomech.2009.09.020
8. van den Borne SW, Diez J, Blankesteijn WM, Verjans J, Hofstra L, Narula J (2010) Myocardial remodeling after infarction: the role of myofibroblasts. Nat Rev Cardiol 7(1):30–37. doi:10.1038/nrcardio.2009.199
9. Turner NA, Porter KE (2013) Function and fate of myofibroblasts after myocardial infarction. Fibrogenesis Tissue Repair 6 (1):5
10. Howard CM, Baudino TA (2014) Dynamic cell-cell and cell-ECM interactions in the heart. J Mol Cell Cardiol 70:19–26
11. Lajiness JD, Conway SJ (2014) Origin, development, and differentiation of cardiac fibroblasts. J Mol Cell Cardiol 70:2–8
12. Maillet M, van Berlo JH, Molkentin JD (2013) Molecular basis of physiological heart growth: fundamental concepts and new players. Nat Rev Mol Cell Biol 14(1):38–48
13. Dobaczewski M, de Haan JJ, Frangogiannis NG (2012) The extracellular matrix modulates fibroblast phenotype and function in the infarcted myocardium. J Cardiovasc Transl Res 5(6):837–847
14. Leask A (2007) TGFbeta, cardiac fibroblasts, and the fibrotic response. Cardiovasc Res 74(2):207–212
15. Davis J, Molkentin JD (2014) Myofibroblasts: trust your heart and let fate decide. J Mol Cell Cardiol 70:9–18
16. Tomasek JJ, Gabbiani G, Hinz B, Chaponnier C, Brown RA (2002) Myofibroblasts and mechano-regulation of connective tissue remodelling. Nat Rev Mol Cell Biol 3(5):349–363
17. Hinz B (2007) Formation and function of the myofibroblast during tissue repair. J Invest Dermatol 127(3):526–537
18. Czubryt MP (2012) Common threads in cardiac fibrosis, infarct scar formation, and wound healing. Fibrogenesis Tissue Repair 5(1):19
19. Lazzerini PE, Capecchi PL, Guideri F, Acampa M, Galeazzi M, Laghi Pasini F (2006) Connective tissue diseases and cardiac rhythm disorders: an overview. Autoimmun Rev 5(5):306–313
20. Brown RD, Ambler SK, Mitchell MD, Long CS (2005) The cardiac fibroblast: therapeutic target in myocardial remodeling and failure. Annu Rev Pharmacol Toxicol 45:657–687
21. Ouzounian M, Lee DS, Liu PP (2008) Diastolic heart failure: mechanisms and controversies. Nat Clin Pract Cardiovasc Med 5(7):375–386
22. Deb A, Ubil E (2014) Cardiac fibroblast in development and wound healing. J Mol Cell Cardiol 70C:47–55
23. Frangogiannis NG (2014) The inflammatory response in myocardial injury, repair, and remodelling. Nat Rev Cardiol 11(5):255–265
24. Rohr S (2009) Myofibroblasts in diseased hearts: new players in cardiac arrhythmias? Heart Rhythm 6(6):848–856. doi:10.1016/j.hrthm.2009.02.038
25. Kisseleva T, Cong M, Paik Y, Scholten D, Jiang C, Benner C, Iwaisako K, Moore-Morris T, Scott B, Tsukamoto H, Evans SM, Dillmann W, Glass CK, Brenner DA (2012) Myofibroblasts revert to an inactive phenotype during regression of liver fibrosis. Proc Natl Acad Sci U S A 109(24):9448–9453
26. Hinz B (2009) Tissue stiffness, latent TGF-beta1 activation, and mechanical signal transduction: implications for the pathogenesis and treatment of fibrosis. Curr Rheumatol Rep 11(2):120–126
27. Ruiz-Ortega M, Rodriguez-Vita J, Sanchez-Lopez E, Carvajal G, Egido J (2007) TGF-beta signaling in vascular fibrosis. Cardiovasc Res 74(2):196–206
28. Grainger DJ (2007) TGF-beta and atherosclerosis in man. Cardiovasc Res 74(2):213–222

29. Varga J, Pasche B (2009) Transforming growth factor beta as a therapeutic target in systemic sclerosis. Nat Rev Rheumatol 5(4):200–206
30. Howell JE, McAnulty RJ (2006) TGF-beta: its role in asthma and therapeutic potential. Curr Drug Targets 7(5):547–565
31. Meier K, Nanney LB (2006) Emerging new drugs for scar reduction. Expert Opin Emerg Drugs 11(1):39–47
32. Wakefield LM, Stuelten C (2007) Keeping order in the neighborhood: new roles for TGFbeta in maintaining epithelial homeostasis. Cancer Cell 12(4):293–295
33. ten Dijke P, Arthur HM (2007) Extracellular control of TGFbeta signalling in vascular development and disease. Nat Rev Mol Cell Biol 8(11):857–869
34. Nishimura SL (2009) Integrin-mediated transforming growth factor-beta activation, a potential therapeutic target in fibrogenic disorders. Am J Pathol 175(4):1362–1370
35. Worthington JJ, Klementowicz JE, Travis MA (2011) TGFbeta: a sleeping giant awoken by integrins. Trends Biochem Sci 36(1):47–54
36. Ikushima H, Miyazono K (2010) TGFbeta signalling: a complex web in cancer progression. Nat Rev Cancer 10(6):415–424. doi:10.1038/nrc2853
37. Yang L, Pang Y, Moses HL (2010) TGF-beta and immune cells: an important regulatory axis in the tumor microenvironment and progression. Trends Immunol 31(6):220–227. doi:10.1016/j.it.2010.04.002
38. Moustakas A, Heldin CH (2009) The regulation of TGFbeta signal transduction. Development 136(22):3699–3714. doi:10.1242/dev.030338
39. Li MO, Wan YY, Sanjabi S, Robertson AK, Flavell RA (2006) Transforming growth factor-beta regulation of immune responses. Annu Rev Immunol 24:99–146
40. Gabbiani G, Ryan GB, Majno G (1971) Presence of modified fibroblasts in granulation tissue and their possible role in wound contraction. Experientia 27(5):549–550
41. Eyden B (2008) The myofibroblast: phenotypic characterization as a prerequisite to understanding its functions in translational medicine. J Cell Mol Med 12(1):22–37
42. Rohr S (2011) Cardiac fibroblasts in cell culture systems: myofibroblasts all along? J Cardiovasc Pharmacol 57(4):389–399
43. Darby I, Skalli O, Gabbiani G (1990) Alpha-smooth muscle actin is transiently expressed by myofibroblasts during experimental wound healing. Lab Invest 63(1):21–29
44. Hinz B, Celetta G, Tomasek JJ, Gabbiani G, Chaponnier C (2001) Alpha-smooth muscle actin expression upregulates fibroblast contractile activity. Mol Biol Cell 12(9):2730–2741
45. Follonier Castella L, Gabbiani G, McCulloch CA, Hinz B (2010) Regulation of myofibroblast activities: calcium pulls some strings behind the scene. Exp Cell Res 316 (15):2390–2401. doi:10.1016/j.yexcr.2010.04.033
46. Arnoldi R, Chaponnier C, Gabbiani G, Hinz B (2012) Heterogeneity of smooth muscle. In: Hill J (ed) Muscle: fundamental biology and mechanisms of disease. Elsevier Inc, Amsterdam, pp 1183–1195
47. Hinz B, Phan SH, Thannickal VJ, Galli A, Bochaton-Piallat ML, Gabbiani G (2007) The myofibroblast: one function, multiple origins. Am J Pathol 170(6):1807–1816
48. Benzonana G, Skalli O, Gabbiani G (1988) Correlation between the distribution of smooth muscle or non muscle myosins and alpha-smooth muscle actin in normal and pathological soft tissues. Cell Motil Cytoskeleton 11(4):260–274
49. Henderson NC, Arnold TD, Katamura Y, Giacomini MM, Rodriguez JD, McCarty JH, Pellicoro A, Raschperger E, Betsholtz C, Ruminski PG, Griggs DW, Prinsen MJ, Maher JJ, Iredale JP, Lacy-Hulbert A, Adams RH, Sheppard D (2013) Targeting of alphav integrin identifies a core molecular pathway that regulates fibrosis in several organs. Nat Med 19(12):1617–1624
50. Mederacke I, Hsu CC, Troeger JS, Huebener P, Mu X, Dapito DH, Pradere JP, Schwabe RF (2013) Fate tracing reveals hepatic stellate cells as dominant contributors to liver fibrosis independent of its aetiology. Nat Commun 4:2823
51. Duffield JS (2014) Cellular and molecular mechanisms in kidney fibrosis. J Clin Invest 124(6):2299–2306

52. Chong JJ, Reinecke H, Iwata M, Torok-Storb B, Stempien-Otero A, Murry CE (2013) Progenitor cells identified by PDGFR-alpha expression in the developing and diseased human heart. Stem Cells Dev 22(13):1932–1943

53. Zhao W, Zhao T, Huang V, Chen Y, Ahokas RA, Sun Y (2011) Platelet-derived growth factor involvement in myocardial remodeling following infarction. J Mol Cell Cardiol 51(5):830–838

54. Ieronimakis N, Hays AL, Janebodin K, Mahoney WM, Jr., Duffield JS, Majesky MW, Reyes M (2013) Coronary adventitial cells are linked to perivascular cardiac fibrosis via TGFbeta1 signaling in the mdx mouse model of Duchenne muscular dystrophy. J Mol Cell Cardiol 63:122–134

55. Olson LE, Soriano P (2009) Increased PDGFRalpha activation disrupts connective tissue development and drives systemic fibrosis. Dev Cell 16(2):303–313

56. Takeda N, Manabe I, Uchino Y, Eguchi K, Matsumoto S, Nishimura S, Shindo T, Sano M, Otsu K, Snider P, Conway SJ, Nagai R (2010) Cardiac fibroblasts are essential for the adaptive response of the murine heart to pressure overload. J Clin Invest 120(1):254–265

57. Acharya A, Baek ST, Huang G, Eskiocak B, Goetsch S, Sung CY, Banfi S, Sauer MF, Olsen GS, Duffield JS, Olson EN, Tallquist MD (2012) The bHLH transcription factor Tcf21 is required for lineage-specific EMT of cardiac fibroblast progenitors. Development 139(12):2139–2149

58. Song K, Nam YJ, Luo X, Qi X, Tan W, Huang GN, Acharya A, Smith CL, Tallquist MD, Neilson EG, Hill JA, Bassel-Duby R, Olson EN (2012) Heart repair by reprogramming non-myocytes with cardiac transcription factors. Nature 485(7400):599–604

59. Braitsch CM, Kanisicak O, van Berlo JH, Molkentin JD, Yutzey KE (2013) Differential expression of embryonic epicardial progenitor markers and localization of cardiac fibrosis in adult ischemic injury and hypertensive heart disease. J Mol Cell Cardiol 65:108–119

60. Squires CE, Escobar GP, Payne JF, Leonardi RA, Goshorn DK, Sheats NJ, Mains IM, Mingoia JT, Flack EC, Lindsey ML (2005) Altered fibroblast function following myocardial infarction. J Mol Cell Cardiol 39(4):699–707

61. Vracko R, Thorning D (1991) Contractile cells in rat myocardial scar tissue. Lab Invest 65(2):214–227

62. von Gise A, Pu WT (2012) Endocardial and epicardial epithelial to mesenchymal transitions in heart development and disease. Circ Res 110(12):1628–1645

63. Zeisberg EM, Tarnavski O, Zeisberg M, Dorfman AL, McMullen JR, Gustafsson E, Chandraker A, Yuan X, Pu WT, Roberts AB, Neilson EG, Sayegh MH, Izumo S, Kalluri R (2007) Endothelial-to-mesenchymal transition contributes to cardiac fibrosis. Nat Med 13(8):952–961

64. Chong JJ, Chandrakanthan V, Xaymardan M, Asli NS, Li J, Ahmed I, Heffernan C, Menon MK, Scarlett CJ, Rashidianfar A, Biben C, Zoellner H, Colvin EK, Pimanda JE, Biankin AV, Zhou B, Pu WT, Prall OW, Harvey RP (2011) Adult cardiac-resident MSC-like stem cells with a proepicardial origin. Cell Stem Cell 9(6):527–540

65. Mollmann H, Nef HM, Kostin S, von Kalle C, Pilz I, Weber M, Schaper J, Hamm CW, Elsasser A (2006) Bone marrow-derived cells contribute to infarct remodelling. Cardiovasc Res 71(4):661–671

66. Haudek SB, Xia Y, Huebener P, Lee JM, Carlson S, Crawford JR, Pilling D, Gomer RH, Trial J, Frangogiannis NG, Entman ML (2006) Bone marrow-derived fibroblast precursors mediate ischemic cardiomyopathy in mice. Proc Natl Acad Sci U S A 103(48):18284–18289

67. van Amerongen MJ, Bou-Gharios G, Popa E, van Ark J, Petersen AH, van Dam GM, van Luyn MJ, Harmsen MC (2008) Bone marrow-derived myofibroblasts contribute functionally to scar formation after myocardial infarction. J Pathol 214(3):377–386

68. Chu PY, Mariani J, Finch S, McMullen JR, Sadoshima J, Marshall T, Kaye DM (2010) Bone marrow-derived cells contribute to fibrosis in the chronically failing heart. Am J Pathol 176(4):1735–1742

69. Souders CA, Bowers SL, Baudino TA (2009) Cardiac fibroblast: the renaissance cell. Circ Res 105(12):1164–1176

70. Espira L, Czubryt MP (2009) Emerging concepts in cardiac matrix biology. Can J Physiol Pharmacol 87(12):996–1008

71. Goldsmith EC, Bradshaw AD, Zile MR, Spinale FG (2014) Myocardial fibroblast-matrix interactions and potential therapeutic targets. J Mol Cell Cardiol 70:92–99
72. Weber KT, Sun Y, Bhattacharya SK, Ahokas RA, Gerling IC (2013) Myofibroblast-mediated mechanisms of pathological remodelling of the heart. Nat Rev Cardiol 10(1):15–26
73. Bowers SL, Banerjee I, Baudino TA (2010) The extracellular matrix: at the center of it all. J Mol Cell Cardiol 48(3):474–482
74. Kong P, Christia P, Frangogiannis NG (2013) The pathogenesis of cardiac fibrosis. Cell Mol Life Sci 71(4):549–574
75. Fomovsky GM, Thomopoulos S, Holmes JW (2010) Contribution of extracellular matrix to the mechanical properties of the heart. J Mol Cell Cardiol 48(3):490–496
76. Shen ZL, Kahn H, Ballarini R, Eppell SJ (2011) Viscoelastic properties of isolated collagen fibrils. Biophys J 100(12):3008–3015
77. Heim AJ, Matthews WG, Koob TJ (2006) Determination of the elastic modulus of native collagen fibrils via radial indentation. Appl Phys Lett 89(18):181902
78. Mishra PK, Givvimani S, Chavali V, Tyagi SC (2013) Cardiac matrix: a clue for future therapy. Biochim Biophys Acta 1832(12):2271–2276
79. Shamhart PE, Meszaros JG (2010) Non-fibrillar collagens: key mediators of post-infarction cardiac remodeling? J Mol Cell Cardiol 48(3):530–537
80. Naugle JE, Olson ER, Zhang X, Mase SE, Pilati CF, Maron MB, Folkesson HG, Horne WI, Doane KJ, Meszaros JG (2006) Type VI collagen induces cardiac myofibroblast differentiation: implications for postinfarction remodeling. Am J Physiol Heart Circ Physiol 290(1):H323–330.
81. Morales MO, Price RL, Goldsmith EC (2005) Expression of Discoidin Domain Receptor 2 (DDR2) in the developing heart. Microsc Microanal 11(3):260–267
82. Wilson CG, Stone JW, Fowlkes V, Morales MO, Murphy CJ, Baxter SC, Goldsmith EC (2011) Age-dependent expression of collagen receptors and deformation of type I collagen substrates by rat cardiac fibroblasts. Microsc Microanal 17(4):555–562
83. Wight TN, Potter-Perigo S (2011) The extracellular matrix: an active or passive player in fibrosis? Am J Physiol Gastrointest Liver Physiol 301(6):G950–955
84. Parker MW, Rossi D, Peterson M, Smith K, Sikstrom K, White ES, Connett JE, Henke CA, Larsson O, Bitterman PB (2014) Fibrotic extracellular matrix activates a profibrotic positive feedback loop. J Clin Invest 124(4):1622–1635
85. Boluyt MO, O'Neill L, Meredith AL, Bing OH, Brooks WW, Conrad CH, Crow MT, Lakatta EG (1994) Alterations in cardiac gene expression during the transition from stable hypertrophy to heart failure. Marked upregulation of genes encoding extracellular matrix components. Circ Res 75(1):23–32
86. Samuel JL, Barrieux A, Dufour S, Dubus I, Contard F, Koteliansky V, Farhadian F, Marotte F, Thiery JP, Rappaport L (1991) Accumulation of fetal fibronectin mRNAs during the development of rat cardiac hypertrophy induced by pressure overload. J Clin Invest 88(5):1737–1746
87. Villarreal FJ, Dillmann WH (1992) Cardiac hypertrophy-induced changes in mRNA levels for TGF-beta 1, fibronectin, and collagen. Am J Physiol 262(6 Pt 2):H1861–1866
88. Klingberg F, Hinz B, White ES (2013) The myofibroblast matrix: implications for tissue repair and fibrosis. J Pathol 229(2):298–309
89. Serini G, Bochaton-Piallat ML, Ropraz P, Geinoz A, Borsi L, Zardi L, Gabbiani G (1998) The fibronectin domain ED-A is crucial for myofibroblastic phenotype induction by transforming growth factor-beta1. J Cell Biol 142(3):873–881
90. White ES, Baralle FE, Muro AF (2008) New insights into form and function of fibronectin splice variants. J Pathol 216(1):1–14
91. Singh P, Carraher C, Schwarzbauer JE (2010) Assembly of fibronectin extracellular matrix. Annu Rev Cell Dev Biol 26:397–419
92. Martino MM, Hubbell JA (2010) The 12th–14th type III repeats of fibronectin function as a highly promiscuous growth factor-binding domain. FASEB J 24(12):4711–4721
93. Fontana L, Chen Y, Prijatelj P, Sakai T, Fassler R, Sakai LY, Rifkin DB (2005) Fibronectin is required for integrin alphavbeta6-mediated activation of latent TGF-beta complexes containing LTBP-1. Faseb J 19(13):1798–1808

94. Wijelath ES, Rahman S, Namekata M, Murray J, Nishimura T, Mostafavi-Pour Z, Patel Y, Suda Y, Humphries MJ, Sobel M (2006) Heparin-II domain of fibronectin is a vascular endothelial growth factor-binding domain: enhancement of VEGF biological activity by a singular growth factor/matrix protein synergism. Circ Res 99(8):853–860

95. Huang G, Zhang Y, Kim B, Ge G, Annis DS, Mosher DF, Greenspan DS (2009) Fibronectin binds and enhances the activity of bone morphogenetic protein 1. J Biol Chem 284(38):25879–25888. doi:10.1074/jbc.M109.024125

96. Rahman S, Patel Y, Murray J, Patel KV, Sumathipala R, Sobel M, Wijelath ES (2005) Novel hepatocyte growth factor (HGF) binding domains on fibronectin and vitronectin coordinate a distinct and amplified Met-integrin induced signalling pathway in endothelial cells. BMC Cell Biol 6(1):8

97. Schellings MW, Pinto YM, Heymans S (2004) Matricellular proteins in the heart: possible role during stress and remodeling. Cardiovascular research 64(1):24–31

98. Frangogiannis NG (2012) Matricellular proteins in cardiac adaptation and disease. Physiol Rev 92(2):635–688

99. Xia Y, Dobaczewski M, Gonzalez-Quesada C, Chen W, Biernacka A, Li N, Lee DW, Frangogiannis NG (2011) Endogenous thrombospondin 1 protects the pressure-overloaded myocardium by modulating fibroblast phenotype and matrix metabolism. Hypertension 58(5):902–911

100. Schellings MW, Vanhoutte D, Swinnen M, Cleutjens JP, Debets J, van Leeuwen RE, d'Hooge J, Van de Werf F, Carmeliet P, Pinto YM, Sage EH, Heymans S (2009) Absence of SPARC results in increased cardiac rupture and dysfunction after acute myocardial infarction. J Exp Med 206(1):113–123

101. Imanaka-Yoshida K (2012) Tenascin-C in cardiovascular tissue remodeling: from development to inflammation and repair. Circ J 76(11):2513–2520

102. Vasquez C, Benamer N, Morley GE (2011) The cardiac fibroblast: functional and electrophysiological considerations in healthy and diseased hearts. J Cardiovasc Pharmacol 57(4):380–388

103. Rosker C, Salvarani N, Schmutz S, Grand T, Rohr S (2011) Abolishing myofibroblast arrhythmogeneicity by pharmacological ablation of alpha-smooth muscle actin containing stress fibers. Circ Res 109(10):1120–1131

104. Kakkar R, Lee RT (2010) Intramyocardial fibroblast myocyte communication. Circ Res 106(1):47–57

105. Thompson SA, Copeland CR, Reich DH, Tung L (2011) Mechanical coupling between myofibroblasts and cardiomyocytes slows electric conduction in fibrotic cell monolayers. Circulation 123(19):2083–2093

106. Engler AJ, Carag-Krieger C, Johnson CP, Raab M, Tang HY, Speicher DW, Sanger JW, Sanger JM, Discher DE (2008) Embryonic cardiomyocytes beat best on a matrix with heart-like elasticity: scar-like rigidity inhibits beating. J Cell Sci 121(Pt 22):3794–3802

107. Majkut S, Idema T, Swift J, Krieger C, Liu A, Discher DE (2013) Heart-specific stiffening in early embryos parallels matrix and myosin expression to optimize beating. Curr Biol 23(23):2434–2439

108. Janmey PA, Georges PC, Hvidt S (2007) Basic rheology for biologists. Methods Cell Biol 83:3–27

109. Goffin JM, Pittet P, Csucs G, Lussi JW, Meister JJ, Hinz B (2006) Focal adhesion size controls tension-dependent recruitment of alpha-smooth muscle actin to stress fibers. J Cell Biol 172(2):259–268

110. Wells RG, Discher DE (2008) Matrix elasticity, cytoskeletal tension, and TGF-b: the insoluble and soluble meet. Sci Signal 1:pe13

111. Berry MF, Engler AJ, Woo YJ, Pirolli TJ, Bish LT, Jayasankar V, Morine KJ, Gardner TJ, Discher DE, Sweeney HL (2006) Mesenchymal stem cell injection after myocardial infarction improves myocardial compliance. Am J Physiol Heart Circ Physiol 290(6):H2196–2203

112. Chen W, Frangogiannis NG (2013) Fibroblasts in post-infarction inflammation and cardiac repair. Biochim Biophys Acta 1833(4):945–953

113. van Nieuwenhoven FA, Turner NA (2013) The role of cardiac fibroblasts in the transition from inflammation to fibrosis following myocardial infarction. Vascul Pharmacol 58(3):182–188
114. Lopez B, Gonzalez A, Hermida N, Valencia F, de Teresa E, Diez J (2010) Role of lysyl oxidase in myocardial fibrosis: from basic science to clinical aspects. Am J Physiol Heart Circ Physiol 299(1):H1–9
115. Georges PC, Hui JJ, Gombos Z, McCormick ME, Wang AY, Uemura M, Mick R, Janmey PA, Furth EE, Wells RG (2007) Increased stiffness of the rat liver precedes matrix deposition: implications for fibrosis. Am J Physiol Gastrointest Liver Physiol 293(6):G1147–1154
116. Klingberg F, Chow ML, Koehler BL, Quinn TM, Genot E, Alman BA, Hinz B (2014) Pre-stress in the extracellular matrix sensitizes latent TGF-b1 for activation. J Cell Biol 207:283–297
117. Wen Q, Janmey PA (2013) Effects of non-linearity on cell-ECM interactions. Exp Cell Res 319(16):2481–2489
118. Wang J, Chen H, Seth A, McCulloch CA (2003) Mechanical force regulation of myofibroblast differentiation in cardiac fibroblasts. Am J Physiol Heart Circ Physiol 285(5):H1871–1881. doi:00387.2003 [pii]
119. Merryman WD, Lukoff HD, Long RA, Engelmayr GC Jr, Hopkins RA, Sacks MS (2007) Synergistic effects of cyclic tension and transforming growth factor-beta1 on the aortic valve myofibroblast. Cardiovasc Pathol 16(5):268–276
120. Follonier Castella L, Buscemi L, Godbout C, Meister JJ, Hinz B (2010) A new lockstep mechanism of matrix remodelling based on subcellular contractile events. J Cell Sci 123:1751–1760
121. Grinnell F (1994) Fibroblasts, myofibroblasts, and wound contraction. J Cell Biol 124(4):401–404
122. Arora PD, Narani N, McCulloch CA (1999) The compliance of collagen gels regulates transforming growth factor-beta induction of alpha-smooth muscle actin in fibroblasts. Am J Pathol 154(3):871–882
123. Hinz B (2006) Masters and servants of the force: the role of matrix adhesions in myofibroblast force perception and transmission. Eur J Cell Biol 85(3/4):175–181
124. Junker JP, Kratz C, Tollback A, Kratz G (2008) Mechanical tension stimulates the transdifferentiation of fibroblasts into myofibroblasts in human burn scars. Burns 34(7):942–946
125. Zhao XH, Laschinger C, Arora P, Szaszi K, Kapus A, McCulloch CA (2007) Force activates smooth muscle alpha-actin promoter activity through the Rho signaling pathway. J Cell Sci 120(Pt 10):1801–1809
126. Godbout C, Follonier Castella L, Smith EA, Talele N, Chow ML, Garonna A, Hinz B (2013) The mechanical environment modulates intracellular calcium oscillation activities of myofibroblasts. PLoS ONE 8(5):e64560
127. Henderson NC, Sheppard D (2013) Integrin-mediated regulation of TGFbeta in fibrosis. Biochim Biophys Acta 1832(7):891–896
128. Biernacka A, Dobaczewski M, Frangogiannis NG (2011) TGF-beta signaling in fibrosis. Growth Factors 29(5):196–202
129. Dobaczewski M, Chen W, Frangogiannis NG (2011) Transforming growth factor (TGF)-beta signaling in cardiac remodeling. J Mol Cell Cardiol 51(4):600–606
130. Li G, Li RK, Mickle DA, Weisel RD, Merante F, Ball WT, Christakis GT, Cusimano RJ, Williams WG (1998) Elevated insulin-like growth factor-I and transforming growth factor-beta 1 and their receptors in patients with idiopathic hypertrophic obstructive cardiomyopathy. A possible mechanism. Circulation 98(19 Suppl):II144-149; discussion II149–150
131. Hao J, Ju H, Zhao S, Junaid A, Scammell-La Fleur T, Dixon IM (1999) Elevation of expression of Smads 2, 3, and 4, decorin and TGF-beta in the chronic phase of myocardial infarct scar healing. J Mol Cell Cardiol 31(3):667–678
132. Takahashi N, Calderone A, Izzo NJ Jr, Maki TM, Marsh JD, Colucci WS (1994) Hypertrophic stimuli induce transforming growth factor-beta 1 expression in rat ventricular myocytes. J Clin Invest 94(4):1470–1476

133. Wunsch M, Sharma HS, Markert T, Bernotat-Danielowski S, Schott RJ, Kremer P, Bleese N, Schaper W (1991) In situ localization of transforming growth factor beta 1 in porcine heart: enhanced expression after chronic coronary artery constriction. J Mol Cell Cardiol 23(9):1051–1062

134. Casscells W, Bazoberry F, Speir E, Thompson N, Flanders K, Kondaiah P, Ferrans VJ, Epstein SE, Sporn M (1990) Transforming growth factor-beta 1 in normal heart and in myocardial infarction. Ann N Y Acad Sci 593:148–160

135. Thompson NL, Bazoberry F, Speir EH, Casscells W, Ferrans VJ, Flanders KC, Kondaiah P, Geiser AG, Sporn MB (1988) Transforming growth factor beta-1 in acute myocardial infarction in rats. Growth Factors 1(1):91–99

136. Hein S, Arnon E, Kostin S, Schonburg M, Elsasser A, Polyakova V, Bauer EP, Klovekorn WP, Schaper J (2003) Progression from compensated hypertrophy to failure in the pressure-overloaded human heart: structural deterioration and compensatory mechanisms. Circulation 107(7):984–991

137. Rosenkranz S, Flesch M, Amann K, Haeuseler C, Kilter H, Seeland U, Schluter KD, Bohm M (2002) Alterations of beta-adrenergic signaling and cardiac hypertrophy in transgenic mice overexpressing TGF-beta(1). Am J Physiol Heart Circ Physiol 283(3):H1253–1262

138. Nakajima H, Nakajima HO, Salcher O, Dittie AS, Dembowsky K, Jing S, Field LJ (2000) Atrial but not ventricular fibrosis in mice expressing a mutant transforming growth factor-beta(1) transgene in the heart. Circ Res 86(5):571–579

139. Ikeuchi M, Tsutsui H, Shiomi T, Matsusaka H, Matsushima S, Wen J, Kubota T, Takeshita A (2004) Inhibition of TGF-beta signaling exacerbates early cardiac dysfunction but prevents late remodeling after infarction. Cardiovasc Res 64(3):526–535

140. Shull MM, Ormsby I, Kier AB, Pawlowski S, Diebold RJ, Yin M, Allen R, Sidman C, Proetzel G, Calvin D, Annunziata N, Doetschman T (1992) Targeted disruption of the mouse transforming growth factor-beta 1 gene results in multifocal inflammatory disease. Nature 359(6397):693–699

141. Bierie B, Moses HL (2006) Tumour microenvironment: TGFbeta: the molecular Jekyll and Hyde of cancer. Nat Rev Cancer 6(7):506–520

142. Goumans MJ, Liu Z, ten Dijke P (2009) TGF-beta signaling in vascular biology and dysfunction. Cell Res 19(1):116–127

143. Annes JP, Chen Y, Munger JS, Rifkin DB (2004) Integrin {alpha}V{beta}6-mediated activation of latent TGF-{beta} requires the latent TGF-{beta} binding protein-1. J Cell Biol 165(5):723–734

144. Jenkins G (2008) The role of proteases in transforming growth factor-beta activation. Int J Biochem Cell Biol 40(6/7):1068–1078

145. Wipff PJ, Hinz B (2008) Integrins and the activation of latent transforming growth factor beta1—an intimate relationship. Eur J Cell Biol 87(8/9):601–615

146. Sheppard D (2005) Integrin-mediated activation of latent transforming growth factor beta. Cancer Metastasis Rev 24(3):395–402

147. Annes JP, Munger JS, Rifkin DB (2003) Making sense of latent TGFbeta activation. J Cell Sci 116(Pt 2):217–224

148. Taipale J, Miyazono K, Heldin CH, Keski-Oja J (1994) Latent transforming growth factor-beta 1 associates to fibroblast extracellular matrix via latent TGF-beta binding protein. J Cell Biol 124(1/2):171–181

149. Wipff PJ, Rifkin DB, Meister JJ, Hinz B (2007) Myofibroblast contraction activates latent TGF-beta1 from the extracellular matrix. J Cell Biol 179(6):1311–1323

150. Sarrazy V, Koehler A, Chow ML, Zimina E, Li CX, Kato H, Caldarone CA, Hinz B (2014) Integrins alphavbeta5 and alphavbeta3 promote latent TGF-beta1 activation by human cardiac fibroblast contraction. Cardiovasc Res 102(3):407–417

151. Zilberberg L, Todorovic V, Dabovic B, Horiguchi M, Courousse T, Sakai LY, Rifkin DB (2012) Specificity of latent TGF-beta binding protein (LTBP) incorporation into matrix: role of fibrillins and fibronectin. J Cell Physiol 227(12):3828–3836

152. Robertson IB, Rifkin DB (2013) Unchaining the beast; insights from structural and evolutionary studies on TGFbeta secretion, sequestration, and activation. Cytokine Growth Factor Rev 24(4):355–372
153. Horiguchi M, Ota M, Rifkin DB (2012) Matrix control of transforming growth factor-beta function. J Biochem 152(4):321–329
154. Shi M, Zhu J, Wang R, Chen X, Mi L, Walz T, Springer TA (2011) Latent TGF-beta structure and activation. Nature 474(7351):343–349
155. Buscemi L, Ramonet D, Klingberg F, Formey A, Smith-Clerc J, Meister JJ, Hinz B (2011) The single-molecule mechanics of the latent TGF-beta1 complex. Curr Biol 21:2046–2054. doi:10.1016/j.cub.2011.11.037
156. Giacomini MM, Travis MA, Kudo M, Sheppard D (2012) Epithelial cells utilize cortical actin/myosin to activate latent TGF-beta through integrin alpha(v)beta(6)-dependent physical force. Exp Cell Res 318(6):716–722
157. Discher D, Dong C, Fredberg JJ, Guilak F, Ingber D, Janmey P, Kamm RD, Schmid-Schonbein GW, Weinbaum S (2009) Biomechanics: cell research and applications for the next decade. Ann Biomed Eng 37(5):847–859
158. Storm C, Pastore JJ, MacKintosh FC, Lubensky TC, Janmey PA (2005) Nonlinear elasticity in biological gels. Nature 435(7039):191–194
159. Manso AM, Kang SM, Ross RS (2009) Integrins, focal adhesions, and cardiac fibroblasts. J Investig Med 57(8):856–860
160. Ross RS, Borg TK (2001) Integrins and the myocardium. Circ Res 88(11):1112–1119
161. Bryant JE, Shamhart PE, Luther DJ, Olson ER, Koshy JC, Costic DJ, Mohile MV, Dockry M, Doane KJ, Meszaros JG (2009) Cardiac myofibroblast differentiation is attenuated by alpha(3) integrin blockade: potential role in post-MI remodeling. J Mol Cell Cardiol 46(2):186–192
162. Burgess ML, Terracio L, Hirozane T, Borg TK (2002) Differential integrin expression by cardiac fibroblasts from hypertensive and exercise-trained rat hearts. Cardiovasc Pathol 11(2):78–87
163. Thibault G, Lacombe MJ, Schnapp LM, Lacasse A, Bouzeghrane F, Lapalme G (2001) Upregulation of alpha(8)beta(1)-integrin in cardiac fibroblast by angiotensin II and transforming growth factor-beta1. Am J Physiol Cell Physiol 281(5):C1457–1467
164. Bouzeghrane F, Mercure C, Reudelhuber TL, Thibault G (2004) Alpha8beta1 integrin is upregulated in myofibroblasts of fibrotic and scarring myocardium. J Mol Cell Cardiol 36(3):343–353
165. Graf K, Neuss M, Stawowy P, Hsueh WA, Fleck E, Law RE (2000) Angiotensin II and alpha(v)beta(3) integrin expression in rat neonatal cardiac fibroblasts. Hypertension 35(4):978–984
166. Balasubramanian S, Quinones L, Kasiganesan H, Zhang Y, Pleasant DL, Sundararaj KP, Zile MR, Bradshaw AD, Kuppuswamy D (2012) Beta3 integrin in cardiac fibroblast is critical for extracellular matrix accumulation during pressure overload hypertrophy in mouse. PLoS ONE 7(9):e45076
167. Talior-Volodarsky I, Connelly KA, Arora PD, Gullberg D, McCulloch CA (2012) Alpha11 integrin stimulates myofibroblast differentiation in diabetic cardiomyopathy. Cardiovascular research 96(2):265–275
168. Hinz B (2013) It has to be the alphav: myofibroblast integrins activate latent TGF-beta1. Nature medicine 19(12):1567–1568
169. Tatler AL, John AE, Jolly L, Habgood A, Porte J, Brightling C, Knox AJ, Pang L, Sheppard D, Huang X, Jenkins G (2011) Integrin alphavbeta5-mediated TGF-beta activation by airway smooth muscle cells in asthma. J Immunol 187(11):6094–6107
170. Horan GS, Wood S, Ona V, Li DJ, Lukashev ME, Weinreb PH, Simon KJ, Hahm K, Allaire NE, Rinaldi NJ, Goyal J, Feghali-Bostwick CA, Matteson EL, O'Hara C, Lafyatis R, Davis GS, Huang X, Sheppard D, Violette SM (2008) Partial inhibition of integrin alpha(v)beta6 prevents pulmonary fibrosis without exacerbating inflammation. Am J Respir Crit Care Med 177(1):56–65

171. Munger JS, Huang X, Kawakatsu H, Griffiths MJ, Dalton SL, Wu J, Pittet JF, Kaminski N, Garat C, Matthay MA, Rifkin DB, Sheppard D (1999) The integrin alpha v beta 6 binds and activates latent TGF beta 1: a mechanism for regulating pulmonary inflammation and fibrosis. Cell 96(3):319–328

172. Hahm K, Lukashev ME, Luo Y, Yang WJ, Dolinski BM, Weinreb PH, Simon KJ, Chun Wang L, Leone DR, Lobb RR, McCrann DJ, Allaire NE, Horan GS, Fogo A, Kalluri R, Shield CF, 3rd, Sheppard D, Gardner HA, Violette SM (2007) Alphav beta6 integrin regulates renal fibrosis and inflammation in Alport mouse. Am J Pathol 170(1):110–125

173. Ma LJ, Yang H, Gaspert A, Carlesso G, Barty MM, Davidson JM, Sheppard D, Fogo AB (2003) Transforming growth factor-beta-dependent and -independent pathways of induction of tubulointerstitial fibrosis in beta6(−/−) mice. Am J Pathol 163(4):1261–1273

174. Wang B, Dolinski BM, Kikuchi N, Leone DR, Peters MG, Weinreb PH, Violette SM, Bissell DM (2007) Role of alphavbeta6 integrin in acute biliary fibrosis. Hepatology 46(5):1404–1412

175. Sheppard D (2000) In vivo functions of integrins: lessons from null mutations in mice. Matrix Biol 19(3):203–209

176. Henderson NC, Arnold TD, Katamura Y, Giacomini MM, Rodriguez JD, McCarty JH, Pellicoro A, Raschperger E, Betsholtz C, Ruminski PG, Griggs DW, Prinsen MJ, Maher JJ, Iredale JP, Lacy-Hulbert A, Adams RH, Sheppard D (2013) Targeting of alpha integrin identifies a core molecular pathway that regulates fibrosis in several organs. Nat Med 19(12):1617–1624

177. Borza CM, Pozzi A (2014) Discoidin domain receptors in disease. Matrix Biol 34:185–192

178. Valiathan RR, Marco M, Leitinger B, Kleer CG, Fridman R (2012) Discoidin domain receptor tyrosine kinases: new players in cancer progression. Cancer Metastasis Rev 31(1–2):295–321

179. Buxboim A, Ivanovska IL, Discher DE (2010) Matrix elasticity, cytoskeletal forces and physics of the nucleus: how deeply do cells 'feel' outside and in? J Cell Sci 123(Pt 3):297–308. doi:10.1242/jcs.041186

180. Janmey PA, Winer JP, Murray ME, Wen Q (2009) The hard life of soft cells. Cell Motil Cytoskeleton 66(8):597–605

181. Winer JP, Janmey PA, McCormick ME, Funaki M (2009) Bone marrow-derived human mesenchymal stem cells become quiescent on soft substrates but remain responsive to chemical or mechanical stimuli. Tissue Eng Part A 15(1):147–154

182. Flanagan LA, Ju YE, Marg B, Osterfield M, Janmey PA (2002) Neurite branching on deformable substrates. Neuroreport 13(18):2411–2415

183. Butcher DT, Alliston T, Weaver VM (2009) A tense situation: forcing tumour progression. Nat Rev Cancer 9(2):108–122

184. Liu F, Mih JD, Shea BS, Kho AT, Sharif AS, Tager AM, Tschumperlin DJ (2010) Feedback amplification of fibrosis through matrix stiffening and COX-2 suppression. J Cell Biol 190(4):693–706

185. Achterberg VF, Buscemi L, Diekmann H, Smith-Clerc J, Schwengler H, Meister JJ, Wenck H, Gallinat S, Hinz B (2014) The nano-scale mechanical properties of the extracellular matrix regulate dermal fibroblast function. J Invest Dermatol 134(7):1862–1872

186. Engler AJ (2004) Myotubes differentiate optimally on substrates with tissue-like stiffness: pathological implications for soft or stiff microenvironments. J Cell Biol 166:877–887

187. Engler AJ, Rehfeldt F, Sen S, Discher DE (2007) Microtissue elasticity: measurements by atomic force microscopy and its influence on cell differentiation. Methods Cell Biol 83:521–545

188. Ho SP, Marshall SJ, Ryder MI, Marshall GW (2007) The tooth attachment mechanism defined by structure, chemical composition and mechanical properties of collagen fibers in the periodontium. Biomaterials 28(35):5238–5245

189. Hinz B (2013) Matrix mechanics and regulation of the fibroblast phenotype. Periodontol 2000 63(1):14–28

Pathogenic Origins of Fibrosis in the Hypertensive Heart Disease that Accompanies Aldosteronism

R. Justin McCullough, Yao Sun, Kevin P. Newman,
Kodangudi B. Ramanathan, Ramareddy V. Guntaka and Karl T. Weber

Abstract Cardiac fibrosis interferes with the structural homogeneity of the myocardium in hypertensive heart disease (HHD). Its morphologic presentations include: widely scattered microscopic scars which have replaced myocytes lost to necrosis; and a perivascular fibrosis of intramural coronary arteries. An animal model of aldosterone/salt treatment has been used to examine the pathogenic origins of myocyte necrosis and coronary vasculopathy. A common cellular/subcellular pathway involving parathyroid hormone-mediated, intracellular Ca^{2+} overload-induced, mitochondrial-derived oxidative stress was identified. Myofibroblasts and their secretome which includes *de novo* generation of angiotensin peptides, are responsible for fibrogenesis at these sites. Cardioprotection includes upstream prevention of myocyte loss and vascular remodeling or downstream ablation of myofibroblasts and ongoing fibrogenesis at these sites.

Keywords Microscopic scars · Perivascular fibrosis · Aldosteronism · Myofibroblasts · Cardioprotection

1 Introduction

Hypertensive heart disease (HHD) is a major etiologic factor contributing to the appearance of heart failure, a global health problem of epidemic proportions. The hypertrophic growth of cardiomyocytes which accompanies HHD is comparable to the increment in left ventricular mass found with athletic training [1]. On the other hand, diastolic and/or systolic function of the hypertrophied myocardium in HHD are each compromised and related to fibrous tissue which progressively accumulates

K. T. Weber (✉) · R. J. McCullough · Y. Sun · K. P. Newman · K. B. Ramanathan
Division of Cardiovascular Diseases, University of Tennessee Health Science Center,
956 Court Ave., Suite A312, Memphis, TN 38163, USA
e-mail: KTWeber@uthsc.edu

R. V. Guntaka
Department of Microbiology, Immunology & Biochemistry,
University of Tennessee Health Science Center, 858 Madison Ave.,
101H Molecular Sciences Bldg., Memphis, TN, USA

© Springer International Publishing Switzerland 2015   93
I.M.C. Dixon, J. T. Wigle (eds.), *Cardiac Fibrosis and Heart Failure: Cause or Effect?*,
Advances in Biochemistry in Health and Disease 13, DOI 10.1007/978-3-319-17437-2_6

throughout the right and left heart [2]. Fibrosis, composed predominantly of stiff type I fibrillar collagen, not only adversely alters myocardial tissue stiffness but it also serves as substrate for reentrant ventricular arrhythmias [reviewed in 3].

Herein, we review the pathogenic origins to cardiac fibrosis associated with the HHD found in a rat model of chronic aldosteronism. We deal specifically with *(i)* the replacement fibrosis which follows cardiomyocyte necrosis and *(ii)* the reactive fibrosis of intramural coronary arteries which is the result of an immunostimulatory state, where inflammatory cells invade these vessels to create a vasculopathy. Finally, we briefly review the role of myofibroblasts in regulating these fibrous tissue responses and several cardioprotective strategies aimed at the prevention of this adverse structural remodeling of myocardium.

2 Hypertensive Heart Disease

2.1 *Human HHD*

HHD includes left ventricular hypertrophy (LVH) which accompanies an elevation in systemic arteriolar vascular resistance and associated rise in cardiomyocyte systolic developed force. In time, a preclinical stage of hypertrophied myocytes gives way to a progressive accumulation of fibrous tissue and where fibrosis is composed principally of stiff fibrillar type I collagen [4]. The morphologic presentation of this pathologic remodeling includes widely scattered microscopic scars found throughout the right and left heart and perivascular fibrosis of intramural coronary arteries having fibrillar extensions into the contiguous interstitial space (see Fig. 1). The structural remodeling of coronary microcirculation in HHD has been reviewed elsewhere [5, 6].

2.2 *An Animal Model of HHD*

An animal model simulating human HHD enables systematic examination of pathogenic origins of myocardial scarring and coronary vasculopathy. Toward this end, we have used uninephrectomized 8-week-old male Sprague-Dawley rats having subcutaneous implantation of an osmotic minipump releasing aldosterone (ALDO; 0.75 µg/h). One percent NaCl is added to drinking water provided *ad libitum*, which is further fortified with 0.4% KCl to prevent hypokalemia and associated cardiac lesions [7]. We refer to this model as aldosterone/salt treatment (ALDOST), or aldosteronism, where elevations in circulating ALDO are inappropriate for dietary Na^+ intake and plasma renin activity and angiotensin II are each suppressed. Over the course of several weeks, arterial pressure rises and is accompanied by gradual left ventricular hypertrophy (LVH) [8]. At week 1 of ALDOST myocardium appears normal by light microscopy. This *preclinical stage* gives way to cardiac pathology at week 4. As seen in Fig. 2, this *pathologic stage* features: *(i)* microscopic

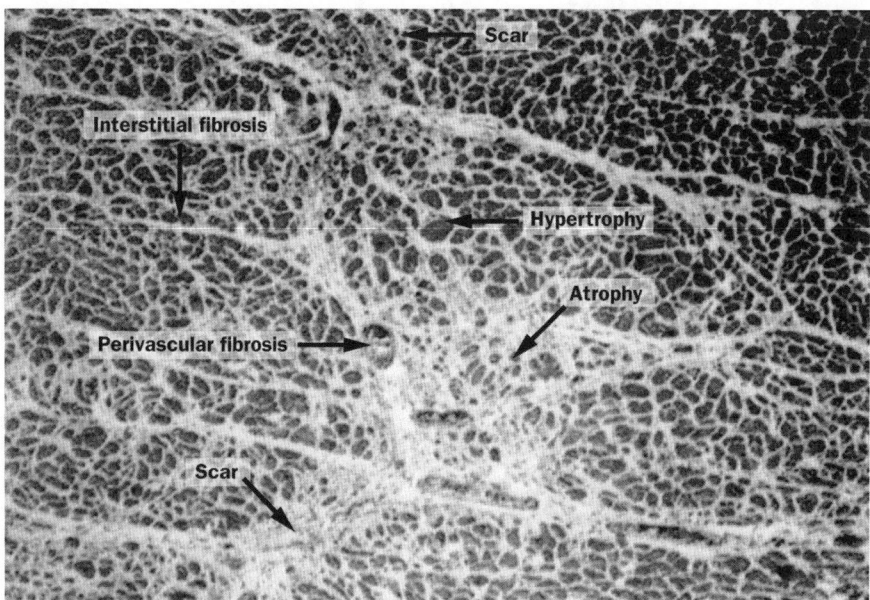

Fig. 1 Postmortem cardiac tissue in human hypertensive heart disease (HHD). Fibrosis presents as widely scattered microscopic scars and perivascular fibrosis of intramural coronary arteries with extensions into the contiguous interstitial space. A heterogeneity in cardiomyocyte size is also evident with hypertrophied as well as atrophied cells bordering on and within scar tissue. (Reprinted from [3])

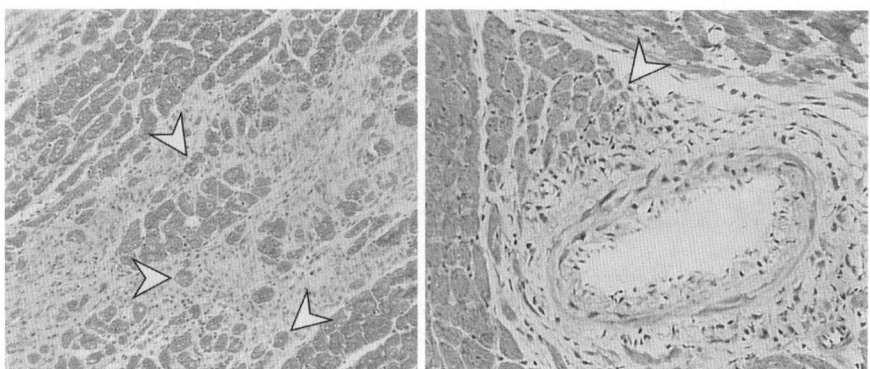

Fig. 2 HHD in rats receiving ALDOST for 4 weeks, where microscopic scars (*left panel*) and perivascular fibrosis (*right panel*) are seen. Atrophic myocytes bordering on and within sites of fibrosis are identified by arrowheads. (Adapted from Kamalov G, et al. *J Cardiovasc Pharmacol.* 2013;62:497–506)

scarring, a *replacement* fibrosis, scattered throughout both the right and left atria and ventricles [9]; and *(ii)* a perivascular/interstitial fibrosis involving the intramural coronary arterial circulation of the right and left heart and which we consider to be a *reactive* fibrosis.

2.3   Remodeling Independent of Hypertension

A series of studies addressed the relevance of hypertension and LVH (vis-à-vis aldosteronism) in contributing to cardiac fibrosis. As reviewed elsewhere [10], a role for hemodynamic factors in promoting cardiac fibrosis was eliminated. These conclusions were based on the following: (*i*) fibrosis was present in the nonpressure-overloaded, nonhypertrophied right atria and ventricle, as well as the pressure overloaded, hypertrophied LV and atrium; (*ii*) the absence of fibrosis when LV pressure overload was created by infrarenal aortic banding, where activation of circulating renin-angiotensin-aldosterone system (RAAS) was absent; and (*iii*) the prevention of fibrosis with either a nondepressor or depressor dose of spironolactone, an aldosterone receptor antagonist. Other confirmatory evidence was seen with an intracerebroventricular infusion of a mineralocorticoid receptor antagonist which prevented hypertension, but not fibrosis [11] and a cardiac-specific upregulation of aldosterone synthase with increased tissue levels of ALDO not accompanied by cardiac fibrosis [12]. Thus, the evidence indicates the adverse structural remodeling of myocardium by fibrous tissue during ALDOST is independent of arterial hypertension and LV hypertrophy. Instead, we hypothesized a circulating factor that accompanies aldosteronism is responsible.

3   Cardiac Myocyte Necrosis as Pathogenic Origin to Microscopic Scarring

3.1   Prooxidant Pathway

Several cellular/subcellular pathways were identified as accounting for cardiomyocyte necrosis and subsequent reparative fibrosis at 4 weeks ALDOST.

Oxidative Stress Evidence of oxidative stress in the myocardium during chronic mineralocorticoidism has been reported by several laboratories [13–17]. A broad spectrum of interventions were used to substantiate the altered redox state. These included: (*i*) the presence of 3-nitrotyrosine, the result of nitrosylation by peroxynitrite and byproduct of the reaction involving superoxide and nitric oxide; (*ii*) an activation of the gp91^{phox} subunit of NADPH oxidase found in inflammatory cells invading the injured myocardium and which contributes to superoxide generation; (*iii*) upregulated redox-sensitive nuclear transcription factor (NF)-κB and a proinflammatory gene cascade it regulates that includes intercellular adhesion molecule (ICAM)-1, monocyte chemoattractant protein (MCP)-1, and tumor necrosis factor (TNF)-alpha; and (*iv*) increased tissue levels of 8-isoprostane and malondialdehyde, biomarkers of lipid peroxidation. Evidence of oxidative stress in blood and urine was found to further support the systemic nature of an altered redox state during ALDOST.

Intracellular Ca^{2+} Overloading A working hypothesis turned to the original concept of Albrecht Fleckenstein: intracellular Ca^{2+} overloading of cardiac myocytes and their mitochondria is an integral pathophysiologic feature of stressor states [18], such as ALDOST. Accordingly, we monitored intracellular Ca^{2+} concentrations in the hearts of rats receiving 1 and 4 weeks ALDOST. Increased Ca^{2+} levels were found in the myocardium as early as week 1 and remained so at week 4, together with biomarker evidence of oxidative stress, such as increased tissue levels of malondialdehyde and 8-isoprostane [17, 19–21]. The underlying circulating factor responsible for intracellular Ca^{2+} overloading during ALDOST, however, remained to be identified.

Calcium and Magnesium Dyshomeostasis and Secondary Hyperparathyroidism (SHPT) Using metabolic studies, rats receiving ALDOST had marked fecal and urinary excretory losses of Ca^{2+} and Mg^{2+} [19]. These urinary and fecal losses of Ca^{2+} and Mg^{2+} led to plasma ionized hypocalcemia and hypomagnesemia. Mediated by the Ca^{2+}-sensing receptor of the parathyroid glands, hypocalcemia prompted increased secretion of parathyroid hormone (PTH) with resultant increased plasma PTH [19]. The accompanying secondary hyperparathyroidism (SHPT) was evidenced by a marked and progressive resorption of bone and ensuing reduction in bone mineral density and bone strength [22].

This led to the working hypothesis that intracellular Ca^{2+} overloading and induction of oxidative stress in ALDOST was PTH-mediated (see Fig. 3), a pathophysiologic scenario embodying the Ca^{2+} paradox of SHPT as suggested by Fujita and Palmieri [23]. Massry and coworkers had earlier demonstrated PTH-mediated intracellular Ca^{2+} overloading of cardiomyocytes. This included: cultured cardiac myocytes incubated with PTH [24]; cells harvested from normal rats receiving a 2-week infusion of PTH; and rats having SHPT with chronic renal failure [25]. In each case, cotreatment with verapamil, a Ca^{2+} channel blocker, prevented the rise in intracellular Ca^{2+}. PTH-mediated intracellular Ca^{2+} overloading is coupled to the induction of oxidative stress in diverse tissues, not only cardiomyocytes. Together, Ca^{2+} overloading and oxidative stress synergistically induce opening of the inner membrane mitochondrial permeability transition pore (mPTP), leading to the structural and functional degeneration of these organelles together with their lost membrane potential and ATP synthesis. Hence, a *mitochondriocentric signal-transducer-effector pathway* is considered the final common pathway to nonischemic cardiomyocyte necrosis [26]. The ensuing replacement fibrosis, or microscopic scarring, is an outcome of subsequent tissue repair. Apoptotic myocyte death does not elicit inflammatory cell or fibroblast responses and therefore is without a morphologic footprint of scar tissue.

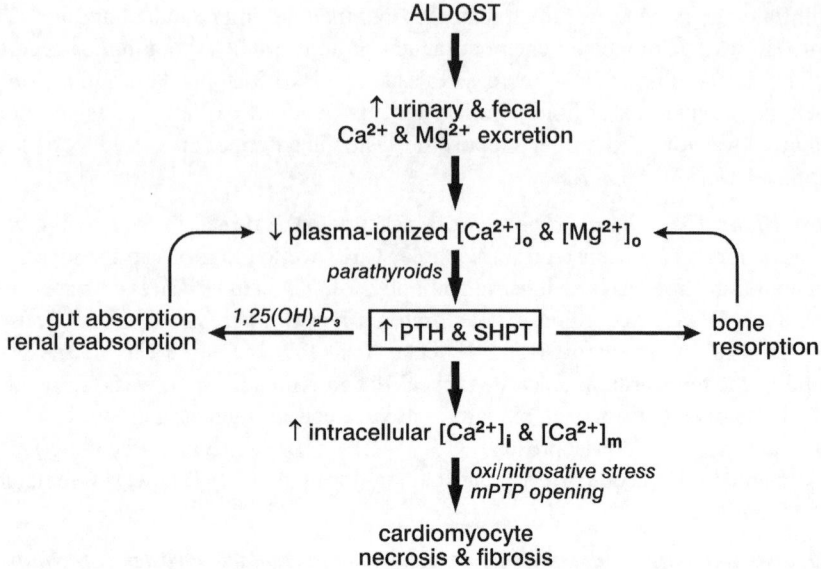

Fig. 3 Aldosterone/salt treatment (ALDOST) is accompanied by marked increments in urinary and fecal excretion of Ca^{2+} and Mg^{2+} which leads to ionized hypocalcemia and hypomagnesemia and incremental secretion of parathyroid hormone (PTH) by the parathyroid glands. Resultant secondary hyperparathyroidism (SHPT) seeks to restore circulating levels of Ca^{2+} and Mg^{2+} through bone resorption and increased absorption and resorption of these cations from the gut and kidneys, respectively. Despite hypocalcemia, PTH promotes intracellular Ca^{2+} overload of cytosolic $[Ca^{2+}]_i$ and mitochondrial $[Ca^{2+}]_m$, and therefore has been termed a Ca^{2+} paradox. Ca^{2+} overload invokes oxidative/nitrosative stress with opening of the mitochondrial inner membrane permeability transition pore (mPTP) leading to structural degeneration of these organelles. Together with lost membrane potential and ATP synthesis cardiomyocyte necrosis follows. Subsequent tissue repair eventuates in fibrosis, or scarring

This pathway was validated by targeted interventions and in so doing pathologic cardiac remodeling by scarring was prevented. They included: (*i*) cotreatment with spironolactone (Spiro), an ALDO receptor antagonist, which attenuated the enhanced urinary and fecal losses of these cations to prevent hypocalcemia and hypomagnesemia and thereby abrogating SHPT [19]; (*ii*) cotreatment with a Ca^{2+} and Mg^{2+}-supplemented diet, fortified with vitamin D_3, to prevent hypocalcemia and SHPT; (*iii*) parathyroidectomy, performed prior to initiating ALDOST [27]; (*iv*) cotreatment with cinacalcet, a calcimimetic that raises the threshold of the parathyroids' Ca^{2+}-sensing receptor to prevent SHPT despite ionized hypocalcemia [28]; (*v*) cotreatment with amlodipine, a Ca^{2+} channel blocker, to prevent intracellular Ca^{2+} overloading [20]; and finally (*vi*) cotreatment with N-acetylcysteine, an antioxidant [16].

Thus, the evidence supports PTH-mediated intracellular Ca^{2+} overloading as the mechanism involved in the induction of oxidative stress during aldosteronism. It is presumed reactive oxygen and nitrogen species overwhelm cellular antioxidant defenses. However, this scenario does not consider whether endogenous antioxidant defenses have been compromised and/or overwhelmed by the overproduction of prooxidants under the pathogenic stimuli leading to intracellular Ca^{2+} overloading.

3.2　Antioxidant Pathways in Cardioprotection

Zinc Dyshomeostasis Chronic mineralocorticoidism is also accompanied by increased enteral and renal losses of Zn^{2+} leading to the appearance of hypozincemia, together with a fall in plasma Cu/Zn-superoxide dismutase (SOD) activity [29]. Also contributory to hypozincemia is a coordinated selective translocation of Zn^{2+} to sites of tissue injury where the upregulation of metallothionein (MT)-1, a Zn^{2+}-binding protein, appears [17, 29].

To systematically address Zn^{2+} kinetics, as antioxidant, in our model of AL-DOST, we used ^{65}Zn as a radioactive tracer. We found a simultaneous fall in plasma ^{65}Zn and a selective accumulation of ^{65}Zn at sites of injury, which included its translocations to injured skin at week 1 that had been freshly incised to implant the minipump and to the injured heart and kidneys at week 4. This pathophysiologic-driven intracellular Zn^{2+} trafficking to injured tissues was accompanied by the temporal upregulation of its binding protein, MT-1 [30]. Thus, rapid translocation of circulating Zn^{2+} to injured tissues also contributes to hypozincemia found with ALDOST, where increased tissue Zn^{2+} is essential to wound healing [31]. Hence, a synchronized dyshomeostasis of Zn^{2+} is another integral feature of myocardial remodeling in aldosteronism. It was therefore necessary to determine whether the rise in cardiac tissue Zn^{2+} involves its cardiac myocytes and mitochondria and antioxidant defenses.

Coupled Ca^{2+} and Zn^{2+} Dyshomeostasis The dyshomeostasis of extra- and intracellular Ca^{2+} and Zn^{2+} which accompanies ALDOST contributes to a contemporaneous dysequilibrium between these pro- and antioxidants. Was the dyshomeostasis of intracellular Ca^{2+} and Zn^{2+} intrinsically coupled in aldosteronism and was the redox state of cardiac myocytes and mitochondria altered? Toward this end, hearts were harvested from rats receiving 4 weeks ALDOST alone or cotreated with Spiro or Amlod. Compared to untreated, age-/sex-matched controls, we found (see Fig. 4) increased cardiomyocyte cytosolic free $[Ca^{2+}]_i$ and $[Zn^{2+}]_i$, together with increased mitochondrial $[Ca^{2+}]_m$ and $[Zn^{2+}]_m$, and each was prevented by Spiro and attenuated by Amlod cotreatment [32].

These iterations in divalent cation composition were accompanied by increased levels of 3-nitrotyrosine and 4-hydroxy-2-nonenal in cardiomyocytes, together with increased H_2O_2 production, malondialdehyde and oxidized glutathione in mitochondria that were also coincident with the increased activities of Cu/Zn-SOD and glutathione peroxidase (GSH-Px) in these cells [17, 26, 32]. Furthermore, these changes in intracellular Zn^{2+} were accompanied by the increased expression of MT-1, Zn^{2+} transporters (Zip1 and ZnT-1) and metal-responsive transcription factor (MTF)-1, an intracellular Zn^{2+} sensor. Thus, in cardiac myocytes and mitochondria from rats with ALDOST, an intrinsically coupled dyshomeostasis of intracellular Ca^{2+} and

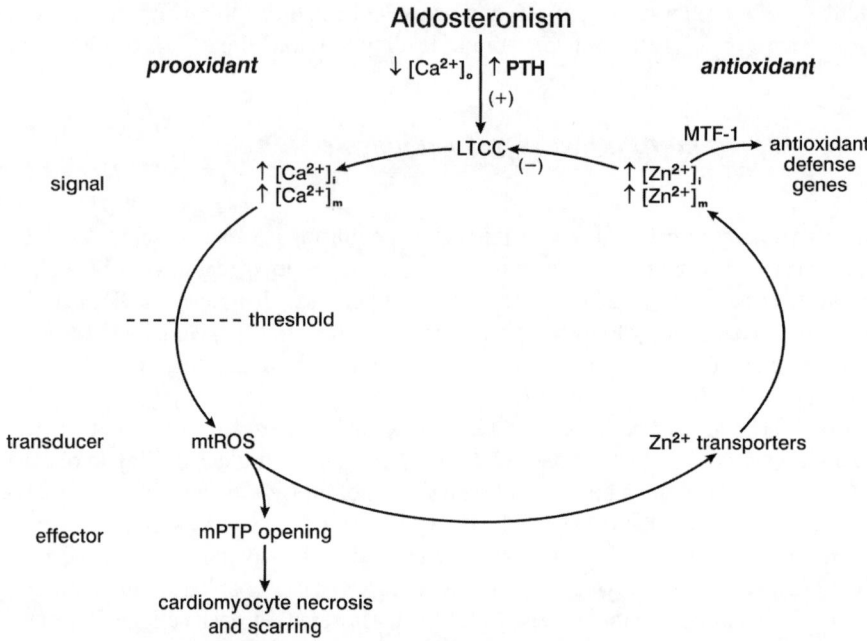

Fig. 4 A coupled dyshomeostasis of Ca^{2+} and Zn^{2+}, representing pro- and antioxidant, respectively, accompanies aldosteronism. (See text for details. Adapted from Cheema Y, et al. *J Cardiovasc Pharmacol.* 2011;58:80–86)

Zn^{2+} appears that alters the redox state of these cells via induction of oxidative stress and generation of antioxidant defenses, respectively. These findings underscore the potential clinical relevance of therapeutic strategies that can uncouple these crucial cations and modulate them in favor of increasing $[Zn^{2+}]_i$ and thereby augmenting antioxidant defenses.

Zinc and Antioxidant Defenses In cardiac myocytes and subsarcolemmal mitochondria harvested by differential centrifugation from the heart at week 4 of ALDOST, increased cytosolic free $[Zn^{2+}]_i$ in cardiac myocytes and total Zn^{2+} concentrations in mitochondria were found [17]. The rise in cardiomyocyte Zn^{2+} was facilitated by the increased expression of membranous Zn^{2+} transporters upregulated by oxidative stress (see Fig. 4). Increased cardiomyocyte $[Zn^{2+}]_i$ serves to enhance antioxidant defenses including their upregulation of MT-1 and activation of MTF-1, which encodes genes related to various antioxidant defenses, such as Cu/Zn-SOD, MT-1, and glutathione synthase [32].

The protective role of raising $[Zn^{2+}]_i$, as natural antagonist to Ca^{2+} entry, using supplemental $ZnSO_4$ or a Zn^{2+} ionophore was next considered [33, 34]. The efficacy of $ZnSO_4$ supplementation in attenuating prooxidant adverse responses, while simultaneously enhancing antioxidant defenses during ALDOST, was addressed. $ZnSO_4$ cotreatment prevented hypozincemia, but not ionized hypocalcemia and the ensuing SHPT. In this context, $ZnSO_4$ attenuated but did not prevent microscopic scarring [17]. Likewise, a Zn^{2+} ionophore alone (pyrrolidine dithiocarbamate, PDTC) [35] was associated with a rise in $[Zn^{2+}]_i$ and a concomitant reduction in $[Ca^{2+}]_i$ in cardiomyocytes. In uncoupling the intrinsically coupled equilibrium between Ca^{2+} and Zn^{2+}, these interventions attenuated oxidative stress in cardiac myocytes and mitochondria and attenuated subsequent necrosis with scarring. Thus, intracellular Ca^{2+} overloading serves as a prooxidant, while increased intracellular Zn^{2+} exerts an antioxidant stimulus with cardiomyocyte survival based on the intrinsic codependency between these two biologically antagonistic divalent cations. The cardioprotective properties of Zn^{2+} have also been found in mice with streptozocin-induced diabetic cardiomyopathy and in rat models of myocardial ischemia/reperfusion and catecholamine-induced injury following isoproterenol administration [17, 33, 34, 36, 37].

4 An Immunostimulatory State as Pathogenic Origin to Perivascular Fibrosis

4.1 The Proinflammatory Vascular Phenotype

An adaptive upregulation of adhesion molecules and chemoattractant chemokines appears within the endothelium of the affected coronary vasculature at week 4 ALDOST. They include: ICAM-1, vascular cell adhesion molecule-1, platelet-endothelial cell adhesion molecule-1; MCP-1; and osteopontin [14, 16, 38–41, 42–44, 45]. MCP-1 is integral to the homing of inflammatory cells into cardiovascular tissue. Within invading inflammatory cells there is evidence of an activation of a redox-sensitive NF-κB and increased expression of a proinflammatory mediator cascade that it regulates, including ICAM-1, MCP-1 and TNF-α. Also, there is an activation of NADPH oxidase, a source of superoxide formation [14–16, 45, 47]. Thus, evidence implicates oxi/nitrosative stress in promoting a proinflammatory vascular phenotype.

4.2 Induction of Oxidative Stress

ALDOST reduces cytosolic free concentrations of $[Mg^{2+}]_i$ in various immune cells, including lymphocytes and monocytes [48, 49]. $[Mg^{2+}]_i$ is the biologically active component of this important divalent cation and, like Zn^{2+}, it is a natural antagonist to Ca^{2+} entry. A reduction in $[Mg^{2+}]_i$ can lead to intracellular Ca^{2+} loading and subsequent induction of oxi/nitrosative stress. Mechanisms responsible for augmented intracellular Ca^{2+} again relate to PTH-mediated Ca^{2+} entry and the presence of SHPT. Evidence in support of Ca^{2+} overload in leading to an altered redox state with activation of immune cells included: (i) reduced $[Mg^{2+}]_i$ in circulating monocytes and lymphocytes (peripheral blood mononuclear cells, PBMC) of rats treated with ALDOST or in man having primary aldosteronism [21, 49, 50]; (*ii*) elevated $[Ca^{2+}]_i$ and total Ca^{2+} concentration of PBMC in response to ALDOST and which occurs prior to tissue invasion, together with increased H_2O_2 production by these PBMC [21]; (*iii*) PTH regulates T-cell activation [51–54]; *(iv)* parathyroidectomy prevents Ca^{2+} overloading of PBMC and vascular lesions [27, 55]; (*v*) upregulated expression of antioxidant defenses in these cells; and (*vi*) prevention of Ca^{2+} loading and oxi/nitrosative stress by co-treatment with either Spiro or an antioxidant [16, 21, 50]. The presence of oxi/nitrosative stress at a systemic level is evidenced by increased serum levels of thiobarbituric acid-reacting substances and reduced activity of plasma α_1-antiproteinase [14, 15, 50]. This early immunostimulatory state featuring PBMC activation is further evidenced by: B cell activation with increased expression of immunoglobulins; an expansion of the B cell lymphocyte subset; an increase in MHC class II-expressing lymphocytes; and increased expression of ICAM-1, integrin-α_1, CC and CXC chemokine proteins and receptors, interleukin-1β and its receptor type 2, and interferon-γ [21, 50]. Evidence of gradual autoreactivity may explain the delayed appearance of vascular remodeling (e.g., first seen at week 4 ALDOST). The prospect that H_2O_2 serves as second messenger to mimic antigen-antigen receptor binding [56] is also raised by these findings given that the heart remains intact, without previous injury, prior to the appearance of these vascular lesions.

4.3 Cardioprotection

In recognizing the pathogenic roles of hormone-induced, redox state-transduced activation of immune cells in leading to the proinflammatory vascular phenotype, the prevention of such adverse structural remodeling could be based on these underlying pathophysiologic mechanisms and where the response in arterial pressure is an indirect outcome to successful immunomodulation [57].

5   Myofibroblasts and Cardiac Fibrosis

Collagen is a stable protein having a half-life of 80–120 days [58]. Usual interstitial fibroblasts are responsible for this gradual turnover of collagen. When active collagen synthesis is invoked at sites of injury, a phenotypically transformed fibroblast-like cell expressing α-smooth muscle actin microfilaments and termed myofibroblast is called into play. Its origins remain controversial. Nonetheless, myofibroblasts are fibrogenic, expressing fibrillar type I collagen at sites of injury and which are regulated in an autocrine manner (see Fig. 5) by angiotensin II, derived *de novo* from these cells [reviewed in 3]. This myofibroblast secretome includes requisites to angiotensin peptide formation, including angiotensin-converting enzyme, and expression of AT_1 receptors. An AT_1 receptor antagonist (e.g., losartan) prevents

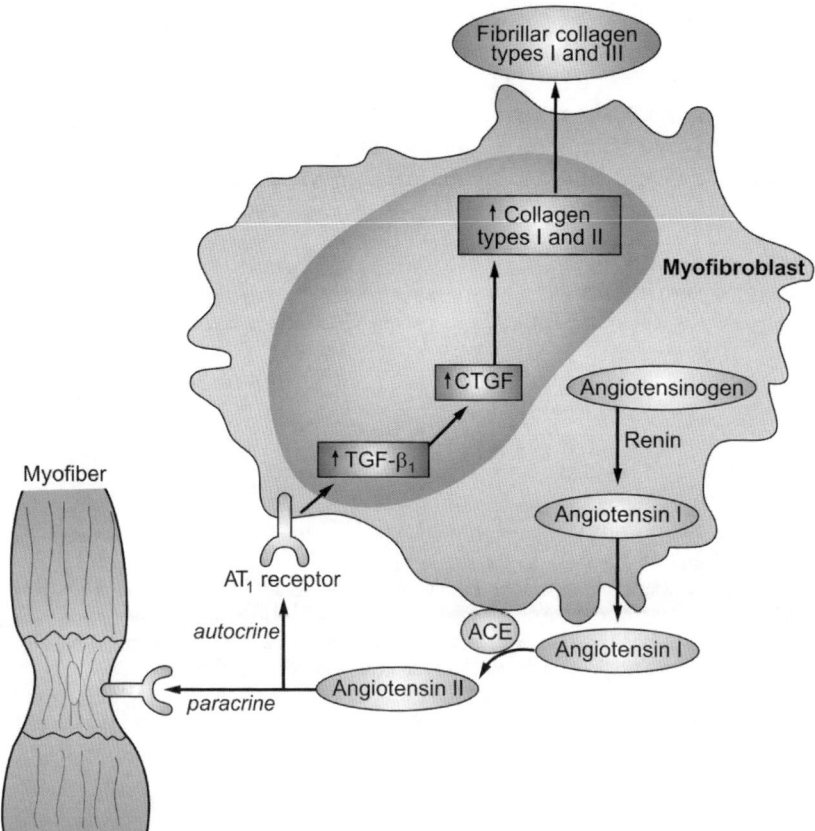

Fig. 5 The myofibroblast secretome includes the *de novo* generation of angiotensin (Ang) peptides. Autocrine properties of tissue AngII regulate collagen turnover by these cells while paracrine signaling involves neighboring myocytes of the myofiber syncytium. (See text. Adapted from [3])

fibrillogenesis at sites of repair, including myocyte necrosis and vasculopathy [59]. Paracrine actions of tissue AngII may regulate the redox state of neighboring cardiomyocytes and where oxidative stress-induced protein degradation by redox-sensitive ligases of the ubiquitin-proteasome system account for their atrophy [60].

6   Summary and Conclusions

Fibrosis disrupts tissue homogeneity. Such is the case in human HHD and in the HHD seen in rats receiving ALDOST, where cardiac fibrosis has its pathologic origins rooted in replacing necrotic myocytes, presenting as microscopic scars, and in the perivascular fibrosis of intramural coronary arteries.

Cellular and subcellular responses related to myocyte necrosis and coronary vasculopathy have a common pathophysiologic origin that includes PTH-mediated, intracellular Ca^{2+} overloading and mitochondrial-based induction of oxidative stress. Cardioprotection from cardiac fibrosis in HHD can be directed at upstream events in preventing myocyte necrosis and immunostimulatory state with activated PBMC or downstream responses aimed at myofibroblast survival and secretome in regulating collagen turnover at sites of injury.

Conflict of Interest This work was supported, in part, by NIH grants R01HL073043, R01HL090867 and R01HL096813 (KTW). Its contents are solely the responsibility of the authors and do not necessarily represent the official views of the NIH. Authors have no conflicts of interest to disclose.

References

1. Shapiro LM, McKenna WJ (1984) Left ventricular hypertrophy: relation of structure to diastolic function in hypertension. Br Heart J 51:637–642
2. Díez J (2009) Towards a new paradigm about hypertensive heart disease. Med Clin North Am 93:637–645
3. Weber KT, Sun Y, Bhattacharya SK, Ahokas RA, Gerling IC (2013) Myofibroblast-mediated mechanisms of pathological remodelling of the heart. Nat Rev Cardiol 10:15–26
4. Weber KT, Janicki JS, Shroff SG, Pick R, Chen RM, Bashey RI (1988) Collagen remodeling of the pressure-overloaded, hypertrophied nonhuman primate myocardium. Circ Res 62:757–765
5. Owens GK (1989) Growth response of aortic smooth muscle cells in hypertension. In: Lee RMKW (ed) Blood vessel changes in hypertension: structure and function. CRC Press, Boca Raton, pp 45–63
6. Cox RH (1989) Mechanical properties of arteries in hypertension. In: Lee RMKW (ed) Blood vessel changes in hypertension: structure and function, vol I. CRC Press, Boca Raton, pp 65–98
7. Darrow DC, Miller HC (1942) The production of cardiac lesions by repeated injections of desoxycorticosterone acetate. J Clin Invest 21:601–611
8. Brilla CG, Pick R, Tan LB, Janicki JS, Weber KT (1990) Remodeling of the rat right and left ventricle in experimental hypertension. Circ Res 67:1355–1364
9. Sun Y, Ramires FJA, Weber KT (1997) Fibrosis of atria and great vessels in response to angiotensin II or aldosterone infusion. Cardiovasc Res 35:138–147
10. Weber KT (2001) Aldosterone in congestive heart failure. N Engl J Med 345:1689–1697
11. Young M, Fullerton M, Dilley R, Funder J (1994) Mineralocorticoids, hypertension, and cardiac fibrosis. J Clin Invest 93:2578–2583
12. Garnier A, Bendall JK, Fuchs S, Escoubet B, Rochais F, Hoerter J, Nehme J, Ambroisine ML, De Angelis N, Morineau G, d'Estienne P, Fischmeister R, Heymes C, Pinet F, Delcayre C (2004) Cardiac specific increase in aldosterone production induces coronary dysfunction in aldosterone synthase-transgenic mice. Circulation 110:1819–1825
13. Somers MJ, Mavromatis K, Galis ZS, Harrison DG (2000) Vascular superoxide production and vasomotor function in hypertension induced by deoxycorticosterone acetate-salt. Circulation 101:1722–1728
14. Pu Q, Neves MF, Virdis A, Touyz RM, Schiffrin EL (2003) Endothelin antagonism on aldosterone-induced oxidative stress and vascular remodeling. Hypertension 42:49–55
15. Virdis A, Neves MF, Amiri F, Viel E, Touyz RM, Schiffrin EL (2002) Spironolactone improves angiotensin-induced vascular changes and oxidative stress. Hypertension 40:504–510

16. Sun Y, Zhang J, Lu L, Chen SS, Quinn MT, Weber KT (2002) Aldosterone-induced inflammation in the rat heart. Role of oxidative stress. Am J Pathol 161:1773–1781
17. Gandhi MS, Deshmukh PA, Kamalov G, Zhao T, Zhao W, Whaley JT, Tichy JR, Bhattacharya SK, Ahokas RA, Sun Y, Gerling IC, Weber KT (2008) Causes and consequences of zinc dyshomeostasis in rats with chronic aldosteronism. J Cardiovasc Pharmacol 52:245–252
18. Fleckenstein A (1967) [Metabolic problems in myocardium insufficiency] [German]. Verh Dtsch Ges Pathol 51:15–30
19. Chhokar VS, Sun Y, Bhattacharya SK, Ahokas RA, Myers LK, Xing Z, Smith RA, Gerling IC, Weber KT (2005) Hyperparathyroidism and the calcium paradox of aldosteronism. Circulation 111:871–878
20. Ahokas RA, Sun Y, Bhattacharya SK, Gerling IC, Weber KT (2005) Aldosteronism and a proinflammatory vascular phenotype. Role of Mg^{2+}, Ca^{2+} and H_2O_2 in peripheral blood mononuclear cells. Circulation 111:51–57
21. Ahokas RA, Warrington KJ, Gerling IC, Sun Y, Wodi LA, Herring PA, Lu L, Bhattacharya SK, Postlethwaite AE, Weber KT (2003) Aldosteronism and peripheral blood mononuclear cell activation. A neuroendocrine-immune interface. Circ Res 93:e124–e135
22. Chhokar VS, Sun Y, Bhattacharya SK, Ahokas RA, Myers LK, Xing Z, Smith RA, Gerling IC, Weber KT (2004) Loss of bone minerals and strength in rats with aldosteronism. Am J Physiol Heart Circ Physiol 287:H2023–H2026
23. Fujita T, Palmieri GM (2000) Calcium paradox disease: calcium deficiency prompting secondary hyperparathyroidism and cellular calcium overload. J Bone Miner Metab 18:109–125
24. Smogorzewski M, Zayed M, Zhang YB, Roe J, Massry SG (1993) Parathyroid hormone increases cytosolic calcium concentration in adult rat cardiac myocytes. Am J Physiol 264:H1998–H2006
25. Perna AF, Smogorzewski M, Massry SG (1989) Effects of verapamil on the abnormalities in fatty acid oxidation of myocardium. Kidney Int 36:453–457
26. Kamalov G, Ahokas RA, Zhao W, Johnson PL, Shahbaz AU, Bhattacharya SK, Sun Y, Gerling IC, Weber KT (2010) Temporal responses to intrinsically coupled calcium and zinc dyshomeostasis in cardiac myocytes and mitochondria during aldosteronism. Am J Physiol Heart Circ Physiol 298:H385–H394
27. Vidal A, Sun Y, Bhattacharya SK, Ahokas RA, Gerling IC, Weber KT (2006) Calcium paradox of aldosteronism and the role of the parathyroid glands. Am J Physiol Heart Circ Physiol 290:H286–H294
28. Selektor Y, Ahokas RA, Bhattacharya SK, Sun Y, Gerling IC, Weber KT (2008) Cinacalcet and the prevention of secondary hyperparathyroidism in rats with aldosteronism. Am J Med Sci 335:105–110
29. Thomas M, Vidal A, Bhattacharya SK, Ahokas RA, Sun Y, Gerling IC, Weber KT (2007) Zinc dyshomeostasis in rats with aldosteronism. Response to spironolactone. Am J Physiol Heart Circ Physiol 293:H2361–H2366
30. Selektor Y, Parker RB, Sun Y, Zhao W, Bhattacharya SK, Weber KT (2008) Tissue [65]zinc translocation in a rat model of chronic aldosteronism. J Cardiovasc Pharmacol 51:359–364
31. Aureli L, Gioia M, Cerbara I, Monaco S, Fasciglione GF, Marini S, Ascenzi P, Topai A, Coletta M (2008) Structural bases for substrate and inhibitor recognition by matrix metalloproteinases. Curr Med Chem 15:2192–2222
32. Kamalov G, Deshmukh PA, Baburyan NY, Gandhi MS, Johnson PL, Ahokas RA, Bhattacharya SK, Sun Y, Gerling IC, Weber KT (2009) Coupled calcium and zinc dyshomeostasis and oxidative stress in cardiac myocytes and mitochondria of rats with chronic aldosteronism. J Cardiovasc Pharmacol 53:414–423
33. Wang J, Song Y, Elsherif L, Song Z, Zhou G, Prabhu SD, Saari JT, Cai L (2006) Cardiac metallothionein induction plays the major role in the prevention of diabetic cardiomyopathy by zinc supplementation. Circulation 113:544–554
34. Karagulova G, Yue Y, Moreyra A, Boutjdir M, Korichneva I (2007) Protective role of intracellular zinc in myocardial ischemia/reperfusion is associated with preservation of protein kinase C isoforms. J Pharmacol Exp Ther 321:517–525

35. Kamalov G, Ahokas RA, Zhao W, Zhao T, Shahbaz AU, Johnson PL, Bhattacharya SK, Sun Y, Gerling IC, Weber KT (2010) Uncoupling the coupled calcium and zinc dyshomeostasis in cardiac myocytes and mitochondria seen in aldosteronism. J Cardiovasc Pharmacol 55:248–254

36. Chvapil M, Owen JA (1977) Effect of zinc on acute and chronic isoproterenol induced heart injury. J Mol Cell Cardiol 9:151–159

37. Singal PK, Dhillon KS, Beamish RE, Dhalla NS (1981) Protective effect of zinc against catecholamine-induced myocardial changes electrocardiographic and ultrastructural studies. Lab Invest 44:426–433

38. Muller DN, Mervaala EM, Schmidt F, Park JK, Dechend R, Genersch E, Breu V, Löffler BM, Ganten D, Schneider W, Haller H, Luft FC (2000) Effect of bosentan on NF-κB, inflammation, and tissue factor in angiotensin II-induced end-organ damage. Hypertension 36:282–290

39. Müller DN, Mervaala EM, Dechend R, Fiebeler A, Park JK, Schmidt F, Theuer J, Breu V, Mackman N, Luther T, Schneider W, Gulba D, Ganten D, Haller H, Luft FC (2000) Angiotensin II (AT₁) receptor blockade reduces vascular tissue factor in angiotensin II-induced cardiac vasculopathy. Am J Pathol 157:111–122

40. Park JK, Muller DN, Mervaala EM, Dechend R, Fiebeler A, Schmidt F, Bieringer M, Schafer O, Lindschau C, Schneider W, Ganten D, Luft FC, Haller H (2000) Cerivastatin prevents angiotensin II-induced renal injury independent of blood pressure- and cholesterol-lowering effects. Kidney Int 58:1420–1430

41. Park JK, Fiebeler A, Muller DN, Mervaala EM, Dechend R, Abou-Rebyeh F, Luft FC, Haller H (2002) Lacidipine inhibits adhesion molecule and oxidase expression independent of blood pressure reduction in angiotensin-induced vascular injury. Hypertension 39(2 Pt 2):685–689

42. Rocha R, Rudolph AE, Frierdich GE, Nachowiak DA, Kekec BK, Blomme EA, McMahon EG, Delyani JA (2002) Aldosterone induces a vascular inflammatory phenotype in the rat heart. Am J Physiol 283:H1802–H1810

43. Muller DN, Dechend R, Mervaala EM, Park JK, Schmidt F, Fiebeler A, Theuer J, Breu V, Ganten D, Haller H, Luft FC (2000) NF-κB inhibition ameliorates angiotensin II-induced inflammatory damage in rats. Hypertension 35(1 Pt 2):193–201

44. Mervaala EM, Müller DN, Park JK, Schmidt F, Löhn M, Breu V, Dragun D, Ganten D, Haller H, Luft FC (1999) Monocyte infiltration and adhesion molecules in a rat model of high human renin hypertension. Hypertension 33(1 Pt 2):389–395

45. Ammarguellat FZ, Gannon PO, Amiri F, Schiffrin EL (2002) Fibrosis, matrix metalloproteinases, and inflammation in the heart of DOCA-salt hypertensive rats: role of ET_A receptors. Hypertension 39(Part 2):679–684

46. Mervaala E, Finckenberg P, Lapatto R, Muller DN, Park JK, Dechend R, Ganten D, Vapaatalo H, Luft FC (2003) Lipoic acid supplementation prevents angiotensin II-induced renal injury. Kidney Int 64:501–508

47. Theuer J, Dechend R, Muller DN, Park JK, Fiebeler A, Barta P, Ganten D, Haller H, Dietz R, Luft FC (2002) Angiotensin II induced inflammation in the kidney and in the heart of double transgenic rats. BMC Cardiovasc Disord 2 (1[Epub]):3

48. Touyz RM, Schiffrin EL (1996) Angiotensin II and vasopressin modulate intracellular free magnesium in vascular smooth muscle cells through Na⁺-dependent protein kinase C pathways. J Biol Chem 271:24353–24358

49. Delva P, Pastori C, Degan M, Montesi G, Brazzarola P, Lechi A (2000) Intralymphocyte free magnesium in patients with primary aldosteronism: aldosterone and lymphocyte magnesium homeostasis. Hypertension 35:113–117

50. Gerling IC, Sun Y, Ahokas RA, Wodi LA, Bhattacharya SK, Warrington KJ, Postlethwaite AE, Weber KT (2003) Aldosteronism: an immunostimulatory state precedes the proinflammatory/fibrogenic cardiac phenotype. Am J Physiol Heart Circ Physiol 285:H813–H821

51. Perry HM, 3rd, Chappel JC, Bellorin-Font E, Tamao J, Martin KJ, Teitelbaum SL (1984) Parathyroid hormone receptors in circulating human mononuclear leukocytes. J Biol Chem 259:5531–5535

52. Klinger M, Alexiewicz JM, Linker-Israeli M, Pitts TO, Gaciong Z, Fadda GZ, Massry SG (1990) Effect of parathyroid hormone on human T cell activation. Kidney Int 37:1543–1551
53. Alexiewicz JM, Gaciong Z, Klinger M, Linker-Israeli M, Pitts TO, Massry SG (1990) Evidence of impaired T cell function in hemodialysis patients: potential role for secondary hyperparathyroidism. Am J Nephrol 10:495–501
54. Ori Y, Korzets A, Malachi T, Gafter U, Breitbart H (1999) Impaired lymphocyte calcium metabolism in end-stage renal disease: enhanced influx, decreased efflux, and reduced response to mitogen. J Lab Clin Med 133:391–400
55. Yang F, Nickerson PA (1988) Effect of parathyroidectomy on arterial hypertrophy, vascular lesions, and aortic calcium content in deoxycorticosterone-induced hypertension. Res Exp Med (Berl) 188:289–297
56. Reth M (2002) Hydrogen peroxide as second messenger in lymphocyte activation. Nat Immunol 3:1129–1134
57. Harrison DG, Guzik TJ, Lob HE, Madhur MS, Marvar PJ, Thabet SR, Vinh A, Weyand CM (2011) Inflammation, immunity, and hypertension. Hypertension 57:132–140
58. Laurent GJ (1987) Dynamic state of collagen: pathways of collagen degradation *in vivo* and their possible role in regulation of collagen mass. Am J Physiol 252:C1–C9
59. Sun Y, Weber KT (1996) Angiotensin converting enzyme and myofibroblasts during tissue repair in the rat heart. J Mol Cell Cardiol 28:851–858
60. Al Darazi F, Zhao W, Zhao T, Sun Y, Marion TN, Ahokas RA, Bhattacharya SK, Gerling IC, Weber KT (2014) Small dedifferentiated cardiomyocytes bordering on microdomains of fibrosis: evidence for reverse remodeling with assisted recovery. J Cardiovasc Pharmacol 64:237–246

Embryological Origin of Valve Progenitor Cells

Michel Pucéat and Thomas Moore-Morris

Abstract The cardiac valves are required for unidirectional blood flow, preventing backflow during diastole. The adult mammalian heart includes four valves: the aortic, pulmonary, mitral and tricuspid valves. Cardiac valves all have common features, notably a stratified structure consisting of three layers of specialized interstitial cells and extracellular matrix. However, the "semilunar" aortic and pulmonary valves and "atrioventricular" tricuspid and mitral valves have notably different embryonic origins. Indeed, several cell lineages, including endocardium, epicardium and neural crest, are valvulogenic. The endocardium, or inner endothelial lining of the heart, makes major contributions to all valves by undergoing endothelial-to-mesenchymal transition. Neural crest and epicardium make secondary contributions to the semilunar and atrioventricular valves, respectively.

The embryonic origins of endocardium, and valve progenitors within, are still not entirely elucidated. The current paradigm stipulates that endocardium is mainly derived from two early embryonic fields, the first and second heart fields. Further delineating the origins of valve progenitors and their specification towards the valve lineages is essential for understanding cardiac congenital defects. Furthermore, it will be essential for developing therapeutic strategies ranging from pharmacological interventions to improving valve replacement. Finally, the biology of valve progenitors is highly relevant to cardiac fibrosis. Indeed, endothelial to mesenchymal transition of endothelium, comparable to that generating valve mesenchyme, is considered to be a major contributor to cardiac fibrosis. However, it has recently been shown that a more likely source of most, if not all, EndoMT derived fibroblasts in heart is EndoMT associated with valvulogenesis.

Keywords Endocardium · Endothelial-mesenchymal transition · Heart · Mouse embryo · Valvulogenesis

M. Pucéat (✉) · T. Moore-Morris
INSERM UMR_S910, Team physiopathology of cardiac development, Medical School La Timone, Aix-Marseille University, 27 Bd Jean Moulin, 13885 Marseille, Cedex 05, France
e-mail: michel.puceat@inserm.fr

© Springer International Publishing Switzerland 2015   109
I.M.C. Dixon, J. T. Wigle (eds.), *Cardiac Fibrosis and Heart Failure: Cause or Effect?*,
Advances in Biochemistry in Health and Disease 13, DOI 10.1007/978-3-319-17437-2_7

1   Introduction

The heart is the first organ to develop in the embryo. Its function of ensuring that blood is distributed to and from other developing organ systems is essential and requires increasing efficiency as the embryo grows. This efficiency is achieved when cardiac valves develop within the contracting myocardium, preventing backflow as blood is pumped by the myocardium's contractions. Oxygenated blood is pumped from the lungs into the left ventricle through the mitral valve, and from the left ventricle through the aortic valve. Deoxygenated blood enters the right ventricle through the tricuspid valve, and is pumped back to the lungs through the pulmonary valve. Valves are stratified structures that consist of up to three extracellular-rich leaflets organized within a fibrous ring, and are firmly anchored to the myocardium. The semilunar aortic and pulmonary valves present a number of differences with the atrioventricular tricuspid and mitral valves. Notably, in terms of structure, the atrioventricular valves are linked to papillary muscles by chordae tendinae, a structural adaptation that prevents prolapsing.

These well characterized structural and positional differences are in contrast with the more elusive origins of the valve progenitors, as well as the morphogenic and biomechanical cues that guide their specification. Indeed, it is currently believed that several early lineages include valve progenitors, and the timing of the specification of these valve progenitors and relative contributions of different lineages is still an area of intense investigation. Furthermore, although much has been revealed on signaling pathways involved in valvulogenesis, much remains to be elucidated, notably in terms of epigenetics or biomechanics (shear stress).

Valves form between the 5th and 8th week of human fetal life, and between E9.5 and E14.5 in mouse. Up to a third of cardiac congenital diseases are characterized by valve malformations [1]. Valves are under intense use, opening more than 2.5 billion times in the life of an adult person. They remodel throughout life, adapting to wear-and-tear, myocardial growth/remodeling and changes in cardiac hemodynamic load. Hence pathological conditions affecting the myocardium lead to adverse remodeling of the valves. Such abnormalities are present in up to 13 % of patients who are 75 years or older [1, 2].

Current therapeutic approaches rely heavily on the replacement of defective valves with mechanical or bioprosthetic valves. Although this has greatly improved the outcome of heart disease for many patients, there are many limitations to this approach. Notably, in the case of congenital malformations in children, replacement valves cannot adapt to the growing myocardium, requiring further interventions. Furthermore, in adults, long term complications can develop linked to limited durablility and thrombogenicity of prostheses.

Hence, the improvement of valve replacement requires these issues to be addressed by using engineered materials with enhanced hemodynamics, mechanical integrity and thromboresistance. Furthermore, cell therapy could also provide a means of giving more functionally integrated replacement valves. Indeed, more in-depth knowledge of the development and characteristics of valve lineages could enable the generating of various human valve cell-types and pave the way for cell therapy.

Valvulogenesis is initiated in the atrio-ventricular canal (AVC) and outflow tract (OFT). These regions become defined when the heart tube elongates and loops, forming the primary ventricle and atrium. Endocardium plays a primary role in early valvulogenesis in both the AVC and OFT. Endocardial cells are separated from the myocardium by a layer of extracellular matrix (ECM) known as cardiac jelly, which is composed of hyaluronan and chondroitin sulfates [3, 4]. Within the AVC and OFT, the myocardium secretes larger amounts of ECM, leading to the formation of "cushions". In response to signals including BMP2 and TGFβ, endocardial cells lining the cushions undergo an endothelial-to-mesenchymal transition (EMT), whereby they delaminate whilst acquiring mesenchymal properties and migrate into the jelly. The cushions become rapidly populated with valve mesenchymal cells characterized by the expression of genes such as Sox9, Sox5, Sox17, Tbx20 and Msxs [4].

Although the endocardium is the major lineage at the origin of valvulogenesis, other lineages have been shown to be essential for normal development of the OFT and AVC valves. Neural crest, a prominent migratory cell population that emerges from the neural tube, makes significant contributions to the valves forming in the OFT. Epicardium also plays a major role in development of the AVC valves. Epicardium is the protective epithelial layer that covers the heart, and undergoes EMT to give rise to cardiac fibroblasts, pericytes, coronary smooth muscle and possibly a subset of endothelial cells and myocytes. Interestingly, epicardium also makes major contributions to the AVC valves, notably generating mesenchyme in the annulus fibrosis and leaflets.

Here we focus on the development of the heterogeneous lineages that contain valve progenitors, as well as the fates of these cells within the maturing valves. Much remains to be determined concerning how and when these subsets of valve progenitors become specified, a point that will be discussed at the end of this section.

2 Early Cardiogenesis and Valve Progenitor Specification

2.1 Overview of Cardiac Development and Key Concepts

Cardiogenesis involves the mobilization of multiple progenitor populations at distinct stages that contribute to specific cardiac compartments. According to the current model, two main populations of progenitors give rise to cardiac myocytes [5]. During gastrulation, a mesodermal cardiac progenitor population, known as the first heart field, emerges from the anterior part of the primitive streak. These cells migrate to the splanchnic mesoderm to form the cardiac crescent. The crescent then fuses at the midline forming a tube-like-structure which elongates on both the arterial and venous poles *via* the addition of progenitor cells originating from the secondary heart field. The latter lies medially and posteriorly to the crescent and is

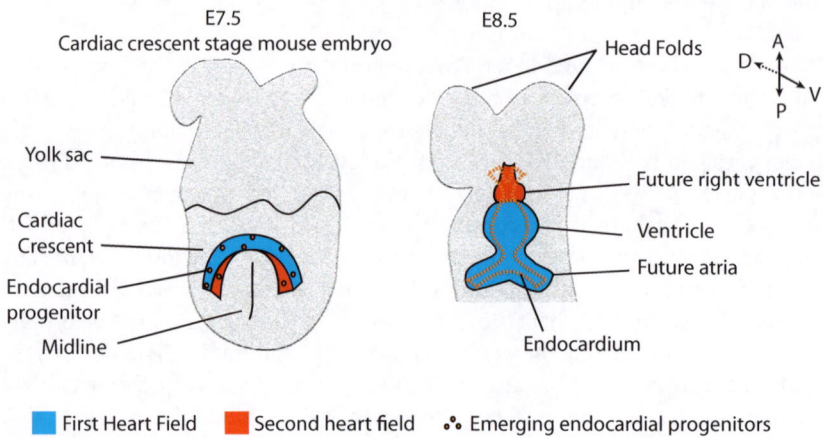

Fig. 1 Emergence of the first valve progenitors occurs in both heart fields. Endocardial cells, including valve progenitors, form by vasculogenesis and give rise to the endocardial tube around which a myocardial layer develops

characterized by the expression of the transcription factor *Isl1* [6]. At this point the heart consists of an outer myocardial layer and an inner endocardial layer (Fig. 1). The latter is formed by de novo vasculogenesis of cells within the cardiac crescent and in the second heart field [7–9]. This process can be recapitulated *in vitro* using human pluripotent stem cells i.e embryonic (HUES) or induced stem cells (iPS), from which both myocardial and endocardial cells can be derived [10, 11].

In order to acquire its definitive form, the heart tube must first undergo rightward looping, whereby the posterior region moves to the anterior allowing segmentation into atrium, an atrioventricular canal, a ventricle and an outflow tract. Intense proliferation of myocardial cells associated with the ventricle and atrium, but not AVC, results in "ballooning" and initial chamber formation [12]. Subsequently, valve formation (valvulogenesis) and septation take place, generating the four cardiac chambers.

2.2   Endocardium Formation

Endocardium forms the inner epithelial lining of the heart, and plays major functions in the development of not only the valves, but also of the formation of the septa, conduction system and trabecular myocardium [13]. In terms of evolution, the development of valves coincides with the separation of cardiac chambers in vertebrates [14, 15]. Although much is known about the development of endocardium, whether the valve progenitors within this lineage i.e. endocardial cells that undergo EMT, represent a distinct population in terms of their embryonic origin is not currently known.

The endocardium develops forming a continuum with the dorsal aorta anteriorly and the cardinal veins posteriorly. Early endocardium is positive for endothelial markers PECAM1, Flk1 and VE-cadherin. However, the origin of endocardial cells is different from other vascular components, although the origin of all endocardial cells has is not completely understood. Early studies using retroviral tracing in chicken and quail embryos have suggested that endocardial and myocardial cells segregate early from a common progenitor in the cardiac field prior to gastrulation [16]. Indeed, retroviral single cell tracking by Mikawa has shown that a population of cells derives from a specific region of the primitive streak migrates to bilateral heart regions, generating either myocardium or endocardium, but not both [17].

Using single-cell tracking in zebrafish, Lee et al. [18] determined that endocardial progenitors were restricted to a specific area of the cardiac field, but could give rise to endothelium, including endocardial cells. This was in agreement with a previous study in chick where a subset of cardiac progenitors was found to co-expressed endothelial (QH-1) and myocardial markers (N-cadherin), suggesting they could give rise to both the myocardial and endocardial lineages [19]. This view has been backed by various studies in mouse, notably showing that these progenitors included a Flk1+ population [20]. Flk1 has been shown to be expressed by endocardium and myocardium as early as E8.5 in mouse heart (Flk1 Lacz reporter) [21]. Genetic lineage tracing in mouse suggests that common myocardial/endocardial progenitors exist relatively late in development, at the cardiac crescent stage, notably being labeled by Nkx2.5-Cre [22] and Isl1-Cre [6] lineage-tracing. Interestingly, an Nkx2.5 response element was identified in the Ets-related protein 71 (Etsrp71), that targets genes required for endothelial/endocardial cell specification of cardiac progenitor cells [23].

Studies performed *in vitro* have put forward the possibility that multipotent cardiac progenitors, notably Flk1+ or Isl+ cells derived from ESCs give rise to myocardium, endothelial (potentially endocardial) and smooth muscle exist [24, 25]. Interestingly, Wnt and BMP signaling that is shown to induce myocardial enrichment in embryoid bodies also promotes the formation of cells with endocardial characteristics, notably the expression of an Nfatc-nuc –LacZ reporter [20].

Fate mapping and cell tracking within quail embryos suggests that some endocardial cells are derived from non-cardiogenic progenitors situated in the second heart field. However, evidence that endocardium represents a distinct lineage from endothelium comes from studies of zebrafish mutants *Cloche* [26] and *Faust*, and hence likely derives from a distinct progenitor pool than other endothelium.

Overall, these studies, looking at expression of precursor markers such as Flk1, or constitutive Cre genetic labeling systems, present limitations. Notably, some of the markers/cre drivers used may not be specific to one lineage throughout cardiac development. Conflicting results, such as the lack of a consensus on the timing of specification of the endocardial progenitors i.e. the existence of early versus late common progenitor for myocardium and endocardium, could be reconciled in the case endocardium is derived from multiple progenitor populations in distinct waves, including a primary wave from vasculogenesis in the first heart field, and subsequent contribtutions from second heart field and possibly other lineages. Future studies could employ approaches such as retrospective clonal analysis, used to provide solid evidence for the presence of two distinct cardiac progenitor populations contributing to specific aspects of the developing heart [5, 27, 28]. Such methods, although time consuming, could provide novel insight into the development of the endocardial lineage. Finally, although controversies have arisen over the extent of the heterogeneity of endocardium, it currently seems likely that endocardial cells share a similar heterogeneous origin to that of the other major early cardiac lineage, the myocytes.

Hence, the embryonic origins of endocardial cells have not been fully elucidated. Key questions remain, including whether distinct developmental origins confers EMT competence to endocardium within the OFT and AVC.

2.3 Epicardium

Epicardium is the protective outer epithelial layer of the heart. In humans, the epicardium is multi-layered and has a sub-epicardial adipose tissue layer, whereas it is formed by a single layer in mouse (and chick). Impaired epicardial development leads to defects in valve development, cardiac myocyte proliferation and alignment as well as conduction system defects.

Epicardial development begins at E9.5 in mouse heart, with the emergence of the pro-epicardium at the venous pole of the heart, a "cauliflower-like" structure (Fig. 2). From E9.5 onwards, pro-epicardial cells begin to migrate and form a sheath

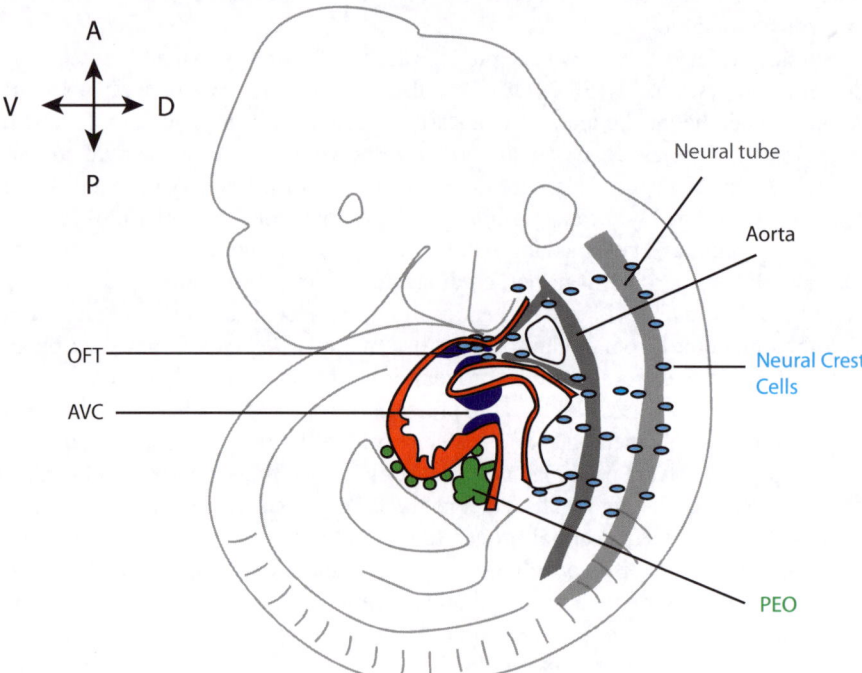

Fig. 2 Illustration of an E9.5 embryo showing the locations of the neural crest and proepicardium/early epicardial cells. Neural crest cells invade the distal parts of the OFT cushions at this stage, whereas epicardial EMT contribution to the AVC cushions/valves occurs at latter embryonic stages

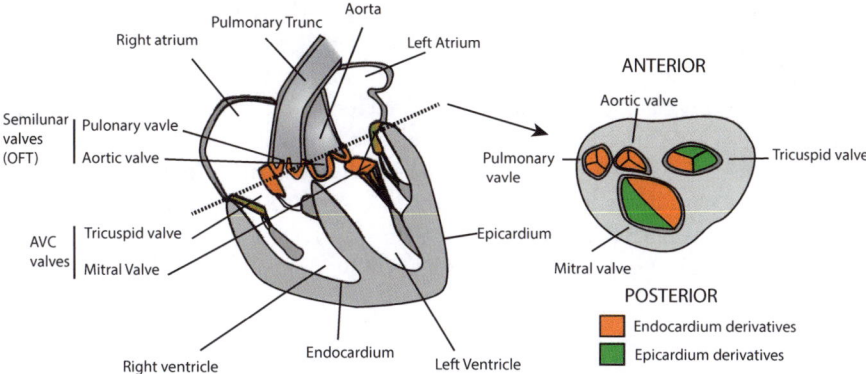

Fig. 3 Illustration summarizing the major contributions of valvulogenic lineages to the adult valve leaflets. Endocardial derivatives make the major contributions to the definitive semilunar valves and the septal and aortic leaflets of the tricuspid and mitral valves. Epicardial derivatives contribute more extensively to the mural leaflets of the tricuspid and mitral valves compared with the septal and aortic leaflets

covering the heart, the epicardium. Early studies by Mikawa [29, 30] using cell-tagging of the propepicardium in avian embryos, show that epicardial cells give rise to independent lineages of coronary smooth muscle and fibroblasts. Subsequent studies using chicken-quail chimeras showed a contribution of epicardial-derived cells (EPDCs) to the AVC cushions and valves, demonstrating that, similarly to endocardium and neural crest, the proepicardium contained valve progenitors [31]. More recent studies using genetic lineage tracing in mouse have provided further details on this contribution from epicardium to the AVC valve leaflets. Interestingly, epicardially-derived mesenchyme has been shown to invest the mural aspects of the AVC valves [32] (Fig. 3). It is probable that this preferential contribution results from the relative proximity of the mural leaflets to the epicardium.

2.4   Neural Crest

The neural crest is a heterogeneous population of cells that originates from the dorsal aspect of the neural tube. These cells arise all along the neural axis and undergo EMT, generating cells that migrate to various locations undergoing ectodermal and mesodermal fates (Fig. 2). These include neurons, glial cells, melanocytes and, at the cephalic level, mesenchymal cells [33].

The cardiac neural crest, a specific subpopulation, plays a key role in morphogenesis of the outflow region of the heart. Initially, understanding the contribution of neural crest to various structures was performed in avian embryos. In particular, neural crest contribution to outflow tract morphogenesis has been well characterized. Removal of a specific portion of the dorsal neural tube between the first and third occipital somites results in a single outflow vessel, or persistent truncus arteriosus, as well as other defects of the pharyngeal arch arteries [34]. Fate mapping, using quail-chick chimeras, showed that neural crest cells first migrated ventrally, covering the caudal pharyngeal arch ateries, and subsequently projected into the aortic sac where they form the aorticopulmonary septum, required for the separation of pulmonary and aortic structues [34, 35]. More recently, genetic lineage tracing in mouse models has confirmed that cardiac neural crest cells first populate the aorticopulmonary septum and conotruncal cushions before septation and contribute to remodeling [36]. This study also demonstrated that very few neural crest derivatives were present in mature semilunar valves, suggesting that the requirement for this cell population is transient.

3 Valve Maturation

The mature SL, tricuspid and mitral valve cusps are complex stratified structures with three layers, each containing composed of specific extracellular matrix (Fig. 4). Notably, the ventricularis (SL)/atrialis (AV) layer is particularly rich in elastin, the intermediate spongiosa rich in proteoglycans and the fibrosa rich in collagen [4]. This constitution provides specific biomechanical properties to the different layers directly in contact with blood flow (ventricularis/atrialis) or providing support. Chordae tendinae provide extra support to the mitral and tricuspid valves, although less prominent equivalent structures also provide support to the SL valves [37].

Valvulogenesis begins in the lumen of the atrio-ventricular canal (AVC) and proximal outflow tract (OFT), where local tissue swellings, termed endocardial cushions, are formed by the accumulation of abundant extracellular matrix in between the endocardium and myocardium. The development of the various leaflets is better characterized in the AVC compared with the OFT. The mural leaflets of the tricuspid and mitral valves i.e. those associated with the ventricular free wall, are generated by the protrusion of atrioventricular myocardium [38, 39]. Myocytes, lost by apoptosis are progressively replaced by mesenchyme that initially forms at the surface. The septal and aortic leaflets of the tricuspid and mitral valves are derived from the inferior and superior AVC cushions, and this is also reportedly the case for the chordinae tendinae [39].

Fig. 4 Illustration depicting the three layers composing the semilunar valves. Each layer has a specific composition in terms of ECM. confering specific biomechanical properties. The mitral and tricuspid valves are also formed of three layers, with an equivalent of the ventricularis, the atrialis, facing the atrial chambers

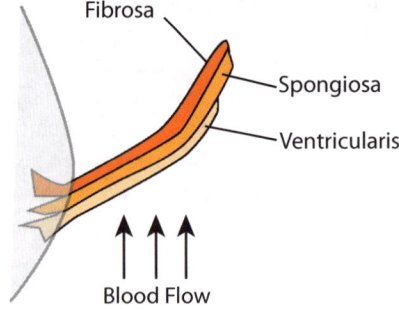

Epithelial to mesenchymal transition, a biological process by which a cell loses its epithelial characteristics and acquires mesenchymal markers and morphology [40, 41], is a key event in early valvulogenesis. EMT of endocardial endothelial valve prospective cells (VECs) is restricted to the cushions, and is brought on by signaling from the underlying myocardium. Key signaling pathways include BMP2 and 4, TGFβ and vascular endothelial growth factor (VEGF) [42]. TGFβs are key modulators of EMT and signaling depends on downstream smads. Notch signaling is required for cells to undergo EMT, and deficient notch signaling results in a lack of EMT [43]. Mechanical forces also play a key role in regulating cushion formation and valve maturation [8]. EMT of VECs gives rise to distinct cell lineages required for valve formation and maturation, including fibroblasts, chondrocytes and more tendinous cells [7, 45].

Other non-endocardial derived cells also contribute to cushion formation. The OFT cushions are specifically populated by mesenchymal cells originating from the neural crest [39]. As the cushions undergo remodeling, epicardium-derived cells (EPDCs) also contribute to the maturing leaflets [32]. EPDCs arise from epithelial-to-mesenchymal transition (EMT) of the epicardium, part of the protective epithelial sheet, or mesothelium, which covers the internal organs. As the valves remodel, more mesenchymal cells are recruited from the hematopoietic lineages [45]. The atrium and ventricle undergo septation in order to form the four cardiac chambers and the AVC divides into left and right ventricular inlets. The OFT separates into left and right ventricular outlets, that are connected to the aorta and pulmonary trunk, respectively. In addition, the AVC endocardial cushions develop into atrio-ventricular (mitral and tricuspid) valves, whereas the OFT endocardial cushions give rise to semilunar (aortic and pulmonic) valves.

4   Valvulopathies, Cardiac Fibrosis and Aortic Stenosis

Cardiac valves are affected in 30 % of cardiac congenital diseases and later in life in ageing people. 2 % of elderly people feature aortic valve undergoing fibrosis and further calcification. Several pathologies often associated with ageing lead to valve fibrosis and calcification. These include hypertension, diabetes, and hypercholesterolemia. The process of calcific aortic stenosis has been the focus of research for more than 60 years [46]. In calcifying valves, the cups slowly thicken and feature fibrosis with a remodeling of the extracellular matrix and ultimately calcification. The mechanical consequence of this pathological process is an increase in valve stiffness and thus a loss in elasticity, which impairs the opening/closing cycle of the valve. The severity of adverse effect can be correlated with the degree of valve calcification.

Valve fibrosis and calcification are linked to myocardial fibrosis. Indeed the loss in valve elasticity imposes an overload to the myocardium that tries to maintain cardiac output. This leads to ventricular hypertrophy as an adaptive phenomenon and then to a decompensation of the myocardium and ventricular fibrosis;

The molecular and cellular mechanisms of valve fibrosis have been also extensively investigated for a few decades. However, both the cells at the origin of fibrosis and/or calcification and the signaling pathways remain incompletely understood. This is somehow reminiscent of the lack of information as to the same processes that occur during development.

While a subset of endothelial cells that might be specifically competent for activation and subsequent osteogenesis do undergo de novo EMT [47, 48], resident valvular interstitial cells (VIC) can also be activated and transformed into myofibroblasts [49] giving rise to osteogenic cells (Fig. 5). The first steps of VIC activation involves an inflammatory response and lipid deposition as suggested by an increase in C-reactive protein in patients with aortic stenosis [50]. The inflammatory cells likely participate in the remodeling of the extracellular matrix of the valve leaflet. This process is very similar to what has been observed in atherosclerosis.

Besides this local cell activation, circulating hematopoietic cells have been proposed to contribute to some osteogenic progenitors [51]. Circulating endothelial

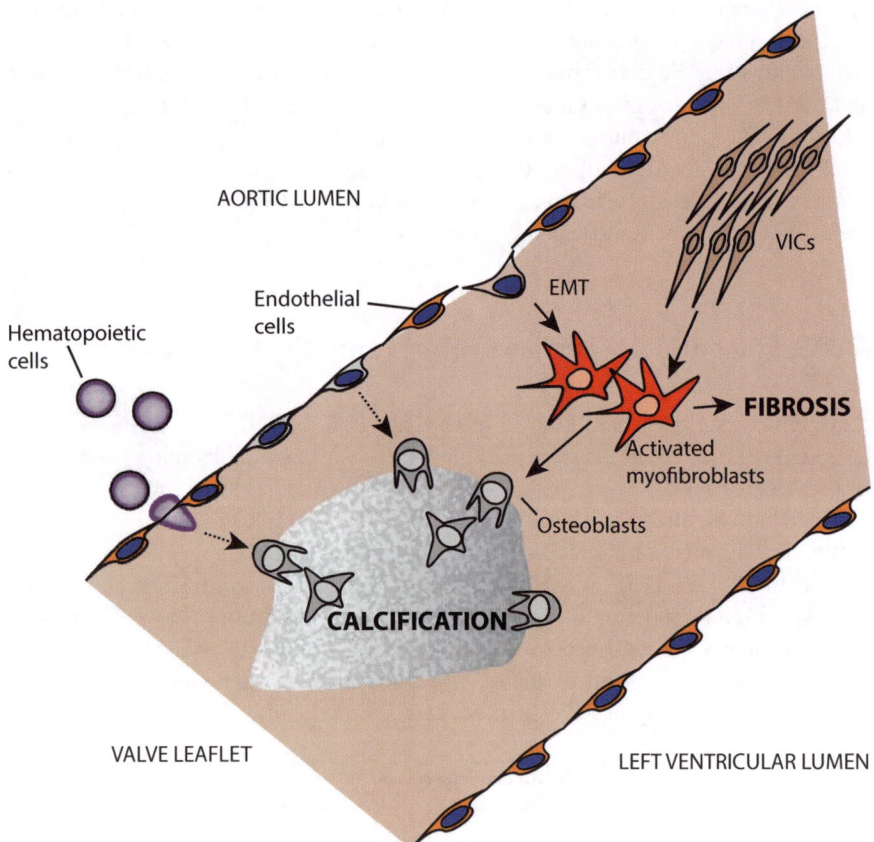

Fig. 5 Cell populations contributing to the calcification of valve leaflets

progenitor cells expressing both endothelial (CD34, KDR) and osteogenic (osteocalcin) markers have been recently found in human calcifying valve [52]. The questions that arise are that of the origin of the progenitors and the signaling pathways that stimulate them. Again the same signaling components active during embryogenesis contribute to the mobilization of progenitor cells. BMP2 and BMP4, secreted by endothelial cells under shear stress [53] and in turn the smad pathway is one of the major morphogens that induces EMT of endocardial cells and a key component of valve calcification.

TGFβ is one of the primary agonist to induce fibrosis in many pathological situations including skin wound healing, liver fibrosis, kidney fibrosis, myocardial fibrosis and an activator of smads.

However TGFβ [54] released together with TNFα and IL1β [56] by T cells infiltrating the endothelium during the inflammatory process does not play a prominent role in valvular fibrosis and calcification. *In vitro* studies have shown that VICs challenged by TGFβ undergo calcification [48, 56]. FGF2 has been shown to counteract TGFβ mediated VIC conversion into myofibroblast [57]. This TGFβ-mediated effect depends on stiffness of the substrate or of the matrix [58, 59]. TGFβ, released by endothelial cells under shear stress, acting through both its canonical signalling pathway in leukocytes and through both canonical and non-canonical signalling in aortic valves of Reversa mice (hypercholesterolemic $Ldlr^{-/-}Apob^{+/+}Mttp^{fl/fl}/Mx$-$1Cre^{+/+}$) fed on a western diet, promotes valve calcification. Wnt/beta catenin signaling is also activated in the process of valve calcification [60].

Wnt signaling might be specifically activated by lipid deposition; indeed intranuclear β-catenin has been observed in hypercholesterolemic mice featuring valve calcification [61]. Wnt is specifically important to drive differentiation of myofibroblasts. Finally, other pathways such as the proinflammatory pathway NFκb and the Runx2/Notch pathways both mediate the calcification process [62, 63].

5   Conclusions

During embryogenesis, the cell lineages at the origin of the cellular components of the valves are still not very well known. This lack of information can also be found in adult as to the cell types that contribute to valve fibrosis and calcification. The same signaling pathways (BMP, TGFβ, Wnt, NFκb…) promote EMT in the AVC and OFT for the formation of cardiac cushions as well as they induce de novo EMT and activate progenitor cells in the adult valve undergoing fibrosis and calcification. Thus developmental biology studies should help in a better understanding of valve disease and should pave the way towards therapeutic approaches. The absence of fibrosis and calcification of the mitral valve in contrast to the aortic valve is still questionable. This could be related to a mechanical issue, the aortic valve being submitted to greater stress than the mitral valve. Another and non-mutually exclusive hypothesis could be a different cellular participation and in turn extracellular matrix composition in both valves as it could be anticipated from different embryological origins of cells that either early contribute or migrate into each valve. That further points to the requirement of more lineage tracing studies in the mouse embryo.

Fibrosis is a general phenomenon observed in many diseases. Aortic fibrosis and myocardial fibrosis are interrelated phenomenon. Any therapeutic approach against valve fibrosis should thus help in reducing myocardial fibrosis. The embryonic origin, the activation process and the cellular physiology and function of valvular fibroblasts are key phenomenon that require to be fully understood in order to develop anti-fibrotic and calcification therapies.

Great progress has been accomplished in the last decades to better understand valve biology and function. There is still much to investigate to get a clear understanding of valve formation and diseases.

References

1. Nkomo VT, Gardin JM, Skelton TN, Gottdiener JS, Scott CG, Enriquez-Sarano M (2006) Burden of valvular heart diseases: a population-based study. Lancet 368:1005–1011
2. Roger VL, Go AS, Lloyd-Jones DM, Benjamin EJ, Berry JD, Borden WB, Bravata DM, Dai S, Ford ES, Fox CS, Fullerton HJ, Gillespie C, Hailpern SM, Heit JA, Howard VJ, Kissela BM, Kittner SJ, Lackland DT, Lichtman JH, Lisabeth LD, Makuc DM, Marcus GM, Marelli A, Matchar DB, Moy CS, Mozaffarian D, Mussolino ME, Nichol G, Paynter NP, Soliman EZ, Sorlie PD, Sotoodehnia N, Turan TN, Virani SS, Wong ND, Woo D, Turner MB (2012) Heart disease and stroke statistics–2012 update: a report from the American Heart Association. Circulation 125:e2–e220
3. Person AD, Klewer SE, Runyan RB (2005) Cell biology of cardiac cushion development. Int Rev Cytol 243:287–335
4. Combs MD, Yutzey KE (2009) Heart valve development: regulatory networks in development and disease. Circ Res 105:408–421
5. Buckingham M, Meilhac S, Zaffran S (2005) Building the mammalian heart from two sources of myocardial cells. Nat Rev Genet 6:826–835
6. Cai CL, Liang X, Shi Y, Chu PH, Pfaff SL, Chen J, Evans S (2003) Isl1 identifies a cardiac progenitor population that proliferates prior to differentiation and contributes a majority of cells to the heart. Dev Cell 5:877–889
7. de Vlaming A, Sauls K, Hajdu Z, Visconti RP, Mehesz AN, Levine RA, Slaugenhaupt SA, Hagege A, Chester AH, Markwald RR, Norris RA (2012) Atrioventricular valve development: new perspectives on an old theme. Differentiation 84:103–116
8. Pucéat M (2012) Embryological origin of the endocardium and derived valve progenitor cells: from developmental biology to stem cell-based valve repair. Biochim Biophys Acta 1833(4):917–922
9. Milgrom-Hoffman M, Harrelson Z, Ferrara N, Zelzer E, Evans SM, Tzahor E (2011) The heart endocardium is derived from vascular endothelial progenitors. Development (Cambridge, England) 138:4777–4787
10. Blin G, Nury D, Stefanovic S, Neri T, Guillevic O, Brinon B, Bellamy V, Rucker-Martin C, Barbry P, Bel A, Bruneval P, Cowan C, Pouly J, Mitalipov S, Gouadon E, Binder P, Hagege A, Desnos M, Renaud JF, Menasche P, Puceat M (2010) A purified population of multipotent cardiovascular progenitors derived from primate pluripotent stem cells engrafts in postmyocardial infarcted nonhuman primates. J Clin Invest 120:1125–1139
11. Neri T, Van Vliet P, Hiriart E, Norris R, Faure E, Zaffran S, Faustino R, Sugi Y, Levine R, De la Pompa J, Terzic A, Evans S, Markwald R (2014) Human embryonic stem cells recapitulate early cardiac valvulogenesis (submitted)

12. Christoffels VM, Habets PE, Franco D, Campione M, de Jong F, Lamers WH, Bao ZZ, Palmer S, Biben C, Harvey RP, Moorman AF (2000) Chamber formation and morphogenesis in the developing mammalian heart. Dev Biol 223:266–278
13. Stankunas K, Hang CT, Tsun ZY, Chen H, Lee NV, Wu JI, Shang C, Bayle JH, Shou W, Iruela-Arispe ML, Chang CP (2008) Endocardial Brg1 represses ADAMTS1 to maintain the microenvironment for myocardial morphogenesis. Dev Cell 14(2):298–311
14. Moorman AF, Christoffels VM (2003) Cardiac chamber formation: development, genes, and evolution. Physiol Rev 83:1223–1267
15. Simoes-Costa MS, Vasconcelos M, Sampaio AC, Cravo RM, Linhares VL, Hochgreb T, Yan CY, Davidson B, Xavier-Neto J (2005) The evolutionary origin of cardiac chambers. Dev Biol 277:1–15
16. Sugi Y, Markwald RR (1996) Formation and early morphogenesis of endocardial endothelial precursor cells and the role of endoderm. Dev Biol 175:66–83
17. Cohen-Gould L, Mikawa T (1996) The fate diversity of mesodermal cells within the heart field during chicken early embryogenesis. Dev Biol 177:265–273
18. Lee RK, Stainier DY, Weinstein BM, Fishman MC (1994) Cardiovascular development in the zebrafish. II. Endocardial progenitors are sequestered within the heart field. Development 120(12):3361–3366
19. Linask KK, Lash JW (1993) Early heart development: dynamics of endocardial cell sorting suggests a common origin with cardiomyocytes. Dev Dyn (an official publication of the American Association of Anatomists) 196:62–69
20. Misfeldt AM, Boyle SC, Tompkins KL, Bautch VL, Labosky PA, Baldwin HS (2009) Endocardial cells are a distinct endothelial lineage derived from Flk1+ multipotent cardiovascular progenitors. Dev Biol 333:78–89
21. Ema M, Takahashi S, Rossant J (2006) Deletion of the selection cassette, but not cis-acting elements, in targeted Flk1-lacZ allele reveals Flk1 expression in multipotent mesodermal progenitors. Blood 107:111–117
22. Stanley EG, Biben C, Elefanty A, Barnett L, Koentgen F, Robb L, Harvey RP (2002) Efficient Cre-mediated deletion in cardiac progenitor cells conferred by a 3'UTR-ires-Cre allele of the homeobox gene Nkx2-5. Int J Dev Biol 46:431–439
23. Ferdous A, Caprioli A, Iacovino M, Martin CM, Morris J, Richardson JA, Latif S, Hammer RE, Harvey RP, Olson EN, Kyba M, Garry DJ (2009) Nkx2-5 transactivates the Ets-related protein 71 gene and specifies an endothelial/endocardial fate in the developing embryo. Proc Natl Acad Sci U S A 106:814–819
24. Kattman SJ, Huber TL, Keller GM (2006) Multipotent flk-1+ cardiovascular progenitor cells give rise to the cardiomyocyte, endothelial, and vascular smooth muscle lineages. Dev Cell 11:723–732
25. Moretti A, Caron L, Nakano A, Lam JT, Bernshausen A, Chen Y, Qyang Y, Bu L, Sasaki M, Martin-Puig S, Sun Y, Evans SM, Laugwitz KL, Chien KR (2006) Multipotent embryonic isl1+ progenitor cells lead to cardiac, smooth muscle, and endothelial cell diversification. Cell 127:1151–1165
26. Stainier DY, Weinstein BM, Detrich HW, 3rd, Zon LI, Fishman MC (1995) Cloche, an early acting zebrafish gene, is required by both the endothelial and hematopoietic lineages. Development 121:3141–3150
27. Meilhac SM, Esner M, Kelly RG, Nicolas JF, Buckingham ME (2004) The clonal origin of myocardial cells in different regions of the embryonic mouse heart. Dev Cell 6:685–698
28. Meilhac SM, Kelly RG, Rocancourt D, Eloy-Trinquet S, Nicolas JF, Buckingham ME (2003) A retrospective clonal analysis of the myocardium reveals two phases of clonal growth in the developing mouse heart. Development 130:3877–3889
29. Mikawa T, Fischman DA (1992) Retroviral analysis of cardiac morphogenesis: discontinuous formation of coronary vessels. Proc Natl Acad Sci U S A 89:9504–9508
30. Mikawa T, Gourdie RG (1996) Pericardial mesoderm generates a population of coronary smooth muscle cells migrating into the heart along with ingrowth of the epicardial organ. Dev Biol 174:221–232

31. Gittenberger-de Groot AC, Vrancken Peeters MP, Mentink MM, Gourdie RG, Poelmann RE (1998) Epicardium-derived cells contribute a novel population to the myocardial wall and the atrioventricular cushions. Circ Res 82:1043–1052

32. Wessels A, van den Hoff MJ, Adamo RF, Phelps AL, Lockhart MM, Sauls K, Briggs LE, Norris RA, van Wijk B, Perez-Pomares JM, Dettman RW, Burch JB (2012) Epicardially derived fibroblasts preferentially contribute to the parietal leaflets of the atrioventricular valves in the murine heart. Dev Biol 366:111–124

33. Le Douarin NM, Dupin E (2012) The neural crest in vertebrate evolution. Curr Opin Genet Dev 22:381–389

34. Kirby ML, Gale TF, Stewart DE (1983) Neural crest cells contribute to normal aorticopulmonary septation. Science (New York, NY) 220:1059–1061

35. Waldo K, Miyagawa-Tomita S, Kumiski D, Kirby ML (1998) Cardiac neural crest cells provide new insight into septation of the cardiac outflow tract: aortic sac to ventricular septal closure. Dev Biol 196:129–144

36. Jiang X, Rowitch DH, Soriano P, McMahon AP, Sucov HM (2000) Fate of the mammalian cardiac neural crest. Development 127:1607–1616

37. Hinton RB Jr, Lincoln J, Deutsch GH, Osinska H, Manning PB, Benson DW, Yutzey KE (2006) Extracellular matrix remodeling and organization in developing and diseased aortic valves. Circ Res 98:1431–1438

38. Chin C, Gandour-Edwards R, Oltjen S, Choy M (1992) Fate of the atrioventricular endocardial cushions in the developing chick heart. Pediatr Res 32:390–393

39. de Lange FJ, Moorman AF, Anderson RH, Manner J, Soufan AT, de Gier-de Vries C, Schneider MD, Webb S, van den Hoff MJ, Christoffels VM (2004) Lineage and morphogenetic analysis of the cardiac valves. Circ Res 95:645–654

40. Hay ED (1995) An overview of epithelio-mesenchymal transformation. Acta Anat 154:8–20

41. Kalluri R, Weinberg RA (2009) The basics of epithelial-mesenchymal transition. J Clin Invest 119:1420–1428

42. Eisenberg LM, Markwald RR (1995) Molecular regulation of atrioventricular valvuloseptal morphogenesis. Circ Res 77:1–6

43. Timmerman LA, Grego-Bessa J, Raya A, Bertran E, Perez-Pomares JM, Diez J, Aranda S, Palomo S, McCormick F, Izpisua-Belmonte JC, de la Pompa JL (2004) Notch promotes epithelial-mesenchymal transition during cardiac development and oncogenic transformation. Genes Dev 18:99–115

44. Butcher JT, Markwald RR (2007) Valvulogenesis: the moving target. Philos Trans R Soc Lond B Biol Sci 362:1489–1503

45. Visconti RP, Ebihara Y, LaRue AC, Fleming PA, McQuinn TC, Masuya M, Minamiguchi H, Markwald RR, Ogawa M, Drake CJ (2006) An in vivo analysis of hematopoietic stem cell potential: hematopoietic origin of cardiac valve interstitial cells. Circ Res 98:690–696

46. Sprague HB, Mallory TB, Chapman EM (1947) (Calcific aortic stenosis). N Engl J Med 237:958–960

47. Wylie-Sears J, Aikawa E, Levine RA, Yang JH, Bischoff J (2011) Mitral valve endothelial cells with osteogenic differentiation potential. Arterioscler Thromb Vasc Biol 31:598–607

48. Paranya G, Vineberg S, Dvorin E, Kaushal S, Roth SJ, Rabkin E, Schoen FJ, Bischoff J (2001) Aortic valve endothelial cells undergo transforming growth factor-beta-mediated and non-transforming growth factor-beta-mediated transdifferentiation in vitro. Am J Pathol 159:1335–1343

49. Liu AC, Joag VR, Gotlieb AI (2007) The emerging role of valve interstitial cell phenotypes in regulating heart valve pathobiology. Am J Pathol 171:1407–1418

50. Galante A, Pietroiusti A, Vellini M, Piccolo P, Possati G, De Bonis M, Grillo RL, Fontana C, Favalli C (2001) C-reactive protein is increased in patients with degenerative aortic valvular stenosis. J Am Coll Cardiol 38:1078–1082

51. Khosla S, Eghbali-Fatourechi GZ (2006) Circulating cells with osteogenic potential. Ann N Y Acad Sci 1068:489–497

52. Gossl M, Khosla S, Zhang X, Higano N, Jordan KL, Loeffler D, Enriquez-Sarano M, Lennon RJ, McGregor U, Lerman LO, Lerman A (2012) Role of circulating osteogenic progenitor cells in calcific aortic stenosis. J Am Coll Cardiol 60:1945–1953

53. Sun L, Rajamannan NM, Sucosky P (2013) Defining the role of fluid shear stress in the expression of early signaling markers for calcific aortic valve disease. PloS One 8:e84433

54. Jian B, Narula N, Li QY, Mohler ER 3rd, Levy RJ (2003) Progression of aortic valve stenosis: TGF-beta1 is present in calcified aortic valve cusps and promotes aortic valve interstitial cell calcification via apoptosis. Ann Thorac Surg 75:457–465; discussion 465–456

55. Kaden JJ, Dempfle CE, Grobholz R, Tran HT, Kilic R, Sarikoc A, Brueckmann M, Vahl C, Hagl S, Haase KK, Borggrefe M (2003) Interleukin-1 beta promotes matrix metalloproteinase expression and cell proliferation in calcific aortic valve stenosis. Atherosclerosis 170:205–211

56. Walker GA, Masters KS, Shah DN, Anseth KS, Leinwand LA (2004) Valvular myofibroblast activation by transforming growth factor-beta: implications for pathological extracellular matrix remodeling in heart valve disease. Circ Res 95:253–260

57. Cushing MC, Mariner PD, Liao JT, Sims EA, Anseth KS (2008) Fibroblast growth factor represses Smad-mediated myofibroblast activation in aortic valvular interstitial cells. FASEB J 22:1769–1777

58. Gould ST, Matherly EE, Smith JN, Heistad DD, Anseth KS (2014) The role of valvular endothelial cell paracrine signaling and matrix elasticity on valvular interstitial cell activation. Biomaterials 35:3596–3606

59. Yip CY, Chen JH, Zhao R, Simmons CA (2009) Calcification by valve interstitial cells is regulated by the stiffness of the extracellular matrix. Arterioscler Thromb Vasc Biol 29:936–942

60. Caira FC, Stock SR, Gleason TG, McGee EC, Huang J, Bonow RO, Spelsberg TC, McCarthy PM, Rahimtoola SH, Rajamannan NM (2006) Human degenerative valve disease is associated with up-regulation of low-density lipoprotein receptor-related protein 5 receptor-mediated bone formation. J Am Coll Cardiol 47:1707–1712

61. Miller JD, Weiss RM, Serrano KM, Castaneda LE, Brooks RM, Zimmerman K, Heistad DD (2010) Evidence for active regulation of pro-osteogenic signaling in advanced aortic valve disease. Arterioscler Thromb Vasc Biol 30:2482–2486

62. Kaden JJ, Bickelhaupt S, Grobholz R, Haase KK, Sarikoc A, Kilic R, Brueckmann M, Lang S, Zahn I, Vahl C, Hagl S, Dempfle CE, Borggrefe M (2004) Receptor activator of nuclear factor kappaB ligand and osteoprotegerin regulate aortic valve calcification. J Mol Cell Cardiol 36:57–66

63. Garg V, Muth AN, Ransom JF, Schluterman MK, Barnes R, King IN, Grossfeld PD, Srivastava D (2005) Mutations in NOTCH1 cause aortic valve disease. Nature 437(7056):270–274

Diverse Cellular Origins of Cardiac Fibroblasts

Fahmida Jahan and Jeffrey T. Wigle

Abstract Cardiac fibroblasts synthesize and remodel the extracellular matrix (ECM) of the heart and are key players in the development of cardiac fibrosis. They are closely associated with cardiac development, function and disease. Cardiac fibroblasts exist as three distinct phenotypes: a non-contractile fibroblast, an intermediate phenotype called a protomyofibroblast, and a contractile myofibroblast. During embryonic development, cardiac fibroblasts are derived mainly from the proepicardium and the cardiac endothelium via the process of epithelial/endothelial to mesenchymal transition (EMT and EndMT). In the adult heart, fibroblasts were previously thought to be a homogeneous cell population and to be derived mainly from the proliferation of resident fibroblasts. However, recent evidence suggests that in diseases such as cardiac fibrosis, activated fibroblasts or myofibroblasts can be derived from multiple sources from the close proximity of the wound area such as from epithelial and endothelial cells via EMT and EndMT, and from circulating bone marrow progenitor cells or from fibrocytes and pericytes. Due to persistent pathological stimuli, the resulting uncontrolled proliferation of fibroblast and their phenoconversion to myofibroblast can lead to ECM remodeling and fibrosis which is associated with various clinically important conditions such as hypertension, atherosclerosis, ischemia, dilated cardiomyopathies, valvular diseases, arrhythmias and heart failure. At present there is no effective therapy for cardiac fibrosis or fibroblast associated pathologies; thus making it a necessity to have a better understanding of the sources of fibroblasts and the underlying mechanisms that lead to their formation in order to limit aberrant fibroblast and myofibroblast generation during pathological conditions.

Keywords Cardiac fibroblasts · Myofibroblasts · Phenoconversion · Extracellular matrix · Epithelial/endothelial-mesenchymal transition

J. T. Wigle (✉) · F. Jahan
Institute of Cardiovascular Sciences, St. Boniface Hospital Research Centre, 351 Tache Ave, Winnipeg, MB R2H 2A6, Canada
e-mail: jwigle@sbrc.ca

F. Jahan · J. T. Wigle
Department of Biochemistry and Medical Genetics, University of Manitoba, Winnipeg, MB, Canada

© Springer International Publishing Switzerland 2015
I.M.C. Dixon, J. T. Wigle (eds.), *Cardiac Fibrosis and Heart Failure: Cause or Effect?*, Advances in Biochemistry in Health and Disease 13, DOI 10.1007/978-3-319-17437-2_8

1 Introduction

Fibroblasts are metabolically active cells that are found in most tissues of the body. They are critical for regulating both the composition and the turnover of the extra-cellular matrix (ECM) components, fluid volume and pressure, and tissue repair [1]. During wound healing, fibroblasts are vital to the precise control of the inflammatory response. Various inflammatory cytokines and growth factors promote the recruitment of fibroblasts to the wounded area. Activated fibroblasts (myofibroblasts) contribute to the wound healing process by producing ECM proteins and responding to and synthesizing cytokines, chemokines, and other inflammatory mediators [2–4]. These cells express myosin and smooth muscle actin, thus they provide a contractile force, which decreases the lesion size. Consequently dysregulation of the wound healing process leads to aberrant fibroblast recruitment and function. Abnormal remodeling of ECM due to excessive or inadequate secretion of matrix components alters organ architecture, impairs function and ultimately leads to organ failure [5–7].

By cell number, fibroblasts constitute the largest cell population in the heart [8]. Cardiac fibroblasts are multi-functional and play crucial roles in both cardiac development and in normal adult physiology [9]. They are crucial for maintaining the structural, biochemical, mechanical and electrophysiological properties of the myocardium. Their primary function is ECM remodeling (synthesis and degradation), which provides a 3D scaffold for myocytes and non-myocytes in the heart. Furthermore, they direct cardiomyocyte growth, cardiac vessel formation and maintain electrophysiological properties to ensure proper cardiac form and function [1, 8–11].

Historically, cardiac fibroblasts were considered to be a homogeneous cell population. However, recently it has been revealed that cardiac fibroblasts are instead a complex heterogeneous population of cells having diverse origins and functions [12, 13]. In the normal heart, fibroblast numbers are largely maintained or increased by the proliferation of resident fibroblasts. But during pathological conditions, fibroblast number can be dramatically increased from various sources such as: (1) proliferation of resident myocardial fibroblasts, (2) recruitment and differentiation of circulating bone marrow progenitor cells, or (3) by the phenotypic conversion of endothelial and epithelial cells into fibroblasts and myofibroblasts (Fig. 1); [13]. Therefore in disease conditions, their aberrant proliferation, activation and recruitment leads to ECM remodeling, tissue fibrosis, myocardial stiffening and dysfunction and thus eventually leads to heart failure [14, 15]. Currently there are no effective therapies that specifically address pathologies associated with fibroblast dysfunction. Hence, a better understanding of their origin, sources and function is the prerequisite to develop an effective therapeutic strategy.

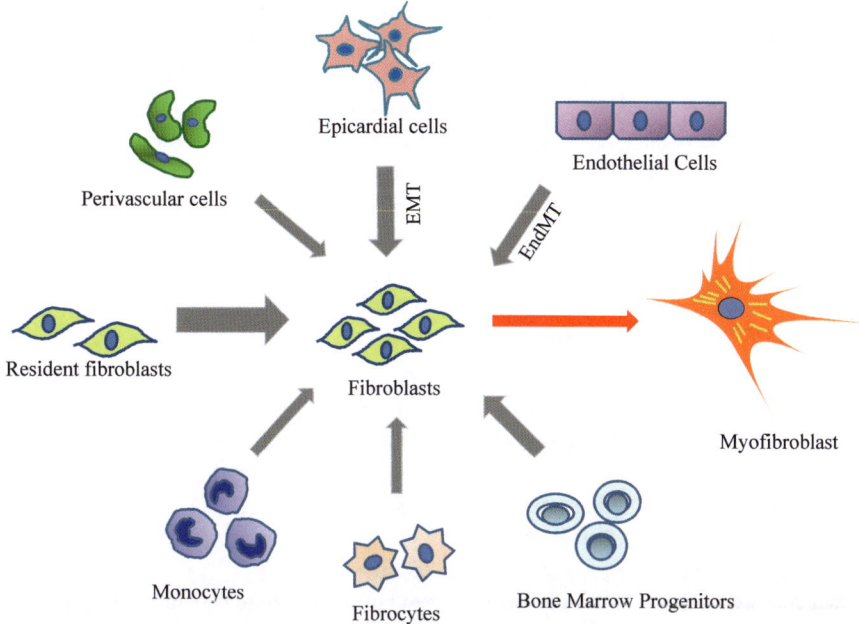

Fig. 1 Diverse sources of cardiac fibroblasts during disease progression. During pathological conditions, a dramatic increase in fibroblast number can be derived from various sources such as proliferation of resident fibroblasts, mesenchymal transition of *epithelial* (*EMT*)/*endothelial* (*EndMT*) cells; and recruitment and differentiation of circulating bone marrow progenitor cells, *monocytes*, *fibrocytes* and *perivascular cells*. These fibroblasts can further phenoconvert into a super mature activated *myofibroblast* that actively synthesizes extracellular matrix proteins and forms stress fibers. *EMT* epithelial to mesenchymal transition, *EndMT* endothelial to mesenchymal transition

2   Phenotypes

Fibroblasts typically exhibit three distinct phenotypes: a non-contractile fibroblast, an intermediate phenotype called a protomyofibroblast [16], and a contractile myofibroblast.

Characteristics of these different cellular phenotypes are described below.

2.1   Fibroblasts

Fibroblasts are generally flat, spindle-shaped cells which may have multiple projections. They populate all of the connective tissues in the body [9, 17]. Although they were considered to be a homogeneous cell population; it is now apparent that these fibroblasts can arise from multiple origins and fibroblasts from different tissues exhibit differential properties and functions [9]. Among the various cell types in the myocardium, the cardiac fibroblasts are distinguished by their lack of a basement membrane [9]. Although the heart is comprised of various cell types including cardiomyocytes, cardiac fibroblasts, endothelial cells, smooth muscle cells and pericytes; cardiac fibroblasts represent the largest population in the heart by cell number [8]. They account for approximately two-thirds of the total cell population, whereas cardiomyocytes comprise about two-thirds of the total volume due to their larger size. However, this ratio may vary from species to species [8].

2.2   Protomyofibroblasts

Fibroblasts can differentiate into an intermediate phenotype called protomyofibroblasts under stress conditions. Protomyofibroblasts are characterized by the synthesis of stress fibres connected to cytoplasmic actins that form fibronexus adhesion complexes under mechanical stress. They also express the ED-A splice variant of fibronectin to form a special organization of cellular fibronectin at the cell surface. These cells are capable of generating contractile force [16].

Mechanical tension is crucial for the induction of contractile force by a protomyofibroblast. However, the exact mechanism behind protomyofibroblast formation *in vivo* is not well understood. *In vitro* studies have shown that fibroblasts extracted from a range of organs and tissues and plated on plastic tissue-culture dishes spontaneously undergo protomyofibroblast phenoconversion. These cells gain the ability to rapidly synthesize stress fibres, focal adhesion molecules and fibronectins- characteristic of a protomyofibroblast phenotype [16]. It is thought that during the tissue repair process, fibroblasts rapidly move to the site of injury and produce a collagen and fibronectin rich ECM [16]. These fibroblasts acquire the ability to form stress fibres and focal adhesions. When wound closure is completed, fibroblasts occupying the granulation tissue in the scar become mechanically stressed due to the tractional force generated by various collagen fibres and stress fibres. Moreover, these fibroblasts also alter the processing of fibronectin splice variants and re-express those of earlier developmental stages. In addition to mechanical tension, growth factors can also play a significant role in this phenoconversion in early developmental stages and during wound healing in the adult. For example, alveolar protomyofibroblasts were absent in platelet-derived growth factor (PDGF)-null mice suggesting that PDGF plays a key role in the generation of protomyofibroblasts [18].

2.3   Myofibroblasts

Protomyofibroblasts further phenoconvert into myofibroblasts in the presence of persistent mechanical stress [16]. It is thought that the protomyofibroblast is an intermediate continuous form when a fibroblast is transitioning into a myofibroblast.

During this transition, a protomyofibroblast switches its phenotype from being pro-liferative and migratory to a hypoproliferative, less migratory and synthetic phase of super mature myofibroblast phenotype [19, 20]. Expression of α-smooth muscle actin (α-SMA), formation of more complex and organized stress fibres and fibron-exus adhesion complexes or super mature focal adhesion molecules are all charac-teristic features of myofibroblasts [16, 19, 20]. Soluble factors including hormones, such as angiotensin II, endothelin I and pro-fibrotic cytokines such as transform-ing growth factor-β (TGF-β), connective tissue growth factor (CCN2/CTGF) and PDGF can induce the generation of myofibroblasts [16, 21]. Among these factors, TGF-β is considered to be the critical driver of myofibroblast phenoconversion. Although TGF-β acts as an anti-proliferative factor in most cells, it can induce fi-broblasts to proliferate and produce ECM during normal tissue repair [22]. TGF-β is also known to elevate ED-A fibronectin expression levels [22]. Both of these effects under stress condition are critical for inducing the phenoconversion from protomyofibroblasts into mature myofibroblasts. TGF-β potentially induces colla-gen synthesis by fibroblastic cells and increases the levels of plasminogen activator inhibitor-1 (PAI-1) and α-SMA expression. It also reduces matrix metalloprotein-ase (MMP) activity by activating the production of their inhibitors and ultimately stimulates myofibroblasts to induce rapid ECM turnover and restoration of injured tissue during the healing process [16, 23]. TGF-β is secreted during tissue injury from a variety of cell sources such as white blood cells, particularly macrophages; platelets and parenchymal cells. TGF-β can also be produced and secreted by fi-broblasts to function in an autocrine manner; making it another important mode for activation and maintenance of myofibroblast when the external inflammatory signal is lost. Damaged epithelial cells are also known to produce and secret TGF-β, and thereby contribute to myofibroblast generation in a paracrine fashion [19]. Blocking the interaction of ED-A fibronectin with the cell surface leads to the attenuation of TGF-β mediated myofibroblast phenoconversion. Moreover it has been shown that when mechanical tension is lost, the myofibroblast phenotype fails to persist even in the presence of TGF-β and ED-A fibronectin. Inhibiting TGF-β signalling under mechanical stress condition prevents myofibroblast formation [24]. Therefore, all of these factors are crucial for the generation of myfibroblast from fibroblast but further investigations are required to discern their possible interplay and underlying mechanisms.

3   Origin of Cardiac Fibroblasts During Development

Fibroblasts play a crucial role during heart development. Depending on the stage of development, their point of origin can vary greatly. During embryogenesis, fi-broblasts are considered to be of mesenchymal origin and participate in the forma-tion of the heart. The proepicardial organ and the epithelial–mesenchymal transition (EMT) during cardiac valve formation are considered to be the two principal sourc-es of the cardiac fibroblast population [17]. However, developing bone marrow and their circulating progenitors, neural crest cells and differentiation from vascular walls can also give rise to fibroblasts during embryonic development [8]. Moreover, it has also been reported that fibroblasts can arise from mesoangioblasts. These are multipotent progenitor cells that differentiate into either vascular endothelial cells or mesodermal fibroblasts. These progenitors originate from the bone marrow hematopoietic stem cells [17].

Cardiac interstitial and annulus fibroblasts are thought to be derived from the embryonic proepicardial mesenchymal cells [11]. The proepicardial cells cover the surface of the embryonic heart and form the epicardium [25]. In the developing heart, the epicardium is the principal source of cardiac progenitor cells. These progenitor cells give rise to two cell types: (1) coronary vascular smooth muscle cells (cVSMCs) and (2) cardiac fibroblasts. These cells contribute to coronary vasculature, cardiac wall, subendocardium, the atrioventricular (AV) cushions and valves, and the fibrous skeleton of the heart [8, 25, 26]. The presence of growth factors, including PDGF, fibroblast growth factor (FGF), and TGF-β, stimulates the epicardium to undergo EMT and form epicardium-derived cells (EPDC). These cells subsequently differentiate into a fibroblast phenotype to form the fibrous heart skeleton. Using chicken-quail chimera, quail derived EPDCs were found in the subendocardium, myocardium, and AV cushions. These EPDCs at the fibrous annulus region were found to be positive for procollagen-I indicating a cardiac fibroblast lineage and thus revealed their role in the formation of the fibrous heart skeleton. Adventitial fibroblasts were also found to be derived from the epicardium [27]. The Tallquist group has shown the role of a E-box binding basic helix-loop-helix (bHLH) transcription factor, Tcf21, in epicardial cell fate determination and cardiac fibroblast development. Using a tamoxifen-inducible Cre expressed from the *Tcf21* locus, they showed that Tcf21-expressing epicardial cells are largely fated to become the cardiac fibroblast lineage specific via EMT. Moreover, *Tcf21* knockout mice lacked cardiac fibroblasts, and fate mapping study showed attenuation of epicardial EMT [28].

Cardiac endothelium may also give rise to valvular fibroblasts which are also known as valvular interstitial cells (VICs) [29]. Endothelial cells detach to form the cardiac cushion and undergo endothelial to mesenchymal transition (EndMT) that is stimulated by various cytokines including TGF-β, PDGF and Wnt. These mesenchymal cells enter the cardiac jelly and acquire a fibroblast phenotype. Various signaling pathways and genes such as vascular endothelial growth factor (VEGF), nuclear factor of activated T-cells, cytoplasmic 1 (NFATc1), Notch, Wnt/β-catenin, bone morphogenetic protein (BMP)/TGF-β, ErbB, and neurofibromatosis 1 (NF1)/ Ras play important and selective roles in regulating endothelial cell proliferation and differentiation during valve development and maturation [30]. VEGF levels need to be strictly controlled during normal heart development as increased VEGF expression can inhibit EndMT [31]. Tissue explants studies have revealed that hypoxia prevents cardiac cushion EndMT by inducing a 10-fold increase in VEGF levels. This finding indicates the possibility that fetal hypoxia may lead to congenital heart defects in the cardiac valves and interatrial septum [32]. Cushion endothelial cells exposed to hyperglycemic conditions during developmental stages had persistent CD31 expression and were unable to correctly initiate the EndMT program [30, 33]. On the other hand cardiac cushions derived from mice lacking CD31 were able to undergo EndMT, even in hyperglycemic conditions [30, 33]. Studies have also indicated that NFATc1 is down-regulated during endocardial EndMT and that NFATc1-expressing cells do not undergo EndMT [34]. Notch signalling is essential for regulation of endocardial cushion EndMT. In $Notch1^{-/-}$ mice, cardiac cushions were found to be hypoplastic suggesting that the endocardium failed to undergo EndMT [35]. Moreover, disruption of Notch signaling particularly diminished TGF-β_2 expression in the heart. In agreement with this result, the expression levels of Snail, a mesenchymal transcription factor, was found to be significantly decreased in the absence of Notch signaling. This finding suggests that Notch signaling may increase the expression of TGF-β_2 in the heart; and TGF-β_2 eventually promotes EndMT in the endocardium [35]. Using a homozygous adenomatous polyposis coli (APC) truncation mutant, nuclear β-catenin was detected throughout the heart indicating that a large population of endocardial cells are able to undergo EndMT [30, 36]. Studies have also indicated that Wnt/β-catenin signaling may play a crucial role in valve development by regulating EndMT. β-catenin can activate the expression of genes required for the mesenchymal transition program. It is speculated that upon CD31 repression, β-catenin levels in the cytosol increase and trigger proliferation of cells undergoing EndMT; thus suggesting that β-catenin may activate the EndMT program which generates the fibroblast population in the cardiac jelly [30]. Using a BMP-2 deficient chick model, mesenchymal cells were shown to fail to invade through a collagen lattice. Similarly by using mouse AV explants, BMP-2 treatment was found to be sufficient to induce EndMT in the myocardium. Additionally, AV endothelial explants were found to synthesize and secret TGF-β_2 in an autocrine mechanism upon BMP-2 treatment [30].

Thus during embryonic developmental stages, cardiac fibroblasts can be derived from a number of sources, especially from proepicardium and cardiac endothelium. However fate mapping studies largely depend on the susceptibility of Cre-dependent reporter and on the efficiency of Cre recombinase, which may represent a limitation of the current findings. Moreover, the lack of specific fibroblast markers makes it challenging to trace their origin. Refining existing fate mapping strategies and identifying appropriate marker genes will enhance our knowledge of cardiac fibroblast heterogeneity during development.

4   Sources of Fibroblast Generation in Cardiac Pathology

In disease, fibroblasts are recruited to the injured area from different sources. Fibroblast generation from nearby sources such as epithelial and endothelial cells via mesenchymal transition (EMT and EndMT), or from fibrocytes, pericytes or from circulating bone marrow progenitor cells is thought to be an effective means to enable the rapid recruitment of fibroblasts to the injured area since recruitment from more distant sites would require migration, activation, and proliferation of these cells (Fig. 1). For instance, during wound healing in epithelial rich areas such as the skin, EMT is a vital process for fibroblast recruitment [13]. Recently there has been increasing evidence suggesting the heterogeneity of cell sources for the fibroblast population found in the damaged heart.

5   Resident Fibroblasts

Previously, it was generally thought that activated fibroblasts or myofibroblasts in fibrotic hearts are derived mainly from the proliferation of existing fibroblasts in the heart called resident fibroblasts [13]. This hypothesis is supported by the fact that cardiac fibroblasts are highly responsive to circulating cues in the surrounding microenvironment that can influence their proliferation and recruitment to the site of pathological inflammation [9]. Initial studies in a pressure-overload mouse model showed that inhibition of TGF-β receptor decreased collagen synthesis and deposition; and inhibited the proliferation and activation of myofibroblasts that resulted in dilated myopathy and dysfunction [37]. During replacement fibrosis in the heart, such stimuli trigger resident fibroblasts to synthesize extracellular matrix components at the site of injury to heal the damaged area [9, 37, 38]. However, increasing evidence suggests that during reactive interstitial fibrosis proliferating pro-fibrotic cells not only originate from the resident fibroblast population, but also are rapidly recruited from multiple nearby sources to enhance cardiac healing [9, 39, 40].

6   Epithelial to Mesenchymal Transition (EMT)

EMT includes a cascade of events through which epithelial cells lose cell-cell contact, cell polarity and acquire migratory and invasive properties and differentiate into mesenchymal cells (Fig. 2) [41]. The structural integrity of epithelial cells are maintained by cell-cell contacts that involve tight junctions, cadherin junctions linked to the actin cytoskeleton, gap junctions to allow direct intercellular communication, desmosomes which are attached to the intermediate filaments and integrin mediated interaction between cell and matrix. During EMT, epithelial cells lose their structural integrity and cell-ECM interactions are modified and migrate into the surrounding tissue. The cellular cytoskeleton is rearranged to enable these cells to invade the surrounding three-dimensional matrix [24]. This process requires a cascade of mesenchymal transcriptional factors to be activated in order to trigger and maintain the mesenchymal features. Mesenchymal cells are spindle-shaped, lack polarity, cell-cell junctions and interaction to basal lamina. Mesenchymal cells remodel the ECM by synthesizing and secreting ECM components. Epithelial cells

Epicardial epithelial cells

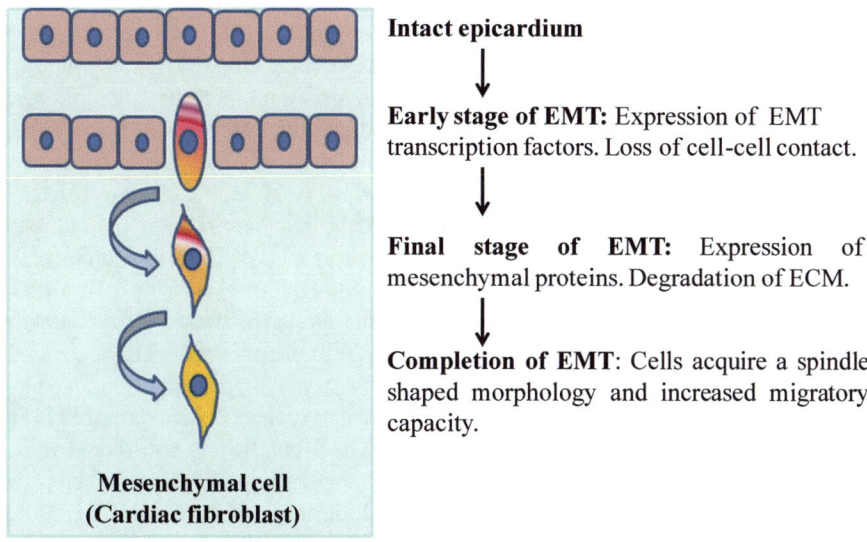

Intact epicardium

↓

Early stage of EMT: Expression of EMT transcription factors. Loss of cell-cell contact.

↓

Final stage of EMT: Expression of mesenchymal proteins. Degradation of ECM.

↓

Completion of EMT: Cells acquire a spindle shaped morphology and increased migratory capacity.

**Mesenchymal cell
(Cardiac fibroblast)**

Fig. 2 Epithelial to mesenchymal transition (*EMT*) in the epicardium. During the first phase of EMT, the expression of transcriptional regulators such as- SNAI1/Snail 1, SNAI2/Snail 2 (Slug), Zeb1 and Zeb2 are up-regulated in epicardial epithelial cells. Their expression leads to the attenuation of E-cadherin, β-catenin, Desmoplakin, Muc-1, Syndecan-1 and Cytokeratin-18 expression; and thus results in the loss of cell-cell contact. During the second phase of EMT, characteristic mesenchymal proteins and growth factors are synthesized such as FSP1, TGF-β, FGF-1,-2,-8, MMP-2, MMP-9, Vimentin, α-SMA, Fibronectin, Collagen type I and Collagen type III. In total, these changes lead to a loss of cell polarity, degradation of ECM and acquisition of a spindle shaped morphology and migratory phenotype that is characteristic of a cardiac fibroblast

characteristically express E-cadherin, in contrast, mesenchymal cells express N-cadherin, fibronectin and vimentin [24]. Consequently, the lack of E-cadherin expression is a signature of EMT. EMT transcription factors such as SNAI1/Snail 1, SNAI2/Snail 2 (Slug), Zeb1, Zeb2 (SIP1), E47 and KLF8 function by repressing transcription of E-cadherin directly. These transcription factors can bind to the promoter region of E-cadherin and repress its expression. Others such as Twist, Goosecoid, E2.2 (TCF4), homeobox protein SIX1 and FOXC2 have been shown to repress E-cadherin indirectly [39, 42–44]. These transcription factors also inhibit the expression of other junctional molecules such as claudins and desmosomes to initiate EMT. Conversely factors such as grainyhead-like protein 2 homologue (GRHL2), *E74-like factor 3* (ELF3) and *E74-like factor 5* (ELF5) are down-regulated in EMT and can drive a reverse process termed mesenchymal-epithelial transition (MET) upon over-expression in mesenchymal cells [44, 45].

Signaling pathways that trigger and control EMT are- TGF-β, FGF, epidermal growth factor (EGF), hepatocyte growth factor (HGF), Wnt/beta-catenin, Notch and hypoxia. In the first phase of EMT, Ras-MAPK pathway up-regulates Snail and Slug expression which leads to desmosomal disruption, cell spreading, and breakage of cell-cell contact, thus initiating the primary essential phase of the EMT program [46, 47]. The second phase involves attainment of cell motility, decreased cytokeratin production, and expression of vimentin [46, 47]. Snail and Slug can also repress p63, a transcription factor which is essential for the development of the epithelial cell lining and maintenance of their structural integrity. Disruption of p63 expression inhibits cell-cell contacts and induces migration in cancer cells [42, 48, 49]. Previously it was reported that Zeb2 down-regulates E-cadherin expression by binding to its promoter. In the epithelial Madin-Darby canine kidney (MDCK) cell line, down-regulation of E-cadherin expression resulted in disrupted cell-cell contact and acquisition of migratory and invasive properties [50]. Recently, it has been shown that Zeb2 is further linked to promoting fibroblast to myofibroblast phenoconversion. Zeb2 expression level has been found to be higher in first passage cardiac myofibroblasts indicating its potential role in the phenoconversion process during cardiac injury and fibrotic condition [51]. Ectopic expression of Ski, a negative regulator of TGF-β_1 signaling, has been shown to repress Zeb2 expression [51].

Among different signalling pathways, the TGF-β pathway is considered to be one of the major drivers of EMT. It activates expression of EMT transcription factors such as Snail and Zeb to mediate EMT during cardiac development and pathological conditions such as cancer and fibrosis [43, 48]. The p53 tumor suppressor represses EMT by increasing transcription of certain microRNAs—such as miR-200 and miR-34 that block Zeb and Snail protein synthesis, and maintains the epithelial structural integrity [52]. The Wnt signaling pathway has also been reported to control EMT during gastrulation, heart valve development and cancer pathogenesis. Activation of the Wnt pathway induces Snail expression which in turn up-regulates vimentin expression [26, 30].

EMT is a crucial process for the development of many tissues and organs, and for various developmental processes such as gastrulation, neural crest development, cardiac valve formation, palate and muscle formation. EMT has also been implicated in pathologies such as wound healing, organ fibrosis and cancer metastasis [53]. EMT is a key process in the production of activated fibroblasts in different organs including the heart, lung, liver, and kidney [47]. The process of EMT has been extensively studied recently and suggests that the epicardial cells can contribute to the fibroblast population by undergoing an EMT [11, 54]. Epithelial cells can differentiate into fibroblasts and promote lung fibrosis in mice by making up the majority of the emerging population of fibroblasts in the fibrotic lesion. This suggests that EMT could be a major source of fibroblasts in pathological conditions [55]. Studies showed that alveolar epithelial cells (AECs) can undergo EMT both *ex vivo* and *in vivo*. It indicates that these cells are may be the progenitors for fibroblasts and contribute to the emerging fibroblast population [55]. Moreover, lineage-tracing and bone marrow transplant studies in mice showed that in kidney fibrosis approximately 12% of fibroblasts were found to arise from the bone marrow and 30% comes from tubular epithelial cells via EMT. Another study has shown that in kidney fibrosis approximately 35% of fibroblasts were derived from endothelial cells via EndMT [48]. However, the extent of contribution of EMT derived fibroblasts in case of injury is yet unknown. It is speculated that the percentage of fibroblasts coming from distinct sources may vary significantly depending on the factors such as the phase of fibrosis, the organ, and the model being used. Moreover, some studies on kidney, liver and lung fibrosis models failed to show the role of EMT in myofibroblast generation [56–58]. This indicates that apart from becoming a mature myofibroblast, EMT may give rise to a range of mesenchymal phenotypes as a strategy to initiate wound healing and promote cell survival in response to injury. More detailed information on EMT and its role in fibrosis has been described in Kalluri and Neilson [47].

7   EMT Derived Fibroblasts in Cardiac Diseases

Although in normal adult heart epicardial cells do not actively undergo EMT, *in vivo* studies of myocardial injuries have shown that epicardium-specific genes expressed during development are re-activated in response to injury [59]. Initially this reactivation is global, but eventually it is confined to the injury area, indicating a role of epicardium in healing cardiac insults (Fig. 2). Subsequently, these epicardial cells undergoing EMT proliferate and migrate into the sub-epicardium and form a thick epicardial cap consisting of epicardial derived cells (EPDCs). For instance, using a $Wt1^{CreERT2}$ mouse model, it was previously shown that epicardial cells did not undergo EMT in normal adult heart. However during cardiac injury, adult mouse epicardium was found to secrete paracrine factors [54, 60]. After myocardial infarction (MI), fetal epicardial genes were shown to be reactivated and led epicardial cells to proliferate and initiated EMT to form a thick layer of mesenchymal cells to protect from myocardial injury [54, 61]. Using a tamoxifen inducible $Wt1CreERT2;Rosa26mTmG$ mouse model, epicardial cells were selectively labelled with mGFP to visualize cell fate after cardiac injury. Experimental MI was performed by left arterial ligation. mGFP labelled EPDCs in the infarct region were found to be spindle shaped and expressed myofibroblast and fibroblast markers [54]. In contrast to the human heart, zebrafish hearts regenerate and repair by epicardial activation, proliferation and by undergoing EMT after cardiac injury. In zebrafish, blocking FGF and PDGF signaling prevented epicardial marker expression and inhibited epicardial EMT during cardiac injury repair [26]. *In vitro* studies have shown that PDGF activates stress fiber production and triggers EMT in epicardial cells. Using an explant model, recombinant PDGF-BB treatment induced cell migration and a fibroblast-like phenotype in an epicardial monolayer from sham operated zebrafish [62]. Using RT-PCR, the expression of EMT markers Snail and Twist genes was found to be up-regulated as well which indicates that fibroblasts can arise from epicardial cells during heart regeneration. The expression patterns of Snail2 and Twist1b were found to be increased at the site of injury by *in situ* hybridization [62]. The role of Wt1-lineage specific cells in healing of the infarct area were analyzed in a mouse model of MI. The epicardial layer over the ischemic area was found to be disrupted immediately after the injury but began to regenerate within 3 days of MI. In the regenerated epicardium, the fetal epicardial genes (Wt1, Tbx18, Raldh) were transiently re-activated and induced epicardial proliferation. Within 2 weeks post-MI, Wt1-lineage positive subepicardial mesenchymal cells were observed. These cells replaced the cardiomyocytes lost and contributed to the fibroblasts, myofibroblast and the coronary endothelium, and later also contributed to the cardiomyocyte population. In epicardium adjacent to the site of injury, the EMT marker Snail1 and the myofibroblast marker α-SMA were expressed and mesenchyme populated the subepicardial space [59]. Using tamoxifin inducible *Wt1CreERT2* mice, another study showed that cells derived from adult epicardium after MI differentiated into myofibroblast/fibroblast and smooth muscle lineages and mainly populated a perivascular niche in the thickened epicardium [26]. Moreover, Wnt1 and Notch signalling pathways have been found to regulate the activation of epicardial EMT and contribute to fibrosis repair in a mouse model of MI and ventricular pressure overload [63].

Therefore, reactivated epicardium and EMT plays a pivotal role in the repair process of injured heart by contributing to the fibroblast pool. However, constitutive reactivation of EMT leads to chronic remodeling of the heart. Thus, blocking the reactivation of this evolutionarily conserved pathway may represent a novel approach to combat fibrosis.

8 Endothelial to Mesenchymal Transition (EndMT)

Studies have shown that vascular endothelial cells can also phenoconvert and acquire a fibroblast-like phenotype upon stimulation with pro-fibrotic signals such as TGF-β_1 or hypoxia, a process called endothelial to mesenchymal transition (EndMT) [64, 65]. These mesenchymal cells leave the microvascular bed and mature to become interstitial fibroblasts. EndMT may have two possible roles in cardiac fibrosis by both giving rise to the fibroblast population and also by decreasing microvasculature density [12]. However, under normal physiological conditions the contribution of this process to form the resident fibroblast population was not found to be significant [12]. This indicates that only after pathological damage or injury cardiac fibroblasts are generated from endothelial cells. Recently, Zeisberg *et al.* showed for the first time that adult cardiac endothelial cells can undergo EndMT. A fate mapping study using endothelial-specific reporter gene Tie1 has shown that between 27–30% of all fibroblasts were derived from endothelial origin via EndMT in a pressure overload–induced cardiac fibrosis model [12]. Similarly, another group observed that in the hearts of diabetic wild-type mice, 15–20% of fibroblasts co-expressed both the endothelial marker CD31 and the fibroblast specific marker S100A4/FSP1, whereas CD31/S100A4 double positive cells were only rarely detected in the diabetic *ET-1f/f;Tie2-Cre* mice hearts. This result indicates that endothelial cell derived ET-1 plays a crucial role in regulating EndMT. However, CD31/S100A4 double labeling, as well as the presence of CD31/α-SMA and CD31/vimentin cells, only allows for the characterization of the intermediate stage of EndMT [66]. Thus, these observations may under-represent the actual number of fibroblasts that are of endothelial origin since in the later stages of diabetes mellitus, EndMT-derived fibroblasts may migrate to the surrounding area and lose expression of endothelial markers [66]. Likewise in EMT, TGF-β is considered to be a major signaling pathway that stimulates EndMT in cardiac injury. Recombinant BMP7 has been shown to inhibit TGF-β function in endothelial cells and thus inhibit End-MT and reduce cardiac fibrosis in aortic banding models [12]. Although TGF-β has been previously shown to induce PAI-1 expression, another study demonstrated that PAI-1can block TGF-β signaling in the normal heart and PAI-1 mutants showed activated TGF-β signaling with a corresponding increase in fibroblast population and developed cardiac fibrosis [23, 67, 68]. Further investigation showed increased TGF-β mediated mesenchymal marker expression in PAI-1 null endothelial cells suggesting that these cells are more likely to undergo EndMT. However, this study did not show any direct evidence of EndMT through lineage tracing experiments [67]. Canonical Wnt signaling was also implicated in the EndMT-mediated cardiac fibrosis following MI. Using a TOPGAL Wnt reporter mouse, canonical Wnt signaling was shown to be increased in the evolving infarct scar and this activity was mainly confined to endothelial and α-SMA-expressing cells. Genetic lineage tracing experiments showed that approximately 40 % of these cells arose from endothelial precursors. Moreover, activation of canonical Wnt signaling triggered morphological and molecular changes to induce a mesenchymal phenotype. Other studies have also reported activation of Wnt expression in the epicardium and in fibroblasts after ischemic myocardial injury [26, 69].

Although recent studies have achieved remarkable success in giving insights into cardiac fibroblast generation from endothelial origin, understanding how these cells are differentiated exclusively during pathological conditions needs to be further investigated.

9   Bone Marrow-Derived (BMD) Progenitor Cells

Bone marrow-derived progenitor cells are another important source of fibroblasts in the diseased heart (Fig. 1). It was initially speculated by the Yano group that cardiac myofibroblasts can arise from bone marrow (BM) during pathological condition, since bone marrow-derived (BMD) cells have been found to contribute to both fibroblast and myofibroblast populations in other organs such as the kidney, lung, liver, stomach, small and large intestines and cancerous tumours [70]. Later on GFP labelled bone marrow transplant studies showed the presence of GFP positive fibroblasts and especially myofibroblasts in the scar and the nearby remnant area after myocardial injury [71]. A similar study showed that 57 % of the myofibroblasts were of bone marrow origin at day 7 post-MI, which decreased to 32 % on day 21 post-MI [72]. This finding raised the debate whether these cells may be simply contributing to the inflammatory response following injury. A significant population of these cells are actually playing role as a specific type of inflammatory cell during acute healing phase and are not contributing to the formation of the chronic lesion. Similar results were obtained using two different mice models- BM transgenic EGFP reporter, or BM cells expressing two reporter genes: luciferase and β-galactosidase under the control of collagen I (α2 chain) gene promoter. After inducing MI by permanent coronary artery ligation technique, α-SMA and collagen I positive myofibroblasts were found to be increased significantly at the site of injury until day 14 and persisted afterwards. A significant number of EGFP-positive BMD cells were observed during the first week post-MI, however the number gradually declined after day 28. Around 21 % of the BMD cells in the infarct area were found to be myofibroblasts. The highest number of BMD myofibroblasts (EGFP and α-SMA double positive) was found on day 7 post-MI. These cells constituted 24 % of all myofibroblasts present in the scar. These BMD myofibroblasts actively synthesized collagen I and were confined to the site of injury [73]. Using a transgenic *Mst1* mouse model, bone marrow-derived cells were shown to contribute to about 17 % of all fibroblasts present in the failing heart [74]. In aortic banded mouse model, bone marrow transplantation experiments showed that 13.4 % FSP positive fibroblasts and 21.1 % α-SMA positive myofibroblasts in the heart were bone marrow derived [12]. These findings suggest that BMD cells may play an important role in myofibroblast generation and can accelerate healing after MI.

10   Monocytes

Monocytes are also considered to be a novel source of pathology-associated fibroblasts (Fig. 1). Using a mouse model of fibrotic ischemia/reperfusion cardiomyopathy (I/RC), it has been demonstrated that 3 % of all non-myocytes in the I/RC heart expressed monocytic markers- CD45, CD11b, CD35 and myofibroblast markers S100A4, αSMA [75]. Moreover the inhibition of monocyte recruitment by injecting CL2MDP liposomes after cryoinjury, dramatically decreased fibroblast number in the infarct area and attenuated myocardial remodeling [76]. CL2MDP liposomes are termed as 'suicide' liposomes that encapsulate clodronate molecules. These liposomes are engulfed by macrophages via endocytosis and then fuse with lysosomes. Phospholipase in the lysosome disrupt the liposomes and causes release of clodronate into the cytoplasm which eventually leads to apoptosis [77]. The above mentioned findings indirectly suggest that monocytes could be a pivotal source for fibroblast and myofibroblast generation and recruitment during injury.

11   Fibrocytes

Fibrocytes are blood-borne mesenchymal progenitor cells that have unique characteristics of expressing cell surface markers of leukocytes, hematopoietic progenitor cells as well as of fibroblasts such as CD11b, CD13, CD34, CD45RO, MHC class II, CD86, collagen I and vimentin. During wound healing circulating fibrocytes rapidly enter the injury site which indicates their potential critical role in the repair process [78]. Data suggests that they are an important source of fibroblasts during pathological conditions (Fig. 1) [79, 80]. They are of hematopoietic origin and exhibit leukocyte-like characteristics [81]. One study showed that 10 % of all cells within the scar area of nephrotic wound were spindle-shaped and positive for procollagen and CD34 markers. In the presence of profibrotic cytokines such as TGF-β or endothelin-1 *in vitro*, fibrocytes can produce fibroblast and myofibroblast markers (fibronectin, collagen and α-SMA) and differentiate into myofibroblasts [79]. In a wound healing model, profibrotic cytokines- IL-4 and IL-13 have also been found to promote fibrocyte differentiation into myofibroblast. On the other hand, the antifibrotic cytokine, IFN-γ, inhibits fibrocyte differentiation [80]. In a rabbit atherosclerotic model, CD34-positive cells were also found to be positive for α-SMA [82].

12 Perivascular Cells

Perivascular cells of the cardiac vessels may present another source of fibroblasts following injury (Fig. 1). In a dermal scarring model, pericytes were shown to phenoconvert into fibroblast-like cells and contributed in collagen production [83]. *In vitro* studies have also shown that retinal pericytes acquire a functionally active fibroblast-like phenotype [84]. Fate mapping studies have demonstrated that CD73+ pericytes can serve as a potential source of fibroblasts during kidney fibrosis [56]. Moreover during lung fibrosis, activated fibroblasts or myofibroblasts have been found to arise from perivascular and peribronchial areas [85]. However, due to lack of pericyte and fibroblast specific markers and difficulty in designing specific fate mapping techniques, it is currently difficult to conclude on the significance of the pericyte's contribution to cardiac fibrosis.

13 Benefits of Understanding the Sources of Fibroblasts

Fibroblasts are intimately involved in cardiac development, maintenance, and disease. They are often implicated as being the major contributors to cardiac fibrosis. Their unique features make them attractive cellular targets for reducing pathological remodeling. Although cardiac fibrosis and fibroblast-associated pathologies are among the largest groups of diseases, at present there are no effective therapies available. Activated cardiac fibroblasts (myofibroblasts) contribute to fibrosis by chronic ECM remodeling and the development of scar tissue which subsequently results in myocardial stiffening, cardiac dysfunction and eventually to heart failure [14, 15]. Thus, an improved therapeutic option would be to restrict their generation or recruitment from other sources. Both EndMT/EMT play key roles in various chronic cardiovascular diseases such as heart failure, pulmonary hypertension, ischemia, hypertrophy and other chronic vascular diseases. Therefore, blocking fibroblast generation from these sources could be a highly effective therapeutic strategy since there is significant evidence showing that EMT and EndMT are specifically linked to various cardiomyopathies. Furthermore, these processes are not activated in the normal adult heart. For instance, treatment with BMP-7 has been shown to inhibit EndMT and thus prevents cardiac fibrosis in experimental mice without affecting the normal homeostasis of the adult heart [12]. Since both EndMT and EMT are key processes in the early development of the heart, understanding the mechanisms behind generation of fibroblast in the adult heart may provide a better insight into developing novel therapeutic approaches [11, 12, 54, 86]. Moreover, studying the mechanisms of fibroblast generation from other sources such as bone-marrow progenitor cells, fibrocytes, monocytes and perivascular cells can also aid in identifying effective new drug targets. Recent studies also implicated importance of fibroblasts in the formation and maintenance of blood vessels which opens up a new avenue for cardiac therapies. Fibroblasts are prominent modifiers of cancer progression and invasion. Like organ fibrosis, these fibroblasts within the tumour stroma stay persistently active. They acquire an activated phenotype as seen during wound healing [87]. These phenoconverted fibroblasts at the site of the tumour are referred to as peritumoral or reactive stromal fibroblasts, or tumour-associated fibroblasts that contribute to cancer progression [87]. Moreover, dysregulation of fibroblast and myofibroblast function and differentiation leads to diseases such as emphysema, asthma, interstitial and chronic obstructive pulmonary diseases in lung; rheumatoid arthritis and osteoarthritis in bones and joints; scleroderma, hypertrophic scars, lipodermatosclerosis in skin; and diseases in other organs including fibrosis related to kidney, liver, eye, nervous system and brain etc [88]. Most importantly, fibrosis in the heart due to uncontrolled proliferation of fibroblasts and their phenoconversion to myofibroblasts is closely associated with clinically important conditions such as hypertension, atherosclerosis, ischemia, dilated cardiomyopathies, valvular diseases, arrhythmias and heart failure; thus making it a major health problem worldwide [89]. Hence, a better understanding of their origin and underlying mechanisms will aid in developing successful therapies for improving patient outcome in all aspects of diseases linked to fibroblast dysregulation.

References

1. Bowers SL, Banerjee I, Baudino TA (2010) The extracellular matrix: at the center of it all. J Mol Cell Cardiol 48:474–482
2. Eckes B, Kessler D, Aumailley M, Krieg T (1999) Interactions of fibroblasts with the extracellular matrix: implications for the understanding of fibrosis. Springer Semin Immunopathol 21:415–429
3. Gabbiani G, Ryan GB, Majne G (1971) Presence of modified fibroblasts in granulation tissue and their possible role in wound contraction. Experientia 27:549–550
4. Martin P (1997) Wound healing—aiming for perfect skin regeneration. Science 276:75–81
5. Weber KT (1997) Monitoring tissue repair and fibrosis from a distance. Circulation 96:2488–2492

6. Eghbali M, Weber KT (1990) Collagen and the myocardium: fibrillar structure, biosynthesis and degradation in relation to hypertrophy and its regression. Mol Cell Biochem 96:1–14
7. Micallef L, Vedrenne N, Billet F, Coulomb B, Darby IA, Desmoulière A (2012) The myofibroblast, multiple origins for major roles in normal and pathological tissue repair. Fibrogenesis Tissue Repair 5:S5
8. Camelliti P, Borg TK, Kohl P (2005) Structural and functional characterisation of cardiac fibroblasts. Cardiovasc Res 65:40–51
9. Krenning G, Zeisberg EM, Kalluri R (2010) The origin of fibroblasts and mechanism of cardiac fibrosis. J Cell Physiol 225:631–637
10. Corda S, Samuel JL, Rappaport L (2000) Extracellular matrix and growth factors during heart growth. Heart Fail Rev 5:119–130
11. Zhou B, von Gise A, Ma Q, Hu YW, Pu WT (2010) Genetic fate mapping demonstrates contribution of epicardium-derived cells to the annulus fibrosis of the mammalian heart. Dev Biol 338:251–261
12. Zeisberg EM, Tarnavski O, Zeisberg M, Dorfman AL, McMullen JR, Gustafsson E, Chandraker A, Yuan X, Pu WT, Roberts AB, Neilson EG, Sayegh MH, Izumo S, Kalluri R (2007) Endothelial-to-mesenchymal transition contributes to cardiac fibrosis. Nat Med 13:952–961
13. Zeisberg EM, Kalluri R (2010) Origins of cardiac fibroblasts. Circ Res 107:1304–1312
14. Freed DH, Cunnington RH, Dangerfield AL, Sutton JS, Dixon IM (2005) Emerging evidence for the role of cardiotrophin-1 in cardiac repair in the infarcted heart. Cardiovasc Res 65:782–792
15. Willems IE, Havenith MG, De Mey JG, Daemen MJ (1994) The alpha-smooth muscle actin-positive cells in healing human myocardial scars. Am J Pathol 145:868–875
16. Gabbiani G (2003) The myofibroblast in wound healing and fibrocontractive diseases. J Pathol 200:500–503
17. Souders CA, Bowers SL, Baudino TA (2009) Cardiac fibroblast: the renaissance cell. Circ Res 105:1164–1176
18. Bostrom H, Willetts K, Pekny M, Leveen P, Lindahl P, Hedstrand H, Pekna M, Hellström M, Gebre-Medhin S, Schalling M, Nilsson M, Kurland S, Törnell J, Heath JK, Betsholtz C (1996) PDGF-A signaling is a critical event in lung alveolar myofibroblast development and alveogenesis. Cell 85:863–873
19. Tomasek JJ, Gabbiani G, Hinz B, Chaponnier C, Brown RA (2002) Myofibroblasts and mechano-regulation of connective tissue remodelling. Nat Rev Mol Cell Biol 3:349–363
20. Turner NA, Porter KE (2013) Function and fate of myofibroblasts after myocardial infarction. Fibrogenesis Tissue Repair 6:5
21. Petrov VV, Fagard RH, Lijnen PJ (2002) Stimulation of collagen production by transforming growth factor-beta1 during differentiation of cardiac fibroblasts to myofibroblasts. Hypertension 39:258–263
22. Leask A, Abraham DJ (2004) TGF-beta signaling and the fibrotic response. FASEB J 18:816–827
23. Kutz SM, Hordines J, McKeown-Longo PJ, Higgins PJ (2001) TGF-beta1-induced PAI-1 gene expression requires MEK activity and cell-to-substrate adhesion. J Cell Sci 114:3905–3914
24. Thiery JP, Sleeman JP (2006) Complex networks orchestrate epithelial-mesenchymal transitions. Nat Rev Mol Cell Biol 7:131–142
25. Lie-Venema H, van den Akker NM, Bax NA, Winter EM, Maas S, Kekarainen T, Hoeben RC, deRuiter MC, Poelmann RE, Gittenberger-de Groot AC (2007) Origin, fate, and function of epicardium-derived cells (EPDCs) in normal and abnormal cardiac development. ScientificWorldJournal 7:1777–1798
26. von Gise A, Pu WT (2012) Endocardial and epicardial epithelial to mesenchymal transitions in heart development and disease. Circ Res 110:1628–1645
27. Gittenberger-de Groot AC, Vrancken Peeters MP, Mentink MM, Gourdie RG, Poelmann RE (1998) Epicardium-derived cells contribute a novel population to the myocardial wall and the atrioventricular cushions. Circ Res 82:1043–1052

28. Acharya A, Baek ST, Huang G, Eskiocak B, Goetsch S, Sung CY, Banfi S, Sauer MF, Olsen GS, Duffield JS, Olson EN, Tallquist MD (2012) The bHLH transcription factor Tcf21 is required for lineage-specific EMT of cardiac fibroblast progenitors. Development 139:2139–2149

29. de Lange FJ, Moorman AF, Anderson RH, Manner J, Soufan AT, de Gier-de Vries C, Schneider MD, Webb S, van den Hoff MJ, Christoffels VM (2004) Lineage and morphogenetic analysis of the cardiac valves. Circ Res 95:645–654

30. Armstrong EJ, Bischoff J (2004) Heart valve development: endothelial cell signaling and differentiation. Circ Res 95:459–470

31. Miquerol L, Gertsenstein M, Harpal K, Rossant J, Nagy A (1999) Multiple developmental roles of VEGF suggested by a LacZ-tagged allele. Dev Biol 212:307–322

32. Dor Y, Klewer SE, McDonald JA, Keshet E, Camenisch TD (2003) VEGF modulates early heart valve formation. Anat Rec A Discov Mol Cell Evol Biol 271:202–208

33. Enciso JM, Gratzinger D, Camenisch TD, Canosa S, Pinter E, Madri JA (2003) Elevated glucose inhibits VEGF-A-mediated endocardial cushion formation: modulation by PECAM-1 and MMP-2. J Cell Biol 160:605–615

34. de la Pompa JL, Timmerman LA, Takimoto H, Yoshida H, Elia AJ, Samper E, Potter J, Wakeham A, Marengere L, Langille BL, Crabtree GR, Mak TW (1998) Role of the NF-ATc transcription factor in morphogenesis of cardiac valves and septum. Nature 392:182–186

35. Timmerman LA, Grego-Bessa J, Raya A, Bertran E, Perez-Pomares JM, Díez J, Aranda S, Palomo S, McCormick F, Izpisúa-Belmonte JC, de la Pompa JL (2004) Notch promotes epithelial-mesenchymal transition during cardiac development and oncogenic transformation. Genes Dev 18:99–115

36. Hurlstone AF, Haramis AP, Wienholds E, Begthel H, Korving J, Van Eeden F, Cuppen E, Zivkovic D, Plasterk RH, Clevers H (2003) The Wnt/beta-catenin pathway regulates cardiac valve formation. Nature 425:633–637

37. Lucas JA, Zhang Y, Li P, Gong K, Miller AP, Hassan E, Hage F, Xing D, Wells B, Oparil S, Chen YF (2010) Inhibition of transforming growth factor-beta signaling induces left ventricular dilation and dysfunction in the pressure-overloaded heart. Am J Physiol Heart Circ Physiol 298:H424–H432

38. Fredj S, Bescond J, Louault C, Potreau D (2005) Interactions between cardiac cells enhance cardiomyocyte hypertrophy and increase fibroblast proliferation. J Cell Physiol 202:891–899

39. Ljungqvist A, Unge G (1973) The proliferative activity of the myocardial tissue in various forms of experimental cardiac hypertrophy. Acta Pathol Microbiol Scand A 81:233–240

40. Mandache E, Unge G, Appelgren LE, Ljungqvist A (1973) The proliferative activity of the heart tissues in various forms of experimental cardiac hypertrophy studied by electron microscope autoradiography. Virchows Arch B Cell Pathol 12:112–122

41. Kong D, Li Y, Wang Z, Sarkar FH (2011) Cancer stem cells and epithelial-to-mesenchymal transition (EMT)-phenotypic cells: are they cousins or twins? Cancers (Basel) 3:716–729

42. Peinado H, Olmeda D, Cano A (2007) Snail, Zeb and bHLH factors in tumour progression: an alliance against the epithelial phenotype? Nat Rev Cancer 7:415–428

43. Yang J, Weinberg RA (2008) Epithelial-mesenchymal transition: at the crossroads of development and tumor metastasis. Dev Cell 14:818–829

44. De Craene B, Berx G (2013) Regulatory networks defining EMT during cancer initiation and progression. Nat Rev Cancer 13:97–110

45. Chakrabarti R, Hwang J, Andres Blanco M, Wei Y, Lukacisin M, Romano RA, Smalley K, Liu S, Yang Q, Ibrahim T, Mercatali L, Amadori D, Haffty BG, Sinha S, Kang Y (2012) Elf5 inhibits the epithelial-mesenchymal transition in mammary gland development and breast cancer metastasis by transcriptionally repressing Snail2. Nat Cell Biol 14:1212–1222

46. Savagner P, Yamada KM, Thiery JP (1997) The zinc-finger protein slug causes desmosome dissociation, an initial and necessary step for growth factor-induced epithelial-mesenchymal transition. J Cell Biol 137:1403–1419

47. Kalluri R, Neilson EG (2003) Epithelial-mesenchymal transition and its implications for fibrosis. J Clin Invest 112:1776–1784

48. Kalluri R, Weinberg RA (2009) The basics of epithelial-mesenchymal transition. J Clin Invest 119:1420–1428
49. Herfs M, Hubert P, Suarez-Carmona M, Reschner A, Saussez S, Berx G, Savagner P, Boniver J, Delvenne P (2010) Regulation of p63 isoforms by snail and slug transcription factors in human squamous cell carcinoma. Am J Pathol 176:1941–1949
50. Vandewalle C, Comijn J, De Craene B, Vermassen P, Bruyneel E, Andersen H, Tulchinsky E, Van Roy F, Berx G (2005) SIP1/ZEB2 induces EMT by repressing genes of different epithelial cell-cell junctions. Nucleic Acids Res 33:6566–6578
51. Cunnington RH, Northcott JM, Ghavami S, Filomeno KL, Jahan F, Kavosh MS, Davies JJ, Wigle JT, Dixon IM (2014) The Ski-Zeb2-Meox2 pathway provides a novel mechanism for regulation of the cardiac myofibroblast phenotype. J Cell Sci 127:40–49
52. Chang CJ, Chao CH, Xia W, Yang JY, Xiong Y, Li CW, Yu WH, Rehman SK, Hsu JL, Lee HH, Liu M, Chen CT, Yu D, Hung MC (2011) p53 regulates epithelial-mesenchymal transition and stem cell properties through modulating miRNAs. Nat Cell Biol 13:317–323
53. Lim J, Thiery JP (2012) Epithelial-mesenchymal transitions: insights from development. Development 139:3471–3486
54. Zhou B, Pu WT (2011) Epicardial epithelial-to-mesenchymal transition in injured heart. J Cell Mol Med 15:2781–2783
55. Kim KK, Kugler MC, Wolters PJ, Robillard L, Galvez MG, Brumwell AN, Sheppard D, Chapman HA (2006) Alveolar epithelial cell mesenchymal transition develops in vivo during pulmonary fibrosis and is regulated by the extracellular matrix. Proc Natl Acad Sci U S A 103:13180–13185
56. Humphreys BD, Lin SL, Kobayashi A, Hudson TE, Nowlin BT, Bonventre JV, Valerius MT, McMahon AP, Duffield JS (2010) Fate tracing reveals the pericyte and not epithelial origin of myofibroblasts in kidney fibrosis. Am J Pathol 176:85–97
57. Chu AS, Diaz R, Hui JJ, Yanger K, Zong Y, Alpini G, Stanger BZ, Wells RG (2011) Lineage tracing demonstrates no evidence of cholangiocyte epithelial-to-mesenchymal transition in murine models of hepatic fibrosis. Hepatology 53:1685–1695
58. Rock JR, Barkauskas CE, Cronce MJ, Xue Y, Harris JR, Liang J, Noble PW, Hogan BL (2011) Multiple stromal populations contribute to pulmonary fibrosis without evidence for epithelial to mesenchymal transition. Proc Natl Acad Sci U S A 108:E1475–E1483
59. van Wijk B, Gunst QD, Moorman AF, van den Hoff MJ (2012) Cardiac regeneration from activated epicardium. PLoS ONE 7:e44692
60. Lepilina A, Coon AN, Kikuchi K, Holdway JE, Roberts RW, Burns CG, Poss KD (2006) A dynamic epicardial injury response supports progenitor cell activity during zebrafish heart regeneration. Cell 127:607–619
61. Zhou B, Honor LB, He H, Ma Q, Oh JH, Butterfield C, Lin RZ, Melero-Martin JM, Dolmatova E, Duffy HS, Gise Av, Zhou P, Hu YW, Wang G, Zhang B, Wang L, Hall JL, Moses MA, McGowan FX, Pu WT (2011) Adult mouse epicardium modulates myocardial injury by secreting paracrine factors. J Clin Invest 121:1894–1904
62. Kim J, Wu Q, Zhang Y, Wiens KM, Huang Y, Rubin N, Shimada H, Handin RI, Chao MY, Tuan TL, Starnes VA, Lien CL (2010) PDGF signaling is required for epicardial function and blood vessel formation in regenerating zebrafish hearts. Proc Natl Acad Sci U S A 107:17206–17210
63. Kovacic JC, Mercader N, Torres M, Boehm M, Fuster V (2012) Epithelial-to-mesenchymal and endothelial-to-mesenchymal transition: from cardiovascular development to disease. Circulation 125:1795–1808
64. Krenning G, Moonen JR, van Luyn MJ, Harmsen MC (2008) Vascular smooth muscle cells for use in vascular tissue engineering obtained by endothelial-to-mesenchymal transdifferentiation (EnMT) on collagen matrices. Biomaterials 29:3703–3711
65. Moonen JR, Krenning G, Brinker MG, Koerts JA, van Luyn MJ, Harmsen MC (2010) Endothelial progenitor cells give rise to pro-angiogenic smooth muscle-like progeny. Cardiovasc Res 86:506–515
66. Widyantoro B, Emoto N, Nakayama K, Anggrahini DW, Adiarto S, Iwasa N, Yagi K, Miyagawa K, Rikitake Y, Suzuki T, Kisanuki YY, Yanagisawa M, Hirata K (2010) Endothelial

cell-derived endothelin-1 promotes cardiac fibrosis in diabetic hearts through stimulation of endothelial-to-mesenchymal transition. Circulation 121:2407–2418

67. Ghosh AK, Bradham WS, Gleaves LA, De Taeye B, Murphy SB, Covington JW, Vaughan DE (2010) Genetic deficiency of plasminogen activator inhibitor-1 promotes cardiac fibrosis in aged mice: involvement of constitutive transforming growth factor-beta signaling and endothelial-to-mesenchymal transition. Circulation 122:1200–1209

68. Yoshimatsu Y, Watabe T (2011) Roles of TGF-beta signals in endothelial-mesenchymal transition during cardiac fibrosis. Int J Inflam 2011:724080

69. Aisagbonhi O, Rai M, Ryzhov S, Atria N, Feoktistov I, Hatzopoulos AK (2011) Experimental myocardial infarction triggers canonical Wnt signaling and endothelial-to-mesenchymal transition. Dis Model Mech 4:469–483

70. Yano T, Miura T, Ikeda Y, Matsuda E, Saito K, Miki T, Kobayashi H, Nishino Y, Ohtani S, Shimamoto K (2005) Intracardiac fibroblasts, but not bone marrow derived cells, are the origin of myofibroblasts in myocardial infarct repair. Cardiovasc Pathol 14:241–246

71. Kania G, Blyszczuk P, Stein S, Valaperti A, Germano D, Miki T, Kobayashi H, Nishino Y, Ohtani S, Shimamoto K (2009) Heart-infiltrating prominin-1+/CD133+ progenitor cells represent the cellular source of transforming growth factor beta-mediated cardiac fibrosis in experimental autoimmune myocarditis. Circ Res 105:462–470

72. Mollmann H, Nef HM, Kostin S, von Kalle C, Pilz I, Weber M, Schaper J, Hamm CW, Elsässer A (2006) Bone marrow-derived cells contribute to infarct remodelling. Cardiovasc Res 71:661–671

73. van Amerongen MJ, Bou-Gharios G, Popa E, van Ark J, Petersen AH, van Dam GM, van Luyn MJ, Harmsen MC (2008) Bone marrow-derived myofibroblasts contribute functionally to scar formation after myocardial infarction. J Pathol 214:377–386

74. Chu PY, Mariani J, Finch S, McMullen JR, Sadoshima J, Marshall T, Kaye DM (2010) Bone marrow-derived cells contribute to fibrosis in the chronically failing heart. Am J Pathol 176:1735–1742

75. Haudek SB, Xia Y, Huebener P, Lee JM, Carlson S, Pilling D, Gomer RH, Trial J, Frangogiannis NG, Entman ML (2006) Bone marrow-derived fibroblast precursors mediate ischemic cardiomyopathy in mice. Proc Natl Acad Sci U S A 103:18284–18289

76. van Amerongen MJ, Harmsen MC, van Rooijen N, Petersen AH, van Luyn MJ (2007) Macrophage depletion impairs wound healing and increases left ventricular remodeling after myocardial injury in mice. Am J Pathol 170:818–829

77. Wang H, Peters T, Sindrilaru A, Scharffetter-Kochanek K (2009) Key role of macrophages in the pathogenesis of CD18 hypomorphic murine model of psoriasis. J Invest Dermatol 129:1100–1114

78. Abe R, Donnelly SC, Peng T, Bucala R, Metz CN (2001) Peripheral blood fibrocytes: differentiation pathway and migration to wound sites. J Immunol 166:7556–7562

79. Bucala R (2008) Circulating fibrocytes: cellular basis for NSF. J Am Coll Radiol 5:36–39

80. Strieter RM, Keeley EC, Burdick MD, Mehrad B (2009) The role of circulating mesenchymal progenitor cells, fibrocytes, in promoting pulmonary fibrosis. Trans Am Clin Climatol Assoc 120:49–59

81. Ogawa M, LaRue AC, Drake CJ (2006) Hematopoietic origin of fibroblasts/myofibroblasts: its pathophysiologic implications. Blood 108:2893–2896

82. Zulli A, Buxton BF, Black MJ, Hare DL (2005) CD34 Class III positive cells are present in atherosclerotic plaques of the rabbit model of atherosclerosis. Histochem Cell Biol 124:517–522

83. Sundberg C, Ivarsson M, Gerdin B, Rubin K (1996) Pericytes as collagen-producing cells in excessive dermal scarring. Lab Invest 74:452–466

84. Covas DT, Panepucci RA, Fontes AM, Silva WA Jr, Orellana MD, Freitas MC, Neder L, Santos AR, Peres LC, Jamur MC, Zago MA (2008) Multipotent mesenchymal stromal cells obtained from diverse human tissues share functional properties and gene-expression profile with CD146+ perivascular cells and fibroblasts. Exp Hematol 36:642–654

85. Hung C, Linn G, Chow YH, Kobayashi A, Mittelsteadt K, Altemeier WA, Gharib SA, Schnapp LM, Duffield JS (2013) Role of lung pericytes and resident fibroblasts in the pathogenesis of pulmonary fibrosis. Am J Respir Crit Care Med 188:820–830

86. van Tuyn J, Atsma DE, Winter EM, van der Velde-van Dijke I, Pijnappels DA, Bax NA, Knaän-Shanzer S, Gittenberger-de Groot AC, Poelmann RE, van der Laarse A, van der Wall EE, Schalij MJ, de Vries AA (2007) Epicardial cells of human adults can undergo an epithelial-to-mesenchymal transition and obtain characteristics of smooth muscle cells *in vitro*. Stem Cells 25:271–278

87. Kalluri R, Zeisberg M (2006) Fibroblasts in cancer. Nat Rev Cancer 6:392–401

88. McAnulty RJ (2007) Fibroblasts and myofibroblasts: their source, function and role in disease. Int J Biochem Cell Biol 39:666–671

89. Khan R, Sheppard R (2006) Fibrosis in heart disease: understanding the role of transforming growth factor-beta in cardiomyopathy, valvular disease and arrhythmia. Immunology 118:10–24

Non-Canonical Regulation of TGF-β₁ Signaling: A Role for Ski/Sno and YAP/TAZ

Matthew R. Zeglinski, Natalie M. Landry and Ian M. C. Dixon

Abstract "Cardiovascular disease (CVD) is a growing epidemic and the leading cause of mortality worldwide. More than one in three Americans are living with some form of CVD. Despite the vast diversity of CVD forms, many disease states are associated with maladaptive remodeling of the myocardial interstitium. Elevated fibrillar collagen expression is considered to be the primary contributor to altered cardiac function based on its adverse influence on electrical signal transduction, myocardial stiffness, and cardiac dysfunction. Amongst the various mechanisms responsible for the production of collagen fibres, transforming growth factor-β_1 (TGF-β_1) has been identified as a critical mediator of the fibrotic response in the injured myocardium. Herein we describe a potential role for the TGF-β_1 negative regulators Ski/Sno, and the TGF-β_1 transcriptional regulators YAP/TAZ in cardiac fibroblast and myofibroblast phenotype and function through modulation of TGF-β_1 signaling."

Keywords TGF-β_1 · Ski/Sno · YAP/TAZ · Fibroblast · Fibrosis

Cardiovascular disease (CVD) is a growing epidemic and the leading cause of mortality worldwide [1, 2]. More than one in three Americans are living with some form of CVD [3]. Despite the vast diversity of CVD, many disease states are associated with maladaptive remodeling of the myocardial interstitium [4]. Elevated fibrillar collagen expression is considered to be the primary contributor to altered cardiac function based on its adverse influence on electrical signal transduction, myocardial stiffness, and cardiac dysfunction [5, 6]. Amongst the various mechanisms responsible for the production of collagen fibres, transforming growth factor-β_1 (TGF-β_1) has been identified as a critical mediator of the fibrotic response in the injured myocardium. Herein we describe a potential role for the TGF-β_1 negative regulators

I. M. C. Dixon (✉)
Department of Physiology and Pathophysiology, St. Boniface Hospital Research Centre, University of Manitoba, R3038 351 Tache Avenue, Winnipeg, MB R2H 2A6, Canada
e-mail: idixon@sbrc.ca

M. R. Zeglinski · N. M. Landry
Department of Physiology and Pathophysiology, St. Boniface Research Centre, Institute of Cardiovascular Sciences, University of Manitoba, Winnipeg, MB, Canada

© Springer International Publishing Switzerland 2015
I.M.C. Dixon, J. T. Wigle (eds.), *Cardiac Fibrosis and Heart Failure: Cause or Effect?*, Advances in Biochemistry in Health and Disease 13, DOI 10.1007/978-3-319-17437-2_9

Ski/Sno, and the TGF-β_1 transcriptional regulators YAP/TAZ in cardiac fibroblast and myofibroblast phenotype and function through modulation of TGF-β_1 signaling.

1 TGF-β Signaling

The TGF-β superfamily of genes and their repressors are phylogenetically ancient and are evolutionarily conserved [7, 8]. While the majority of cytokines are produced as biologically-active molecules, TGF-β ligands (TGF-$\beta_{1/2/3}$) are synthesized and secreted in a latent form that requires extracellular activation, dimerization, and receptor binding to initiate an intracellular signaling cascade. TGF-β_1 signals through a pair of membrane-bound serine/threonine kinase receptors (TβR-II and TβR-I),which activate intracellular Smad proteins that are able to translocate to the nucleus and influence gene expression [9]. Cardiac fibroblasts secrete TGF-β_1 into their surrounding environment, where it is incorporated into the extracellular matrix (ECM) in its inactive form by the TGF-β_1-binding protein (LTBP-1) and latency-associated peptide (LAP) (Fig. 1) [10]. Release of TGF-β_1 from this inhibitory

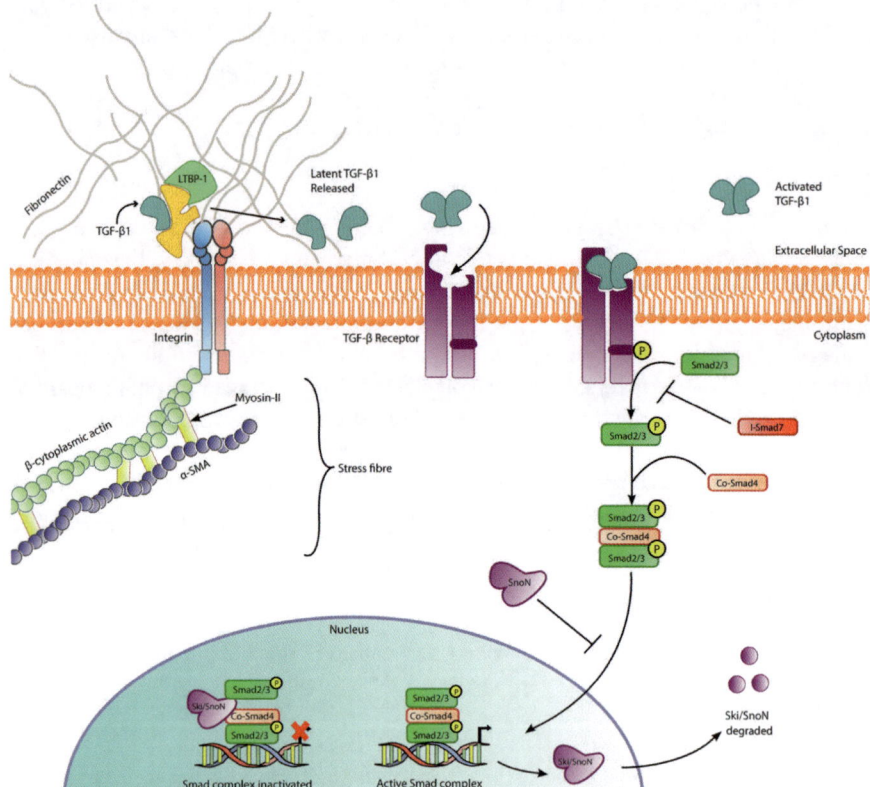

Fig. 1 Canonical TGF-β Signaling and regulation by Ski/Sno TGF-β1, transforming growth factor-β 1; LAP, latency-associated peptide; LTBP-1, latent transforming growth factor β-binding protein 1; *α-SMA* α-smooth muscle actin, *I-Smad7* inhibitory Smad 7, *Co-Smad4* co-mediator Smad 4, *P* phosphorylated, *SnoN* Ski-related novel protein N

complex can occur through a variety of mechanisms including proteolytic cleavage, oxidative stress, low pH, and mechanical stress [9, 11–13]. Ultimately, the effects of TGF-β_1 depend on numerous factors including cell type, cross-talk with other signaling pathways (e.g. Hippo, Wnt/β-catenin), stage of pathogenesis, and the microenvironment [14, 15]. These factors illustrate the pleiotropic nature of the pathway and the requirement for strict regulation of signal transduction downstream from the ligand-receptor interaction.

2   TGF-β_1 and Mechanical Stress

The involvement TGF-β_1 in the pathogenesis of cardiac fibrosis has been well-characterized [16–18]. Most notably, in the infarcted myocardium TGF-β_1 signaling results in a phenoconversion event where relatively inactive resident fibroblasts become hyper-synthetic/secretory, contractile myofibroblasts characterized by marked expression of proteins including alpha-smooth muscle actin (α-SMA) [19], embryonic smooth muscle myosin (SMemb), and extracellular domain-A (ED-A) fibronectin [20].

The entire complement of factors involved in the induction of TGF-β-responsive gene expression during cardiac remodeling has yet to be fully described; however recent studies have demonstrated that cardiac fibrosis is the result of a mechanosensory response by fibroblasts that is related to substrate stiffness [21]. Mechanistically, it is thought that distinctive membrane-bound integrins (αvβ5 and αvβ3) on mesenchymal cells (such as cardiac fibroblasts) are responsible for TGF-β_1 activation [15]. Recent work conducted by Hinz and colleagues have shown that release of TGF-β_1 from the matrix bound latency complex is initiated by mechanical stress, whereby cytoskeletal contractions are transmitted to the ECM *via* integrins [15]. This finding not only provides an elegant biomechanical link between TGF-β_1 and cardiac fibrosis, but further implicates chronic stressors, such as hypertension, to the myocardium with heart failure.

3   The Ski/Sno Superfamily

In canonical TGF-β_1 signaling, Smad proteins act in a cascade of serial phosphorylation events that either promote (i.e. via Smad2/3 and Smad4) or inhibit (i.e. via Smad7) signal transduction [12]. Ski and the Ski-related novel gene (Sno) encode structurally and functionally similar proteins that negatively regulate TGF-β_1 signaling. While Sno is expressed as several isoforms in human cells (SnoN, SnoN2, SnoA, and SnoI) that arise due to alternative gene splicing, Ski mRNA has not been shown to be spliced and exists in a singular form [22, 23].

Ski and Sno have been shown to contain a highly-conserved region of homology at the N-terminus [24] that consists of a unique Dachshund Homology Domain (DHD) and a structural motif similar to a SAND (derived from Sp100, AIRE-1, NucP41/75,

DEAF-1) domain. The Ski/Sno DHD does not appear to bind directly to DNA despite its similarity to the winged-helix class of DNA-binding domains [25] but may be required for protein-protein interactions with co-factors including C184M, Nuclear hormone Co-Repressor (N-CoR), and Ski-interacting protein (Skip) [26, 27].

SAND domains are generally associated with DNA-binding and transcriptional repression, but Ski/Sno proteins use the structure's I-loop to interact with the Smad L3 loop, rather than with genomic DNA [28]. The effect of Ski/Sno binding to Smad4 essentially leads to the dissociation of the functional Smad2/3-Smad4 heterocomplex, thus repressing TGF-β_1 signaling at the nuclear level. Furthermore, the Ski/Sno SAND-like domain has been implicated in binding to other transcription factors, such as the tumor suppressor retinoblastoma (Rb) [29], which may provide a link to Ski's involvement in TGF-β-dependent cell-cycle arrest.

The combined properties of the Ski/Sno DHD and SAND domains allow these proteins to interact with nuclear co-factors that suppress transcription of TGF-β_1-responsive genes in addition to Smad proteins. SAND-Smad interactions within the nucleus has been reported to be coupled with the disruption of the Smad complex, along with the physical blocking of p300 and CREB-binding protein (p300/CBP), both of which are co-activators to Smad3 [30]. This initial contact promotes the formation of a transcriptional-silencing complex where Ski/Sno recruits N-CoR via its DHD, which then associates with chromatin-modifying proteins such as histone deacetylase (HDAC), methyl-CpG binding protein 2 (MeCP2), mSin3A and/or homeodomain-interacting protein kinase 2 (HIPK2) [31]. While several protein-protein interactions have been identified for Ski, other members of the Ski/Sno superfamily such as Fussel-15 and Fussel-18 are known to bind Smad2 and Smad3 in a similar fashion however their expression has been limited to neuronal tissues [32].

In contrast to the N-terminus, the C-terminal region of Ski/Sno is much less conserved. A small region near the C-terminus has been associated with homo- and heterodimerization of Ski and SnoN [33] and variability in the degree of TGF-β_1 repression. Heterodimers of Ski and SnoN possess greater oncogenic potential and are more thermodynamically stable than homodimers of either protein [34]. Although the dimerization of Ski/Sno proteins plays some role in the efficacy of the co-repression, the overall importance of the event has yet to be fully characterized.

4   TGF-β Regulation of Ski/Sno

While both Ski and SnoN are affected by TGF-β_1 post-translationally, the expression of SnoN is directly modified by TGF-β_1 signaling whereas Ski is not. The expression of Smad-mediated ubiquitination regulatory factor 2 (Smurf2) is induced shortly after stimulation by TGF-β_1 which subsequently labels Smad2, Ski, and SnoN for degradation via ubiquitination [35–37]. The effect is two-fold for nuclear SnoN, whose interaction with Smad2 enhances the ability for Smurf2 to recruit the ubiquitin-dependent proteasome [38]. In addition, a marked increase in SnoN approximately 2h after TGF-β_1 stimulation results from the binding of a Smad2/

Smad4 complex to a TGF-β-responsive element (or Smad binding element) in the SnoN promoter [39]. Thus SnoN expression is a component of negative feedback in TGF-β_1 signaling. While the significance of the bi-phasic nature of Smurf2 in fibrotic heart tissue has yet to be determined, it has been linked to fibroblast trans-formation in human cancer cells [39].

5   Ski/Sno Regulation of TGF-β_1 Signaling

Ski and Sno are capable of regulating TGF-β signaling through the disruption of Smad signaling at both the cytoplasmic and nuclear levels [40, 41]. As the study of Ski/Sno proteins was initially conducted in cancer and immortalized cells, the belief that these proteins acted solely as nuclear signaling moieties was established [22, 42]. Subsequent studies demonstrated that the sub-cellular distribution of Ski and SnoN depends on a variety of factors, including cell type and cytokine environment both within and surrounding the cell [40, 43]. It has become evident that the distri-bution of sub-cellular components may underpin prominent pathogenetic changes within the progression of cardiac fibrosis.

Under normal physiological conditions, SnoN is primarily localized to the cy-toplasm [40]. This is in stark contrast to transformed cells where SnoN is uniquely nuclear. Accumulation of nuclear SnoN increases dramatically upon altered cellu-lar state including cell cycle arrest, differentiation, and phenotypic adaptation [40]. While nuclear SnoN is subject to TGF-β-induced proteosomal degradation, cytoplas-mic SnoN is far more resistant to this mode of removal and functions to prevent the nuclear translocation of Smad2/3-Smad4 heterotrimers, essentially repressing the TGF-β_1 signaling cascade. Experiments with a truncated variant of SnoN indi-cated impaired proteosomal removal of the variant, which in turn, abrogated TGF-β-induced cell cycle arrest and was associated with increased cell proliferation and eventual transformation [44]. While these observations have not yet been verified in myocardial cells, the argument that, based on its main function as a repressor protein of TGF-β_1 signaling, modulation of SnoN distribution in fibroblast cells of healing myocardial infarcts could influence myofibroblast trans-differentiation and ultimately the progression of matrix dysfunction, chronic wound healing, and eventual fibrosis.

In contrast to SnoN, the intracellular distribution of Ski and its post-translational modifications eg, phosphorylation and ubiquitination, are better understood. We suggest that Ski activation may play a complex role in post-myocardial infarction (MI) wound healing. In this context, it has been shown that cytosolic Ski negatively regulates TGF-β_1 signaling [45]. Furthermore, a recent *in vitro* study in primary rat cardiac fibroblasts indicates that acute TGF-β_1 stimulation induces a marked shut-tling of Ski into the nucleus [46]. The same study showed that cytosolic levels of Ski are significantly increased in post-MI cardiac tissue, *in vivo*. Thus a decrease in nuclear Ski could be responsible for unimpeded TGF-β_1 signaling, resulting in improper wound healing of the heart post-MI. While nuclear Ski has a molecular weight of 95 kDa, the cytosolic isoform has a molecular weight of 105 kDa, suggesting that an undefined post-translational modification had occurred [46]. Previous investigations showed that Ski accumulates in the cytoplasm of epithelial cells when proteasomal activity is inhibited [47]. While Ski is known to be phosphorylated [48], the significant variation in molecular weight leads us to suggest that ubiquitination and proteasome activity may provide an alternative modality for control of function. Further elucidation of the physiological significance of Ski and its undefined isoforms in the healthy and post-MI heart, as well as its role(s) in wound healing is of major interest and significance.

6   Implications of Ski/Sno Function in the Pathology of Cardiac Fibrosis

Preliminary investigations into the role of Ski in the regulation of the cardiac myofibroblast phenotype have shown that Ski overexpression promotes the reversal or "dialing back" of the myofibroblast phenotype to a fibroblast-like state *in vitro* [46]. Parallel *in vivo* analyses using a rat model of myocardial infarction (MI) also revealed the up-regulation of cytosolic Ski with a corresponding decrease in nuclear localization in the peri-infarct region of myocardium surrounding the infarct scar, as well as in myofibroblasts in the scar. With this in mind, the regulatory mechanisms underlying Ski/Sno signaling in fibroblast and myofibroblast cells are likely of considerable importance with respect to cardiac fibrosis.

As Ski/Sno proteins have been implicated in the induction and progression of cardiac fibrosis, it is of interest to investigate potential downstream targets with an emphasis on controlling the fibroblast phenotype in damaged or fibrosing myocardium. While the full complement of Ski-targeted proteins has been only partially described, those identified thus far have been associated with development, differentiation, and transcriptional regulation. For example, the Zeb2 E-box binding protein which regulates the expression of Mesenchyme homeobox 2 (Meox2 or Gax) is a downstream target of Ski signaling [49, 50]. While Meox2 was originally associated with muscle development, recent analyses provided by Cunnington et al. revealed that Ski-mediated repression of Zeb2 in cardiac fibroblasts up-regulates the expression of Meox2 [49, 51]. In addition, rat ventricular myofibroblasts overexpressing Meox2 showed increased proliferation. However down-regulation of nuclear Meox2 corresponds to significantly decreased expression of myofibroblast markers. It was also observed that the expression of Zeb2 is markedly increased in cardiac myofibroblasts and that this expression is down-regulated subsequent to overexpression of Ski [49]. Thus, the differential expression of Meox2 and Zeb2, mediated by Ski is a viable mechanism by which myofibroblast phenoconversion occurs. Subsequent findings have shown that the sub-cellular localization of Zeb2 and Meox2 also varies between the fibroblast and myofibroblast phenotypes. In the quiescent fibroblast phenotype, Meox2 appears to be confined to the nucleus, whereas in activated myofibroblasts, Meox2 is distributed throughout the cytoplasm. Conversely, it was found that Zeb2 is more concentrated in the nucleus of myofibroblasts, presenting an interesting dichotomy of sub-cellular localization with respect to Ski, which is mainly cytoplasmic in the post-MI myofibroblast [46]. It was suggested that nuclear Ski maintains elevated levels of Meox2 to sustain fibroblast quiescence. Additional elucidation of this putative mechanism would bring a greater understanding to the mechanisms underlying myofibroblast phenoconversion. Further study into the downstream effects of Ski/Sno and related proteins in cardiac fibroblasts is also warranted. However, studies to address the regulatory factors controlling the expression of Ski and SnoN also merits attention to better elucidate their activity in both the healthy and pathological state.

As the effects of intracellular TGF-β_1 signaling are diverse, cross-talk with other signaling pathways are not uncommon. One particular Smad-independent TGF-β_1 response involving the p38 mitogen-activated protein kinase (p38 or p38 MAPK) pathway is of interest as its activation, which results in cell cycle arrest, has been linked to the dysregulation of Ski [52]. Work carried out by Li et al. demonstrated that the decreased levels of Ski in vascular smooth muscle cells (VSMCs) after injury to rat arteries was correlated with significant VSMC proliferation [52]. Using *in vivo* gene therapy to up-regulate Ski expression after injury, the degree of VSMC propagation was markedly decreased. It was found that Smad3 phosphorylation, normally indicative of TGF-β_1 signaling, was decreased and was coupled with evidence that p38 signaling was activated. While these observations explain the inhibitory effects of Ski, they are not sufficient to elaborate as to how Ski levels are down-regulated in injured vascular tissue. Subsequent investigations found that Ski gene expression is negatively regulated by microRNA-21 (miR-21) in VSMCs [53]. A putative miR-21 recognition sequence was identified at the 3'-UTR of the Ski mRNA transcript in rats. Furthermore, it was found that an increase in miR-21 transcripts stimulated VMSC proliferation and inhibited p38-p21-p27 signaling resulting in a complete reversal of the effects of Ski up-regulation. Prior to these findings, miR-21 had already been identified as a key contributor to the onset of myocardial disease by inducing MAPK signaling in cardiac fibroblasts, thus promoting interstitial fibrosis [54, 55]. Despite these findings, the precise mechanism by which miR-21 induces a pro-fibrotic state remains unclear. Although confirmation of miR-21 regulation of Ski in cardiac fibroblasts is still required, it presents an appealing target for fine-tuning Ski expression. Indeed, miR-21 has already piqued the interest of the pharmaceutical industry as a potential target for kidney fibrosis; several others (i.e.: miR-15 family and miR-29) have also been noted for their putative value in treatment of cardiac remodeling [56]. Future examinations of Ski/Sno proteins in cardiac fibroblasts and myofibroblasts should include analyses of relevant miRNA expression to broaden the therapeutic arsenal with which effective treatment for cardiac fibrosis could be formulated.

Evidently, the Ski/Sno superfamily is of paramount importance to the management of cellular responses to multiple factors. The complex, environment-specific nature of Ski/Sno functionality presents an interesting paradigm when considering Ski/Sno proteins as potential therapeutic targets for TGF-β_1-induced fibrosis. Therefore, the modulation and sub-levels of control of Ski/Sno activity, along with crosstalk to other signaling pathways, are of principal interest when the targeted down-regulation of TGF-β_1 in myofibroblasts is the ultimate goal of future treatment, and perhaps prevention, of cardiac fibrosis.

7 The Hippo Pathway

Another pathway that has been identified to regulate TGF-β_1 signaling through non-canonical mechanisms is the recently identified Hippo pathway. The Hippo signaling pathway was originally discovered as a central regulator of organ size and tumor development in *Drosophila melanogaster* (*D. melanogaster*) [30, 57–61]. Its role has since grown in scope to include stem cell function, development, and disease including cancer and fibrotic diseases. In the early 2000's a search for modulators of tissue overgrowth lead to the identification of a core kinase cascade responsible for regulating tissue size during development [58, 60, 62–64]. This pathway, later coined the Hippo pathway, was comprised of the Hippo and Warts kinase, the adaptor proteins Salvador and Mob, the transcription factor Scalloped, and the transcriptional regulator Yorkie. Activation of Salvador and Mob by Hippo lead to Warts phosphorylation of Yorkie and Yorkie's exclusion from the nucleus [57, 61, 65]. Within the nucleus Yorkie promotes cell proliferation and inhibits apoptosis through its association with the transcription factor Scalloped [61, 65].

As the Hippo pathway became well known for its role in regulating fruit fly development, studies into other organisms found Hippo signaling to be highly conserved within eukaryotes. The core components of mammalian Hippo signaling include mammalian sterile 20-like (Mst1/2), salvador-like homolog (Sav1), large tumor suppressor (Lats1/2), MOBs, TEA-domain family member (TEAD), and Yes-associated protein/transcriptional co-activator with a PDZ-binding domain (YAP/TAZ) which are the orthologs to Hippo, Salvador, Warts, Mats, Scalloped, and Yorkie in *D. melanogaster* respectively [64, 66]. YAP and TAZ are paralogs that share approximately 45% identity and are functionally similar. Structurally, both contain a C-terminal PDZ-binding motif, an N-terminal TEAD binding domain, and a WW-domain (two tryptophan residues separated by 20–23 amino acids) that recognizes a PPXY (proline, proline, any, tyrosine) motif in proteins which regulate their sub-cellular localization and activity levels [64, 67, 68]. In this discussion we will consider YAP and TAZ as one (YAP/TAZ) due to their regulational and functional similarities, except where otherwise stated.

In the mammalian Hippo signaling pathway, phosphorylation of Mst1/2 recruits and binds Sav1 [69, 70]. This complex then phosphorylates Lats1/2 which in turn binds to MOB1 and induces its auto-phosphorylation. Finally Lats1/2 phosphorylates YAP/TAZ and prevents their nuclear translocation and/or targets nuclear YAP/TAZ for nuclear exportation through association with 14-3-3 protein chaperons [68]. Ultimately, phosphorylation of YAP and TAZ results in cytoplasmic retention and inhibition of proliferative and survival signals (Fig. 2) [70]. When de-phosphorylated, YAP is found within the nucleus where it interacts with TEAD transcription factors to regulate the transcription of genes that control tissue growth, and cell survival [64, 71, 72]. Activation of the canonical Hippo pathway in mammals is

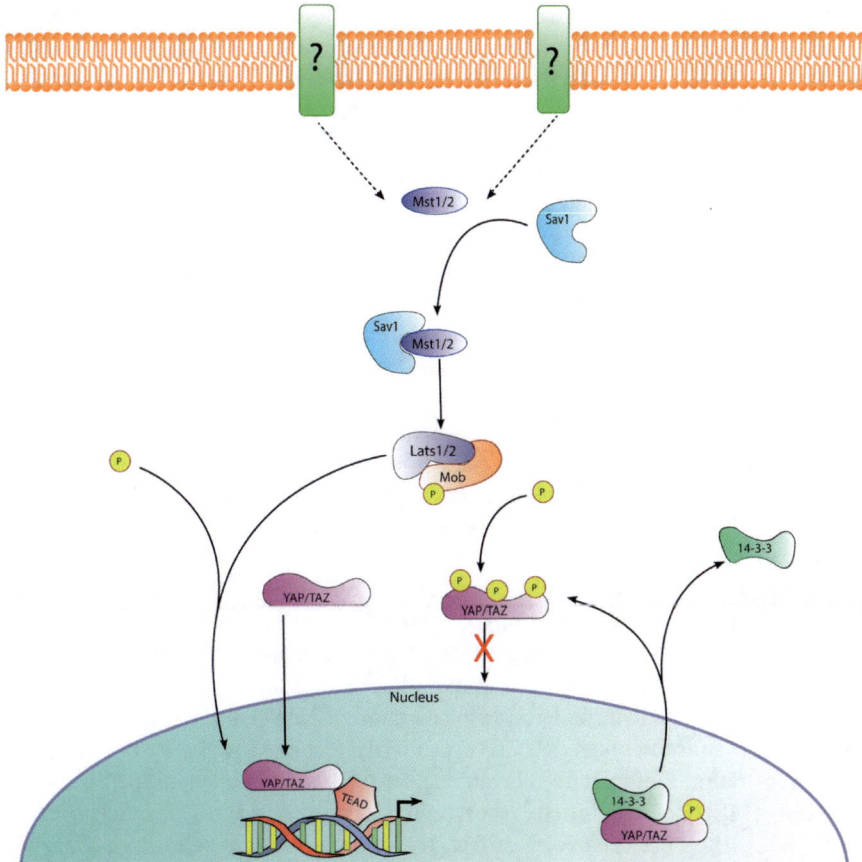

Fig. 2 Canonical Mammalian Hippo Signaling. *Mst1/2* macrophage stimulating 1 or 2, *Sav1* Salvador homolog 1, *Lats1/2* large tumor suppressor kinase 1 or 2, *Mob* Mps one binder, *YAP* Yes-associated protein, *TAZ* Transcriptional co-activator with PDZ binding motif, *TEAD* transcriptional enhancer factor domain family member, *14-3-3* 14-3-3 proteins, *P* phosphorylated

not well understood; however, FAT4 and CD44 cell surface receptors have been demonstrated to regulate the core components of Hippo signaling through some yet to be determined factors [67]. In addition to canonical signaling, YAP/TAZ can be activated through a variety of other stimuli including G-protein coupled receptors [73], cell-to-cell contact [74], cell polarity[64, 75], adhesion and junction proteins [71], and mechanical stress [76–78].

8   YAP/TAZ and Mechanical Stress

As discussed above, activation of the myofibroblast phenotype can be induced by mechanical stress [13, 15, 21, 79]. In cell culture, 2-dimensional plating of fibroblastic cells onto plastic, a relatively stiff substrate, biomechanically activates their phenoconversion from the fibroblast to the myofibroblast phenotype. Conversely, those cells cultured onto relatively soft or compressible silicon substrates (5-kPa compressibility), which is comparable to the "soft" physiological stiffness of 3-dimensional matrices *in vivo*, retain their quiescent phenotype [13, 21, 79]. Recent studies have provided evidence that extracellular matrix stiffness can regulate YAP/TAZ sub-cellular localization and transcriptional activity independent of upstream Hippo components [76–78, 80]. A study by Dupont et al. used a bioinformatics approach to identify which signaling pathways were activated when cells were under mechanical stress [76]. They found that the YAP/TAZ transcriptional regulators were relatively over-represented in those cells under mechanical stress. Their functional studies demonstrated that mammary epithelial cells (MECs) grown on fibronectin-coated acrylamide hydrogels ranging in Young's modulus (a measure of elasticity) from 0.7 to 40 kPa demonstrated a near linear range of YAP activation levels. Cells grown on stiff (40 kPa) hydrogels, representative of scar tissue, had an activity level similar to that of MECs grown on hard plastic [76], whereas MECs grown on soft (0.7 kPa) hydrogels, representing physiologically healthy tissue, inhibited YAP/TAZ activation to levels that were comparable to that of siRNA knockdown studies.

As the Hippo cascade is an inhibitor of YAP/TAZ function, the "gold standard" for measuring the activity level of the Hippo pathway has been to identify the sub-cellular localization of YAP/TAZ. Further evidence that mechanical stress can influence YAP/TAZ activity came from immunofluorescent studies. Those cells grown on stiff substrates had strong nuclear localization of YAP/TAZ, whereas those cultured on soft substrates had YAP/TAZ primarily localized to the cytoplasm [76]. To demonstrate a biologically relevant aspect of this phenomenon, mesenchymal stem cells (MSC) were cultured on soft (cytoplasmic YAP) and stiff (nuclear YAP) matrices as MSCs are capable of differentiating into different cell linages depending on their environment. MSCs cultured on stiff substrates undergo osteogenic differentiation,which could be prevented through inhibition (by phosphorylation) of YAP and TAZ. Conversely, MSCs cultured on soft substrates undergo adipogenesis [76]. These phenoconversion events were found to be independent of the upstream Hippo signaling cascade as knockdown of Lats1/2 (an inhibitory kinase for YAP/TAZ) in osteoblast differentiated MSCs could not rescue the phenotype [76].

How is it that the stiffness of the ECM is capable of influencing intracellular signaling pathways? Transduction of these mechanical cues is mediated through the cytoskeleton. Formation of stress fibers is characteristic of the myofibroblast phenotype. Both F-actin and Rho (regulators of the cytoskeleton and stress fiber formation) appear to play important roles in the regulation of YAP/TAZ [74, 76, 77, 80–84]. Treatment of NIH-3T3 cells with anti-actin drugs resulted in a reduction in the amount of stress fiber formation along with cytoplasmic retention of YAP [74]. To demonstrate that F-actin was the specific cytoskeletal element required for YAP regulation, microtubules were disrupted with the drug nocodazole. Despite microtubule disorganization, there was no difference in YAP sub-cellular localization between nocodazole-treated and untreated controls [74]. These experiments are supported by a similar study conducted in *D. melanogaster* where cytochalasin

D, an F-actin destabilizing drug, reduced the activity level of Yorkie (the *D. melanogaster* homolog of YAP) [80]. To determine that F-actin was the cytoskeletal regulator of Yorkie activity, treatment with the myosin inhibitors blebbistatin, ML-7, and Y27623 resulted in a phenotype similar to that which was seen with the anti-actin drugs in NIH-3T3 cells, which included reduced stress fiber formation and increased cytoplasmic YAP.

It is clear that the extracellular environment that a cell is surrounded by in three dimensions plays a significant role in modulating its phenotype and function. Transduction of extracellular mechanical forces to the cell plays a critical role in determining the cells response to that environment. In addition to activation TGF-β₁ (discussed in detail above), mechanical stress can induce the activation and nuclear accumulation of YAP/TAZ leading to cell proliferation, differentiation, and survival. To date, the role that YAP/TAZ play within the cardiac fibroblast is unknown, however it is likely that YAP/TAZ play a significant role in regulating the myofibroblast phenotype, particularly during chronic wound healing.

9   Cross-Talk Between the Hippo and TGF-β Pathways

Both TGF-β/Smad and YAP/TAZ/Hippo signaling have been shown to play important roles in regulating cell phenotype and function. Recently, Hippo signaling has been demonstrated to play a significant role in regulating Smad-mediated TGF-β₁ signaling [64, 75, 84]. Both YAP and TAZ have been shown to play a role in dictating Smad complex sub-cellular localization and activation levels [75, 85]. In a 2008 study by Varelas et al. TAZ was found to bind to Smad2/4 and Smad3/4 complexes following TGF-β₁ treatment in Mv1Lu cells [85]. To lend a functional aspect to their findings, they used a reporter system and found that TGF-β₁ responsive Smad reporters were significantly inhibited in TAZ knockout cells, indicating that TAZ may be a critical factor in Smad-mediated gene transcription at the level of the gene promoter [85]. Additionally, their studies found that TAZ was a critical factor for Smad2 nuclear accumulation in human embryonic stem cells (hESC). Upon TGF-β₁ stimulation, Smad2 was primarily nuclear, however when TAZ expression was lost, Smad2 was redistributed amongst the entire cell despite TGF-β₁ stimulation [85]. These studies have since been duplicated in HepG2, Cos7, and NIH-3T3 cells demonstrating the broad range of Hippo/TGF-β interaction in a variety of cell types. Interestingly, over-expression of TAZ leads to its accumulation in the cytoplasm which prevented Smad nuclear localization. Thus, sub-cellular localization of TAZ has a direct impact on TGF-β/Smad dependent signaling.

As YAP and TAZ are functionally similar, a 2011 study by Varelas et al. demonstrated YAP was also able to directly interact with Smads and regulate TGF-β₁ mediated gene signaling [75]. In addition to their previous study, the Varelas group demonstrated that cell density plays a major role in YAP/Smad localization and function. At low cell densities YAP and Smad-complexes co-localized to the nucleus, whereas at high densities YAP and Smad-complexes were exclusively colocalized to the cytoplasm [71, 74, 75]. In addition, they also showed that polarized epithelial cells displayed cytoplasmic YAP which restricted R-Smad/Co-Smad nuclear localization [75]. Epithelial cells that had lost their polarization display increased YAP/Smad nuclear localization and an increased sensitivity to TGF-β_1 stimulation. The effects of YAP/TAZ on Smad signaling were found to be regulated by upstream members of the Hippo pathway, specifically Mst1/2 and Lats1/2. Single and/or double knockout of either Mst1/2 or Lats1/2 resulted in an increase of YAP/TAZ nuclear accumulation, Smad2/3 nuclear accumulation and induction of TGF-β_1 regulated genes [75, 85]. While studies aimed at describing a specific role for YAP/TAZ in the regulation of TGF-β_1 in health and disease are relatively few, the potential for these proteins to play a significant role in the pathogenicity of numerous diseases, including cardiac fibrosis, is significant due to the major role that TGF-β/Smad signaling plays in the myofibroblast.

10   The Effects of YAP in the Heart

Several studies have described a significant role for YAP/TAZ in the developing and diseased heart [64, 86–90]. Specifically, YAP/TAZ has been shown to have a major influence on cardiomyocyte proliferation and survival [87–92]. During mouse embryogenesis, conditional knockout of YAP *in vivo* is embryonically lethal by day 10.5 (E10.5) due to myocardial hypoplasia and contractile deficiency as a result of reduced cardiomyocyte proliferation [90]. Additionally, conditional knockout of Sav-1 at E9.5 showed a significant reduction in phospho-YAP levels but no change in total YAP levels. Unlike YAP conditional knockouts, Sav1 conditional knockout animals survived to term but died soon after due to cardiomegaly [90]. Over-expression of YAP has been shown to stimulate myocyte proliferation and increased heart size (cardiomegaly) [86, 89, 90]. Transgenic mice over-expressing YAP displayed a thicker myocardium and an increased heart weight/tibia length ratio [90]. Following injury, such as MI, forced expression of constitutively active YAP (YAP[Con+]) by changing Serine112 to an Alanine (S112A) to prevent Lats1/2 phosphorylation of YAP promoted cardiac regeneration by inducing cardiomyocyte proliferation, thereby reducing scar size, and improving cardiac function [86, 89]. Some studies have shown that transgenic mice with YAP[Con+] had complete cardiac regeneration after injury with little to no scaring [90]. Deletion of upstream regulators of YAP, such as Sav1, Mst1/2 or Lats2 (which regulate the phosphorylation and inhibition of YAP by sequestering it in the cytoplasm), lead to a phenotype that was phenotypically equivalent to YAP[Con+] [87]. Conversely, cardiomyocyte specific knockdown of YAP in post-natal mice hearts demonstrated gross cardiac deficiencies as deletion suppressed myocyte proliferation during development [86, 92]. Post-natal mice demonstrated functional deficits by 6 weeks and significant wall thinning with dilated cardiomyopathy as early as 9 weeks which worsened with age and resulted in cardiac fibrosis and lethal heart failure [90].

A role for YAP/TAZ signaling in cardiac fibroblasts has yet to be described. As YAP/TAZ induces cell proliferation and survival, we suggest a role for YAP/TAZ in promoting the myofibroblast phenotype in the stressed/damaged heart. Following cardiac injury, such as MI, cardiac fibroblasts undergo a phenoconversion to become hyper-synthetic cardiac myofibroblasts. Acutely, their presence is beneficial as they secrete collagen fibres which form scar tissue to provide tensile and structural strength to the injured myocardium. When scar formation is near completion, the majority of cells undergo apoptosis and are removed by macrophages. However, myofibroblasts have been shown to persist within the infarcted region for years following the initial injury, continually producing scar tissue. Chronic and excessive deposition of scar tissue will lead to go global cardiac stiffening and ultimately result in heart failure. We suggest that the chronic deposition of scar tissue provides a niche to the myofibroblast that allows for constitutive activation of nuclear YAP/TAZ, which supresses genes associated with cell death and promotes those associated with survival and proliferation. This may be a novel mechanism by which myofibroblasts are able to "escape" the normal wound healing response and induce pathological remodeling of the myocardium. While this theory merits exploration, the therapeutic potential of targeting YAP/TAZ and the Hippo pathway in cardiac disease is an attractive target for pharmaceuticals for those patients suffering from heart failure.

11 Synopsis

Regulation of TGF-β₁ signaling is complex and involves a variety of factors, some of which may have yet to be discovered. TGF-β₁ signaling has been demonstrated to play a significant role in the pathogenesis of numerous fibrotic diseases including MI. In the healthy heart, TGF-β₁ signaling is the sum of stimulatory (R-Smad, YAP/TAZ) and inhibitory (I-Smad7, Ski/Sno, YAP/TAZ) factors. During disease, this balance is lost and the scale is tipped and promotes the synthesis and secretion of fibrillar collagens. Two pairs of structurally and functionally similar proteins have recently been identified to play a role in TGF-β₁ regulation of the myocardium. Although Ski/Sno have been shown to inhibit TGF-β₁ signaling through disruption of Smad complexes, YAP/TAZ has been demonstrate to either promote or inhibit Smad signaling depending on its phosphorylation status.

Our group has recently described a role for Ski in cardiac myofibroblasts, demonstrating its ability to revert the myofibroblast to a proto-myofibroblast phenotype through its ability to disrupt Smad signaling [46]. Unlike Ski, YAP/TAZ has yet to be characterized in playing a role in the cardiac myofibroblast during myocardial remodeling. It has, however, been shown to play a significant role in regulating cardiomyocyte proliferation and survival, thus it is likely that YAP/TAZ play an important role in regulating the fibrotic response, possibly through fibroblast-to-myofibroblast phenoconversion. We believe that Ski/Sno and YAP/TAZ are in a steady balance in the healthy heart which is lost following cardiac injury and pro-

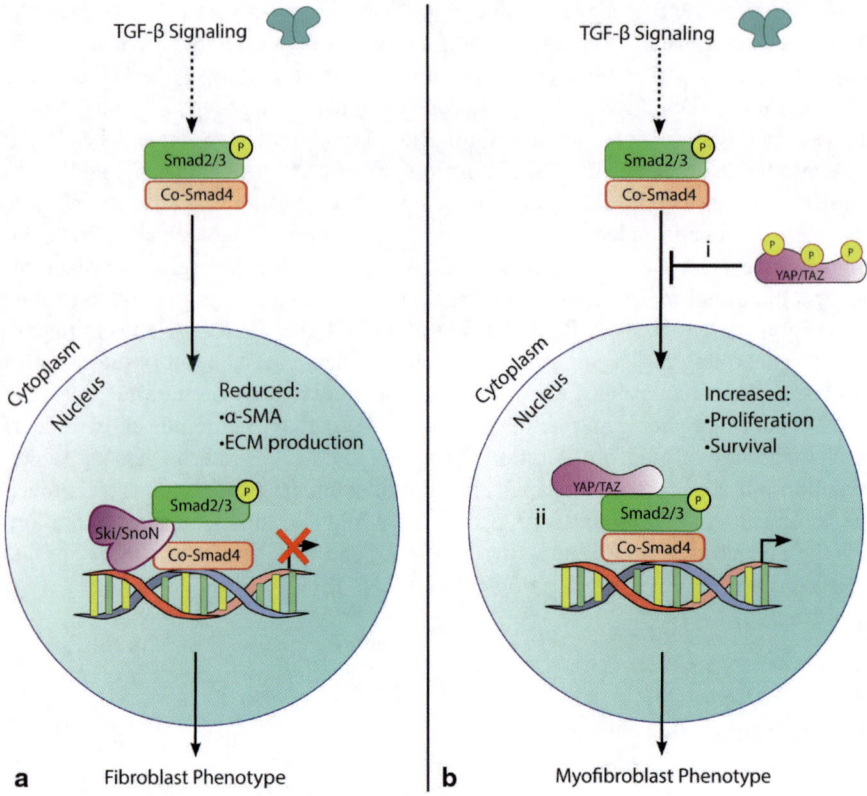

Fig. 3 Hypothetical regulation of the myofibroblast phenotype by YAP/TAZ and Ski/Sno. (**a**) Antagonism of fibroblast-to-myofibroblast phenoconversion by nuclear Ski/SnoN by disrupting the active phosphorylated R-Smad/Co-Smad complex at TGF-β-reponsive gene promoters. (**b**) Proposed inhibitory mechanism (*i*) by which phosphorylated YAP/TAZ disrupts TGF-β signaling in fibroblasts, along with putative mechanism by which nuclear YAP/TAZ is involved in enhancing (*ii*) the effects of Smad-mediated TGF-β signaling. *TGF-β* transforming growth factor-β, *R-Smad* receptor-regulated Smad, *Co-Smad* co-mediator Smad, *YAP* Yes-associated protein, *TAZ* transcriptional co-activator with PDZ binding motif, *α-SMA* α-smooth muscle actin, *ECM* extracellular matrix, *SnoN* Ski-related novel protein N, *P* phosphorylated

motes a fibrotic environment (Fig. 3). While this hypothesis requires exploration, it provides an attractive means to regulate the myofibroblast phenotype and function.

References

1. Mathers CD, Loncar D (2006) Projections of global mortality and burden of disease from 2002 to 2030. PLoS Med 3:e442
2. Mendis SPP, Norrving B (2011) Global atlas on cardiovascular disease prevention and control. World Health Organization. Geneva. ISBN 978 924 156437 3

3. Roger VL, Go AS, Lloyd-Jones DM, Benjamin EJ, Berry JD, Borden WB, Bravata DM, Dai S, Ford ES, Fox CS, Fullerton HJ, Gillespie C, Hailpern SM, Heit JA, Howard VJ, Kissela BM, Kittner SJ, Lackland DT, Lichtman JH, Lisabeth LD, Makuc DM, Marcus GM, Marelli A, Matchar DB, Moy CS, Mozaffarian D, Mussolino ME, Nichol G, Paynter NP, Soliman EZ, Sorlie PD, Sotoodehnia N, Turan TN, Virani SS, Wong ND, Woo D, Turner MB, American Heart Association Statistics C, Stroke Statistics S (2012) Heart disease and stroke statistics–2012 update: A report from the American heart association. Circulation 125:e2–e220
4. Dobaczewski M, de Haan JJ, Frangogiannis NG (2012) The extracellular matrix modulates fibroblast phenotype and function in the infarcted myocardium. J Cardiovasc Transl Res 5:837–847
5. Popovic AD, Neskovic AN, Pavlovski K, Marinkovic J, Babic R, Bojic M, Tan M, Thomas JD (1997) Association of ventricular arrhythmias with left ventricular remodelling after myocardial infarction. Heart 77:423–427
6. John Sutton M St, Lee D, Rouleau JL, Goldman S, Plappert T, Braunwald E, Pfeffer MA (2003) Left ventricular remodeling and ventricular arrhythmias after myocardial infarction. Circulation 107:2577–2582
7. Newfeld SJ, Wisotzkey RG, Kumar S (1999) Molecular evolution of a developmental pathway: phylogenetic analyses of transforming growth factor-beta family ligands, receptors and Smad signal transducers. Genetics 152:783–795
8. Graff JM, Bansal A, Melton DA (1996) Xenopus Mad proteins transduce distinct subsets of signals for the TGF beta superfamily. Cell 85:479–487
9. Annes JP, Munger JS, Rifkin DB (2003) Making sense of latent TGFbeta activation. J Cell Sci 116:217–224
10. Worthington JJ, Klementowicz JE, Travis MA (2011) TGFbeta: a sleeping giant awoken by integrins. Trends Biochem Sci 36:47–54
11. Ahamed J, Burg N, Yoshinaga K, Janczak CA, Rifkin DB, Coller BS (2008) *In vitro* and in vivo evidence for shear-induced activation of latent transforming growth factor-beta1. Blood 112:3650–3660
12. Derynck R, Zhang YE (2003) Smad-dependent and Smad-independent pathways in TGF-beta family signalling. Nature 425:577–584
13. Wipff PJ, Rifkin DB, Meister JJ, Hinz B (2007) Myofibroblast contraction activates latent TGF-beta1 from the extracellular matrix. J Cell Biol 179:1311–1323
14. Amini-Nik S, Cambridge E, Yu W, Guo A, Whetstone H, Nadesan P, Poon R, Hinz B, Alman BA (2014) Beta-catenin-regulated myeloid cell adhesion and migration determine wound healing. J Clin Invest 124:2599–2610
15. Sarrazy V, Koehler A, Chow ML, Zimina E, Li CX, Kato H, Caldarone CA, Hinz B (2014) Integrins $\alpha v\beta 5$ and $\alpha v\beta 3$ promote latent TGF-beta1 activation by human cardiac fibroblast contraction. Cardiovasc Res 102:407–417
16. Rosenkranz S (2004) TGF-beta1 and angiotensin networking in cardiac remodeling. Cardiovasc Res 63:423–432
17. Bujak M, Frangogiannis NG (2007) The role of TGF-beta signaling in myocardial infarction and cardiac remodeling. Cardiovasc Res 74:184–195
18. Saltis J, Agrotis A, Bobik A (1996) Regulation and interactions of transforming growth factor-beta with cardiovascular cells: implications for development and disease. Clin Exp Pharmacol Physiol 23:193–200
19. Desmouliere A, Geinoz A, Gabbiani F, Gabbiani G (1993) Transforming growth factor-beta 1 induces alpha-smooth muscle actin expression in granulation tissue myofibroblasts and in quiescent and growing cultured fibroblasts. J Cell Biol 122:103–111
20. Serini G, Bochaton-Piallat ML, Ropraz P, Geinoz A, Borsi L, Zardi L, Gabbiani G (1998) The fibronectin domain ED-A is crucial for myofibroblastic phenotype induction by transforming growth factor-beta1. J Cell Biol 142:873–881
21. Wipff PJ, Hinz B (2008) Integrins and the activation of latent transforming growth factor beta1—an intimate relationship. Eur J Cell Biol 87:601–615

22. Pearson-White S (1993) SnoI, a novel alternatively spliced isoform of the ski protooncogene homolog, sno. Nucleic Acids Res 21:4632–4638
23. Pearson-White S, Crittenden R (1997) Proto-oncogene sno expression, alternative isoforms and immediate early serum response. Nucleic Acids Res 25:2930–2937
24. Zheng G, Teumer J, Colmenares C, Richmond C, Stavnezer E (1997) Identification of a core functional and structural domain of the v-ski oncoprotein responsible for both transformation and myogenesis. Oncogene 15:459–471
25. Wilson JJ, Malakhova M, Zhang R, Joachimiak A, Hegde RS (2004) Crystal structure of the dachshund homology domain of human SKI. Structure 12:785–792
26. Dahl R, Wani B, Hayman MJ (1998) The Ski oncoprotein interacts with Skip, the human homolog of drosophila Bx42. Oncogene 16:1579–1586
27. Kokura K, Kim H, Shinagawa T, Khan MM, Nomura T, Ishii S (2003) The Ski-binding protein C184M negatively regulates tumor growth factor-beta signaling by sequestering the smad proteins in the cytoplasm. J Biol Chem 278:20133–20139
28. Wu JW, Krawitz AR, Chai J, Li W, Zhang F, Luo K, Shi Y (2002) Structural mechanism of Smad4 recognition by the nuclear oncoprotein ski: insights on Ski-mediated repression of TGF-beta signaling. Cell 111:357–367
29. Tokitou F, Nomura T, Khan MM, Kaul SC, Wadhwa R, Yasukawa T, Kohno I, Ishii S (1999) Viral ski inhibits retinoblastoma protein (rb)-mediated transcriptional repression in a dominant negative fashion. J Biol Chem 274:4485–4488
30. Akiyoshi S, Inoue H, Hanai J, Kusanagi K, Nemoto N, Miyazono K, Kawabata M (1999) C-Ski acts as a transcriptional co-repressor in transforming growth factor-beta signaling through interaction with smads. J Biol Chem 274:35269–35277
31. Nomura T, Khan MM, Kaul SC, Dong HD, Wadhwa R, Colmenares C, Kohno I, Ishii S (1999) Ski is a component of the histone deacetylase complex required for transcriptional repression by mad and thyroid hormone receptor. Genes Dev 13:412–423
32. Arndt S, Poser I, Schubert T, Moser M, Bosserhoff AK (2005) Cloning and functional characterization of a new Ski homolog, fussel-18, specifically expressed in neuronal tissues. Lab Invest 85:1330–1341
33. Heyman HC, Stavnezer E (1994) A carboxyl-terminal region of the ski oncoprotein mediates homodimerization as well as heterodimerization with the related protein snon. J Biol Chem 269:26996–27003
34. Cohen SB, Zheng G, Heyman HC, Stavnezer E (1999) Heterodimers of the SnoN and Ski oncoproteins form preferentially over homodimers and are more potent transforming agents. Nucleic Acids Res 27:1006–1014
35. Lin X, Liang M, Feng XH (2000) Smurf2 is a ubiquitin E3 ligase mediating proteasome-dependent degradation of Smad2 in transforming growth factor-beta signaling. J Biol Chem 275:36818–36822
36. Ohashi N, Yamamoto T, Uchida C, Togawa A, Fukasawa H, Fujigaki Y, Suzuki S, Kitagawa K, Hattori T, Oda T, Hayashi H, Hishida A, Kitagawa M (2005) Transcriptional induction of Smurf2 ubiquitin ligase by TGF-beta. FEBS Lett 579:2557–2563
37. Stroschein SL, Bonni S, Wrana JL, Luo K (2001) Smad3 recruits the anaphase-promoting complex for ubiquitination and degradation of SnoN. Genes Dev 15:2822–2836
38. Bonni S, Wang HR, Causing CG, Kavsak P, Stroschein SL, Luo K, Wrana JL (2001) TGF-beta induces assembly of a Smad2-Smurf2 ubiquitin ligase complex that targets SnoN for degradation. Nat Cell Biol 3:587–595
39. Zhu Q, Pearson-White S, Luo K (2005) Requirement for the SnoN oncoprotein in transforming growth factor beta-induced oncogenic transformation of fibroblast cells. Mol Cell Biol 25:10731–10744
40. Krakowski AR, Laboureau J, Mauviel A, Bissell MJ, Luo K (2005) Cytoplasmic snon in normal tissues and nonmalignant cells antagonizes TGF-beta signaling by sequestration of the Smad proteins. Proc Natl Acad Sci U S A 102:12437–12442
41. Xu W, Angelis K, Danielpour D, Haddad MM, Bischof O, Campisi J, Stavnezer E, Medrano EE (2000) Ski acts as a co-repressor with Smad2 and Smad3 to regulate the response to type beta transforming growth factor. Proc Natl Acad Sci U S A 97:5924–5929

42. Boyer PL, Colmenares C, Stavnezer E, Hughes SH (1993) Sequence and biological activity of chicken snoN cDNA clones. Oncogene 8:457–466
43. Reed JA, Bales E, Xu W, Okan NA, Bandyopadhyay D, Medrano EE (2001) Cytoplasmic localization of the oncogenic protein ski in human cutaneous melanomas in vivo: functional implications for transforming growth factor beta signaling. Cancer Res 61:8074–8078
44. Edmiston JS, Yeudall WA, Chung TD, Lebman DA (2005) Inability of transforming growth factor-beta to cause snon degradation leads to resistance to transforming growth factor-beta-induced growth arrest in esophageal cancer cells. Cancer Res 65:4782–4788
45. Prunier C, Pessah M, Ferrand N, Seo SR, Howe P, Atfi A (2003) The oncoprotein Ski acts as an antagonist of transforming growth factor-beta signaling by suppressing Smad2 phosphorylation. J Biol Chem 278:26249–26257
46. Cunnington RH, Wang B, Ghavami S, Bathe KL, Rattan SG, Dixon IM (2011) Antifibrotic properties of c-Ski and its regulation of cardiac myofibroblast phenotype and contractility. Am J Physiol Cell Physiol 300:C176–C186
47. Nagata M, Goto K, Ehata S, Kobayashi N, Saitoh M, Miyoshi H, Imamura T, Miyazawa K, Miyazono K (2006) Nuclear and cytoplasmic c-Ski differently modulate cellular functions. Genes Cells 11:1267–1280
48. Sutrave P, Kelly AM, Hughes SH (1990) Ski can cause selective growth of skeletal muscle in transgenic mice. Genes Dev 4:1462–1472
49. Cunnington RH, Northcott JM, Ghavami S, Filomeno KL, Jahan F, Kavosh MS, Davies JJ, Wigle JT, Dixon IM (2014) The Ski-Zeb2-Meox2 pathway provides a novel mechanism for regulation of the cardiac myofibroblast phenotype. J Cell Sci 127:40–49
50. Chen Y, Banda M, Speyer CL, Smith JS, Rabson AB, Gorski DH (2010) Regulation of the expression and activity of the antiangiogenic homeobox gene GAX/MEOX2 by ZEB2 and microRNA-221. Mol Cell Biol 30:3902–3913
51. Skopicki HA, Lyons GE, Schatteman G, Smith RC, Andres V, Schirm S, Isner J, Walsh K (1997) Embryonic expression of the Gax homeodomain protein in cardiac, smooth, and skeletal muscle. Circ Res 80:452–462
52. Li J, Li P, Zhang Y, Li GB, Zhou YG, Yang K, Dai SS (2013) C-Ski inhibits the proliferation of vascular smooth muscle cells via suppressing Smad3 signaling but stimulating p38 pathway. Cell Signal 25:159–167
53. Li J, Zhao L, He X, Yang T, Yang K (2014) Mir-21 inhibits c-Ski signaling to promote the proliferation of rat vascular smooth muscle cells. Cell Signal 26:724–729
54. Patrick DM, Montgomery RL, Qi X, Obad S, Kauppinen S, Hill JA, van Rooij E, Olson EN (2010) Stress-dependent cardiac remodeling occurs in the absence of microRNA-21 in mice. J Clin Invest 120:3912–3916
55. Thum T, Gross C, Fiedler J, Fischer T, Kissler S, Bussen M, Galuppo P, Just S, Rottbauer W, Frantz S, Castoldi M, Soutschek J, Koteliansky V, Rosenwald A, Basson MA, Licht JD, Pena JT, Rouhanifard SH, Muckenthaler MU, Tuschl T, Martin GR, Bauersachs J, Engelhardt S (2008) MicroRNA-21 contributes to myocardial disease by stimulating MAP kinase signalling in fibroblasts. Nature 456:980–984
56. Hennessy EJ, Moore KJ (2013) Using microrna as an alternative treatment for hyperlipidemia and cardiovascular disease: cardio-miRs in the pipeline. J Cardiovasc Pharmacol 62:247–254
57. Huang J, Wu S, Barrera J, Matthews K, Pan D (2005) The hippo signaling pathway coordinately regulates cell proliferation and apoptosis by inactivating Yorkie, the drosophila homolog of YAP. Cell 122:421–434
58. Pantalacci S, Tapon N, Leopold P (2003) The salvador partner Hippo promotes apoptosis and cell-cycle exit in drosophila. Nat Cell Biol 5:921–927
59. Tapon N, Harvey KF, Bell DW, Wahrer DC, Schiripo TA, Haber D, Hariharan IK (2002) Salvador promotes both cell cycle exit and apoptosis in drosophila and is mutated in human cancer cell lines. Cell 110:467–478
60. Wu S, Huang J, Dong J, Pan D (2003) Hippo encodes a Ste-20 family protein kinase that restricts cell proliferation and promotes apoptosis in conjunction with salvador and warts. Cell 114:445–456

61. Wu S, Liu Y, Zheng Y, Dong J, Pan D (2008) The TEAD/TEF family protein scalloped medi-
     ates transcriptional output of the Hippo growth-regulatory pathway. Dev Cell 14:388–398
62. Jia J, Zhang W, Wang B, Trinko R, Jiang J (2003) The drosophila Ste20 family kinase dMST
     functions as a tumor suppressor by restricting cell proliferation and promoting apoptosis.
     Genes Dev 17:2514–2519
63. Udan RS, Kango-Singh M, Nolo R, Tao C, Halder G (2003) Hippo promotes proliferation
     arrest and apoptosis in the Salvador/Warts pathway. Nat Cell Biol 5:914–920
64. Varelas X (2014) The Hippo pathway effectors TAZ and YAP in development, homeostasis
     and disease. Development 141:1614–1626
65. Zhang L, Ren F, Zhang Q, Chen Y, Wang B, Jiang J (2008) The TEAD/TEF family of transcrip-
     tion factor scalloped mediates Hippo signaling in organ size control. Dev Cell 14:377–387
66. Lin Z, Pu WT (2014) Harnessing Hippo in the heart: Hippo/yap signaling and applications
     to heart regeneration and rejuvenation. Stem Cell Res 13(3 PtB):571–586. doi:10.1016/j.
     scr.2014.04.010
67. Nishio M, Otsubo K, Maehama T, Mimori K, Suzuki A (2013) Capturing the mammalian
     Hippo: elucidating its role in cancer. Cancer Sci 104:1271–1277
68. Kanai F, Marignani PA, Sarbassova D, Yagi R, Hall RA, Donowitz M, Hisaminato A, Fuji-
     wara T, Ito Y, Cantley LC, Yaffe MB (2000) TAZ: a novel transcriptional co-activator regu-
     lated by interactions with 14-3-3 and PDZ domain proteins. EMBO J 19:6778–6791
69. Callus BA, Verhagen AM, Vaux DL (2006) Association of mammalian sterile twenty kinases,
     Mst1 and Mst2, with hsalvador via C-terminal coiled-coil domains, leads to its stabilization
     and phosphorylation. FEBS J 273:4264–4276
70. Zhao B, Tumaneng K, Guan KL (2011) The Hippo pathway in organ size control, tissue
     regeneration and stem cell self-renewal. Nat Cell Biol 13:877–883
71. Gumbiner BM, Kim NG (2014) The Hippo-Yap signaling pathway and contact inhibition of
     growth. J Cell Sci 127:709–717
72. Nishioka N, Inoue K, Adachi K, Kiyonari H, Ota M, Ralston A, Yabuta N, Hirahara S, Ste-
     phenson RO, Ogonuki N, Makita R, Kurihara H, Morin-Kensicki EM, Nojima H, Rossant J,
     Nakao K, Niwa H, Sasaki H (2009) The Hippo signaling pathway components Lats and Yap
     pattern tead4 activity to distinguish mouse trophectoderm from inner cell mass. Dev Cell
     16:398–410
73. Yu FX, Zhao B, Panupinthu N, Jewell JL, Lian I, Wang LH, Zhao J, Yuan H, Tumaneng K, Li
     H, Fu XD, Mills GB, Guan KL (2012) Regulation of the Hippo-YAP pathway by G-protein-
     coupled receptor signaling. Cell 150:780–791
74. Wada K, Itoga K, Okano T, Yonemura S, Sasaki H (2011) Hippo pathway regulation by cell
     morphology and stress fibers. Development 138:3907–3914
75. Varelas X, Samavarchi-Tehrani P, Narimatsu M, Weiss A, Cockburn K, Larsen BG, Rossant
     J, Wrana JL (2010) The crumbs complex couples cell density sensing to hippo-dependent
     control of the TGF-β-SMAD pathway. Dev Cell 19:831–844
76. Dupont S, Morsut L, Aragona M, Enzo E, Giulitti S, Cordenonsi M, Zanconato F, Digabel J
     L, Forcato M, Bicciato S, Elvassore N, Piccolo S (2011) Role of YAP/TAZ in mechanotrans-
     duction. Nature 474:179–183
77. Hao J, Zhang Y, Wang Y, Ye R, Qiu J, Zhao Z, Li J (2014) Role of extracellular matrix and
     YAP/TAZ in cell fate determination. Cell Signal 26:186–191
78. Tschumperlin DJ, Liu F, Tager AM (2013) Biomechanical regulation of mesenchymal cell
     function. Curr Opin Rheumatol 25:92–100
79. Balestrini JL, Chaudhry S, Sarrazy V, Koehler A, Hinz B (2012) The mechanical memory of
     lung myofibroblasts. Integr Biol 4:410–421
80. Sansores-Garcia L, Bossuyt W, Wada K, Yonemura S, Tao C, Sasaki H, Halder G (2011)
     Modulating F-actin organization induces organ growth by affecting the Hippo pathway.
     EMBO J 30:2325–2335
81. Low BC, Pan CQ, Shivashankar GV, Bershadsky A, Sudol M, Sheetz M (2014) YAP/TAZ as
     mechanosensors and mechanotransducers in regulating organ size and tumor growth. FEBS
     Lett 588(16): 2663–2670. doi:10.1016/j.febsfet.2014.04.012

82. Matsui Y, Lai ZC (2013) Mutual regulation between Hippo signaling and actin cytoskeleton. Protein Cell 4:904–910

83. Reddy P, Deguchi M, Cheng Y, Hsueh AJ (2013) Correction: actin cytoskeleton regulates Hippo signaling. PloS One 8(9):e73763. doi:10.1371/journal.pone.0073763

84. Reddy P, Deguchi M, Cheng Y, Hsueh AJ (2013) Actin cytoskeleton regulates Hippo signaling. PloS One 8:e73763

85. Varelas X, Sakuma R, Samavarchi-Tehrani P, Peerani R, Rao BM, Dembowy J, Yaffe MB, Zandstra PW, Wrana JL (2008) TAZ controls Smad nucleocytoplasmic shuttling and regulates human embryonic stem-cell self-renewal. Nat Cell Biol 10:837–848

86. Del Re DP, Yang Y, Nakano N, Cho J, Zhai P, Yamamoto T, Zhang N, Yabuta N, Nojima H, Pan D, Sadoshima J (2013) Yes-associated protein isoform 1 (Yap1) promotes cardiomyocyte survival and growth to protect against myocardial ischemic injury. J Biol Chem 288:3977–3988

87. Heallen T, Zhang M, Wang J, Bonilla-Claudio M, Klysik E, Johnson RL, Martin JF (2011) Hippo pathway inhibits Wnt signaling to restrain cardiomyocyte proliferation and heart size. Science 332:458–461

88. Lin Z, von Gise A, Zhou P, Gu F, Ma Q, Jiang J, Yau AL, Buck JN, Gouin KA, van Gorp PR, Zhou B, Chen J, Seidman JG, Wang DZ, Pu WT (2014) Cardiac-specific YAP activation improves cardiac function and survival in an experimental murine myocardial infarction model. Circ Res 115(3):354–363. doi:10.1161/CircResAHA.115.303632

89. von Gise A, Lin Z, Schlegelmilch K, Honor LB, Pan GM, Buck JN, Ma Q, Ishiwata T, Zhou B, Camargo FD, Pu WT (2012) YAP1, the nuclear target of Hippo signaling, stimulates heart growth through cardiomyocyte proliferation but not hypertrophy. Proc Natl Acad Sci U S A 109:2394–2399

90. Xin M, Kim Y, Sutherland LB, Murakami M, Qi X, McAnally J, Porrello ER, Mahmoud AI, Tan W, Shelton JM, Richardson JA, Sadek HA, Bassel-Duby R, Olson EN (2013) Hippo pathway effector Yap promotes cardiac regeneration. Proc Natl Acad Sci U S A 110:13839–13844

91. Liu JY, Li YH, Lin HX, Liao YJ, Mai SJ, Liu ZW, Zhang ZL, Jiang LJ, Zhang JX, Kung HF, Zeng YX, Zhou FJ, Xie D (2013) Overexpression of YAP 1 contributes to progressive features and poor prognosis of human urothelial carcinoma of the bladder. BMC Cancer 13:349

92. Xin M, Kim Y, Sutherland LB, Qi X, McAnally J, Schwartz RJ, Richardson JA, Bassel-Duby R, Olson EN (2011) Regulation of insulin-like growth factor signaling by Yap governs cardiomyocyte proliferation and embryonic heart size. Sci Signal 4:ra70

Molecular Mechanisms of Smooth Muscle and Fibroblast Phenotype Conversions in the Failing Heart

Christina Pagiatakis, Dr. John C. McDermott

Abstract The mechanisms of gene regulation in cardiac hypertrophy and fibrosis are important in understanding the regulation of pathological gene expression in the heart. Cardiac hypertrophy is characterized by enlargement of the heart as a result of an increase in cardiomyocyte size and also enhanced fibrosis due primarily to phenotypic conversion of fibroblasts to myofibroblasts. Also, atherosclerosis, a disease characterized by formation of plaque within the arterial wall, and restenosis, which is the process of arterial wall healing in response to vascular injury, are highly affected by vascular remodelling. Vascular smooth muscle cells thus play a key role in vascular remodelling, as they modulate their phenotype in response to vascular injury and are a significant source of extracellular matrix components of the vessel wall. In view of the profound effects of both the fibroblast to myofibroblast conversion and also the role of vascular smooth muscle cells in vascular remodelling, we review the activation of the smooth muscle actin gene in these contexts to examine the common and non-overlapping molecular circuitry underlying these cellular processes in the cardiovascular system.

Keywords Vascular smooth muscle cell · Arterial wall · MEF2 · TGFβ · Myocardin

Cardiac hypertrophy is characterized by the enlargement of the heart as a result of an increase in cardiomyocyte size and enhanced fibrosis due primarily to phenotypic conversion of fibroblasts to myofibroblasts. This is typically a result of increased biomechanical stress. There are two types of cardiac hypertrophy: physiological hypertrophy and pathological hypertrophy [1]. The former occurs during normal growth and development, and also in response to exercise and pregnancy, whereas the latter occurs typically due to loss of cardiomyocytes following myocardial infarction or as a result of arterial hypertension. Physiological hypertrophy is not associated with adverse cardiac function, fibrosis or heart failure, whereas pathological hypertrophy results in congestive heart failure, arrhythmia and mortality

J. C. McDermott (✉) · C. Pagiatakis
Department of Biology, York University, 4700 Keele Street, Toronto, ON M3J 1P3, Canada
e-mail: jmcderm@yorku.ca

C. Pagiatakis
e-mail: cpag@yorku.ca

© Springer International Publishing Switzerland 2015   167
I.M.C. Dixon, J. T. Wigle (eds.), *Cardiac Fibrosis and Heart Failure: Cause or Effect?*,
Advances in Biochemistry in Health and Disease 13, DOI 10.1007/978-3-319-17437-2_10

[2]. It should be noted that cardiac hypertrophy can also have a genetic basis due to mutations in contractile proteins of the cardiac muscle sarcomere, such as β-myosin heavy chain, myosin light chain, troponin, actin, myosin binding protein C and α-tropomyosin [3]. However, most of the studies reviewed here are based on hypertrophic adaptations to hypertension or experimental model systems that mimic it.

During pathological hypertrophy in the adult, there is an up-regulation of genes normally associated with embryonic and fetal development concurrently with a down-regulation of adult myocardial genes. There are various alterations in cardiac gene expression which result in both apoptosis and fibrosis [4]. This phenomenon is termed 'fetal gene activation' and involves increased expression of β-MHC, atrial natriuretic factor, SM22, smooth muscle and skeletal muscle α-actin. Furthermore, cardiomyocytes decrease their overall oxidative capacity and rely on anaerobic glucose metabolism; it appears that physiological hypertrophy plays an adaptive role to increased cardiac wall stress. Conversely, pathological hypertrophy will ultimately result in congestive heart failure [5].

The mechanisms of gene regulation in cardiac hypertrophy are important to understands the regulation of pathological gene expression in the heart. For example, forced expression of activated calcineurin induces hypertrophy, fetal gene activation and heart failure in the transgenic mouse [1]. CaMKs phosphorylate class IIa HDACs to relieve their repressive effects on transcription, and forced expression of CaMKIV in the heart also induces hypertrophy, concomitanly with increased expression of ANF and down-regulation of α-MHC [6]. CAMKII is also an important factor in cardiac excitation-contraction coupling in response to β-adrenergic signalling [7], but also plays an important role in pathological cardiac remodelling in response to endothelin-1 [8]. Interestingly, CaMKII targets HDAC4 specifically, to promote fetal gene activation following α-adrenergic agonist treatment [9]. Mice deficient in CaMKIIδ are protected from pathological hypertrophy and fetal gene activation [10], however targeted deletion of HDAC5 and HDAC9 results in cardiac hypertrophy and increased pressure overload resulting in cardiac remodelling and increased fetal gene activation [11].

Cardiac remodelling, a phenomenon which occurs in response to heamodynamic load and/or injury, is characterized by a physical alteration in the hearts' dimension, mass or shape, and there are several molecular pathways which regulate cardiac remodelling. There are several agonists of cardiac remodelling, including Angiotensin II (AngII), Endothelin-1 (ET-1) and α-adrenergic stimulation [12]. Many studies have shown that these agonists that activate Gq-coupled receptors target specific downstream targets such as PKC and PKD1, which are important for the nuclear export of HDAC5 in cardiomyocytes [13]. Reduced expression of PKD1 prevents agonist-induced hypertrophy in cardiomyocytes, whereas conditional deletion of PKD1 in mice shows improved cardiac function and reduced hypertrophy [14].

1 Atherosclerosis and Restenosis

Atherosclerosis is a disease that is characterized by the formation of a plaque (also termed atheroma), within the arterial intima and media. Following the formation of the atheroma, the lumen will eventually narrow to cause ischemia. Disruption of the atheroma results in thrombus formation, which is the leading cause of angina, myocardial infarction and even cerebral infarction in the brain [15].

Atherogenesis is a process that occurs in response to chronic injury such as shear stress or oxidative stress. Examples of sources of endothelial injury are free radicals produced from cigarette smoking, hypertension, diabetes mellitus, oxidized LDL and elevated homocysteine [15]. Free radicals and reactive oxygen species not only contribute to intimal injury, but also neutralize the protective effects of nitric oxide produced by the endothelium on the vasculature [16]. Injury to the intima results in changes to the endothelium's capacity to regulate its adhesiveness and perme-ability to various circulating factors. Importantly, upon injury, the endothelium will increase its production of vasoconstrictors (such as Ang-II and ET-1), which re-sult in activation of cytokines and growth factors and subsequent internalization of oxidized LDL within the vessel wall. Internalization of LDL within the vessel wall stimulates conversion of macrophages into foam cells, which form the initial lesion during atherogenesis. Secretion of cytokines, chemokines and growth factors at the arterial lesion result in an inflammatory response which promotes proliferation and migration of vascular smooth muscle cells from the media to the lesion within the vessel wall [15].

As the VSMCs are being activated and migrate towards the site of lesion for-mation, the formation of the atherosclerotic plaque continues, while the VSMCs form a fibrous cap over the lesion. At this stage, the lesion is considered as an ad-vanced plaque, which will continue to develop as a result of increase proliferation of VSMCs, macrophages and T-cells. As the plaque grows larger, degradation of the fibrous cap is promoted by secretion of MMPs from the activated macrophages, resulting in instability of the plaque, and subsequent hemorrhage and rupture of the plaque [15] (Fig. 1).

Angiotensin II is a factor that plays a major role in not only formation of the plaque, but also its instability and rupture. Under oxidative stress conditions, Ang-II promotes vasoconstriction, inflammation and vessel remodelling. Vessel

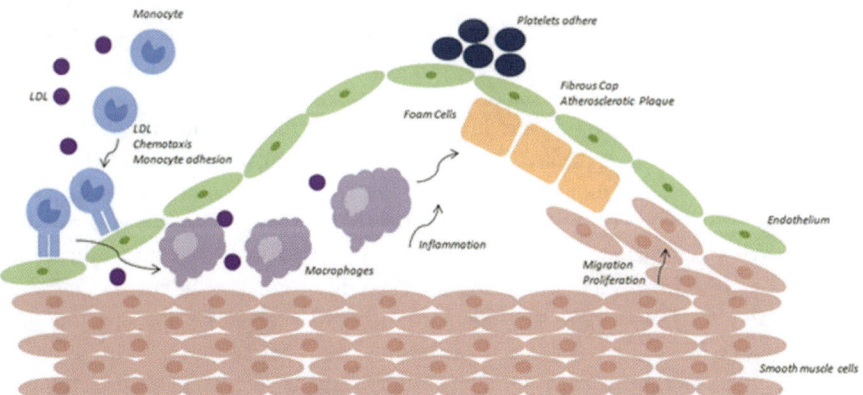

Fig. 1 Summary of atherogenesis: macrophages are recruited to consume oxidezed LDL and become foam cells within the arterial wall. Platelets adhere to the dysfunctional endothelium to release growth factors, causing VSMCs to alter their phenotype and proliferate and migrate to the site of plaque formation

constriction is a result of angiotensin stimulating the release of endothelin-1 and noradrenaline, and also stimulates expression of monocyte chemoattractant protein 1 (MCP-1) and tumor necrosis factor (TNFα). Angiotensin also activates NADH, which in turn promotes vascular oxidative stress, and induces expression of various growth factors, including PDGF, bFGF and IFG-1, which contribute to the vascular inflammatory response [16].

Restenosis is the process of the arterial wall healing in response to mechanical injury. It is comprised of two stages: neointimal hyperplasia and vessel remodelling. Neointimal hyperplasia is a result of platelet aggregation and inflammatory cell infiltration as a result of release of cytokines and growth factors which stimulate recruitment of activated VSMCs to the site of injury. The resulting neointima is comprised of synthetic VSMCs, extracellular matrix components and macrophages. As a result of vessel remodelling, production of extracellular matrix components increases, and VSMCs at the site of neointimal formation exhibit down-regulated expression of SM-MHC isoforms post-injury, but unaltered expression of SMα-actin. Interestingly, 6 months post arterial injury, expression of SM-MHC is recovered [17].

2   Role of VSMCs in Atherosclerosis

The role of vascular smooth muscle cells (VSMCs) has been shown to be a key component of development, as they are the major source of extracellular matrix components of vessel walls [18]. During development, VSMCs will both proliferate and differentiate to form components of the vasculature. In development, the proliferative phenotype of VSMCs (synthetic phenotype) refers to migration and proliferation of these cells to specific sites to form the vasculature. On the other hand, differentiated VSMCs (contractile phenotype) line vessel walls to regulate blood flow. The earliest VSMC differentiation marker is smooth muscle α-actin (SMαA), whose expression can be detected as early as smooth muscle precursors are recruited into the vessel wall. Following SMαA induction, other smooth muscle-marker genes are sequentially induced: SM22, calponin, SM-MHC I, and finally SM-MHC II [18].

Post-natally, VSMCs modulate their phenotype in response to various extracellular signals, and, unlike striated muscle, do not terminally differentiate. This phenotypic modulation and expression of either the synthetic or contractile phenotypes is not mutually exclusive. Differentiated VSMCs in mature vessels express matrix components and proliferate at low levels. However, following vascular injury, contractile VSMCs down-regulate muscle-specific differentiation genes and increase proliferation, to contribute to the vascular regenerative response and promote vessel healing [19]. This phenotypic modulation, from contractile to proliferative, is key to maintaining the integrity of the vascular system, but also plays an important role in many vascular diseases, such as atherosclerosis and restenosis following angioplasty [20].

VSMCs lie in the media of vessels, and are considered mature contractile cells that regulate the integrity of the vasculature, and also blood flow. In areas of turbulent flow, as well as arterial bifurcations, where differences in pressure exist, there is a higher chance of developing atherosclerotic lesions [21]. Under shear stress or mechanical strain, VSMCs in these areas modify their phenotype as a result of atherogenic stimuli (fibronectin, collagen, PDGF and reactive oxygen species), to a synthetic one. VSMCs residing in the media will migrate into the developing lesion. The synthetic phenotype is characterized by increased DNA synthesis and expression of cell-cycle markers and a decreased expression of smooth muscle marker genes such as SM-MHC and SMα-actin. There is also a morphological change to the cells, whereby myofilaments are replaced with rough endoplasmic reticulum and golgi, as well as a change in the cell shape from a more elongated shape to a rounder one [20] (Fig. 2).

Activated VSMCs contribute to plaque formation and size, not only by migrating to the site of injury, but also by affecting lipid uptake through LDL receptors, by contributing to inflammatory cytokine production and by altering the production of extracellular matrix components [21].

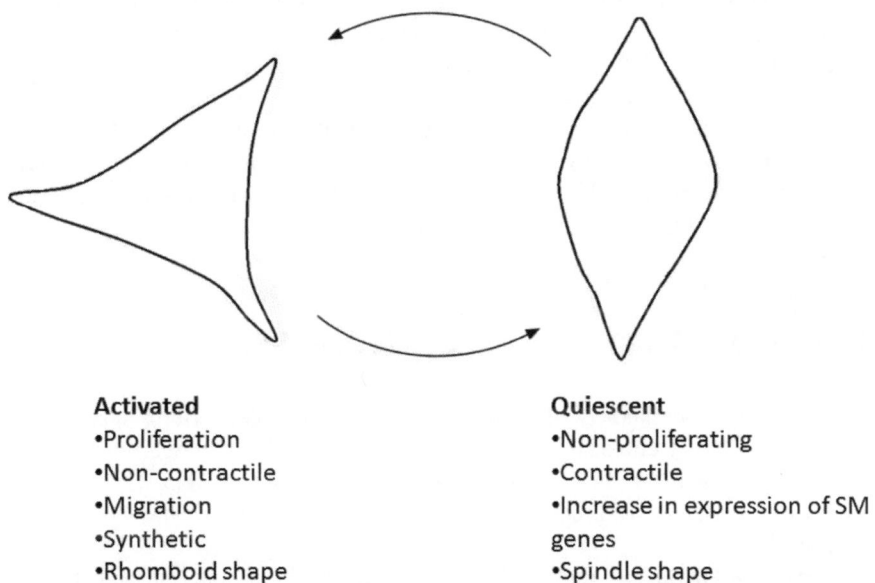

Activated
•Proliferation
•Non-contractile
•Migration
•Synthetic
•Rhomboid shape

Quiescent
•Non-proliferating
•Contractile
•Increase in expression of SM genes
•Spindle shape

Fig. 2 Schematic representation of the phenotypic switch of vascular smooth muscle cells between a quiescent/contractile and a proliferative/synthetic state

3   Transcriptional Regulation of Smooth Muscle Phenotype Conversions

There are several transcription factors as well as transcriptional co-regulators that have been shown to regulate the smooth muscle phenotype. GATA-6, a transcription factor of zinc finger motif DNA-binding domain proteins, has been shown to play a key role in regulating SMC-specific promoters. In quiescent (or contractile) smooth muscle cells, GATA-6 is expressed, however in response to injury, it is down-regulated [22]. GATA-6 has been shown to be a key player in the induction of the differentiated smooth muscle phenotype; upregulation promotes withdrawal from the cell cycle. In a vascular injury model, the phenotypic switch of VSMCs into the de-differentiated or synthetic phenotype is associated with a downregulation of GATA6 [23].

The smooth muscle myosin heavy chain promoter has been shown to be regulated by GATA6 through specific regulatory elements [24]. Apart from GATA6, there are various transcription factors that function in a combinatorial manner to regulate transcription of smooth muscle specific genes, and thus the quiescent, differentiated phenotype. It has been shown that GATA6, SRF and TAAT binding sites in the promoter regions of smooth muscle-specific genes act in concert to promote transcription [25]. For example, myocardin, a potent co-activator of SRF in both cardiac and smooth muscle cells functions through its interaction with SRF, which binds to *cis* elements termed CArG boxes, which are found in the promoters of muscle specific genes. Interestingly, CArG boxes are also found in the promoters of serum inducible genes, which regulate proliferation. Thus it is evident how smooth muscle cells can fluctuate between a proliferated and differentiated state, as a result of which co-activators are regulating SRF on the CArG boxes of various genes. Therefore, binding of myocardin to SRF on muscle-specific promoters induces transcriptional activation and a differentiated phenotype [26].

4   Fibrosis

A key factor in the majority of heart disease is the presence of fibrosis, an excess production of extracellular matrix proteins which alters the structure, shape of the heart. These changes to the heart, brought on by cardiovascular injury, have severe effects on ventricular contractility, valvular function and electrical conduction of the heart [27].

One of the key factors that has been implicated in fibrosis is TGFβ1. Although TGFβ1 is known to promote collagen production, little is known about the mechanism by which it induces fibrosis. Studies have shown that TGFβ1 inhibition can attenuate fibrosis in the heart. Interestingly, genetic studies on TGFβ1 gene polymorphism and dysregulation have been shown them to be factors in having a predisposition to heart disease. Therefore pharmacologic or targeted gene therapies are potentially important therapeutic approaches to treating fibrosis [28].

To understand fibrosis, it is important to understand the structure and composition of the heart. The heart is not an organ that is comprised solely of muscle cells. In fact, in the heart, the number of fibroblast cells actually outnumbers that of cardiac myocytes. It is primarily the fibroblasts in the heart that give rise to the extracellular matrix and allows for fibrosis in the myocardium [28]. The connective and elastic tissue in the heart is important for the maintenance of structure and architecture of various components of the heart. However, fibrotic tissue in excess will give rise to cardiac pathologies. Increased levels of collagen within the myocardium have an effect on ventricular elasticity [29]. Stiffening of the ventricle will then have an effect on myocyte contraction and relaxation, resulting in aberrant ventricular filling and thus increased pressure [30]. Presence of fibrotic tissues also has a detrimental effect on systolic function, due to increased collagen concentration in the myocardium. Although extracellular matrix proteins usually function as a repair mechanism, in fibrotic conditions where there is a reduction of muscle tissue, the outcome is poor ventricular contraction and reduced cardiac output. The fibrotic heart is not able to produce adequate pressures for systemic perfusion, as a result of increased collagen concentration and changes in ventricular geometry [31]. Overall, the presence and upregulation of fibrotic proteins results in a change in ventricular size and shape, which negatively influences heart function.

5   Fibroblast to Myofibroblast Transition

Fibroblasts are spindle shaped cells that reside in the majority of tissues and organs of the body that are associated with extracellular matrix molecules. They are characterized by expression of vimentin and absence of expression of desmin and SMα-actin. Activated fibroblasts are associated with synthesis and secretion of ECM molecules such as collagens, proteoglycans and fibronectin. Fibroblast cells originate from the mesenchyme and portray a diverse phenotypic variability such as non-contractile fibroblast, protomyofibroblast and contractile myofibroblast. Myofibroblasts are distinguished from fibroblasts by their expression of SMα-actin in stress fibres and various ECM proteins. Although myofibroblasts express SMα-actin, they can be distinguished from actual smooth muscle cells by their lack of desmin and smooth muscle myosin expression (Fig. 3). The origin of myofibroblasts is uncertain. They may arise from transdifferentiation of fibroblasts and smooth muscle cells, however, whether the populations of myofibroblasts derived from fibroblasts or smooth muscle cells form distinct or similar populations is unknown. Whether or not fibroblasts can differentiate into smooth muscle cells or *vice versa* also remains unclear, however it is possible that fibroblasts can differentiate into myofibroblast-like cells, whose protein expression pattern resembles that of smooth muscle cells [32, 33].

Recent studies have shown that following fibrosis or injury, the recruited fibroblasts or myofibroblasts may arise from different sources. Such sources could be dedifferentiated epithelial cells by epithelial-mesenchymal transition (EMT), bone marrow derived mesenchymal stem cells or tissue derived mesenchymal stem cells [34].

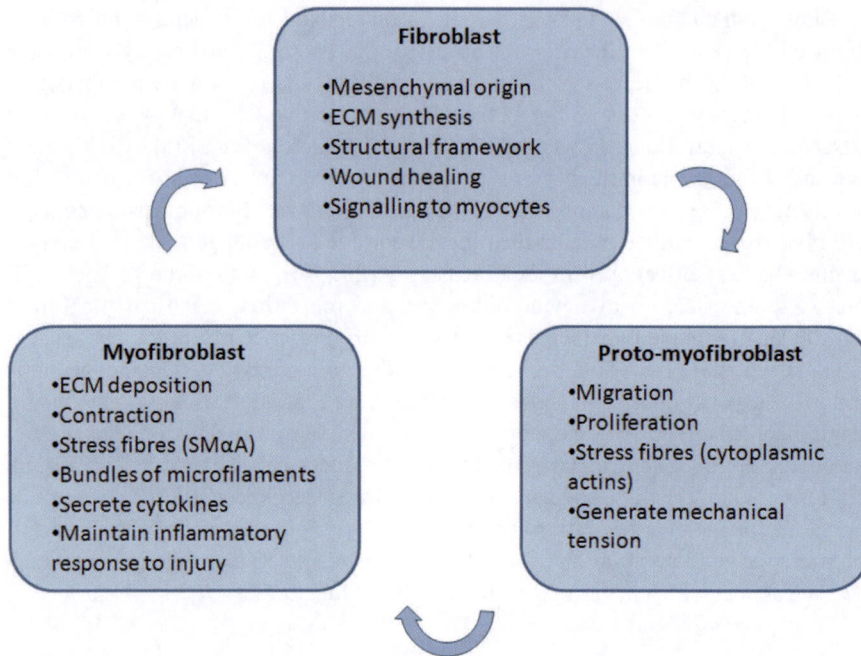

Fig. 3 Schematic representation of the basic characteristics of fibroblast to myofibroblast transition

Following tissue injury, such as myocardial infarction, the balance between collagen synthesis and degradation is regulated by myofibroblasts. The origin of these cells is mainly from cardiac fibroblasts and have the ability to respond to various mechanical, paracrine and autocrine factors. In response to mechanical stretch or pro-inflammatory cytokines, myofibroblasts increase synthesis and deposition of ECM proteins to replace necrotic myocardial tissue; this process results in scar formation [35]. Myofibroblasts play a key role in the formation of stress fibres, expression of smooth muscle genes and collagen synthesis and deposition. Normal myofibroblast function involves stabilizing the infarcted area and promotes scar tussue formation and contraction. However, abnormal amounts of myofibroblasts as a result of persistent signal elevation can result in abnormal myocardial stiffness and impairment of ventricular function due to the excessive fibrotic deposition [36].

6 Molecular Regulation of the Smooth Muscle Actin Gene in Phenotype Conversion

In normal arteries, VSMCs regulate vascular tone, and are quiescent, expressing high levels of contractile, smooth muscle-specific genes. Upon arterial injury, VSMCs lose expression of the contractile genes and proliferate. This is termed phenotypic modulation [37]. Fully differentiated smooth muscle cells upregulate genes encoding proteins that are involved in smooth muscle contraction, such as α-actin, myosin heavy chain, myosin light chain, caldesmon, vinculin, calponin, SM22 and metavinculin. However, many of these genes are also expressed in other cell types; for example, myosin light chain, caldesmon, vinculin and metavinculin smooth muscle isoforms are products of alternatively spliced genes that are expressed in a variety of cell types. The most abundant of the smooth muscle-specific genes is smooth muscle α-actin, which is exclusively expressed in smooth muscle, and smooth muscle-related cells in normal adults. Although it is expressed transiently in cardiac and skeletal muscle during development, and also in myofibroblasts in tumors, wounds, and in proliferating smooth muscle cells in atherosclerotic lesions. Thus, because of its essential nature in VSMCs, the transcriptional regulation of the smooth muscle α-actin gene locus has become a paradigm for understanding the molecular regulation of differentiation and phenotypic conversions in smooth muscle cells [38].

The smooth muscle α-actin gene proximal promoter region contains several conserved regulatory elements that are essential in its regulation. One such element is the CArG box (CC[A/T_6]GG), which was first identified as the serum response element (SRE) in the promoter region of the immediate early gene *c-fos*. It was first identified as playing a role in inducing promoter activity in response to growth factor stimulation. The CArG box is a consensus binding site for the serum response factor (SRF), which binds to DNA as a homodimer to promote transcription of a variety of genes, including smooth muscle-specific genes. Although it is ubiquitously expressed, SRF is essential in the control of the smooth muscle α-actin promoter. Interestingly, it has been shown to regulate two opposing events: activation of muscle-specific genes to promote differentiation, and activation of immediate early genes to promote proliferation [39].

The key to SRF-dependent regulation of smooth muscle-specific gene expression was found to be through the co-factor myocardin, which has been shown to be essential for smooth muscle cell differentiation. Myocardin transactivates multiple smooth muscle genes in a CArG dependent manner, but interestingly fails to activate *c-fos,* in the same manner, indicating that its role in differentiation is dependent on SRF. Unlike SRF, Myocardin is not ubiquitously expressed, and its expression is restricted to cardiac and smooth muscle tissue, it is evident that it plays a key role in the regulation of smooth muscle phenotypic regulation. Furthermore, studies have shown that activation of smooth muscle specific genes through myocardin occurs as a result of the interaction of myocardin and SRF, and not due to direct binding of myocardin to the promoter region of these genes [39]. Furthermore, the transcription factor Myocyte Enhancer Factor 2 (MEF2), has been shown to play a critical role in the phenotypic modulation of smooth muscle cells, and like SRF, regulates both immediate early genes and smooth muscle marker genes. Calcium signalling has been implicated in the control of this phenotypic switching by controlling two distinct signalling pathways. It has been shown that induction of immediate-early genes occurs via de-repression of MEF2 from HDAC4 in a calcium/calmodulin-dependent manner [19], whereas the induction of smooth muscle specific genes occurs via a MEF2-dependent RhoA/ROCK dependent signalling pathway. MEF2 has been shown to be genetically upstream of myocardin, and recent studies have documented that the RhoA/ROCK pathway is functioning through MEF2 and myocardin to regulate calcium sensitivity in smooth muscle cells. This pathway involves the de-repression of MEF2 from PP1, the catalytic subunit of myosin light chain phosphatase which regulates contraction in smooth muscle cells, by the PP1 inhibitor CPI-17 (PKC-potentiated protein phosphatase inhibitor of 17 kDa). Activation of the RhoA/ROCK pathway induces phosphorylation of CPI-17, which physically interacts with PP1 to relieve its repressive effects on MEF2, thus inducing expression of myocardin, resulting in its interaction with SRF and concomitant upregulation of smooth muscle marker genes, including smooth muscle α-actin (Fig. 4) [40].

Recent studies have also implicated transforming growth factor β(TGFβ) in the phenotypic transition of smooth muscle cells. Little is known about the exact mechanism by which TGFβ functions to potently up-regulate smooth muscle specific genes, however studies have shown that in neural crest cells and fibroblasts, TGFβ induces smooth muscle α-actin, as well as other smooth-muscle specific genes, potentially through canonical Smad signalling, and the RhoA/ROCK pathway [41, 42].

Therefore, it is important to identify the mechanisms regulating the fibroblast to myofibroblast transition so as to control aberrant activation and cardiac pathologies associated with excessive myofibroblast activity. Studies have demonstrated the hormone relaxin is produced in the heart to stimulate mouse neonatal cardiomyocyte growth. Interestingly relaxin has been shown to inhibit TGFβ1-induced fibroblast to myofibroblast transition; this was indicated by a downregulation of

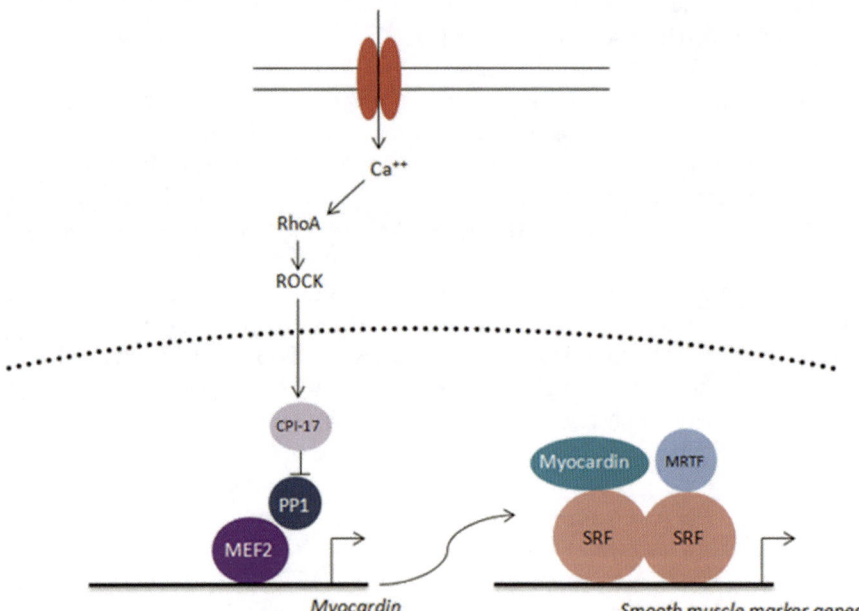

Fig. 4 Summary of the regulation of myocardin/SRF smooth muscle marker gene expression through a RhoA/ROCK-MEF2-CPI-17-dependent mechanism [40]

smooth muscle α-actin and type I collagen expression. It was found that the Notch-1 signalling pathway is involved in this pathway, and inhibition of Notch-1 potentiated TGFβ1 induced myofibroblast differentiation and abrogated the inhibitory effects of relaxin. Thus Notch appears to also play an important role by downregulating TGFβ-dependent fibroblast to myofibroblast transitions, providing another potential therapeutic target [43, 44].

7   Concluding Remarks

Cardiovascular disease is one of the leading causes of death worldwide. A characteristic of patients presenting with hypertension and heart failure is cardiac fibrosis, due to constant activation of the tissue repair program and persistent activation of fibroblast migration to the site of injury. Initially, this mechanism serves to synthesize new extracellular matrix, however prolonged activation results in excess scar tissue formation leading to fibrosis. Although the mechanisms underlying fibrosis are being characterized, there is still much to learn concerning the underlying molecular etiology of fibrosis in order to allow advances in therapeutic interventions. Basic studies have implicated a complex milieu of hormones and signalling pathways that contribute to the control of gene expression and ultimately the fibrotic phenotype. One important challenge for cardiovascular disease will be to develop novel therapeutic approaches aimed at these molecular pathways.

References

1. Frey N, Olson EN (2003) Cardiac hypertrophy: the good, the bad, and the ugly. Annu Rev Physiol 65:45–79. doi:10.1146/annurev.physiol.65.092101.142243
2. Sadoshima J, Izumo S (1997) The cellular and molecular response of cardiac myocytes to mechanical stress. Annu Rev Physiol 59:551–571. doi:10.1146/annurev.physiol.59.1.551
3. Michele DE, Metzger JM (2000) Contractile dysfunction in hypertrophic cardiomyopathy: elucidating primary defects of mutant contractile proteins by gene transfer. Trends Cardiovasc Med 10(4):177–182
4. Czubryt MP, Olson EN (2004) Balancing contractility and energy production: the role of myocyte enhancer factor 2 (MEF2) in cardiac hypertrophy. Recent Prog Horm Res 59:105–124
5. Finck BN, Kelly DP (2007) Peroxisome proliferator-activated receptor gamma coactivator-1 (PGC-1) regulatory cascade in cardiac physiology and disease. Circulation 115(19):2540–2548. doi:10.1161/CIRCULATIONAHA.107.670588
6. Passier R, Zeng H, Frey N, Naya FJ, Nicol RL, McKinsey TA, Overbeek P, Richardson JA, Grant SR, Olson EN (2000) CaM kinase signaling induces cardiac hypertrophy and activates the MEF2 transcription factor in vivo. J Clin Invest 105(10):1395–1406. doi:10.1172/JCI8551
7. Maier LS, Bers DM (2007) Role of Ca2 +/calmodulin-dependent protein kinase (CaMK) in excitation-contraction coupling in the heart. Cardiovasc Res 73(4):631–640. doi:10.1016/j.cardiores.2006.11.005
8. Wu X, Zhang T, Bossuyt J, Li X, McKinsey TA, Dedman JR, Olson EN, Chen J, Brown JH, Bers DM (2006) Local InsP3-dependent perinuclear Ca2 + signaling in cardiac myocyte excitation-transcription coupling. J Clin Invest 116(3):675–682. doi:10.1172/JCI27374

9. Backs J, Song K, Bezprozvannaya S, Chang S, Olson EN (2006) CaM kinase II selectively signals to histone deacetylase 4 during cardiomyocyte hypertrophy. J Clin Invest 116(7):1853–1864. doi:10.1172/JCI27438

10. Ling H, Zhang T, Pereira L, Means CK, Cheng H, Gu Y, Dalton ND, Peterson KL, Chen J, Bers D, Brown JH (2009) Requirement for Ca2 +/calmodulin-dependent kinase II in the transition from pressure overload-induced cardiac hypertrophy to heart failure in mice. J Clin Invest 119(5):1230–1240. doi:10.1172/JCI38022

11. Chang S, McKinsey TA, Zhang CL, Richardson JA, Hill JA, Olson EN (2004) Histone deacetylases 5 and 9 govern responsiveness of the heart to a subset of stress signals and play redundant roles in heart development. Mol Cell Biol 24(19):8467–8476. doi:10.1128/MCB.24.19.8467-8476.2004

12. Yusuf S, Sleight P, Pogue J, Bosch J, Davies R, Dagenais G (2000) Effects of an angiotensin-converting-enzyme inhibitor, ramipril, on cardiovascular events in high-risk patients. The Heart Outcomes Prevention Evaluation Study Investigators. N Engl J Med 342(3):145–153. doi:10.1056/NEJM200001203420301

13. Vega RB, Harrison BC, Meadows E, Roberts CR, Papst PJ, Olson EN, McKinsey TA (2004) Protein kinases C and D mediate agonist-dependent cardiac hypertrophy through nuclear export of histone deacetylase 5. Mol Cell Biol 24(19):8374–8385. doi:10.1128/MCB.24.19.8374-8385.2004

14. Harrison BC, Kim MS, van Rooij E, Plato CF, Papst PJ, Vega RB, McAnally JA, Richardson JA, Bassel-Duby R, Olson EN, McKinsey TA (2006) Regulation of cardiac stress signaling by protein kinase d1. Mol Cell Biol 26(10):3875–3888. doi:10.1128/MCB.26.10.3875-3888.2006

15. Ross R (1999) Atherosclerosis–an inflammatory disease. N Engl J Med 340(2):115–126. doi:10.1056/NEJM199901143400207

16. Dzau VJ (2001) Theodore Cooper Lecture: Tissue angiotensin and pathobiology of vascular disease: a unifying hypothesis. Hypertens 37(4):1047–1052

17. Costa MA, Simon DI (2005) Molecular basis of restenosis and drug-eluting stents. Circulation 111(17):2257–2273. doi:10.1161/01.CIR.0000163587.36485.A7

18. Moretti A, Caron L, Nakano A, Lam JT, Bernshausen A, Chen Y, Qyang Y, Bu L, Sasaki M, Martin-Puig S, Sun Y, Evans SM, Laugwitz KL, Chien KR (2006) Multipotent embryonic isl1 + progenitor cells lead to cardiac, smooth muscle, and endothelial cell diversification. Cell 127(6):1151–1165. doi:10.1016/j.cell.2006.10.029

19. Gordon JW, Pagiatakis C, Salma J, Du M, Andreucci JJ, Zhao J, Hou G, Perry RL, Dan Q, Courtman D, Bendeck MP, McDermott JC (2009) Protein kinase A-regulated assembly of a MEF2{middle dot}HDAC4 repressor complex controls c-Jun expression in vascular smooth muscle cells. J Biol Chem 284(28):19027–19042. doi:10.1074/jbc.M109.000539

20. Owens GK, Kumar MS, Wamhoff BR (2004) Molecular regulation of vascular smooth muscle cell differentiation in development and disease. Physiological Rev 84(3):767–801. doi:10.1152/physrev.00041.2003

21. Doran AC, Meller N, McNamara CA (2008) Role of smooth muscle cells in the initiation and early progression of atherosclerosis. Arteriosclerosis Thromb Vasc Biol 28(5):812–819. doi:10.1161/ATVBAHA.107.159327

22. Mano T, Luo Z, Malendowicz SL, Evans T, Walsh K (1999) Reversal of GATA-6 downregulation promotes smooth muscle differentiation and inhibits intimal hyperplasia in balloon-injured rat carotid artery. Circulation Res 84(6):647–654

23. Morrisey EE (2000) GATA-6: the proliferation stops here: cell proliferation in glomerular mesangial and vascular smooth muscle cells. Circulation Res 87(8):638–640

24. Kakita T, Hasegawa K, Morimoto T, Kaburagi S, Wada H, Sasayama S (1999) p300 protein as a coactivator of GATA-5 in the transcription of cardiac-restricted atrial natriuretic factor gene. J Biol Chem 274(48):34096–34102

25. Nishida W, Nakamura M, Mori S, Takahashi M, Ohkawa Y, Tadokoro S, Yoshida K, Hiwada K, Hayashi K, Sobue K (2002) A triad of serum response factor and the GATA and NK families governs the transcription of smooth and cardiac muscle genes. J Biol Chem 277(9):7308–7317. doi:10.1074/jbc.M111824200

26. Pipes GC, Creemers EE, Olson EN (2006) The myocardin family of transcriptional coactivators: versatile regulators of cell growth, migration, and myogenesis. Genes Dev 20(12):1545–1556. doi:10.1101/gad.1428006
27. Helske S, Lindstedt KA, Laine M, Mayranpaa M, Werkkala K, Lommi J, Turto H, Kupari M, Kovanen PT (2004) Induction of local angiotensin II-producing systems in stenotic aortic valves. J Am Coll Cardiol 44(9):1859–1866. doi:10.1016/j.jacc.2004.07.054
28. Nag AC (1980) Study of non-muscle cells of the adult mammalian heart: a fine structural analysis and distribution. Cytobios 28(109):41–61
29. MacKenna DA, Vaplon SM, McCulloch AD (1997) Microstructural model of perimysial collagen fibers for resting myocardial mechanics during ventricular filling. Am J Physiol 273(3 Pt 2):H1576–1586
30. Covell JW (1990) Factors influencing diastolic function. Possible role of the extracellular matrix. Circulation 81(2 Suppl):III155–158
31. Deten A, Holzl A, Leicht M, Barth W, Zimmer HG (2001) Changes in extracellular matrix and in transforming growth factor beta isoforms after coronary artery ligation in rats. J Mol Cell cardiol 33 (6):1191–1207. doi:10.1006/jmcc.2001.1383
32. Desmouliere A, Darby IA, Gabbiani G (2003) Normal and pathologic soft tissue remodeling: role of the myofibroblast, with special emphasis on liver and kidney fibrosis. Lab Invest 83(12):1689–1707
33. Eyden B (2005) The myofibroblast: a study of normal, reactive and neoplastic tissues, with an emphasis on ultrastructure. J Submicrosc Cytol Pathol:37(2):109–204
34. Chambers RC, Leoni P, Kaminski N, Laurent GJ, Heller RA (2003) Global expression profiling of fibroblast responses to transforming growth factor-beta1 reveals the induction of inhibitor of differentiation-1 and provides evidence of smooth muscle cell phenotypic switching. Am J Pathol 162(2):533–546
35. Fan D, Takawale A, Lee J, Kassiri Z (2012) Cardiac fibroblasts, fibrosis and extracellular matrix remodeling in heart disease. Fibrogenesis Tissue Repair 5(1):15. doi:10.1186/1755-1536-5-15
36. Kolonin MG, Evans KW, Mani SA, Gomer RH (2012) Alternative origins of stroma in normal organs and disease. Stem cell Res 8(2):312–323. doi:10.1016/j.scr.2011.11.005
37. Grabski AD, Shimizu T, Deou J, Mahoney WM Jr, Reidy MA, Daum G (2009) Sphingosine-1-phosphate receptor-2 regulates expression of smooth muscle alpha-actin after arterial injury. Arterioscler Thromb Vasc Biol 29(10):1644–1650. doi:10.1161/ATVBAHA.109.191965
38. Shimizu RT, Blank RS, Jervis R, Lawrenz-Smith SC, Owens GK (1995) The smooth muscle alpha-actin gene promoter is differentially regulated in smooth muscle versus non-smooth muscle cells. J Biol Chem 270(13):7631–7643
39. Hendrix JA, Wamhoff BR, McDonald OG, Sinha S, Yoshida T, Owens GK (2005) 5' CArG degeneracy in smooth muscle alpha-actin is required for injury-induced gene suppression in vivo. J Clin Invest 115(2):418–427. doi:10.1172/JCI22648
40. Pagiatakis C, Gordon JW, Ehyai S, McDermott JC (2012) A novel RhoA/ROCK-CPI-17-MEF2C signaling pathway regulates vascular smooth muscle cell gene expression. J Biol Chem 287(11):8361–8370. doi:10.1074/jbc.M111.286203
41. Tang Y, Yang X, Friesel RE, Vary CP, Liaw L (2011) Mechanisms of TGF-beta-induced differentiation in human vascular smooth muscle cells. J Vasc Res 48(6):485–494. doi:10.1159/000327776
42. Chen S, Crawford M, Day RM, Briones VR, Leader JE, Jose PA, Lechleider RJ (2006) RhoA modulates Smad signaling during transforming growth factor-beta-induced smooth muscle differentiation. J Biol Chem 281(3):1765–1770. doi:10.1074/jbc.M507771200
43. Sassoli C, Chellini F, Pini A, Tani A, Nistri S, Nosi D, Zecchi-Orlandini S, Bani D, Formigli L (2013) Relaxin prevents cardiac fibroblast-myofibroblast transition via notch-1-mediated inhibition of TGF-beta/Smad3 signaling. PloS one 8(5):e63896. doi:10.1371/journal.pone.0063896
44. Fahed AC, Gelb BD, Seidman JG, Seidman CE (2013) Genetics of congenital heart disease: the glass half empty. Circulation Res 112(4):707–720. doi:10.1161/CIRCRESAHA.112.300853C. Pagiatakis and J. C. McDermott

Current and Future Strategies for the Diagnosis and Treatment of Cardiac Fibrosis

Patricia L. Roche and Michael P. Czubryt

Abstract Cardiac fibrosis ensues from a mismatch between extracellular matrix production and degradation, resulting in increased and excessive deposition of matrix components including fibrillar collagen types I and III. Increased collagen synthesis and cross-linking strengthens the myocardium but also increases wall stiffness and thereby negatively impact both diastolic (filling) and systolic (contractile) function. Myocardial fibrosis occurs secondary to a host of cardiovascular diseases, including hypertension, coronary artery disease with or without myocardial infarction, myocarditis of various origins, systemic fibrotic diseases (sclerosis), and congenital heart defects (including dilated and hypertrophic cardiomyopathies) (Chaturvedi et al., Circulation 121(8):979–988, 2010). Fibrosis is a significant contributor to the pathogenesis of heart failure (HF), which causes the majority of hospitalizations in patients over 65, and thus represents a considerable but largely ignored clinical concern (Biernacka and Frangogiannis, Aging Dis 2(2):158–173, 2011).

While fibrosis has long been recognized as a common sequela to numerous cardiovascular diseases, until recently diagnosis without biopsy was challenging, and thus the incidence and prevalence of cardiac fibrosis have been difficult to accurately assess. Even more challenging has been the clinical management of cardiac fibrosis, which has largely been limited to the treatment of co-morbidities and remains elusive. Here we review the presentation, diagnosis and treatment of cardiac fibrosis. The potential for future treatments and therapies is also examined, including how research into fibrotic diseases in other tissues may provide promising new avenues of pursuit for ameliorating fibrosis in cardiac patients.

Keywords Therapeutics · Cardiac fibrosis · miRNAs · Anti-fibrotics

M. P. Czubryt (✉) · P. L. Roche
Department of Physiology and Pathophysiology, Institute of Cardiovascular Sciences,
University of Manitoba, R4008, St. Boniface Hospital Research Centre, 351 Tache Avenue,
Winnipeg, MB, Canada R2H 2A6

© Springer International Publishing Switzerland 2015   181
I.M.C. Dixon, J. T. Wigle (eds.), *Cardiac Fibrosis and Heart Failure: Cause or Effect?*,
Advances in Biochemistry in Health and Disease 13, DOI 10.1007/978-3-319-17437-2_11

1 Clinical Presentation of Cardiac Fibrosis

Heart failure frequently occurs secondary to myocardial infarction or chronic arterial hypertension, both of which typically present with clinically significant cardiac fibrosis. In hypertension, increased afterload and stress on the myocardium leads to hypertrophy of the left ventricle (LV) in an attempt to compensate for increased hemodynamic demand. Myocyte hypertrophy and progressive fibrosis lead to dysfunction of the LV, eventually resulting in heart failure and the development of cardiac arrhythmias which increase the risk of sudden cardiac death [3]. Maintenance of fibrosis and progression to heart failure is further augmented by activation of neurohormonal and cytokine pathways in response to increased cardiac stress and injury. Multiple mechanisms thus converge to induce significant changes in cardiomyocytes and interstitial fibroblasts that worsen the degree of LV remodeling and myocardial rigidity. These structural and functional changes at the cellular, tissue, and organ levels are the catalyst for the inevitable decline in cardiac output and diastolic function that leads to the inability of the heart to sufficiently meet the body's demands.

Excess collagen production is mediated primarily by an activated fibroblast called the myofibroblast. Activation of myofibroblasts, which may also be derived from non-fibroblast sources, occurs in response to a variety of local and systemic pro-fibrotic factors [4]. The key players in cardiac fibrosis include injury-induced growth factors like transforming growth factor-β (TGF-β), connective tissue growth factor (CTGF), and platelet-derived growth factor (PDGF), neurohormonal agents such as the renin-angiotensin-aldosterone system (RAAS), proteins related to vascular damage such as endothelin-1, and hemodynamic factors (pressure and volume overload) which alter the physical forces acting upon and throughout the myocardium [5–9]. The interplay of these and other factors in driving the fibrotic process is complex and dependent upon the relative degree of activation of these various pathways, etiology of fibrosis, the presence and severity of disease in other organs, and patient genetics.

In general cardiac fibrosis typically presents in two forms, although the precise nomenclature of these forms is subject to some debate [10, 11]. Reactive fibrosis occurs as an adaptive response to increased cardiac stress (via pressure or volume overload) characterized by cardiomyocyte hypertrophy. Spreading from perivascular spaces, reactive fibrosis presents as a diffuse deposition of excess collagen throughout the myocardium [12]. With increased dilatation, reactive fibrosis may eventually lead to the second form of fibrosis, replacement fibrosis, which is associated with cardiomyocyte loss. In myocardial infarction, a post-injury inflammatory phase induces expression of wound healing genes that initiate the infarct scar formation process. Replacement fibrosis occurs post-myocardial infarct directly succeeding the inflammatory phase and can completely bypass reactive fibrosis [13]. The presentation of the fibrotic lesion also differs, depending upon the type of fibrosis present. Macroscopic scars may arise from replacement and reactive fibrosis, are often more pronounced over the atrial-ventricular septum, may range from small patches to large transmural scars, and are often readily visible using non-invasive imaging techniques [14]. It is noteworthy that, for reasons that remain unclear, fibrosis distal to the initial infarcted region frequently follows the initial healing phase in otherwise apparently healthy areas of the myocardium.

The most common primary disease associated with the development of cardiac fibrosis and related heart failure is hypertensive heart disease, which affects over 1 billion people worldwide [15]. Approximately one quarter of cases of heart failure can be directly attributed to hypertension, and it precedes heart failure in 90% of patients [16, 17]. In fact, hypertension alone accounts for the majority of cardiovascular events in the elderly [18]. Hypertensive heart disease is associated with an increase in the collagen volume fraction (CVF) of the myocardium [19]. While occurring most commonly with left ventricular hypertrophy, fibrosis may be associated with a variety of cardiovascular diseases. For example, familial mutation of a cardiac muscle sarcomere gene results in congenital hypertrophic cardiomyopathy (HCM), which affects about 1 in 500 North Americans [10]. The hallmark histological features of HCM include myocyte hypertrophy, fibrosis and small-vessel disease. Though the type of fibrosis manifested varies between individuals, the degree of fibrosis is associated with small vessel disease, cardiac mass, age, septal thickness, and progression to end-stage heart failure or sudden cardiac death [20, 21]. An unusually strong correlation is observed between small vessel disease and myocardial scarring.

2   Diagnosing Cardiac Fibrosis

A number of obstacles hinder the diagnosis and treatment of cardiac fibrosis. For example, its symptoms bear a strong similarity to those of constrictive pericarditis, and fibrosis may present atypically in women, the obese, and the elderly. Additionally, in reactive fibrosis, the distribution of fibrotic tissue is often diffuse and requires invasive techniques for proper characterization. Prior to the advent of full-body MRI scans, detecting the presence of cardiac fibrosis was strictly limited to histological examination of post-mortem patient biopsies for collagen content. In response to this dilemma, Burwell et al. noted a list of common characteristics which could aid in categorizing cardiac fibrosis. These included elevated venous pressure, congestive hepatomegaly, peritoneal and peripheral edema, cardiomegaly, and poor response to heart failure treatments (digitalis, diuretics, and low-salt diet) [22]. They also employed fluoroscopy, a live X-ray imaging technique that posed significant risk of carcinogenesis to both the physician and the patient, to assess changes in cardiac function. Today, clinicians utilize both functional and structural characteristics to assess fibrosis-related cardiac dysfunction. Upon presenting with symptoms of heart failure, routine testing is used to determine the underlying etiology, including blood tests for serum levels of cardiovascular disease biomarkers, and echocardiography for assessment of heart function [23]. Common functional abnormalities observed in patients with cardiac fibrosis include reductions in cardiac output, ejection fraction, and coronary flow, as well as conduction abnormalities detected by echocardiogram [1]. Due to the increased stiffness of the myocardium, about 50% of patients will retain systolic function and a normal ejection fraction, yet will display diastolic (filling) dysfunction and reduced net cardiac output [24].

The subjective, error-prone and highly invasive use of histological assessment of cardiac biopsies from living patients has resulted in the development of less invasive, more sensitive and more accurate methods for identifying and quantifying cardiac fibrosis [25]. At present, late-enhancement magnetic resonance imaging (MRI) is the gold standard for assessing fibrotic lesions, often in conjunction with blood work for key cardiac protein and enzyme levels (discussed below). Unlike biopsies, MRI is non-invasive and possesses high spatial and temporal resolution, enhanced by the infusion of tissue contrast agents into the bloodstream which are then taken up into the tissue but not absorbed by cells. Due to paramagnetism, these chelated metal ions (e.g. gadolinium) reduce the T1 relaxation time of nearby protons and increase the rate of stimulated emission, resulting in a higher contrast image [26]. The use of T1 mapping techniques, such as modified Look-Locker inversion recovery (MOLLI), is a relatively recent advancement in MRI technology [27, 28]. Late gadolinium enhancement (LGE) is useful in marking chronic myocardial injury due to the expansion of fibrotic interstitium by increased collagen deposition, which results in increased gadolinium concentration in affected areas [29]. The degree of LGE can be correlated with the degree of fibrotic tissue in the heart. The presence of multiple right ventricular wall aneurysms, non-vascular distribution of contrast agent (i.e. interstitial enhancement), and absence of an increase in enhancement in a T2-weighted signal are indicators of fibrotic lesions [26, 30]. These properties make LGE-MRI a useful tool for quantifying ischemic and non-ischemic fibrosis, though its application is limited in that it cannot be used to visualize diffuse, microscopic lesions [30, 31]. In addition to gadolinium, imaging research has generated nanoparticle-enhanced MRI probes that are targeted specifically for angiotensin II-converting enzyme (ACE) overexpression or fibrillar collagen [32–34]. Collagen-avid peptides show increased sensitivity, reducing the amount of probe required for visualization of interstitial collagen, but have not yet made the leap from animal model to the clinic [35].

In addition to MRI, a number of other molecular imaging techniques have been proposed as potential tools in assessment of interstitial collagen deposition. Such approaches show promise for molecular imaging in conditions of diffuse fibrosis, and are relatively fast, sensitive, and inexpensive [36]. Compared to LGE-MRI, imaging techniques such as single photon emission computed tomography and photon emission tomography have increased sensitivity to enhancement agents (in the nanomolar range), though their spatial resolution is limited [35]. Near-infrared fluorescence imaging is another emerging visualization technique with increased sensitivity due to reduced tissue absorption and scattering [37]. However, its use is presently limited to detecting tumours close to the skin surface, and improvements in this modality are required for its use in deeper tissues such as the heart [36].

3   Biomarkers of Cardiac Fibrosis

Examination of fibrotic biomarkers in the serum is an emerging focus of current cardiovascular research, and has the potential to provide an additional level of confidence in diagnosing and assessing the severity of cardiac fibrosis. Candidate

Table 1 Diagnostic serum biomarkers for indications of cardiac fibrosis.

Biomarker	Indication
Pro-collagen type I C-terminal peptide (PICP)	Synthesis of type I collagen
Pro-collagen type III N-terminal peptide (PINP)	Synthesis of type III collagen
Collagen type I C-terminal telopeptide (ICTP)	Degradation of type I collagen
Galectin-3	Myocardial fibrosis
Syndecan-1	Myocardial fibrosis and remodeling
High sensitivity cardiac troponin T (hs-cTnT)	Myocardial fibrosis
Brain natriuretic peptide (BNP)	Myocardial remodeling, fibrosis in some cases
Matrix metalloproteinase 9 (MMP)	Correlation with prolidase activity

biomarkers include indicators of matrix synthesis and degradation (collagen propeptides and telopeptides, matrix metalloproteinases (MMPs) and tissue inhibitors of MMPs (TIMPs)), as well as relevant ratios indicative of the fibrotic process (collagen type I:III, MMP:TIMP) (Table 1). Serum levels of the collagen metabolism marker pro-collagen type I C-terminal peptide (PICP) were positively correlated with the degree of fibrotic tissue (assessed histologically) in hypertensive heart disease [38, 39]. The N-terminal peptide of pro-collagen III (PIIINP) is a marker of type III collagen synthesis, and is significantly increased in the serum of patients with HF, hypertrophic and dilated cardiomyopathy [40]. The serum level of type I collagen telopeptide (ICTP) can be used as a marker of collagen degradation, and may be useful as a therapeutic index for treatment. PICP and PIIINP, or a combination, are the most commonly-used markers for identifying collagen products in serum [40]. Abnormal serum levels of MMP2, MMP9 and TIMP1 have been reported in heart failure populations, including hypertrophic and dilated cardiomyopathy [41, 42]. MMP9 serum levels are positively correlated with the activity of prolidase, a key regulatory enzyme in metabolism of the collagen component hydroxyproline [42]. Other potential biomarkers include matricellular proteins (e.g. syndecan-1), pro-fibrotic inflammatory mediators such as galectin-3, natriuretic peptides such as brain natriuretic peptide, high-sensitivity cardiac troponin T, and neurohormonal RAAS peptides, which have been correlated with clinical endpoints [43–46].

To date there is no single biomarker that directly correlates with the degree of myocardial fibrosis, and the usefulness of serum marker quantification relies on determining the most representative combination of markers. Additionally, correlations between collagen products or proteases are often not observed in circulating blood samples, thus requiring relatively invasive transcardiac blood sampling [47]. The clinical efficacy of these measures is also dependent upon their ability to identify at-risk patients, their association with clinical endpoints, and their ability to consistently predict responses to treatment [48–50]. The development of noninvasive assessment methods such as imaging and biomarkers will permit larger clinical trials for assessing the efficacy of treatments for cardiac fibrosis.

4 Current Treatments

Cardiac fibrosis is not typically the direct target of patient treatment regimens. Rather, treatment strategies are aimed at alleviating symptoms associated with the underlying pathology such as chronic hypertension and heart failure, and improving the quality of life of the patient (Table 2). For this reason, most drugs indicated for the treatment of heart failure accompanied by fibrosis involve reducing the burden on the heart by decreasing cardiac workload. Early treatment strategies for fibrosis-related heart failure, secondary to hypertension or hypertrophy, were limited to the positive inotrope digitalis, diuretics, and decreased dietary salt intake [22]. Current treatment approaches focus on diuresis (primarily via targeting of RAAS), vasodilation, and negative inotropic agents such as beta blockers. However, due in part to the inability of these treatments to reduce or reverse cardiac fibrosis and diastolic dysfunction, heart failure frequently worsens to the point that transplant is the only option for patient survival. Hence, many emerging avenues of research are focused on not only attenuating, but actively reversing cardiac remodeling and myocardial fibrosis, which in combination with treatments targeting the primary disease, hold promise in returning the failing heart to a functional state.

4.1 Targeting the Renin-Angiotensin-Aldosterone System

One of the major advances in cardiovascular medicine in the last 50 years was the development of drugs targeting the renin-angiotensin-aldosterone system, a critical mediator of blood volume, arterial pressure, and cardiovascular function [51]. Inappropriate activation of RAAS not only contributes to hypertension and cardiac hypertrophy, but also cardiac fibrosis. Although the exact mechanisms by which RAAS stimulates fibrosis of the heart are still under scrutiny, it is likely that these hormones act by creating a highly pro-fibrotic environment via increased inflammation, oxidative stress, and cell death. Activation of RAAS is triggered by various factors, including sympathetic nervous system stimulation, baroreflex, decreased sodium delivery to the distal tubule of the kidney, and renal artery hypotension [51]. By reducing systemic blood pressure, RAAS-targeting drugs attenuate the volume and pressure burdens placed on the heart, preventing further hypertrophy and slowing the progression of fibrosis. In addition, some RAAS components may have direct pro-fibrotic actions that can be inhibited with the use of targeted agents, further augmenting their anti-fibrotic potential.

4.1.1 Angiotensin-Converting Enzyme Inhibitors

Angiotensin-converting enzyme (ACE) inhibitors are one of the drug classes of choice in delaying the progression to heart failure in patients with LV pressure and volume overload. In two different rat models of hypertension, captopril prevented

Table 2 Current cardiovascular therapeutics demonstrating impact on cardiac fibrosis.

Agent	Target	Primary Indication(s)
Aliskerin	Renin inhibitor	Hypertension
Candesartan	Angiotensin II AT1 receptor antagonist	Hypertension, hypertrophy
Losartan	Angiotensin II AT1 receptor antagonist	Hypertension, hypertrophy
Lisinopril	Angiotensin-converting enzyme inhibitor	Hypertension, CHF
Eplerenone	Aldosterone (mineralocorticoid) receptor antagonist	Left ventricular systolic dysfunction, post-infarct CHF
Spironolactone	Aldosterone (mineralocorticoid) receptor antagonist	Refractory edema
Torasemide	Diuretic	Heart failure-associated edema, hypertension
Carvedilol	Beta-adrenergic receptor blocker	Mild to moderate congestive heart failure
Nifedipine	L-type calcium channel blocker	Hypertension, chronic stable angina
Atorvastatin	HMG-CoA reductase inhibitor	Cholesterol reduction
Provastatin	HMG-CoA reductase inhibitor	Cholesterol reduction
Tranilast	Anti-inflammatory	Anti-allergen (Japan, South Korea)
Pirfenidone	Anti-inflammatory	Treatment of idiopathic pulmonary fibrosis

the induction of type I collagen gene expression, and in deoxycorticosterone acetate (DOCA)-salt rats, reduced perivascular and interstitial fibrosis, resulting in decreased LV stiffness [52, 53]. In a transverse aortic constriction (TAC) model of pressure overload, captopril reduced interstitial collagen (relative to myocyte number) to sham control levels [54]. Nephrectomy-induced LV hypertrophy in male rats was regressed with captopril treatment, as was fibrosis [55]. In a mouse model of chronic viral myocarditis, captopril significantly reduced CVF from~14 to 9.4% [56]. Conversely, captopril was unable to effectively prevent fibrosis in rat models of hypertrophy induced by isoproterenol, N-nitro-L-arginine methyl ester or hyperaldosteronism [57–59]. Thus captopril may be useful for altering cardiac fibrosis only in specific situations.

More promising is the modest regression of fibrosis in hypertensive heart disease patients by related ACE inhibitors such as lisinopril, which was effective at reducing fibrosis in the absence of reductions in blood pressure or LV hypertrophy, suggesting a potential direct effect on fibroblast function [3]. Endomyocardial biopsy samples from these patients showed a decrease in CVF from 6.9 to 6.3%, as well as reduced hydroxyproline concentration (a major component of collagen). These changes were accompanied by an improvement in diastolic function. Lisinopril was also shown to attenuate fibrosis in rat models of hypertension and aortic stenosis [60–62]. Enalapril regressed fibrosis in rats with chronic renal failure (CVF reduced from 2.6 to 1.6%), spontaneous hypertension (59% reduction in replacement fibrosis), chronic hypoxia-induced right ventricular hypertrophy (myocardial hydroxyproline reduced by 26%) and atrial fibrillation-induced HF (CVF reduced from 11.2 to 8.3%) [63–66]. A modest reduction in cardiac fibrosis (assessed by echocardiography) was observed with enalapril treatment in hypertensive patients with LV hypertrophy [67]. Enalapril was unable to reduce fibrosis in calcific aortic valve disease, and its effect on myocardial fibrosis was not examined in patients with LV dysfunction or aortic stenosis [68–70]. The efficacy of ACE inhibitors is counterbalanced by the potential of a patient to develop 'aldosterone escape', in which ACE inhibition is insufficient to reliably repress the release of downstream aldosterone [71]. In many cases, there is conflicting evidence of the effect of ACE inhibitors on cardiac fibrosis, and their effects are often dependent upon dual therapy, of which some combinations have shown increased risk of adverse events [72, 73].

4.1.2 Angiotensin II Receptor Antagonists

Another class of drugs used in the treatment of hypertension and related complications are the AT1 receptor blockers/antagonists (ARBs). ARBs directly induce vasodilation, reduce secretion of vasopressin, and decrease the downstream production and secretion of aldosterone. ARBs have also been shown to interfere with the progression of fibrosis. For example, after 12 months of treatment, losartan decreased CVF from 5.7 to 3.7%, an effect associated with a reduction in myocardial stiffness in patients with hypertensive heart disease independent of changes in blood pressure or LV mass [74]. Similarly, losartan decreased late gadolinium enhancement by 23% (compared to a 31% *increase* with placebo) in a small group of 20 patients with nonobstructive hypertrophic cardiomyopathy, and was also found to reduce myocardial fibrosis (assessed by echocardiography) in hypertensive patients with type 2 diabetes and LV hypertrophy [75, 76]. In a trial of patients with end-stage renal disease, losartan significantly reduced serum PICP levels after 6 months [77]. In post-MI patients, losartan was also found to reduce PIIINP serum levels and improve LV function, in combination with the ACE inhibitor perindopril [78]. The anti-fibrotic action of losartan may be due at least in part to its inhibition of type I collagen synthesis [79]. Candesartan has also been shown to be effective in reducing myocardial fibrosis in a clinical trial of hypertensive patients with LV hypertrophy, as assessed by echocardiography [67]. In patients with atrial fibrillation, candesartan reduced serum PIIINP levels after 24 months, though the relationship with fibrosis in this setting is unclear [80]. Irbesartan, which has dual actions of AT1 receptor blockade and peroxisome proliferator-activated receptor-γ (PPARγ) activation, has been shown to be useful in reducing fibrosis in transgenic mice over-expressing AngII. In these mice, irbesartan decreased CVF, reduced expression of TGF-β, CTGF, and collagens, and reduced phosphorylation of the growth factor pathway mediator extracellular signal-related kinase (ERK) [81]. Since irbesartan also increases PPARγ expression, these receptors may be involved in preventing the progression of fibrosis. This is further supported by the observation that fenofibrate, an agonist of the related PPARα, also decreased fibrosis in hypertensive rat models [82–85]. Of the ARBs currently in use, losartan shows the greatest potential for the directed treatment of cardiac fibrosis.

4.1.3 Aldosterone Antagonists or Anti-mineralocorticoids

Targeting the effects of aldosterone has proven to be effective in improving LV function. Initial studies by Weber et al. showed that chronic infusion of aldosterone augmented myocardial fibrosis, and fibrosis is evident in hyperaldosteronism [86, 87]. Tissue-specific activation of aldosterone has been observed in the infarcted myocardium, and a positive feedback loop appears to exist between aldosterone and ACE-mediated AngII production [88].

Aldosterone antagonists are diuretic drugs whose mechanism of action involves blockade of aldosterone binding to its mineralocorticoid receptors in the kidney, and thus are also called anti-mineralocorticoids or MRAs. As with ACE inhibitors and ARBs, MRAs are commonly used in the treatment of chronic heart failure to reduce peripheral and peritoneal edema as well as cardiac workload. In patients with heart failure or myocardial infarction (MI), spironolactone was found to reduce serum levels of PIIINP and improve LV ejection fraction [89–92]. Similarly, PICP levels were reduced with spironolactone treatment in stroke survivors after only one month and in patients with metabolic syndrome after 6 months [93, 94]. Both PIIINP and PICP levels were reduced by spironolactone after 6 months in obese patients with impaired LV function [95]. In subclinical diabetic cardiomyopathy, spironolactone did not produce significant changes in serum levels of PIIINP or PICP, though LV function was improved [96]. Spironolactone drastically reduced mortality in NYHA class II to III heart failure, while concomitantly increasing nitric oxide bioavailability, which may contribute to its anti-fibrotic effects [97]. A recent *in vitro* experiment demonstrated a direct anti-fibrotic effect of spironolactone in rat cardiac fibroblasts [98]. Spironolactone ameliorated increased levels of hydroxyproline, lysyl oxidase (LOX), CTGF, and microRNA-21 stimulated by AngII treatment in these cells. However, spironolactone was incapable of preventing LOX- or Ras homolog gene family member A (RhoA)-induced up-regulation of CTGF expression, suggesting that this drug plays a pathway-specific role in preventing cardiac fibrosis [98]. Canrenone, an active metabolite of spironolactone, was shown to reduce perivascular and interstitial collagen (by 47 and 34%, respectively) in a rat post-MI heart failure model at a high dose (28 mg/kg/day), but had no effect when administered at a low dose (8 mg/kg/day) [99]. Further research into the mechanisms of spironolactone and canrenone will identify differences in their molecular targets. In aldosterone/salt-induced hypertensive rats, the related MRA, eplerenone, negated the development of perivascular fibrotic lesions [100].

Overall, MRAs circumvent the problem of aldosterone escape observed in some cases of ACE inhibition and ARB therapy, and greatly improve morbidity and mortality in advanced cases of heart failure. Spironolactone and canrenone represent potentially useful tools in directed treatment of cardiac fibrosis. Clinical trials examining the use of MRAs in combination with other promising RAAS-targeting agents would provide further insight into the potential of these drugs in treating cardiac fibrosis.

4.1.4   Direct Renin Inhibitors

A more recent approach in treating HF and associated conditions such as fibrosis is the direct targeting of renin, the rate-limiting enzyme in RAAS activation. The direct renin inhibitor (DRI) aliskerin was found to attenuate collagen deposition by over 40% in a non-hypertrophic mouse model of fibrosis, even at levels that were suboptimal for blood pressure reduction, and reduced expression of collagen type I in cultured fibroblasts [101]. Aliskerin was also able to rescue the phenotype of transgenic tissue renin (Ren2) over-expressing mice by reducing hypertrophy and myocardial fibrosis [102]. However, in a rat DOCA-salt model of hypertension, aliskerin failed to reduce CVF [103].

Evidence presented by some of the aforementioned studies suggests that there may be some additional effects of certain ACE inhibitors and ARBs that act independently of blood pressure to reduce, and even regress, the progression of myocardial fibrosis. However, the mechanisms of action by which these effects are achieved have yet to be determined and will require further study on a molecular level, as current mechanistic knowledge is largely limited to vascular effects. Dual therapy of RAAS-targeting drugs is also a common trend in treatment, with over 200,000 patients in the USA treated with dual blockade, the majority of which include an ACE inhibitor and an ARB [104]. However, due to the variety of diuretic and vasodilatory effects of these drugs, the risks of this type of therapy may in some cases outweigh the potential anti-fibrotic benefits due to increased incidence of hyperkalemia, hypotension, and renal failure compared to monotherapy [105].

4.2   Diuretics

Despite increased extracellular fluid volume accompanying heart failure, inappropriate RAAS activation results in increased tubular reabsorption of sodium and water, causing hypertension and edema [106]. Different classes of diuretics exist which form the basis of pharmacological treatment for decreasing edema in heart failure [23, 106]. *In vivo* studies have provided evidence of anti-fibrotic actions of diuretics independent of their blood pressure-lowering effects [23].

The loop diuretic torasemide has been shown to reduce fibrosis in both myocardial biopsies from heart failure patients (NYHA class II to IV) as well as in rat models of HF [107–109]. In a small clinical trial of 36 patients with NYHA class II to IV congestive heart failure, torasemide, but not furosemide, was able to reduce CVF by nearly 50% (from 8.0 to 4.5%) and reduce serum PICP after 8 months [107]. These results were substantiated by the much larger Torasemide (Prolonged Release) on Myocardial Fibrosis in Patients With Heart Failure (TORAFIC) study, which demonstrated a significant decrease in serum PICP levels in 77 patients with mild (NYHA class II) HF—an effect not observed with furosemide [110]. These studies highlight torasemide, which is already used clinically in HF management, as an attractive novel therapeutic agent for cardiac fibrosis.

4.3   The Sympathetic Nervous System

Chronic activation of the sympathetic nervous system causes significant alterations in cardiac function and structure, and is implicated in a number of cardiovascular diseases. Inappropriate sympathetic activation in heart failure is evidenced by increased sympathetic outflow, increased levels of plasma norepinephrine (NE) and NE spillover from activated cardiorenal sympathetic nerve terminals to the circulation [111, 112]. In addition to its vasopressor effects through $\alpha 1$ and $\alpha 2$ adrenergic receptors, NE induces increased synthesis of fibrillar collagens I and III in both rat hearts and in primary cardiac fibroblasts, and acts synergistically with TGF-β via the mitogen-activated protein kinase (MAPK) pathway to induce expression of ECM remodeling genes (including collagen I, fibronectin, and plasminogen activator-inhibitor 1) [113, 114]. Further evidence of the role of sympathetic activation in fibrosis is provided by the observation that sympathectomy of both spontaneously hypertensive rats and rats with pressure induced overload prevents LV hypertrophy and fibrosis [115, 116]. However, blocking sympathetic activation does not consistently prevent fibrosis, and may in fact promote its progression through unknown mechanisms.

4.3.1   Beta-Adrenergic Blockade

In addition to tachycardia and cardiac inotropy, chronic sympathetic activation in HF acts in a positive feedback loop with RAAS to worsen the progression of hypertrophy and fibrosis [117]. The release of renin is stimulated through a $\beta 1$-receptor-

mediated mechanism, and β receptor activation has been shown to increase angiotensin II release in mesenteric arteries [118, 119]. Consequently, AngII facilitates
the natriuretic effects of renal sympathetic nerves [120]. Sympathetic activation also
augments RAAS indirectly via modulation of other peptide mediators, including en-
dothelin-1, atrial natriuretic peptide (which exerts anti-fibrotic effects), and adrenomedullin [121–124].

Three generations of beta blockers are used clinically. First generation beta
blockers such as propranolol are non-selective and poorly tolerated, and thus are
rarely used currently in treating HF [125]. The majority of animal studies with propranolol have failed to demonstrate any anti-fibrotic effects [52, 126, 127]. Second
generation beta blockers are selective for β1 receptors, and are well tolerated [128,
129]. Metoprolol is a second generation beta blocker that has been shown to have
varying effects in animal models of cardiac remodeling and was unable to inhibit
the expression of pro-fibrotic genes in cultured fibroblasts [130–139]. Third generation beta blockers such as carvedilol and bucindolol are non-selective but have additional properties (including vasodilation) that increase their efficacy and tolerability
in hypertension-related fibrosis [140]. Carvedilol is the preferred beta-blocker for
patients with chronic HF, and evidence suggests it may have anti-fibrotic properties
[141, 142]. Numerous animal models of post-MI remodeling, hypertension, dilated
cardiomyopathy, and LV hypertrophy have demonstrated reductions in myocardial
fibrosis with carvedilol treatment [133, 137, 139, 143–147]. In pressure-overloaded
rats, carvedilol reduced myocardial expression of fibronectin and fibrillar collagens,
as well as reduced the proliferation of fibroblasts cultured from these hearts [139].
In fibroblasts from the hearts of post-MI rats, carvedilol was again found to reduce
fibrillar collagen expression via Smad3 inhibition, and also reduced α-smooth muscle
actin (α-SMA) expression, while up-regulating microRNA-29b [148]. In human cardiac fibroblasts, carvedilol, but not metoprolol or propranolol, decreased stimulation
of platelet-derived growth factor (PDGF) receptors by recombinant PDGF-BB [149].
In addition, carvedilol is able to block L-type calcium channel activation and prevent
endothelin-1 and PDGF-induced proliferation in vascular smooth muscle cells [150,
151].

The divergent effects of some beta blockers on cardiac fibrosis may be due to their
ability to act as 'biased ligands.' In addition to blocking activation of β-adrenergic
receptors, beta blockers are also capable of inducing responses through G-protein-
independent and β-arrestin-dependent pathways that may contribute to cardiac fibrosis [152]. Thus, the majority of evidence does not support beta blockade for the
directed treatment of cardiac fibrosis, and its use may obscure the efficacy of other
HF treatments [153]. However, the evidence of beneficial effects with carvedilol is
considerable, and it appears to amplify the anti-fibrotic effects of dual therapy with
RAAS-targeting agents [137, 154–157]. Carvedilol may thus indeed be useful in
treating cardiac fibrosis, especially as part of dual therapy.

4.3.2 Alpha-Adrenergic Blockade

Studies examining the role of α1-adrenergic receptors in fibrosis are restricted to
animal models which have produced conflicting results. In transgenic ABKO mice
lacking the cardiac α1-adrenergic receptor, TAC-induced hypertrophy resulted in an

approximately 3-fold increase in LV interstitial collagen (26%) compared to wildtype controls (7–10%), as well as an increased number of fibroblasts [158]. However, the degree of perivascular and right ventricular fibrosis was not significantly different in the ABKO TAC and wild type TAC mice. Conversely, mice over-expressing cardiac α1 receptors showed an increase in cardiac mRNA levels of pro-fibrotic factors thrombospondin-1, tenascin C and CTGF, though TGF-β levels remained unchanged [159]. The α1 receptor antagonist doxazosin was found to reduce cardiac collagen deposition in TAC-induced pressure overload, and reduced aortic collagen synthesis in spontaneously hypertensive rats [116, 160]. Obviously, further research is needed to better define the effect of α-adrenergic activity on cardiac fibrosis.

4.4 Calcium Channel Blockers

Calcium channel blockers (CCBs) are commonly used in the treatment of hypertension, acting mainly as vasodilators to reduce systemic blood pressure. Recent evidence suggests they may also be effective in reducing myocardial fibrosis. Beyond their well-demonstrated roles in cardiomyocytes and vascular smooth muscle cells, calcium channels also appear to regulate critical fibroblast functions, including AngII-induced proliferation [161]. The L-type CCB amlodipine has been found to modestly reduce CVF (by ~3%) in the LV of hypertensive rat models, and CVF was also reduced by amlodipine in hamsters with dilated cardiomyopathy [162–164]. In animal MI models, reduction of scar collagen by amlodipine was limited to pre-treatment with the drug prior to induction of MI. Treatment post-MI actually increased CVF, perivascular fibrosis, scar length, and wall thickness [165–168]. In rats treated with amlodipine post-MI, there was decreased myofibroblast apoptosis, concomitant with a decrease in activation of the pro-apoptotic B-cell lymphoma-2 (Bcl-2) family member, Bcl-2-associated death promoter (Bad) [168]. Clinically, amlodipine has been unsuccessful in reducing cardiac fibrosis or type I collagen metabolism in patients with hypertension and LV hypertrophy [74]. Patients with mild to moderate hypertension showed no changes in serum levels of PICP, PINP, or aldosterone with amlodipine treatment [169]. Similarly, in hypertensive patients with type 2 diabetes and LV hypertrophy, amlodipine was unable to reduce myocardial fibrosis (assessed by echocardiography) [76, 170]. However, although amlodipine is ineffective as a monotherapy, it has shown promising results for treating LV hypertrophy and fibrosis in combination with ACE inhibitors and statins [171–176]. Other L-type CCBs have also demonstrated anti-fibrotic effects [177]. Administration of nifedipine in Dahl salt-sensitive hypertensive rats reduced cardiac collagen I:III mRNA ratio, MMP2 activity, as well as interstitial and perivascular fibrosis by almost 50%, even in the absence of blood pressure reduction [178]. Similarly, in a rat model of renal hypertension, nifedipine reduced CVF in the LV (from 12.3 to 5.6%) and right ventricle (from 20.6 to 10.4%) [179]. Similar results were achieved in other studies with animal models of hypertension [180–183]. Studies with non-selective CCBs are limited to animal models of MI and pressure overload-induced hypertrophy, but of these, mibefradil has shown potential in terms of cardiac CVF and hydroxyproline reduction [74, 165, 184–186]. However, mibefradil was removed from the market in 1998 due to potentially harmful interactions with other drugs.

The anti-fibrotic potential of CCBs in animal models of hypertension and MI may be due to a direct effect of these drugs on myocardial fibroblast and myofibroblast activities. Examination of tissue sections from the ventricles of AngII/aldosterone-infused hypertensive rats showed a decrease in α-SMA with mibefradil treatment [74]. Similarly, nifedipine reduced cardiac fibroblast proliferation and conversion to the myofibroblast phenotype in AngII-treated cells [187]. In this setting, nifedipine also decreased the production of reactive oxygen species and prevented induction of fibronectin and CTGF expression. The anti-fibrotic effects of nifedipine in cardiac fibroblasts may be a result of inhibited activation of the profibrotic pathway mediators ERK1/2 and c-Jun N-terminal kinase (JNK). In combination with olmesartan, nifedipine inhibited MMP9 activation in diabetic mice with hypoxia-induced LV remodeling [188]. Additionally, nifedipine was found to activate PPARγ in cultured vascular smooth muscle cells [182]. Hence, nifedipine likely inhibits fibrosis by regulating various pathways in fibroblasts and other cell types. Overall, nifedipine represents an attractive strategy for the treatment of cardiac fibrosis. In addition to calcium channels, non-selective transient receptor potential channels (TRPCs) have been recently implicated in fibroblast-driven cardiac fibrosis, though no TRPC blocking drugs currently exist [189].

4.5   Statins

Statins are inhibitors of HMG-CoA reductase, and are used to lower serum cholesterol in patients at risk for cardiovascular disease. However, recent evidence suggests a pleiotropic role of statins, including anti-fibrotic effects. Statins have been demonstrated to inhibit AngII-mediated reactive oxygen species (ROS) production and inhibition of atrial myofibroblast proliferation *in vitro* (via reduction of RhoA activation) [190, 191]. In an Ang-II-transgenic mouse model of organ failure, cerivastatin was able to inhibit increased deposition of fibrotic markers such as collagen and fibronectin, alongside reductions in blood pressure and hypertrophy [192]. Similarly, cerivastatin reduced van Gieson staining for myocardial fibrosis by~67% in Dahl salt-sensitive rats fed a high-salt diet, attenuated increases in pro-inflammatory interleukins, and decreased both cardiac hypertrophy and cardiomyocyte apoptosis [193]. In spontaneously hypertensive rats, both simvastatin and pravastatin prevented the induction of reactive fibrosis, as assessed by ventricular collagen staining [194]. Simvastatin also reduced perivascular fibrosis and cardiomyocyte apoptosis in rats heterozygous for adrenomedullin (an anti-fibrotic peptide hormone) treated with AngII and a high-salt diet [195]. In a mouse model of metabolic syndrome, a 3 month treatment with rosuvastatin significantly reduced CVF to wild-type levels, but produced no significant changes in cardiomyocyte hypertrophy. Rosuvastatin treatment was also associated with decreased fibrotic markers in cultured fibroblasts, including TGF-β1, α-SMA, pro-collagen type I, and LOX [196]. Additionally, a recent study in a rat model of post-MI heart failure showed that atorvastatin reduced ventricular fibrosis as well as expression of type I and III collagen mRNA. Interestingly, this study found that although MMP-2 levels were unchanged between untreated and atorvastatin-treated rats, post-MI fibrosis was associated with a significant decrease in MMP-2/TIMP-2 ratio, which was attenuated with atorvastatin [197].

In vitro treatment with atorvastatin and pravastatin demonstrated that statins are capable of reducing AngII- or TGF-β-induced cardiac fibroblast proliferation, collagen production, CTGF expression, and both canonical (Smad-mediated) and non-canonical (MAPK pathway-mediated) TGF-β1 signaling [198–200]. Thus, it appears that statins have an inhibitory effect on cardiac fibrosis via direct regulation of cardiac fibroblast function.

4.6   Inhibition of Transforming Growth Factor-β

The induction of TGF-β expression is a cardinal feature of injury and fibrosis in a variety of tissues, including the heart, lungs, kidneys, and pancreas [201–207]. TGF-β normally exists in a matrix-bound, latent form, and is activated in response to cardiac injury through the activity of extracellular proteases, ROS production, and mechanical disruption of the matrix itself, likely through direct activation of cell-matrix adhesion signaling molecules such as integrins [208–215]. An overwhelming body of evidence has shown that TGF-β is a critical mediator of cardiac fibrosis through various mediators such as CTGF, and induces activation, proliferation, motility, pro-fibrotic gene expression and survival of fibroblasts and myofibroblasts [216–218]. Due to its prolific effects in driving the fibrotic process, many drug companies have synthesized inhibitors of TGF-β as novel treatments for cardiac fibrosis and nearly every component in its pathway has been identified as a potential target. The approaches utilized include anti-ligand antisense oligonucleotides and RNA, ligand-competitive peptides (ligand traps), and antibodies targeting ligand, receptor, or associated proteins. Models of cardiac fibrosis in post-MI rats have shown promise for the inhibition of TGF-β receptor types I and II, which are efficient in preventing cardiac hypertrophy, dysfunction, and the development of fibrosis by reducing myofibroblast accumulation and collagen synthesis [219–221]. However, while some TGFβ ligand-inhibiting antibodies have shown safety and efficacy in systemic sclerosis and fibrosis of other organs, to date there have been no clinical trials testing direct TGF-β inhibition specifically in cardiac fibrosis. In addition, some currently available drugs used in treating heart failure and fibrosis have also been found to inhibit TGF-β signaling, such as the ARBs losartan and candesartan, as well as pirfenidone, which has recently been approved for treatment of idiopathic pulmonary fibrosis in Europe [222]. Conversely some common drugs, such as aspirin, may increase serum levels of TGF-β and contribute to the progression of fibrosis in these diseases [223].

Targeting such a pleiotropic pathway may ultimately result in a higher risk to benefit ratio than is feasible for using such drugs in clinical therapy of cardiac fibrosis, but there may still be potential for drugs targeting more specific downstream mediators of TGF-β and other pro-fibrotic pathways. For example, the p38 MAPK inhibitor RWJ-67657 ameliorated post-MI fibrosis by repressing expression of fibrillar collagens and the myofibroblast marker α-SMA in cultured fibroblasts [224]. Imanitib mesylate, a specific receptor tyrosine kinase (RTK) inhibitor that is implicated in inhibition of PDGF and TGF-β pathways, reduced perivascular and interstitial fibrosis, collagen I and III mRNA, and PDGF receptor β activation in spontaneously hypertensive rats [225]. On the other hand, the RTK inhibitor PF-04254644, which specifically targets the hepatocyte growth factor receptor pathway, was found to increase myocardial fibrosis and dysfunction with repeated dosing [226]. Rho kinase inhibition by fasudil reduces interstitial fibrosis in mouse models of pressure overload (TAC) and MI via inhibition of the TGF-β1-TAK (TGF-β-activated kinase) pathway [227]. Fasudil was also able to attenuate fibrosis in an excessive endurance exercise-induced rat model of cardiac hypertrophy, concomitant with improvements in LV function [228].

In addition to TGF-β, other growth factors such as CTGF and PDGF have been proposed as potential targets for anti-fibrotic drugs [8, 9, 229–232]. Although these may be less pleiotropic than TGF-β, large gaps remain in our understanding of the full spectrum of their effects in the heart and other organs, thus the potential existence of significant, adverse off-target effects cannot be ignored.

4.7 Anti-inflammatory Agents

The inflammatory response is a common phenomenon in cardiac injury, and has been shown to contribute to fibrogenesis. However, the inflammatory phase of cardiac injury post-MI, which is characterized by cardiomyocyte necrosis, leukocyte infiltration, and the release of pro-inflammatory cytokines and chemokines, precedes the initial healing phase that eventually progresses to a fibrotic state. Nonetheless, the anti-allergic pharmacological agent tranilast has been demonstrated to reduce hydroxyproline content in rat models of renal hypertension and diabetic cardiomyopathy [233, 234]. In a canine model of atrial fibrosis, tranilast significantly reduced atrial expression of TIMP-1 and TGF-β1, and decreased fibrotic tissue area to 1.4% compared to 9.3% in untreated animals [235]. Pirfenidone is a small molecule drug which is highly promising for the treatment of idiopathic pulmonary fibrosis (IPF), and is the first drug licensed for its treatment [236, 237]. Both in vivo (TAC-induced pressure overload) and in vitro (rat cardiac fibroblast) studies with pirfenidone indicate that it reduced fibrosis via reduction of NLRP3 (Nod-like receptor pyrin domain containing 3) inflammasome activation, which is implicated in the development of fibrosis [238, 239]. These results indicate a direct anti-inflammatory effect of pirfenidone in the reduction of myocardial fibrosis secondary to pressure overload.

5 Future Directions

The treatment strategies for cardiac fibrosis and related heart failure are presently aimed at treating the underlying causes (i.e. hypertension and hypertrophy), and there currently exist no effective methods for eliminating or reversing the remodeling process in the cardiac extracellular matrix, although RAAS-targeting drugs (specifically lisinopril and losartan) show promise by mechanisms that remain poorlydefined. Currently, patients with heart failure concomitant with cardiac fibrosis will be scheduled to receive heart transplants, for which they must wait several months. A heart transplant is a highly invasive and risky procedure, especially considering the demographic of these patients. In addition, despite the availability of effective anti-hypertensive drugs used in heart failure patients with LV hypertrophy, many patients still have uncontrolled high blood pressure which is refractory to treatment. Controlling the underlying condition, in addition to inhibiting or reversing the fibrotic process, is the ultimate goal for research and clinical development in the treatment of cardiac fibrosis in heart failure. In this regard, a number of current avenues of research are showing promise for the development of novel, targeted anti-fibrotics in a variety of tissues, and which may be effective in the heart (Table 3). It should be noted that the research directions described here do not comprise an exhaustive list; due to space limitations, factors and pathways for which only single studies currently exist have been excluded, but this does not diminish their potential consideration for anti-fibrotic therapeutic development in future.

5.1 Endothelin-1

Endothelin-1 (ET-1) is a highly potent and long-lasting vasoconstrictor up-regulated in a variety of cardiovascular diseases. In addition to the heart, ET-1 is up-regulated in other fibrotic tissues including the lung and kidney [240, 241]. Increased levels of ET-1 are also seen in response to tissue damage, hypoxia, and induction of pro-fibrotic TGF-β and AngII. ET-1 is a highly pro-fibrotic agent, enhancing collagen expression in myocardial fibroblasts and smooth muscle cells, and inducing myofibroblast phenoconversion from resident fibroblasts, bone marrow-derived monocytes, vascular pericytes, as well as epithelial and endothelial mesenchymal transition [242–254]. In addition to promoting the myofibroblast phenotype, ET-1 has also been shown to increase myofibroblast resistance to apoptosis in the setting of post-MI scarring [255]. Molecular studies indicate that ET-1 mediates fibrosis by altering the expression and activity of genes involved in cell adhesion and by increasing expression of connective tissue growth factor (CTGF) and TGF-β1 [256]. However, although endothelin antagonism in animal models of hypertension and MI initially showed promise in reducing cardiac fibrosis, in HF patients ET-1 receptor blockers of various types have failed to alter LV remodelling [257, 258].

Table 3 Novel prospective drug targets in cardiac fibrosis

Pro-fibrotic targets	Proposed mechanism
Lysyl oxidase	Up-regulation of CTGF; collagen maturation
Connective tissue growth factor (CTGF)	Unclear, but can act via integrin signaling to promote a pro-fibrotic environment
Platelet derived growth factor (PDGF)	Cardiac fibroblast proliferation, myofibroblast activation, increased type I collagen synthesis, TGF-β synthesis, decreased MMP:TIMP ratio
Nod-like receptor protein-3 (NLRP3) inflammasome	Caspase-1 activation
Endothelin-1 (ET-1)	Altered cell adhesion, increased expression of TGF-β and CTGF
microRNA-21	Sprouty-1 inhibition, ERK/MAPK pathway activation
Scleraxis	Increases synthesis of fibrillar collagens; synergy with TGF-β/Smad signaling
Transient receptor potential channels (TRPCs)	ERK pathway activation
Anti-fibrotic targets	*Proposed mechanism*
MicroRNA-29b	Reduced α-SMA and collagen type I & III synthesis
Cyclic adenosine monophosphate (cAMP)	Reduced myofibroblast phenoconversion, TGF-β signaling
Estrogen receptor-β	Increased cAMP signaling
Peroxisome proliferator-activated receptor-γ	Unknown
Atrial natriuretic peptide (ANP)	Inhibition of expression of collagen type III, CTGF, PAI-1, ET-1, tenascin-C

5.2   MicroRNAs

MicroRNAs (miRs) are small, non-coding RNAs that regulate post-transcriptional gene expression. Several miRs have been identified as regulators of proteins involved in myocardial fibrosis, often targeting multiple players in a given pathway [259–262]. Inhibition of miRs is accomplished by the use of antagomiRs—complementary oligonucleotides that block micro-RNA binding to its target transcripts. The temporal and spatial specificity gene regulation by miRs has made these molecules an attractive new target for the development of fibrotic therapies. However, the modality by which these miRs are delivered is a critical factor in drug development, as tiny LNA (locked nucleic acid) inhibitors are quickly cleared by the kidney, and miR networks likely have compensatory mechanisms in the case of miR-specific knockout animals that may confound results [263].

The majority of miRs implicated in fibrogenesis are usually down-regulated in disease, suggesting that they act as physiological repressors of pro-fibrotic genes. In the heart, miR-21 is expressed mainly by cardiac fibroblasts and its expression is augmented in fibrosis. In transgenic mice over-expressing the β1-adrenergic receptor, treatment with an antagomiR against miR-21 prevented the development of fibrosis. It was determined that miR-21 augments the ERK/MAPK pathway in these mice via inhibition of mRNA translation of the Sprouty1 mRNA, whose product is an inhibitor of this pro-fibrotic pathway [264]. The miR-29 family members have been validated as direct repressors of fibrotic genes such as collagens via *in vitro* luciferase assays. Increased miR-29b expression induced repression of fibrillar collagens and α-SMA in post-MI rat cardiac fibroblasts [148]. Systemic antagomiR-29b infusion has been shown to increase expression of fibrillar collagens type I and III, as well as the matrix proteins fibrillin-1 and elastin-1 [265]. Loss of miR-29b has been associated with the development of fibrosis in the heart, lungs, and kidneys via a Smad3-dependent mechanism of TGF-β signaling [266–268]. In AngII-induced hypertension in mice, knockdown of miR-29b enhanced, while over-expression inhibited, cardiac fibrosis [266]. Over-expression of miR-29b was associated with ~50% reduction in CVF, as well as decreased mRNA and protein levels of α-SMA and type I collagen compared to wild-type controls.

Antagonism of other miRs, including miR-24, -132, and -214, has been shown to reduce cardiac fibrosis following MI or TAC, whereas over-expression of miRs 101, 206, and miR-1 was found to decrease fibrosis post-insult. The role of other cardiac microRNAs such as miR-24 remains unclear due to conflicting results in knockdown and over-expression studies [269].

5.3   Targeting the Extracellular Matrix

A highly attractive prospect in reducing or reversing cardiac fibrosis is targeting of the myofibroblast phenotype. Since mechanical strain and tissue stiffness are key physical factors mediating the appearance of the myofibroblast in the stressed myocardium, it has been proposed that altering the activity of matrix-degrading enzymes may aid in regressing the myofibroblast phenotype in fibrosis [218, 270]. It is also possible that by reducing the rigidity of the cardiac matrix, blood flow may be improved, allowing the influx of molecules that would further the regression of fibrotic lesions and improve cardiomyocyte survival. Identifying direct modulators of myofibroblast genes like α-SMA, the ED-A splice variant of fibronectin, vimentin, and the embryonic form of smooth muscle myosin heavy chain (SMemb), as well as fibrillar collagens type I and III, will allow for the development of highly specific drug targets [4, 271–275]. One such putative target is the pro-fibrotic transcription factor scleraxis, which directly transactivates the promoter of collagen Iα2 [276, 277]. Matrix disorganization and excess collagen deposition resulting from dysregulation of this protein in other tissues supports the potential for an important role for scleraxis in mediating the fibrotic process [278–281].

6   Summary

The secret to ameliorating cardiac fibrosis may perhaps lie in the treatment of fibrosis of other organs in the body. Although there exists a great deal of tissue heterogeneity in terms of cell phenotype and pathophysiology of fibrosis, there is evidence to suggest that similar mechanisms may underlie the pathology in multiple organ systems. Common features of fibrogenesis in various tissue types include epithelial responses to injury (release of pro-inflammatory cytokines and chemokines such as interleukins 13 and 17), increased TGF-β signaling, and the central and requisite role of the myofibroblast [282]. The second messenger cyclic AMP has demonstrated anti-fibrotic effects in various tissues, and its increased expression has been directly correlated with preventing the phenoconversion of fibroblasts to myofibroblasts [283–285]. Thus, it is possible that treatments for one disease may be applicable to other tissues, such as the aforementioned pirfenidone. When selecting targets for therapy of any disease, researchers must also consider the potential effects it may have on other organs and body systems. An important area that is lacking in our current animal models that likely inhibits translation to humans is the comparison of gender differences—an issue that is brought to the forefront when considering the efficacy of female sex hormones in preventing disease. For example, the estrogen receptor-β (ER-β) agonist, 17β-estradiol, attenuated AngII-induced hypertrophy and fibrosis, an effect which was abrogated in ER-β knockout mice [286].

Although some of the current treatments for hypertension, hypertrophy, and heart failure have been shown to ameliorate the progression of fibrosis, there still remains no actual therapy directed specifically at interfering with the cellular and molecular mechanisms responsible for cardiac fibrosis. Drugs that have been shown to provide beneficial effects in reducing fibrosis in other organs, such as pirfenidone, are now on the radar. However the mechanisms of action of most of these pharmacological agents remain unknown and the true extent of their effects on the heart and other organs remains to be determined. Additionally, the salutary effects on fibrosis have been relatively modest to date, and an ideal drug target should result in reversal of the condition rather than simply slowing its progression. The increasing number of elderly patients and incidence of heart disease calls for a targeted and highly efficacious treatment for cardiac fibrosis. Some candidates for drug targets that appear to be specific to the fibrotic process are making their way down the pipeline, including microRNAs and direct transcriptional regulators such as scleraxis, and the development of more sensitive and explicit biomarkers and imaging techniques will be invaluable for the assessment of novel therapeutic strategies. As noted by Friedman et al., we are finally "nearing the starting line" in developing effective treatments for fibrosis—but the future looks promising indeed [282].

Acknowledgements This work was supported by an Open Operating Grant (MOP-106671) to MPC from the Canadian Institutes for Health Research, and by a graduate studentship to PLR from the Manitoba Health Research Council.

References

1. Chaturvedi RR, Herron T, Simmons R, Shore D, Kumar P, Sethia B, Chua F, Vassiliadis E, Kentish JC (2010) Passive stiffness of myocardium from congenital heart disease and implications for diastole. Circulation 121(8):979–988
2. Biernacka A, Frangogiannis NG (2011) Aging and cardiac fibrosis. Aging Dis 2(2):158–173
3. Brilla CG, Funck RC, Rupp H (2000) Lisinopril-mediated regression of myocardial fibrosis in patients with hypertensive heart disease. Circulation 102(12):1388–1393
4. Hinz B, Gabbiani G (2010) Fibrosis: recent advances in myofibroblast biology and new therapeutic perspectives. F1000 Biol Rep 2:78
5. Weber KT (2000) Targeting pathological remodeling: concepts of cardioprotection and reparation. Circulation 102(12):1342–1345
6. Sanderson JE, Lai KB, Shum IO, Wei S, Chow LT (2001) Transforming growth factor-beta(1) expression in dilated cardiomyopathy. Heart 86(6):701–708
7. Daniels A, van Bilsen M, Goldschmeding R, van der Vusse GJ, van Nieuwenhoven FA (2009) Connective tissue growth factor and cardiac fibrosis. Acta Physiol (Oxf) 195(3):321–338
8. Zhao W, Zhao T, Huang V, Chen Y, Ahokas RA, Sun Y (2011) Platelet-derived growth factor involvement in myocardial remodeling following infarction. J Mol Cell Cardiol 51(5):830–838
9. Zhao T, Zhao W, Chen Y, Li VS, Meng W, Sun Y (2013) Platelet-derived growth factor-D promotes fibrogenesis of cardiac fibroblasts. Am J Physiol Heart Circ Physiol 304(12):H1719–H1726
10. Anderson KR, Sutton MG, Lie JT (1979) Histopathological types of cardiac fibrosis in myocardial disease. J Pathol 128(2):79–85
11. Weber KT (2000) Fibrosis and hypertensive heart disease. Curr Opin Cardiol 15(4):264–272
12. Isoyama S, Nitta-Komatsubara Y (2002) Acute and chronic adaptation to hemodynamic overload and ischemia in the aged heart. Heart Fail Rev 7(1):63–69
13. Hasenfuss G (1998) Animal models of human cardiovascular disease, heart failure and hypertrophy. Cardiovasc Res 39(1):60–76
14. Maron BJ, Epstein SE, Roberts WC (1979) Hypertrophic cardiomyopathy and transmural myocardial infarction without significant atherosclerosis of the extramural coronary arteries. Am J Cardiol 43(6):1086–1102
15. Kearney PM, Whelton M, Reynolds K, Muntner P, Whelton PK, He J (2005) Global burden of hypertension: analysis of worldwide data. Lancet 365(9455):217–223

16. Kannel WB, Cobb J (1992) Left ventricular hypertrophy and mortality-results from the Framingham Study. Cardiology 81(4/5):291–298
17. Chobanian AV, Bakris GL, Black HR, Cushman WC, Green LA, Izzo JL Jr, Jones DW, Materson BJ, Oparil S, Wright JT Jr, Roccella EJ (2003) The seventh report of the joint national committee on prevention, detection, evaluation, and treatment of high blood pressure: the JNC 7 report. JAMA 289(19):2560–2572
18. Yamasaki N, Kitaoka H, Matsumura Y, Furuno T, Nishinaga M, Doi Y (2003) Heart failure in the elderly. Intern Med 42(5):383–388
19. Rossi MA (1998) Pathologic fibrosis and connective tissue matrix in left ventricular hypertrophy due to chronic arterial hypertension in humans. J Hypertens 16(7):1031–1041
20. Arteaga E, de Araujo AQ, Bernstein M, Ramires FJ, Ianni BM, Fernandes F, Mady C (2009) Prognostic value of the collagen volume fraction in hypertrophic cardiomyopathy. Arq Bras Cardiol 92(3):210–214, 216–220
21. Almaas VM, Haugaa KH, Strom EH, Scott H, Dahl CP, Leren TP, Geiran OR, Endresen K, Edvardsen T, Aakhus S, Amlie JP (2013) Increased amount of interstitial fibrosis predicts ventricular arrhythmias, and is associated with reduced myocardial septal function in patients with obstructive hypertrophic cardiomyopathy. Europace 15(9):1319–1327
22. Burwell CS, Robin ED (1959) Diagnosis of diffuse myocardial fibrosis. Circulation 20:606–614
23. McKelvie RS, Moe GW, Ezekowitz JA, Heckman GA, Costigan J, Ducharme A, Estrella-Holder E, Giannetti N, Grzeslo A, Harkness K, Howlett JG, Kouz S, Leblanc K, Mann E, Nigam A, O'Meara E, Rajda M, Steinhart B, Swiggum E, Le VV, Zieroth S, Arnold JM, Ashton T, D'Astous M, Dorian P, Haddad H, Isaac DL, Leblanc MH, Liu P, Rao V, Ross HJ, Sussex B (2013) The 2012 Canadian Cardiovascular Society heart failure management guidelines update: focus on acute and chronic heart failure. Can J Cardiol 29(2):168–181
24. Burlew BS, Weber KT (2002) Cardiac fibrosis as a cause of diastolic dysfunction. Herz 27(2):92–98
25. Diez J (2008) Diagnosis and treatment of myocardial fibrosis in hypertensive heart disease. Circ J 72(Suppl A):A8–A12
26. Plastiras SC, Kelekis N, Tzelepis GE (2006) Magnetic resonance imaging for the detection of myocardial fibrosis in scleroderma. N Engl J Med 354(20):2194–2196
27. Stuckey DJ, McSweeney SJ, Thin MZ, Habib J, Price AN, Fiedler LR, Gsell W, Prasad SK, Schneider MD (2014) T1 mapping detects pharmacological retardation of diffuse cardiac fibrosis in mouse pressure-overload hypertrophy. Circ Cardiovas Imaging 7(2):240–249
28. Won S, Davies-Venn C, Liu S, Bluemke DA (2013) Noninvasive imaging of myocardial extracellular matrix for assessment of fibrosis. Curr Opin Cardiol 28(3):282–289
29. Vohringer M, Mahrholdt H, Yilmaz A, Sechtem U (2007) Significance of late gadolinium enhancement in cardiovascular magnetic resonance imaging (CMR). Herz 32(2):129–137
30. Moon JC, Prasad SK (2005) Cardiovascular magnetic resonance and the evaluation of heart failure. Curr Cardiol Rep 7(1):39–44
31. Tandri H, Saranathan M, Rodriguez ER, Martinez C, Bomma C, Nasir K, Rosen B, Lima JA, Calkins H, Bluemke DA (2005) Noninvasive detection of myocardial fibrosis in arrhythmogenic right ventricular cardiomyopathy using delayed-enhancement magnetic resonance imaging. J Am Coll Cardiol 45(1):98–103
32. Shirani J, Dilsizian V (2007) Molecular imaging in heart failure. Curr Opin Biotechnol 18(1):65–72
33. Ghann WE, Aras O, Fleiter T, Daniel MC (2012) Syntheses and characterization of lisinopril-coated gold nanoparticles as highly stable targeted CT contrast agents in cardiovascular diseases. Langmuir 28(28):10398–10408
34. Femia FJ, Maresca KP, Hillier SM, Zimmerman CN, Joyal JL, Barrett JA, Aras O, Dilsizian V, Eckelman WC, Babich JW (2008) Synthesis and evaluation of a series of 99mTc(CO)3 + lisinopril complexes for *in vivo* imaging of angiotensin-converting enzyme expression. J Nucl Med 49(6):970–977

35. de Haas HJ, Arbustini E, Fuster V, Kramer CM, Narula J (2014) Molecular imaging of the cardiac extracellular matrix. Circ Res 114(5):903–915
36. Hadjipanayis CG, Jiang H, Roberts DW, Yang L (2011) Current and future clinical applications for optical imaging of cancer: from intraoperative surgical guidance to cancer screening. Semin Oncol 38(1):109–118
37. Chen J, Tung CH, Allport JR, Chen S, Weissleder R, Huang PL (2005) Near-infrared fluorescent imaging of matrix metalloproteinase activity after myocardial infarction. Circulation 111(14):1800–1805
38. Lopez B, Gonzalez A, Beaumont J, Querejeta R, Larman M, Diez J (2007) Identification of a potential cardiac antifibrotic mechanism of torasemide in patients with chronic heart failure. J Am Coll Cardiol 50(9):859–867
39. Querejeta R, Varo N, Lopez B, Larman M, Artinano E, Etayo JC, Martinez Ubago JL, Gutierrez-Stampa M, Emparanza JI, Gil MJ, Monreal I, Mindan JP, Diez J (2000) Serum carboxyterminal propeptide of procollagen type I is a marker of myocardial fibrosis in hypertensive heart disease. Circulation 101(14):1729–1735
40. Lijnen PJ, Maharani T, Finahari N, Prihadi JS (2012) Serum collagen markers and heart failure. Cardiovasc Hematol Disord Drug Targets 12(1):51–55
41. Lombardi R, Betocchi S, Losi MA, Tocchetti CG, Aversa M, Miranda M, D'Alessandro G, Cacace A, Ciampi Q, Chiariello M (2003) Myocardial collagen turnover in hypertrophic cardiomyopathy. Circulation 108(12):1455–1460
42. Toprak G, Yuksel H, Demirpence O, Islamoglu Y, Evliyaoglu O, Mete N (2013) Fibrosis in heart failure subtypes. Eur Rev Med Pharmacol Sci 17(17):2302–2309
43. Tromp J, van der Pol A, Klip IT, de Boer RA, Jaarsma T, van Gilst WH, Voors AA, van Veldhuisen DJ, van der Meer P (2014) The fibrosis marker syndecan-1 and outcome in heart failure patients with reduced and preserved ejection fraction. Circ Heart Fail 7:457–462
44. Lopez-Andres N, Rossignol P, Iraqi W, Fay R, Nuee J, Ghio S, Cleland JG, Zannad F, Lacolley P (2012) Association of galectin-3 and fibrosis markers with long-term cardiovascular outcomes in patients with heart failure, left ventricular dysfunction, and dyssynchrony: insights from the CARE-HF (Cardiac Resynchronization in Heart Failure) trial. Eur J Heart Fail 14(1):74–81
45. Lin YH, Lin LY, Wu YW, Chien KL, Lee CM, Hsu RB, Chao CL, Wang SS, Hsein YC, Liao LC, Ho YL, Chen MF (2009) The relationship between serum galectin-3 and serum markers of cardiac extracellular matrix turnover in heart failure patients. Clinica Chimica Acta 409(1/2):96–99
46. Kawasaki T, Sakai C, Harimoto K, Yamano M, Miki S, Kamitani T (2013) Usefulness of high-sensitivity cardiac troponin T and brain natriuretic peptide as biomarkers of myocardial fibrosis in patients with hypertrophic cardiomyopathy. Am J Cardiol 112(6):867–872
47. Kaye DM, Khammy O, Mariani J, Maeder MT (2013) Relationship of circulating matrix biomarkers to myocardial matrix metabolism in advanced heart failure. Eur J Heart Fail 15(3):292–298
48. Zannad F (2014) What is measured by cardiac fibrosis biomarkers and imaging? Circ Heart Fail 7(2):239–242
49. Lopez B, Gonzalez A, Querejeta R, Larman M, Diez J (2006) Alterations in the pattern of collagen deposition may contribute to the deterioration of systolic function in hypertensive patients with heart failure. J Am Coll Cardiol 48(1):89–96
50. Zannad F, Alla F, Dousset B, Perez A, Pitt B (2000) Limitation of excessive extracellular matrix turnover may contribute to survival benefit of spironolactone therapy in patients with congestive heart failure: insights from the randomized aldactone evaluation study (RALES). Rales Investigators. Circulation 102(22):2700–2706
51. Sanders GD, Coeytaux R, Dolor RJ, Hasselblad V, Patel UD, Powers B, Yancy WS, Gray RN, Irvine RJ, Kendrick A (2011) Angiotensin-converting enzyme inhibitors (ACEIs), angiotensin ii receptor antagonists (ARBs), and direct renin inhibitors for treating essential hypertension: an update. AHRQ comparative effectiveness reviews. Rockville (MD)

52. Ziegelhoffer-Mihalovicova B, Arnold N, Marx G, Tannapfel A, Zimmer HG, Rassler B (2006) Effects of salt loading and various therapies on cardiac hypertrophy and fibrosis in young spontaneously hypertensive rats. Life Sci 79(9):838–846
53. Brown L, Duce B, Miric G, Sernia C (1999) Reversal of cardiac fibrosis in deoxycorticoste-rone acetate-salt hypertensive rats by inhibition of the renin-angiotensin system. J Am Soc Nephrol JASN 10(Suppl 11):S143–S148
54. Rossi MA, Peres LC (1992) Effect of captopril on the prevention and regression of myocar-dial cell hypertrophy and interstitial fibrosis in pressure overload cardiac hypertrophy. Am Heart J 124(3):700–709
55. Rahman M, Kim SJ, Kim JS, Kim SZ, Lee YU, Kang HS (2010) Myocardial calcification and hypertension following chronic renal failure and ameliorative effects of furosemide and captopril. Cardiology 116(3):194–205
56. Guo C, Wang Y, Liang H, Zhang J (2010) ADAMTS-1 contributes to the antifibrotic effect of captopril by accelerating the degradation of type I collagen in chronic viral myocarditis. Eur J Pharmacol 629(1–3):104–110
57. Gallego M, Espina L, Vegas L, Echevarria E, Iriarte MM, Casis O (2001) Spironolactone and captopril attenuates isoproterenol-induced cardiac remodelling in rats. Pharmacol Res 44(4):311–315
58. Simko F, Pechanova O, Pelouch V, Krajcirovicova K, Celec P, Palffy R, Bednarova K, Vrankova S, Adamcova M, Paulis L (2010) Continuous light and L-NAME-induced left ven-tricular remodelling: different protection with melatonin and captopril. J Hypertens 28(Suppl 1):S13–S18
59. Brilla CG, Matsubara LS, Weber KT (1993) Anti-aldosterone treatment and the prevention of myocardial fibrosis in primary and secondary hyperaldosteronism. J Mol Cell Cardiol 25(5):563–575
60. Brilla CG, Matsubara L, Weber KT (1996) Advanced hypertensive heart disease in spontane-ously hypertensive rats. Lisinopril-mediated regression of myocardial fibrosis. Hypertension 28(2):269–275
61. Liang B, Leenen FH (2007) Prevention of salt induced hypertension and fibrosis by angioten-sin converting enzyme inhibitors in Dahl S rats. Br J Pharmacol 152(6):903–914
62. Goncalves G, Zornoff LA, Ribeiro HB, Okoshi MP, Cordaro FR, Okoshi K, Padovani CR, Aragon FF, Cicogna AC (2005) Blockade of renin-angiotensin system attenuates cardiac re-modeling in rats undergoing aortic stenosis. Arq Bras Cardiol 84(4):304–308
63. Tyralla K, Adamczak M, Benz K, Campean V, Gross ML, Hilgers KF, Ritz E, Amann K (2011) High-dose enalapril treatment reverses myocardial fibrosis in experimental uremic cardiomyopathy. PloS ONE 6(1):e15287
64. Pahor M, Bernabei R, Sgadari A, Gambassi G Jr, Giudice P L, Pacifici L, Ramacci MT, La-grasta C, Olivetti G, Carbonin P (1991) Enalapril prevents cardiac fibrosis and arrhythmias in hypertensive rats. Hypertension 18(2):148–157
65. Pelouch V, Kolar F, Ost'adal B, Milerova M, Cihak R, Widimsky J (1997) Regression of chronic hypoxia-induced pulmonary hypertension, right ventricular hypertrophy, and fibro-sis: effect of enalapril. Cardiovasc Drugs Ther 11(2):177–185
66. Shi Y, Li D, Tardif JC, Nattel S (2002) Enalapril effects on atrial remodeling and atrial fibril-lation in experimental congestive heart failure. Cardiovasc Res 54(2):456–461
67. Ciulla MM, Paliotti R, Esposito A, Cuspidi C, Muiesan ML, Rosei EA, Magrini F, Zanchetti A (2009) Effects of antihypertensive treatment on ultrasound measures of myocardial fibro-sis in hypertensive patients with left ventricular hypertrophy: results of a randomized trial comparing the angiotensin receptor antagonist, candesartan and the angiotensin-converting enzyme inhibitor, enalapril. J Hypertens 27(3):626–632
68. Cote N, Mahmut A, Fournier D, Boulanger MC, Couture C, Despres JP, Trahan S, Bosse Y, Page S, Pibarot P, Mathieu P (2014) Angiotensin receptor blockers are associated with reduced fibrosis and interleukin-6 expression in calcific aortic valve disease. Pathobiology J Immunopathol Mol Cell Biol 81(1):15–24

69. Vermes E, Tardif JC, Bourassa MG, Racine N, Levesque S, White M, Guerra PG, Ducharme A (2003) Enalapril decreases the incidence of atrial fibrillation in patients with left ventricular dysfunction: insight from the studies of left ventricular dysfunction (SOLVD) trials. Circulation 107(23):2926–2931
70. Chockalingam A, Venkatesan S, Subramaniam T, Jagannathan V, Elangovan S, Alagesan R, Gnanavelu G, Dorairajan S, Krishna BP, Chockalingam V (2004) Safety and efficacy of angiotensin-converting enzyme inhibitors in symptomatic severe aortic stenosis: symptomatic cardiac obstruction-pilot study of enalapril in aortic stenosis (SCOPE-AS). Am Heart J 147(4):E19
71. Prakash ES (2005) "Aldosterone escape" or refractory hyperaldosteronism? MedGenMed 7(3):25
72. SoRelle R (2003) Two better than one. Circulation 108(15):e9042–e9043
73. Lakhdar R, Al-Mallah MH, Lanfear DE (2008) Safety and tolerability of angiotensin-converting enzyme inhibitor versus the combination of angiotensin-converting enzyme inhibitor and angiotensin receptor blocker in patients with left ventricular dysfunction: a systematic review and meta-analysis of randomized controlled trials. J Card Fail 14(3):181–188
74. Diez J, Querejeta R, Lopez B, Gonzalez A, Larman M, Martinez Ubago JL (2002) Losartan-dependent regression of myocardial fibrosis is associated with reduction of left ventricular chamber stiffness in hypertensive patients. Circulation 105(21):2512–2517
75. Shimada YJ, Passeri JJ, Baggish AL, O'Callaghan C, Lowry PA, Yannekis G, Abbara S, Ghoshhajra BB, Rothman RD, Ho CY, Januzzi JL, Seidman CE, Fifer MA (2013) Effects of losartan on left ventricular hypertrophy and fibrosis in patients with nonobstructive hypertrophic cardiomyopathy. JACC Heart Fail 1(6):480–487
76. Fogari R, Mugellini A, Destro M, Corradi L, Lazzari P, Zoppi A, Preti P, Derosa G (2012) Losartan and amlodipine on myocardial structure and function: a prospective, randomized, clinical trial. Diabet Med 29(1):24–31
77. Shibasaki Y, Nishiue T, Masaki H, Tamura K, Matsumoto N, Mori Y, Nishikawa M, Matsubara H, Iwasaka T (2005) Impact of the angiotensin II receptor antagonist, losartan, on myocardial fibrosis in patients with end-stage renal disease: assessment by ultrasonic integrated backscatter and biochemical markers. Hypertens Res 28(10):787–795
78. Li L, Liu RY, Zhao XY, Zhang JY, Jia M, Lu PQ (2009) Effects of combination therapy with perindopril and losartan on left ventricular remodelling in patients with myocardial infarction. Clin Exp Pharmacol Physiol 36(7):704–710
79. Lopez B, Querejeta R, Varo N, Gonzalez A, Larman M, Martinez Ubago JL, Diez J (2001) Usefulness of serum carboxy-terminal propeptide of procollagen type I in assessment of the cardioreparative ability of antihypertensive treatment in hypertensive patients. Circulation 104(3):286–291
80. Kawamura M, Ito H, Onuki T, Miyoshi F, Watanabe N, Asano T, Tanno K, Kobayashi Y (2010) Candesartan decreases type III procollagen-N-peptide levels and inflammatory marker levels and maintains sinus rhythm in patients with atrial fibrillation. J Cardiovasc Pharmacol 55(5):511–517
81. Zhang ZZ, Shang QH, Jin HY, Song B, Oudit GY, Lu L, Zhou T, Xu YL, Gao PJ, Zhu DL, Penninger JM, Zhong JC (2013) Cardiac protective effects of irbesartan via the PPAR-gamma signaling pathway in angiotensin-converting enzyme 2-deficient mice. J Transl Med 11:229
82. Ogata T, Miyauchi T, Sakai S, Takanashi M, Irukayama-Tomobe Y, Yamaguchi I (2004) Myocardial fibrosis and diastolic dysfunction in deoxycorticosterone acetate-salt hypertensive rats is ameliorated by the peroxisome proliferator-activated receptor-alpha activator fenofibrate, partly by suppressing inflammatory responses associated with the nuclear factor-kappa-B pathway. J Am Coll Cardiol 43(8):1481–1488
83. Iglarz M, Touyz RM, Viel EC, Paradis P, Amiri F, Diep QN, Schiffrin EL (2003) Peroxisome proliferator-activated receptor-alpha and receptor-gamma activators prevent cardiac fibrosis in mineralocorticoid-dependent hypertension. Hypertension 42(4):737–743

84. Diep QN, Benkirane K, Amiri F, Cohn JS, Endemann D, Schiffrin EL (2004) PPAR alpha activator fenofibrate inhibits myocardial inflammation and fibrosis in angiotensin II-infused rats. J Mol Cell Cardiol 36(2):295–304

85. Ogata T, Miyauchi T, Sakai S, Irukayama-Tomobe Y, Goto K, Yamaguchi I (2002) Stimulation of peroxisome-proliferator-activated receptor alpha (PPAR alpha) attenuates cardiac fibrosis and endothelin-1 production in pressure-overloaded rat hearts. Clin Sci (Lond) 103(Suppl 48):284S–288S

86. Weber KT, Brilla CG (1991) Pathological hypertrophy and cardiac interstitium. Fibrosis and renin-angiotensin-aldosterone system. Circulation 83(6):1849–1865

87. Sanderson JE, Cockram CS, Yu CM, Campbell SE, Weber KT (1996) Myocardial fibrosis and hyperaldosteronism. Eur Heart J 17(11):1761–1762

88. Silvestre JS, Heymes C, Oubenaissa A, Robert V, Aupetit-Faisant B, Carayon A, Swynghedauw B, Delcayre C (1999) Activation of cardiac aldosterone production in rat myocardial infarction: effect of angiotensin II receptor blockade and role in cardiac fibrosis. Circulation 99(20):2694–2701

89. Macdonald JE, Kennedy N, Struthers AD (2004) Effects of spironolactone on endothelial function, vascular angiotensin converting enzyme activity, and other prognostic markers in patients with mild heart failure already taking optimal treatment. Heart 90(7):765–770

90. MacFadyen RJ, Barr CS, Struthers AD (1997) Aldosterone blockade reduces vascular collagen turnover, improves heart rate variability and reduces early morning rise in heart rate in heart failure patients. Cardiovasc Res 35(1):30–34

91. Hayashi M, Tsutamoto T, Wada A, Tsutsui T, Ishii C, Ohno K, Fujii M, Taniguchi A, Hamatani T, Nozato Y, Kataoka K, Morigami N, Ohnishi M, Kinoshita M, Horie M (2003) Immediate administration of mineralocorticoid receptor antagonist spironolactone prevents post-infarct left ventricular remodeling associated with suppression of a marker of myocardial collagen synthesis in patients with first anterior acute myocardial infarction. Circulation 107(20):2559–2565

92. Cicoira M, Zanolla L, Rossi A, Golia G, Franceschini L, Brighetti G, Marino P, Zardini P (2002) Long-term, dose-dependent effects of spironolactone on left ventricular function and exercise tolerance in patients with chronic heart failure. J Am Coll Cardiol 40(2):304–310

93. Wong KY, Wong SY, McSwiggan S, Ogston SA, Sze KY, MacWalter RS, Struthers AD (2013) Myocardial fibrosis and QTc are reduced following treatment with spironolactone or amiloride in stroke survivors: a randomised placebo-controlled cross-over trial. Int J Cardiol 168(6):5229–5233

94. Kosmala W, Przewlocka-Kosmala M, Szczepanik-Osadnik H, Mysiak A, O'Moore-Sullivan T, Marwick TH (2011) A randomized study of the beneficial effects of aldosterone antagonism on LV function, structure, and fibrosis markers in metabolic syndrome. JACC Cardiovasc Imaging 4(12):1239–1249

95. Kosmala W, Przewlocka-Kosmala M, Szczepanik-Osadnik H, Mysiak A, Marwick TH (2013) Fibrosis and cardiac function in obesity: a randomised controlled trial of aldosterone blockade. Heart 99(5):320–326

96. Jellis CL, Sacre JW, Wright J, Jenkins C, Haluska B, Jeffriess L, Martin J, Marwick TH (2014) Biomarker and imaging responses to spironolactone in subclinical diabetic cardiomyopathy. Eur Heart J Cardiovasc Imaging 15:776–786

97. Farquharson CA, Struthers AD (2000) Spironolactone increases nitric oxide bioactivity, improves endothelial vasodilator dysfunction, and suppresses vascular angiotensin I/angiotensin II conversion in patients with chronic heart failure. Circulation 101(6):594–597

98. Lavall D, Selzer C, Schuster P, Lenski M, Adam O, Schaefers HJ, Boehm M, Laufs U (2014) The mineralocorticoid receptor promotes fibrotic remodeling in atrial fibrillation. J Biol Chem

99. Cittadini A, Casaburi C, Monti MG, Di Gianni A, Serpico R, Scherillo G, Saldamarco L, Vanasia M, Sacca L (2002) Effects of canrenone on myocardial reactive fibrosis in a rat model of postinfarction heart failure. Cardiovasc Drugs Ther 16(3):195–201

100. Struthers AD (2005) Pathophysiology of heart failure following myocardial infarction. Heart 91(Suppl 2):ii14–ii16. Discussion ii31, ii43–ii18

101. Zhi H, Luptak I, Alreja G, Shi J, Guan J, Metes-Kosik N, Joseph J (2013) Effects of direct renin inhibition on myocardial fibrosis and cardiac fibroblast function. PloS ONE 8(12):e81612

102. Whaley-Connell A, Habibi J, Rehmer N, Ardhanari S, Hayden MR, Pulakat L, Krueger C, Ferrario CM, DeMarco VG, Sowers JR (2013) Renin inhibition and AT(1)R blockade improve metabolic signaling, oxidant stress and myocardial tissue remodeling. Metabolism 62(6):861–872

103. Ma L, Hua J, He L, Li Q, Zhou J, Yu J (2012) Anti-fibrotic effect of aliskiren in rats with deoxycorticosterone induced myocardial fibrosis and its potential mechanism. Bosn J Basic Med Sci 12(2):69–73

104. Messerli FH, Staessen JA, Zannad F (2010) Of fads, fashion, surrogate endpoints and dual RAS blockade. EurHeart J 31(18):2205–2208

105. Makani H, Bangalore S, Desouza KA, Shah A, Messerli FH (2013) Efficacy and safety of dual blockade of the renin-angiotensin system: meta-analysis of randomised trials. BMJ 346:f360

106. Paul S (2002) Balancing diuretic therapy in heart failure: loop diuretics, thiazides, and aldosterone antagonists. Congest Heart Fail 8(6):307–312

107. Lopez B, Querejeta R, Gonzalez A, Sanchez E, Larman M, Diez J (2004) Effects of loop diuretics on myocardial fibrosis and collagen type I turnover in chronic heart failure. J Am Coll Cardiol 43(11):2028–2035

108. Veeraveedu PT, Watanabe K, Ma M, Thandavarayan RA, Palaniyandi SS, Yamaguchi K, Suzuki K, Kodama M, Aizawa Y (2008) Comparative effects of torasemide and furosemide in rats with heart failure. Biochem Pharmacol 75(3):649–659

109. Veeraveedu PT, Watanabe K, Ma M, Palaniyandi SS, Yamaguchi K, Suzuki K, Kodama M, Aizawa Y (2008) Torasemide, a long-acting loop diuretic, reduces the progression of myocarditis to dilated cardiomyopathy. Eur J Pharmacol 581(1/2):121–131

110. TORAFIC Investigators Group (2011) Effects of prolonged-release torasemide versus furosemide on myocardial fibrosis in hypertensive patients with chronic heart failure: a randomized, blinded-end point, active-controlled study. Clin Ther 33(9):1204–1213 (e1203)

111. Leimbach WN Jr, Wallin BG, Victor RG, Aylward PE, Sundlof G, Mark AL (1986) Direct evidence from intraneural recordings for increased central sympathetic outflow in patients with heart failure. Circulation 73(5):913–919

112. Hasking GJ, Esler MD, Jennings GL, Burton D, Johns JA, Korner PI (1986) Norepinephrine spillover to plasma in patients with congestive heart failure: evidence of increased overall and cardiorenal sympathetic nervous activity. Circulation 73(4):615–621

113. Briest W, Rassler B, Deten A, Zimmer HG (2003) Norepinephrine-induced cardiac hypertrophy and fibrosis are not due to mast cell degranulation. Mol Cell Biochem 252(1/2):229–237

114. Akiyama-Uchida Y, Ashizawa N, Ohtsuru A, Seto S, Tsukazaki T, Kikuchi H, Yamashita S, Yano K (2002) Norepinephrine enhances fibrosis mediated by TGF-beta in cardiac fibroblasts. Hypertension 40(2):148–154

115. Levick SP, Murray DB, Janicki JS, Brower GL (2010) Sympathetic nervous system modulation of inflammation and remodeling in the hypertensive heart. Hypertension 55(2):270–276

116. Perlini S, Palladini G, Ferrero I, Tozzi R, Fallarini S, Facoetti A, Nano R, Clari F, Busca G, Fogari R, Ferrari AU (2005) Sympathectomy or doxazosin, but not propranolol, blunt myocardial interstitial fibrosis in pressure-overload hypertrophy. Hypertension 46(5):1213–1218

117. Bos R, Mougenot N, Findji L, Mediani O, Vanhoutte PM, Lechat P (2005) Inhibition of catecholamine-induced cardiac fibrosis by an aldosterone antagonist. J Cardiovasc Pharmacol 45(1):8–13

118.   van Zwieten PA, de Jonge A (1986) Interaction between the adrenergic and renin-angioten-sin-aldosterone-systems. Postgrad Med J 62(Suppl 1):23–27
119.   Nakamaru M, Jackson EK, Inagami T (1986) Beta-adrenoceptor-mediated release of angio-tensin II from mesenteric arteries. Am J Physiol 250(1 Pt 2):H144–H148
120.   Hirsch AT, Pinto YM, Schunkert H, Dzau VJ (1990) Potential role of the tissue renin-angio-tensin system in the pathophysiology of congestive heart failure. Am J Cardiol 66(11):22D–30D. Discussion 30D-32D
121.   Pacher R, Bergler-Klein J, Globits S, Teufelsbauer H, Schuller M, Krauter A, Ogris E, Rodler S, Wutte M, Hartter E (1993) Plasma big endothelin-1 concentrations in congestive heart failure patients with or without systemic hypertension. Am J Cardiol 71(15):1293–1299
122.   Brandt RR, Wright RS, Redfield MM, Burnett JC Jr (1993) Atrial natriuretic peptide in heart failure. J Am Coll Cardiol 22(4 Suppl A):86A–92A
123.   Nishikimi T, Saito Y, Kitamura K, Ishimitsu T, Eto T, Kangawa K, Matsuo H, Omae T, Mat-suoka H (1995) Increased plasma levels of adrenomedullin in patients with heart failure. J Am Coll Cardiol 26(6):1424–1431
124.   Andreis PG, Neri G, Prayer-Galetti T, Rossi GP, Gottardo G, Malendowicz LK, Nussdorfer GG (1997) Effects of adrenomedullin on the human adrenal glands: an *in vitro* study. J Clin Endocrinol Metab 82(4):1167–1170
125.   Stephen SA (1966) Unwanted effects of propranolol. Am J Cardiol 18(3):463–472
126.   Su JZ, Chen SC, Wu KG, Chen DG, Rui HB, Wang XY, Wang HJ (1999) Effects of perin-dopril, propranolol, and dihydrochlorothiazide on cardiovascular remodelling in spontane-ously hypertensive rats. Zhongguo yao li xue bao 20(10):923–928
127.   Marano G, Palazzesi S, Vergari A, Catalano L, Gaudi S, Testa C, Canese R, Carpinelli G, Podo F, Ferrari AU (2003) Inhibition of left ventricular remodelling preserves chamber systolic function in pressure-overloaded mice. Pflugers Arch 446(4):429–436
128.   Waagstein F, Bristow MR, Swedberg K, Camerini F, Fowler MB, Silver MA, Gilbert EM, Johnson MR, Goss FG, Hjalmarson A (1993) Beneficial effects of metoprolol in idiopathic dilated cardiomyopathy. metoprolol in dilated cardiomyopathy (MDC) trial study group. Lancet 342(8885):1441–1446
129.   A randomized trial of beta-blockade in heart failure. The Cardiac Insufficiency Bisoprolol Study (CIBIS). CIBIS Investigators and Committees (1994). Circulation 90(4):1765–1773
130.   Li Y, Shi J, Yang BF, Liu L, Han CL, Li WM, Dong DL, Pan ZW, Liu GZ, Geng JQ, Sheng L, Tan XY, Sun DH, Gong ZH, Gong YT (2012) Ketamine-induced ventricular structural, sympathetic and electrophysiological remodelling: pathological consequences and protec-tive effects of metoprolol. Br J Pharmacol 165(6):1748–1756
131.   Chan V, Fenning A, Hoey A, Brown L (2011) Chronic beta-adrenoceptor antagonist treat-ment controls cardiovascular remodeling in heart failure in the aging spontaneously hyper-tensive rat. J Cardiovasc Pharmacol 58(4):424–431
132.   Latini R, Masson S, Jeremic G, Luvara G, Fiordaliso F, Calvillo L, Bernasconi R, Torri M, Rondelli I, Razzetti R, Bongrani S (1998) Comparative efficacy of a DA2/alpha2 agonist and a beta-blocker in reducing adrenergic drive and cardiac fibrosis in an experimental model of left ventricular dysfunction after coronary artery occlusion. J Cardiovasc Pharma-col 31(4):601–608
133.   Sun T, Shen LH, Chen H, Li HW, Guo CY, Li ZZ, Tang CS (2008) Effects of carvedilol and metoprolol on cardiac fibrosis in rats with experimental myocardial infarction. Zhonghua xin xue guan bing za zhi 36(1):68–71
134.   Huntgeburth M, Tiemann K, Shahverdyan R, Schluter KD, Schreckenberg R, Gross ML, Modersheim S, Caglayan E, Muller-Ehmsen J, Ghanem A, Vantler M, Zimmermann WH, Bohm M, Rosenkranz S (2011) Transforming growth factor beta(1) oppositely regulates the hypertrophic and contractile response to beta-adrenergic stimulation in the heart. PloS ONE 6(11):e26628
135.   Serpi R, Tolonen AM, Tenhunen O, Pievilainen O, Kubin AM, Vaskivuo T, Soini Y, Kerkela R, Leskinen H, Ruskoaho H (2009) Divergent effects of losartan and metoprolol on cardiac

remodeling, c-kit + cells, proliferation and apoptosis in the left ventricle after myocardial infarction. Clin Trans Sci 2(6):422–430

136. Seeland U, Schaffer A, Selejan S, Hohl M, Reil JC, Muller P, Rosenkranz S, Bohm M (2009) Effects of AT1- and beta-adrenergic receptor antagonists on TGF-beta1-induced fibrosis in transgenic mice. Eur J Clin Invest 39(10):851–859

137. Wei S, Chow LT, Sanderson JE (2000) Effect of carvedilol in comparison with metoprolol on myocardial collagen postinfarction. J Am Coll Cardiol 36(1):276–281

138. Becher PM, Lindner D, Miteva K, Savvatis K, Zietsch C, Schmack B, Van Linthout S, Westermann D, Schultheiss HP, Tschope C (2012) Role of heart rate reduction in the prevention of experimental heart failure: comparison between If-channel blockade and beta-receptor blockade. Hypertension 59(5):949–957

139. Grimm D, Huber M, Jabusch HC, Shakibaei M, Fredersdorf S, Paul M, Riegger GA, Kromer EP (2001) Extracellular matrix proteins in cardiac fibroblasts derived from rat hearts with chronic pressure overload: effects of beta-receptor blockade. J Mol Cell Cardiol 33(3):487–501

140. Feuerstein GZ, Ruffolo RR Jr (1996) Carvedilol, a novel vasodilating beta-blocker with the potential for cardiovascular organ protection. Eur Heart J 17(Suppl B):24–29

141. Keating GM, Jarvis B (2003) Carvedilol: a review of its use in chronic heart failure. Drugs 63(16):1697–1741

142. Kveiborg B, Major-Petersen A, Christiansen B, Torp-Pedersen C (2007) Carvedilol in the treatment of chronic heart failure: lessons from the carvedilol or metoprolol European trial. Vasc Health Risk Manag 3(1):31–37

143. Barone FC, Campbell WG Jr, Nelson AH, Feuerstein GZ (1998) Carvedilol prevents severe hypertensive cardiomyopathy and remodeling. J Hypertens 16(6):871–884

144. Nanjo S, Yamazaki J, Yoshikawa K, Ishii T, Togane Y (2006) Carvedilol prevents myocardial fibrosis in hamsters. Int Heart J 47(4):607–616

145. Watanabe K, Takahashi T, Nakazawa M, Wahed MI, Fuse K, Tanabe N, Kodama M, Aizawa Y, Ashino H, Tazawa S (2002) Effects of carvedilol on cardiac function and cardiac adrenergic neuronal damage in rats with dilated cardiomyopathy. J Nucl Med 43(4):531–535

146. Arozal W, Sari FR, Watanabe K, Arumugam S, Veeraveedu PT, Ma M, Thandavarayan RA, Sukumaran V, Lakshmanan AP, Kobayashi Y, Mito S, Soetikno V, Suzuki K (2011) Carvedilol-afforded protection against daunorubicin-induced cardiomyopathic rats *in vivo:* effects on cardiac fibrosis and hypertrophy. ISRN Pharmacol 2011:430549

147. Chen J, Huang C, Zhang B, Huang Q, Xu L (2013) The effects of carvedilol on cardiac structural remodeling: the role of endogenous nitric oxide in the activity of carvedilol. Mol Med Rep 7(4):1155–1158

148. Zhu JN, Chen R, Fu YH, Lin QX, Huang S, Guo LL, Zhang MZ, Deng CY, Zou X, Zhong SL, Yang M, Zhuang J, Yu XY, Shan ZX (2013) Smad3 inactivation and MiR-29b upregulation mediate the effect of carvedilol on attenuating the acute myocardium infarction-induced myocardial fibrosis in rat. PloS ONE 8(9):e75557

149. Lotze U, Heinke S, Fritzenwanger M, Krack A, Muller S, Figulla HR (2002) Carvedilol inhibits platelet-derived growth factor-induced signal transduction in human cardiac fibroblasts. J Cardiovasc Pharmacol 39(4):576–589

150. Nakajima T, Ma J, Iida H, Iwasawa K, Jo T, Omata M, Nagai R (2003) Inhibitory effects of carvedilol on calcium channels in vascular smooth muscle cells. Jpn Heart J 44(6):963–978

151. Sung CP, Arleth AJ, Ohlstein EH (1993) Carvedilol inhibits vascular smooth muscle cell proliferation. J Cardiovasc Pharmacol 21(2):221–227

152. Nakaya M, Chikura S, Watari K, Mizuno N, Mochinaga K, Mangmool S, Koyanagi S, Ohdo S, Sato Y, Ide T, Nishida M, Kurose H (2012) Induction of cardiac fibrosis by beta-blocker in G protein-independent and G protein-coupled receptor kinase 5/beta-arrestin2-dependent signaling pathways. J Biol Chem 287(42):35669–35677

153. Cavallari LH, Momary KM, Groo VL, Viana MA, Camp JR, Stamos TD (2007) Association of beta-blocker dose with serum procollagen concentrations and cardiac response to spironolactone in patients with heart failure. Pharmacotherapy 27(6):801–812

154. Sia YT, Parker TG, Tsoporis JN, Liu P, Adam A, Rouleau JL (2002) Long-term effects of carvedilol on left ventricular function, remodeling, and expression of cardiac cytokines after large myocardial infarction in the rat. J Cardiovasc Pharmacol 39(1):73–87

155. Khattar RS, Senior R, Soman P, van der Does R, Lahiri A (2001) Regression of left ventricular remodeling in chronic heart failure: comparative and combined effects of captopril and carvedilol. Am Heart J 142(4):704–713

156. Udelson JE (2004) Ventricular remodeling in heart failure and the effect of beta-blockade. Am J Cardiol 93(9A):43B–48B

157. Badheka AO, Tuliani TA, Rathod A, Ali Kizilbash M, Bharadwaj A, Afonso L (2012) Combined use of direct renin inhibitor and carvedilol in heart failure with preserved systolic function. Med Hypotheses 79(4):448–451

158. O'Connell TD, Swigart PM, Rodrigo MC, Ishizaka S, Joho S, Turnbull L, Tecott LH, Baker AJ, Foster E, Grossman W, Simpson PC (2006) Alpha1-adrenergic receptors prevent a maladaptive cardiac response to pressure overload. J Clin Invest 116(4):1005–1015

159. Pe'er J, Levinger S, Chirambo M, Ron N, Okon E (1991) Malignant fibrous histiocytoma of the skin and the conjunctiva in xeroderma pigmentosum. Arch Pathol Lab Med 115(9):910–914

160. Chichester CO, Rodgers RL (1987) Effects of doxazosin on vascular collagen synthesis, arterial pressure and serum lipids in the spontaneously hypertensive rat. J Cardiovasc Pharmacol 10(Suppl 9):S21–S26

161. Wang LP, Wang Y, Zhao LM, Li GR, Deng XL (2013) Angiotensin II upregulates K(Ca)3.1 channels and stimulates cell proliferation in rat cardiac fibroblasts. Biochem Pharmacol 85(10):1486–1494

162. Sevilla MA, Voces F, Carron R, Guerrero EI, Ardanaz N, San Roman L, Arevalo MA, Montero MJ (2004) Amlodipine decreases fibrosis and cardiac hypertrophy in spontaneously hypertensive rats: persistent effects after withdrawal. Life Sci 75(7):881–891

163. Takatsu M, Hattori T, Murase T, Ohtake M, Kato M, Nashima K, Nakashima C, Takahashi K, Ito H, Niinuma K, Aritomi S, Murohara T, Nagata K (2012) Comparison of the effects of cilnidipine and amlodipine on cardiac remodeling and diastolic dysfunction in Dahl salt-sensitive rats. J Hypertens 30(9):1845–1855

164. Watanabe M, Kawaguchi H, Onozuka H, Mikami T, Urasawa K, Okamoto H, Watanabe S, Abe K, Kitabatake A (1998) Chronic effects of enalapril and amlodipine on cardiac remodeling in cardiomyopathic hamster hearts. J Cardiovasc Pharmacol 32(2):248–259

165. Sandmann S, Claas R, Cleutjens JP, Daemen MJ, Unger T (2001) Calcium channel blockade limits cardiac remodeling and improves cardiac function in myocardial infarction-induced heart failure in rats. J Cardiovasc Pharmacol 37(1):64–77

166. Whittaker P, Zhang HP, Kloner RA (2000) Biphasic survival response to amlodipine after myocardial infarction in rats: association with cardiac vascular remodeling. Cardiovasc Pathol 9(2):85–93

167. Jugdutt BI, Menon V, Kumar D, Idikio H (2002) Vascular remodeling during healing after myocardial infarction in the dog model: effects of reperfusion, amlodipine and enalapril. J Am Coll Cardiol 39(9):1538–1545

168. Ogino A, Takemura G, Kanamori H, Okada H, Maruyama R, Miyata S, Esaki M, Nakagawa M, Aoyama T, Ushikoshi H, Kawasaki M, Minatoguchi S, Fujiwara T, Fujiwara H (2007) Amlodipine inhibits granulation tissue cell apoptosis through reducing calcineurin activity to attenuate postinfarction cardiac remodeling. Am J Physiol Heart Circ Physiol 293(4):H2271–H2280

169. Rajzer M, Klocek M, Kawecka-Jaszcz K (2003) Effect of amlodipine, quinapril, and losartan on pulse wave velocity and plasma collagen markers in patients with mild-to-moderate arterial hypertension. Am J Hypertens 16(6):439–444

170. Ishimitsu T, Kobayashi T, Honda T, Takahashi M, Minami J, Ohta S, Inada H, Yoshii M, Ono H, Matsuoka H (2005) Protective effects of an angiotensin II receptor blocker and a long-acting calcium channel blocker against cardiovascular organ injuries in hypertensive patients. Hypertens Res 28(4):351–359

171. Nakamura T, Fukuda M, Kataoka K, Nako H, Tokutomi Y, Dong YF, Yamamoto E, Yasuda O, Ogawa H, Kim-Mitsuyama S (2011) Eplerenone potentiates protective effects of amlodipine against cardiovascular injury in salt-sensitive hypertensive rats. Hypertens Res 34(7):817–824

172. Lu JC, Cui W, Zhang HL, Liu F, Han M, Liu DM, Yin HN, Zhang K, Du J (2009) Additive beneficial effects of amlodipine and atorvastatin in reversing advanced cardiac hypertrophy in elderly spontaneously hypertensive rats. Clin Exp Pharmacol Physiol 36(11):1110–1119

173. Ge CJ, Lu SZ, Chen YD, Wu XF, Hu SJ, Ji Y (2008) Synergistic effect of amlodipine and atorvastatin on blood pressure, left ventricular remodeling, and C-reactive protein in hypertensive patients with primary hypercholesterolemia. Heart Vessels 23(2):91–95

174. Martin-Ventura JL, Tunon J, Duran MC, Blanco-Colio LM, Vivanco F, Egido J (2006) Vascular protection of dual therapy (atorvastatin-amlodipine) in hypertensive patients. J Am Soc Nephrol JASN 17(12 Suppl 3):S189–S193

175. Jukema JW, van der Hoorn JW (2004) Amlodipine and atorvastatin in atherosclerosis: a review of the potential of combination therapy. Expert Opin Pharmacother 5(2):459–468

176. Kang BY, Wang W, Palade P, Sharma SG, Mehta JL (2009) Cardiac hypertrophy during hypercholesterolemia and its amelioration with rosuvastatin and amlodipine. J Cardiovasc Pharmacol 54(4):327–334

177. Susic D, Varagic J, Frohlich ED (1999) Pharmacologic agents on cardiovascular mass, coronary dynamics and collagen in aged spontaneously hypertensive rats. J Hypertens 17(8):1209–1215

178. Yamada T, Nagata K, Cheng XW, Obata K, Saka M, Miyachi M, Naruse K, Nishizawa T, Noda A, Izawa H, Kuzuya M, Okumura K, Murohara T, Yokota M (2009) Long-term administration of nifedipine attenuates cardiac remodeling and diastolic heart failure in hypertensive rats. Eur J Pharmacol 615(1–3):163–170

179. Brilla CG (2000) Regression of myocardial fibrosis in hypertensive heart disease: diverse effects of various antihypertensive drugs. Cardiovasc Res 46(2):324–331

180. Kimpara T, Okabe M, Nishimura H, Hayashi T, Imamura K, Kawamura K (1997) Ultrastructural changes during myocardial hypertrophy and its regression: long-term effects of nifedipine in adult spontaneously hypertensive rats. Heart Vessels 12(3):143–151

181. Amann K, Greber D, Gharehbaghi H, Wiest G, Lange B, Ganten U, Mattfeldt T, Mall G (1992) Effects of nifedipine and moxonidine on cardiac structure in spontaneously hypertensive rats. Stereological studies on myocytes, capillaries, arteries, and cardiac interstitium. Am J Hypertens 5(2):76–83

182. Hashimoto R, Umemoto S, Guo F, Umeji K, Itoh S, Kishi H, Kobayashi S, Matsuzaki M (2010) Nifedipine activates PPARgamma and exerts antioxidative action through Cu/Zn-SOD independent of blood-pressure lowering in SHRSP. J Atheroscler Thromb 17(8):785–795

183. Campbell SE, Turek Z, Rakusan K, Kazda S (1993) Cardiac structural remodelling after treatment of spontaneously hypertensive rats with nifedipine or nisoldipine. Cardiovasc Res 27(7):1350–1358

184. Mulder P, Richard V, Thuillez C (1998) Different effects of calcium antagonists in a rat model of heart failure. Cardiology 89(Suppl 1):33–37

185. Sandmann S, Bohle RM, Dreyer T, Unger T (2000) The T-type calcium channel blocker mibefradil reduced interstitial and perivascular fibrosis and improved hemodynamic parameters in myocardial infarction-induced cardiac failure in rats. Virchows Arch 436(2):147–157

186. Ramires FJ, Sun Y, Weber KT (1998) Myocardial fibrosis associated with aldosterone or angiotensin II administration: attenuation by calcium channel blockade. J Mol Cell Cardiol 30(3):475–483

187. Jia Y, Xu J, Yu Y, Guo J, Liu P, Chen S, Jiang J (2013) Nifedipine inhibits angiotensin II-induced cardiac fibrosis via downregulating Nox4-derived ROS generation and suppressing ERK1/2, JNK signaling pathways. Pharmazie 68(6):435–441

188. Yamashita C, Hayashi T, Mori T, Matsumoto C, Kitada K, Miyamura M, Sohmiya K, Ukimura A, Okada Y, Yoshioka T, Kitaura Y, Matsumura Y (2010) Efficacy of olmesartan and nifedipine on recurrent hypoxia-induced left ventricular remodeling in diabetic mice. Life Sci 86(9/10):322–330

189. Yue Z, Zhang Y, Xie J, Jiang J, Yue L (2013) Transient receptor potential (TRP) channels and cardiac fibrosis. Curr Top Med Chem 13(3):270–282

190. Delbosc S, Cristol JP, Descomps B, Mimran A, Jover B (2002) Simvastatin prevents angiotensin II-induced cardiac alteration and oxidative stress. Hypertension 40(2):142–147

191. Porter KE, Turner NA, O'Regan DJ, Balmforth AJ, Ball SG (2004) Simvastatin reduces human atrial myofibroblast proliferation independently of cholesterol lowering via inhibition of RhoA. Cardiovasc Res 61(4):745–755

192. Dechend R, Fiebeler A, Park JK, Muller DN, Theuer J, Mervaala E, Bieringer M, Gulba D, Dietz R, Luft FC, Haller H (2001) Amelioration of angiotensin II-induced cardiac injury by a 3-hydroxy-3-methylglutaryl coenzyme a reductase inhibitor. Circulation 104(5):576–581

193. Hasegawa H, Yamamoto R, Takano H, Mizukami M, Asakawa M, Nagai T, Komuro I (2003) 3-Hydroxy-3-methylglutaryl coenzyme A reductase inhibitors prevent the development of cardiac hypertrophy and heart failure in rats. J Mol Cell Cardiol 35(8):953–960

194. Bezerra DG, Mandarim-de-Lacerda CA (2005) Beneficial effect of simvastatin and pravastatin treatment on adverse cardiac remodelling and glomeruli loss in spontaneously hypertensive rats. Clin Sci (Lond) 108(4):349–355

195. Yamamoto C, Fukuda N, Jumabay M, Saito K, Matsumoto T, Ueno T, Soma M, Matsumoto K, Shimosawa T (2011) Protective effects of statin on cardiac fibrosis and apoptosis in adrenomedullin-knockout mice treated with angiotensin II and high salt loading. Hypertens Res 34(3):348–353

196. Hermida N, Markl A, Hamelet J, Van Assche T, Vanderper A, Herijgers P, van Bilsen M, Hilfiker-Kleiner D, Noppe G, Beauloye C, Horman S, Balligand JL (2013) HMGCoA reductase inhibition reverses myocardial fibrosis and diastolic dysfunction through AMP-activated protein kinase activation in a mouse model of metabolic syndrome. Cardiovasc Res 99(1):44–54

197. An Z, Yang G, He YQ, Dong N, Ge LL, Li SM, Zhang WQ (2013) Atorvastatin reduces myocardial fibrosis in a rat model with post-myocardial infarction heart failure by increasing the matrix metalloproteinase-2/tissue matrix metalloproteinase inhibitor-2 ratio. Chin Med J (Engl) 126(11):2149–2156

198. Chen J, Mehta JL (2006) Angiotensin II-mediated oxidative stress and procollagen-1 expression in cardiac fibroblasts: blockade by pravastatin and pioglitazone. Am J Physiol Heart Circ Physiol 291(4):H1738–H1745

199. Martin J, Denver R, Bailey M, Krum H (2005) In vitro inhibitory effects of atorvastatin on cardiac fibroblasts: implications for ventricular remodelling. Clin Exp Pharmacol Physiol 32(9):697–701

200. Shyu KG, Wang BW, Chen WJ, Kuan P, Hung CR (2010) Mechanism of the inhibitory effect of atorvastatin on endoglin expression induced by transforming growth factor-beta1 in cultured cardiac fibroblasts. Eur J Heart Fail 12(3):219–226

201. Leask A, Abraham DJ (2004) TGF-beta signaling and the fibrotic response. Faseb J 18(7):816-827.

202. Ihn H (2002) The role of TGF-beta signaling in the pathogenesis of fibrosis in scleroderma. Arch Immunol Ther Exp (Warsz) 50(5):325–331

203. Biernacka A, Dobaczewski M, Frangogiannis NG (2011) TGF-beta signaling in fibrosis. Growth Factors 29(5):196–202

204. Dobaczewski M, Chen W, Frangogiannis NG (2011) Transforming growth factor (TGF)-beta signaling in cardiac remodeling. J Mol Cell Cardiol 51(4):600–606

205. Verrecchia F, Mauviel A (2007) Transforming growth factor-beta and fibrosis. World J Gastroenterol 13(22):3056–3062

206. Fernandez IE, Eickelberg O (2012) The impact of TGF-beta on lung fibrosis: from targeting to biomarkers. Proc Am Thorac Soc 9(3):111–116 (023AW)

207. Tatler AL, Jenkins G (2012) TGF-beta activation and lung fibrosis. Proc Am Thorac Soc 9(3):130–136
208. Shi M, Zhu J, Wang R, Chen X, Mi L, Walz T, Springer TA (2011) Latent TGF-beta structure and activation. Nature 474(7351):343–349
209. Tatti O, Vehvilainen P, Lehti K, Keski-Oja J (2008) MT1-MMP releases latent TGF-beta1 from endothelial cell extracellular matrix via proteolytic processing of LTBP-1. Exp Cell Res 314(13):2501–2514
210. Maeda S, Dean DD, Gomez R, Schwartz Z, Boyan BD (2002) The first stage of transforming growth factor beta1 activation is release of the large latent complex from the extracellular matrix of growth plate chondrocytes by matrix vesicle stromelysin-1 (MMP-3). Calcif Tissue Int 70(1):54–65
211. Jain M, Rivera S, Monclus EA, Synenki L, Zirk A, Eisenbart J, Feghali-Bostwick C, Mutlu GM, Budinger GR, Chandel NS (2013) Mitochondrial reactive oxygen species regulate transforming growth factor-beta signaling. J Biol Chem 288(2):770–777
212. Purnomo Y, Piccart Y, Coenen T, Prihadi JS, Lijnen PJ (2013) Oxidative stress and transforming growth factor-beta1-induced cardiac fibrosis. Cardiovasc Hematol Disord Drug Targets 13(2):165–172
213. Buscemi L, Ramonet D, Klingberg F, Formey A, Smith-Clerc J, Meister JJ, Hinz B (2011) The single-molecule mechanics of the latent TGF-beta1 complex. Curr Biol 21(24):2046–2054
214. Wipff PJ, Hinz B (2008) Integrins and the activation of latent transforming growth factor beta1- an intimate relationship. Eur J Cell Biol 87(8/9):601–615
215. Hyytiainen M, Penttinen C, Keski-Oja J (2004) Latent TGF-beta binding proteins: extracellular matrix association and roles in TGF-beta activation. Crit Rev Clin Lab Sci 41(3):233–264
216. Chen MM, Lam A, Abraham JA, Schreiner GF, Joly AH (2000) CTGF expression is induced by TGF- beta in cardiac fibroblasts and cardiac myocytes: a potential role in heart fibrosis. J Mol Cell Cardiol 32(10):1805–1819. doi:10.1006/jmcc.2000.1215
217. Hinz B (2007) Formation and function of the myofibroblast during tissue repair. J Invest Dermatol 127(3):526–537
218. Hinz B (2010) The myofibroblast: paradigm for a mechanically active cell. J Biomech 43(1):146–155
219. Tan SM, Zhang Y, Connelly KA, Gilbert RE, Kelly DJ (2010) Targeted inhibition of activin receptor-like kinase 5 signaling attenuates cardiac dysfunction following myocardial infarction. Am J Physiol Heart Circ Physiol 298(5):H1415–H1425
220. Ikeuchi M, Tsutsui H, Shiomi T, Matsusaka H, Matsushima S, Wen J, Kubota T, Takeshita A (2004) Inhibition of TGF-beta signaling exacerbates early cardiac dysfunction but prevents late remodeling after infarction. Cardiovasc Res 64(3):526–535
221. Okada H, Takemura G, Kosai K, Li Y, Takahashi T, Esaki M, Yuge K, Miyata S, Maruyama R, Mikami A, Minatoguchi S, Fujiwara T, Fujiwara H (2005) Postinfarction gene therapy against transforming growth factor-beta signal modulates infarct tissue dynamics and attenuates left ventricular remodeling and heart failure. Circulation 111(19):2430–2437
222. Takeda Y, Tsujino K, Kijima T, Kumanogoh A (2014) Efficacy and safety of pirfenidone for idiopathic pulmonary fibrosis. Patient Prefer Adherence 8:361–370
223. Grainger DJ, Kemp PR, Metcalfe JC, Liu AC, Lawn RM, Williams NR, Grace AA, Schofield PM, Chauhan A (1995) The serum concentration of active transforming growth factor-beta is severely depressed in advanced atherosclerosis. Nat Med 1(1):74–79
224. See F, Thomas W, Way K, Tzanidis A, Kompa A, Lewis D, Itescu S, Krum H (2004) p38 mitogen-activated protein kinase inhibition improves cardiac function and attenuates left ventricular remodeling following myocardial infarction in the rat. J Am Coll Cardiol 44(8):1679–1689
225. Jang SW, Ihm SH, Choo EH, Kim OR, Chang K, Park CS, Kim HY, Seung KB (2014) Imatinib mesylate attenuates myocardial remodeling through inhibition of platelet-derived growth factor and transforming growth factor activation in a rat model of hypertension. Hypertension 63:1228–1234

226. Aguirre SA, Heyen JR, Collette W 3rd, Bobrowski W, Blasi ER (2010) Cardiovascular effects in rats following exposure to a receptor tyrosine kinase inhibitor. Toxicol Pathol 38(3):416–428
227. Li Q, Xu Y, Li X, Guo Y, Liu G (2012) Inhibition of Rho-kinase ameliorates myocardial remodeling and fibrosis in pressure overload and myocardial infarction: role of TGF-beta1-TAK1. Toxicol Lett 211(2):91–97
228. Ho TJ, Huang CC, Huang CY, Lin WT (2012) Fasudil, a Rho-kinase inhibitor, protects against excessive endurance exercise training-induced cardiac hypertrophy, apoptosis and fibrosis in rats. Eur J Appl Physiol 112(8):2943–2955
229. Tsoutsman T, Wang X, Garchow K, Riser B, Twigg S, Semsarian C (2013) CCN2 plays a key role in extracellular matrix gene expression in severe hypertrophic cardiomyopathy and heart failure. J Mol Cell Cardiol 62:164–178
230. Gu J, Liu X, Wang QX, Tan HW, Guo M, Jiang WF, Zhou L (2012) Angiotensin II increases CTGF expression via MAPKs/TGF-beta1/TRAF6 pathway in atrial fibroblasts. Exp Cell Res 318(16):2105–2115
231. Ruperez M, Lorenzo O, Blanco-Colio LM, Esteban V, Egido J, Ruiz-Ortega M (2003) Connective tissue growth factor is a mediator of angiotensin II-induced fibrosis. Circulation 108(12):1499–1505
232. Ponten A, Li X, Thoren P, Aase K, Sjoblom T, Ostman A, Eriksson U (2003) Transgenic overexpression of platelet-derived growth factor-C in the mouse heart induces cardiac fibrosis, hypertrophy, and dilated cardiomyopathy. Am J Pathol 163(2):673–682
233. Hocher B, Godes M, Olivier J, Weil J, Eschenhagen T, Slowinski T, Neumayer HH, Bauer C, Paul M, Pinto YM (2002) Inhibition of left ventricular fibrosis by tranilast in rats with renovascular hypertension. J Hypertens 20(4):745–751
234. Kagitani S, Ueno H, Hirade S, Takahashi T, Takata M, Inoue H (2004) Tranilast attenuates myocardial fibrosis in association with suppression of monocyte/macrophage infiltration in DOCA/salt hypertensive rats. J Hypertens 22(5):1007–1015
235. Nakatani Y, Nishida K, Sakabe M, Kataoka N, Sakamoto T, Yamaguchi Y, Iwamoto J, Mizumaki K, Fujiki A, Inoue H (2013) Tranilast prevents atrial remodeling and development of atrial fibrillation in a canine model of atrial tachycardia and left ventricular dysfunction. J Am Coll Cardiol 61(5):582–588
236. Woodcock HV, Maher TM (2014) The treatment of idiopathic pulmonary fibrosis. F1000 Prime Rep 6:16
237. Albera C, Ferrero C, Rindone E, Zanotto S, Rizza E (2013) Where do we stand with IPF treatment? Respir Res 14(Suppl 1):S7
238. Roche P, Czubryt MP (2013) Pirfenidone and the inflammasome: getting to the heart of cardiac remodeling. Cardiology 126(1):59–61
239. Wang Y, Wu Y, Chen J, Zhao S, Li H (2013) Pirfenidone attenuates cardiac fibrosis in a mouse model of TAC-induced left ventricular remodeling by suppressing NLRP3 inflammasome formation. Cardiology 126(1):1–11
240. Kawaguchi Y, Suzuki K, Hara M, Hidaka T, Ishizuka T, Kawagoe M, Nakamura H (1994) Increased endothelin-1 production in fibroblasts derived from patients with systemic sclerosis. Ann Rheum Dis 53(8):506–510
241. Shi-Wen X, Rodriguez-Pascual F, Lamas S, Holmes A, Howat S, Pearson JD, Dashwood MR, du Bois RM, Denton CP, Black CM, Abraham DJ, Leask A (2006) Constitutive ALK5-independent c-Jun N-terminal kinase activation contributes to endothelin-1 overexpression in pulmonary fibrosis: evidence of an autocrine endothelin loop operating through the endothelin A and B receptors. Mol Cell Biol 26(14):5518–5527
242. Guarda E, Katwa LC, Myers PR, Tyagi SC, Weber KT (1993) Effects of endothelins on collagen turnover in cardiac fibroblasts. Cardiovasc Res 27(12):2130–2134
243. Mansoor AM, Honda M, Saida K, Ishinaga Y, Kuramochi T, Maeda A, Takabatake T, Mitsui Y (1995) Endothelin induced collagen remodeling in experimental pulmonary hypertension. Biochem Biophys Res Commun 215(3):981–986

244. Rizvi MA, Katwa L, Spadone DP, Myers PR (1996) The effects of endothelin-1 on collagen type I and type III synthesis in cultured porcine coronary artery vascular smooth muscle cells. J Mol Cell Cardiol 28(2):243–252

245. Lagares D, Busnadiego O, Garcia-Fernandez RA, Kapoor M, Liu S, Carter DE, Abraham D, Shi-Wen X, Carreira P, Fontaine BA, Shea BS, Tager AM, Leask A, Lamas S, Rodriguez-Pascual F (2012) Inhibition of focal adhesion kinase prevents experimental lung fibrosis and myofibroblast formation. Arthritis Rheum 64(5):1653–1664

246. Lagares D, Busnadiego O, Garcia-Fernandez RA, Lamas S, Rodriguez-Pascual F (2012) Adenoviral gene transfer of endothelin-1 in the lung induces pulmonary fibrosis through the activation of focal adhesion kinase. Am J Respir Cell Mol Biol 47(6):834–842

247. Lagares D, Garcia-Fernandez RA, Jimenez CL, Magan-Marchal N, Busnadiego O, Lamas S, Rodriguez-Pascual F (2010) Endothelin 1 contributes to the effect of transforming growth factor beta1 on wound repair and skin fibrosis. Arthritis Rheum 62(3):878–889

248. Nishida M, Onohara N, Sato Y, Suda R, Ogushi M, Tanabe S, Inoue R, Mori Y, Kurose H (2007) Galpha12/13-mediated up-regulation of TRPC6 negatively regulates endothelin-1-induced cardiac myofibroblast formation and collagen synthesis through nuclear factor of activated T cells activation. J Biol Chem 282(32):23117–23128

249. Im J, Gil K (2011) Effect of anaerobic digestion on the high rate of nitritation, treating piggery wastewater. J Environ Sci (China) 23(11):1787–1793

250. Fligny C, Duffield JS (2013) Activation of pericytes: recent insights into kidney fibrosis and microvascular rarefaction. Curr Opin Rheumatol 25(1):78–86

251. Simonson MS, Ismail-Beigi F (2011) Endothelin-1 increases collagen accumulation in renal mesangial cells by stimulating a chemokine and cytokine autocrine signaling loop. J Biol Chem 286(13):11003–11008

252. Jain R, Shaul PW, Borok Z, Willis BC (2007) Endothelin-1 induces alveolar epithelial-mesenchymal transition through endothelin type A receptor-mediated production of TGF-beta1. Am J Respir Cell Mol Biol 37(1):38–47

253. Piera-Velazquez S, Li Z, Jimenez SA (2011) Role of endothelial-mesenchymal transition (EndoMT) in the pathogenesis of fibrotic disorders. Am J Pathol 179(3):1074–1080

254. Sun G, Stacey MA, Bellini A, Marini M, Mattoli S (1997) Endothelin-1 induces bronchial myofibroblast differentiation. Peptides 18(9):1449–1451

255. Horowitz JC, Ajayi IO, Kulasekaran P, Rogers DS, White JB, Townsend SK, White ES, Nho RS, Higgins PD, Huang SK, Sisson TH (2012) Survivin expression induced by endothelin-1 promotes myofibroblast resistance to apoptosis. Int J Biochem Cell Biol 44(1):158–169

256. Rodriguez-Pascual F, Busnadiego O, Gonzalez-Santamaria J (2013) The profibrotic role of endothelin-1: Is the door still open for the treatment of fibrotic diseases? Life Sci 118:156–164

257. Kohan DE, Cleland JG, Rubin LJ, Theodorescu D, Barton M (2012) Clinical trials with endothelin receptor antagonists: what went wrong and where can we improve? Life Sci 91(13/14):528–539

258. Visnagri A, Kandhare AD, Ghosh P, Bodhankar SL (2013) Endothelin receptor blocker bosentan inhibits hypertensive cardiac fibrosis in pressure overload-induced cardiac hypertrophy in rats. Cardiovasc Endocrinol 2:85–97

259. Creemers EE, Pinto YM (2011) Molecular mechanisms that control interstitial fibrosis in the pressure-overloaded heart. Cardiovasc Res 89(2):265–272

260. Dai Y, Khaidakov M, Wang X, Ding Z, Su W, Price E, Palade P, Chen M, Mehta JL (2013) MicroRNAs involved in the regulation of postischemic cardiac fibrosis. Hypertension 61(4):751–756

261. Tijsen AJ, Pinto YM, Creemers EE (2012) Non-cardiomyocyte microRNAs in heart failure. Cardiovasc Res 93(4):573–582

262. Bowen T, Jenkins RH, Fraser DJ (2013) MicroRNAs, transforming growth factor beta-1, and tissue fibrosis. J Pathol 229(2):274–285

263. van Rooij E, Olson EN (2012) MicroRNA therapeutics for cardiovascular disease: opportunities and obstacles. Nat Rev Drug Discov 11(11):860–872
264. Thum T, Gross C, Fiedler J, Fischer T, Kissler S, Bussen M, Galuppo P, Just S, Rottbauer W, Frantz S, Castoldi M, Soutschek J, Koteliansky V, Rosenwald A, Basson MA, Licht JD, Pena JT, Rouhanifard SH, Muckenthaler MU, Tuschl T, Martin GR, Bauersachs J, Engelhardt S (2008) MicroRNA-21 contributes to myocardial disease by stimulating MAP kinase signalling in fibroblasts. Nature 456(7224):980–984
265. van Rooij E, Sutherland LB, Thatcher JE, DiMaio JM, Naseem RH, Marshall WS, Hill JA, Olson EN (2008) Dysregulation of microRNAs after myocardial infarction reveals a role of miR-29 in cardiac fibrosis. Proc Natl Acad Sci U S A 105(35):13027–13032
266. Zhang Y, Huang XR, Wei LH, Chung AC, Yu CM, Lan HY (2014) miR-29b as a therapeutic agent for angiotensin II-induced cardiac fibrosis by targeting TGF-beta/Smad3 signaling. Mol Ther 22:974–985
267. Xiao J, Meng XM, Huang XR, Chung AC, Feng YL, Hui DS, Yu CM, Sung JJ, Lan HY (2012) miR-29 inhibits bleomycin-induced pulmonary fibrosis in mice. Mol Ther 20(6):1251–1260
268. Qin W, Chung AC, Huang XR, Meng XM, Hui DS, Yu CM, Sung JJ, Lan HY (2011) TGF-beta/Smad3 signaling promotes renal fibrosis by inhibiting miR-29. J Am Soc Nephrol 22(8):1462–1474
269. Wijnen WJ, Pinto YM, Creemers EE (2013) The therapeutic potential of miRNAs in cardiac fibrosis: where do we stand? J Cardiovasc Trans Res 6(6):899–908
270. Tomasek JJ, Gabbiani G, Hinz B, Chaponnier C, Brown RA (2002) Myofibroblasts and mechano-regulation of connective tissue remodelling. Nat Rev Mol Cell Biol 3(5):349–363
271. Santiago JJ, Dangerfield AL, Rattan SG, Bathe KL, Cunnington RH, Raizman JE, Bedosky KM, Freed DH, Kardami E, Dixon IM (2010) Cardiac fibroblast to myofibroblast differentiation *in vivo* and *in vitro:* expression of focal adhesion components in neonatal and adult rat ventricular myofibroblasts. Dev Dyn 239(6):1573–1584
272. Serini G, Bochaton-Piallat ML, Ropraz P, Geinoz A, Borsi L, Zardi L, Gabbiani G (1998) The fibronectin domain ED-A is crucial for myofibroblastic phenotype induction by transforming growth factor-beta1. J Cell Biol 142(3):873–881
273. Frangogiannis NG, Michael LH, Entman ML (2000) Myofibroblasts in reperfused myocardial infarcts express the embryonic form of smooth muscle myosin heavy chain (SMemb). Cardiovasc Res 48(1):89–100
274. Shiojima I, Aikawa M, Suzuki J, Yazaki Y, Nagai R (1999) Embryonic smooth muscle myosin heavy chain SMemb is expressed in pressure-overloaded cardiac fibroblasts. Jpn Heart J 40(6):803–818
275. Hinz B, Phan SH, Thannickal VJ, Prunotto M, Desmouliere A, Varga J, De Wever O, Mareel M, Gabbiani G (2012) Recent developments in myofibroblast biology: paradigms for connective tissue remodeling. Am J Pathol 180(4):1340–1355
276. Espira L, Lamoureux L, Jones SC, Gerard RD, Dixon IM, Czubryt MP (2009) The basic helix-loop-helix transcription factor scleraxis regulates fibroblast collagen synthesis. J Mol Cell Cardiol 47(2):188–195
277. Bagchi RA, Czubryt MP (2012) Synergistic roles of scleraxis and Smads in the regulation of collagen 1alpha2 gene expression. Biochim Biophys Acta 1823(10):1936–1944
278. Barnette DN, Hulin A, Ahmed AS, Colige AC, Azhar M, Lincoln J (2013) Tgfbeta-Smad and MAPK signaling mediate scleraxis and proteoglycan expression in heart valves. J Mol Cell Cardiol 65:137–146
279. Levay AK, Peacock JD, Lu Y, Koch M, Hinton RB Jr, Kadler KE, Lincoln J (2008) Scleraxis is required for cell lineage differentiation and extracellular matrix remodeling during murine heart valve formation *in vivo*. Circ Res 103(9):948–956
280. Wang L, Bresee CS, Jiang H, He W, Ren T, Schweitzer R, Brigande JV (2011) Scleraxis is required for differentiation of the stapedius and tensor tympani tendons of the middle ear. J Assoc Res Otolaryngol 12(4):407–421

281. Murchison ND, Price BA, Conner DA, Keene DR, Olson EN, Tabin CJ, Schweitzer R (2007) Regulation of tendon differentiation by scleraxis distinguishes force-transmitting tendons from muscle-anchoring tendons. Development 134(14):2697–2708
282. Friedman SL, Sheppard D, Duffield JS, Violette S (2013) Therapy for fibrotic diseases: nearing the starting line. Science Trans Med 5(167):167sr161
283. Liu X, Sun SQ, Hassid A, Ostrom RS (2006) cAMP inhibits transforming growth factor-beta-stimulated collagen synthesis via inhibition of extracellular signal-regulated kinase 1/2 and Smad signaling in cardiac fibroblasts. Mol Pharmacol 70(6):1992–2003
284. Huang S, Wettlaufer SH, Hogaboam C, Aronoff DM, Peters-Golden M (2007) Prostaglandin E(2) inhibits collagen expression and proliferation in patient-derived normal lung fibroblasts via E prostanoid 2 receptor and cAMP signaling. Am J Physiol Lung Cell Mol Physiol 292(2):L405–L413
285. Lu D, Aroonsakool N, Yokoyama U, Patel HH, Insel PA (2013) Increase in cellular cyclic AMP concentrations reverses the profibrogenic phenotype of cardiac myofibroblasts: a novel therapeutic approach for cardiac fibrosis. Mol Pharmacol 84(6):787–793
286. Pedram A, Razandi M, O'Mahony F, Lubahn D, Levin ER (2010) Estrogen receptor-beta prevents cardiac fibrosis. Mol Endocrinol 24(11):2152–2165

Remodelling of the Cardiac Extracellular Matrix: Role of Collagen Degradation and Accumulation in Pathogenesis of Heart Failure

Abhijit Takawale, Mengcheng Shen, Dong Fan and Zamaneh Kassiri

Abstract The extracellular matrix (ECM) serves a number of functions in every tissue including the myocardium. While the function of ECM as a structural scaffold is well established, it has become increasingly recognized that it plays a number of additional functions including providing a reservoir for growth factors and cytokines allowing their rapid release and activation in response to environmental cues. In addition, components of the ECM are critical in the interstitial transport of numerous molecules and drugs. Therefore, impaired integrity of the ECM would influence multiple aspects of an organ's structural and function. In the myocardium, the primary component of the ECM network structure is the fibrillar collagens I and III. Multiple steps and various enzymes are involved from collagen mRNA synthesis to collagen fibre formation. Alterations in each step can impact collagen fibre production resulting in an uncoupling between collagen mRNA and protein levels. In this chapter, we will provide an overview of the mechanisms involved in myocardial fibrosis, the disease-dependent nature and consequence of different types of fibrosis, clinical biomarkers of collagen turnover, and potential therapeutic approaches in managing myocardial fibrosis.

Keywords Extracellular matrix · Replacement fibrosis · Reactive fibrosis · TGFβ1 · Matrix metalloproteinases · Tissue inhibitor of metalloproteinases

1 Introduction

Prevalence of cardiovascular diseases and the associated morbidities and mortalities continue to rise. It is projected that in North America, the medical cost of cardiovascular disorders will triple by 2030 [1, 2]. Heart function is the result of a complex interaction between the cardiomyocytes and non-cardiomyocytes mediated by

Z. Kassiri (✉) · A. Takawale · M. Shen · D. Fan
Department of Physiology Cardiovascular Research Center, Mazankowski Alberta Heart Institute, University of Alberta, 474 HMRC, Edmonton, AB T6G 2S2, Canada
e-mail: z.kassiri@ualberta.ca

© Springer International Publishing Switzerland 2015
I.M.C. Dixon, J. T. Wigle (eds.), *Cardiac Fibrosis and Heart Failure: Cause or Effect?*,
Advances in Biochemistry in Health and Disease 13, DOI 10.1007/978-3-319-17437-2_12

the extracellular matrix (ECM), a network that connects the different components within the myocardium. The main component of the ECM network structure are collagens (type I and type III), which are produced as pro-collagens and undergo a number of processing steps before collagen fibrils and collagen fibres are deposited in the interstitium. A defect in any of these steps can lead to adverse ECM remodeling.

Regardless of the initiating cause, heart failure (HF) is associated with impaired ECM composition and integrity. Excess accumulation of ECM proteins, also known as fibrosis, has been clearly linked to myocardial stiffness, diastolic and systolic dysfunction [3]. Meanwhile, excess degradation of the ECM can also negatively impact myocardial compliance, cardiac structure and function, with adverse outcomes [4–9]. Therefore, intact ECM structural and integrity is required for optimal cardiac function.

2 Myocardial Extracellular Matrix (ECM)

Myocardial ECM is comprised of structural proteins that form the fibrillar structure of the matrix (mainly collagen type I and III), the basement membrane that forms an interface between the cardiomyocytes and the interstitial space, and non-structural proteins such as proteoglycans. ECM serves a number of functions in the heart, it provides a structural scaffold for different cell types and the vasculature within the myocardium, ensures optimal cardiac architecture, and transmits single myocyte contractility to whole heart pumping function for effective systolic and diastolic functions of the heart [8]. In a healthy heart, the cell-ECM connection is maintained by integrins which allows for transmission of signals across the cell membrane. The ECM can also serve as a reservoir for a number of growth factors and cytokines that remain bound to proteoglycans, the non-structural component of the ECM, and are released in response to a cellular or signaling cue, whereby their activity is tightly regulated beyond their mRNA and protein synthesis [10].

ECM undergoes continuous turnover whereby the existing ECM proteins are degraded by multiple proteinases, and replaced by newly synthesized proteins. In heart disease, adverse ECM remodeling results in excess deposition of ECM proteins leads to fibrosis, or aberrant degradation and compromised integrity of the ECM network structure. Myocardial fibrosis often occurs in response to increased afterload (e.g. pressure overload, aortic stenosis) and increases stiffness of the myocardium which can lead to diastolic dysfunction. Inversely, volume overload leads to disruption of the ECM network, ventricular dilation and systolic dysfunction mainly due to cardiomyocyte slippage and loss of cell-ECM connections. Moreover, it is not uncommon for both types of adverse ECM remodeling to take place within the same heart resulting in a complex systolic and diastolic dysfunction. Cardiac fibroblasts (cFB) are the primary source of ECM proteins [11, 12]. In response to appropriate stimuli, cFBs can differentiate into the more mobile and contractile myofibroblasts (myoFBs) with a greater synthetic ability to produce ECM proteins

[12, 13]. In addition, other cell types such as inflammatory cells and endothelial cells can undergo transformation and become myofibroblasts [12]. Collagen type I and III are the primary fibrillar collagen that comprise the ECM network structure in the heart and alterations in their synthesis and degradation can markedly impact the overall ECM structure.

3   Different Types of Myocardial Fibrosis and Impact on Cardiac Function

Heart failure is a complex progressive syndrome that can be initiated by a number of heart diseases such as hypertensive, valvular defect, viral myocarditis, ischemic or dilated cardiomyopathies [14, 15], among which ischemic and hypertensive heart disease account for the main two contributing factors to HF [15]. A common characteristic among these heart diseases is fibrosis, excess ECM accumulation in the myocardium [12, 16]. However, the underlying molecular mechanism and the functional consequences can be different depending on the type of disease. For instance, after acute MI, necrotic cardiomyocytes are replaced by fibrillar collagen through the process of 'reparative' fibrosis, preserving the integrity of the myocardium by formation of the scar or infarct. The degradation products of the original ECM in the infarcted myocardium serve as chemoattractants that trigger infiltration of inflammatory cells to the site of injury [17]. These inflammatory cells release a number of cytokines as well as MMPs, while a subpopulation of the infiltrated cells transform into myofibroblasts [18–20] that produce collagen and other ECM proteins, thereby playing a critical role in formation of scar (infarction) to replace the necrotic myocardium [21]. However, unlike the prototypical wound healing response in static tissue interfaces such as skin, prolonged collagen degradation induced by expanded population of myofibroblasts can result in expansion of the infarct tissue leading to LV dilation, worsened systolic dysfunction and eventually HF. The ECM in the myocardium remote to the infarct zone also undergoes mainly in response to neurohormonal stimuli triggered by the ischemic injury [22]. After the scar is formed at the injured sites, myofibroblasts are supposed to return to a quiescent phenotype. However, if myofibroblast dispersion via apoptotic cell death fails to occur, the 'reparative' fibrosis becomes 'active' fibrosis, in which myofibroblasts contribute to perpetual matrix formation through their ongoing production of signalling molecules that promote fibrogenesis in an autocrine manner [23]. Therefore, the initial fibrotic response following myocardial infarction is a healing process and its interruption will lead to cardiac rupture with devastating outcomes.

Myocardial ECM remodeling following pressure overload, as in hypertensive heart disease, is often a reactive process where interstitial and perivascular fibrosis occurs in the absence of myocyte necrosis. This process is driven predominantly by neurohormonal stimuli. The role of Ang II, aldosterone, salt loading and nephrectomy have been extensively investigated on collagen deposition in hypertensive and pressure overloaded cardiomyopathy [24–27]. All major components of the renin-angiotensin-system (RAS) exhibit profibrotic activity, and angiotensin converting enzyme inhibitor (ACEi) [28] and angiotensin receptor blockers (AEBs) [29] have proven to have beneficial effects in heart disease patients. Among the components of the RAS system, Ang II is a major contributor to myocardial hypertrophy, fibrosis and remodeling in hypertensive heart disease [30]. In severe cases of hypertensive cardiomyopathy, myocyte necrosis can occur due to catecholamine or Ang II-mediated toxicity resulting in a combination of reparative and reactive fibrosis in the myocardium [30, 31].

Heterogeneous remodeling of the ECM can also occur in the diseased myocardium where fibrotic lesions and disrupted ECM are found in the same heart. While fibrosis increases myocardial stiffness and impairs relaxation, degradation of the ECM network results in systolic dysfunction. Myocardial fibrosis leads to diastolic dysfunction, abnormal left ventricular (LV) filling, reduced LV compliance, and increased diastolic pressure. This may occur in patients with normal LV ejection fraction (EF) and can eventually lead to heart failure with preserved EF (HFpEF) [32]. The hospital re-admission rates of patients with HFpEF are similar to those with HF and reduced EF (HFrEF), and worsening of diastolic dysfunction has been shown to be an independent predictor of mortality [33]. However, different therapeutic approaches are required for each type of HF, and therefore, understanding the molecular mechanisms underlying myocardial fibrosis and the associated diastolic dysfunction is critical in developing effective therapies. Consistently, a thorough assessment of collagen turnover and factors that disturb this balance in HF in animal models can translate into significant contributions to the advancement of clinical care of HF patients.

4 Collagen Synthesis, Assembly and Deposition and Related Biomarkers in the Heart

Collagens make up 2–4 % of a healthy human body [34]. Based on their structure, collagen molecules are divided into two main classes: fibril forming collagens (I, II, III, V, XI) and non-fibril forming collagens including collagen IV and VI which are expressed in the heart [35, 36]. The fibril forming collagens have long continuous triple helix structures while non-fibril forming collagens are more heterogeneous and further classified based on molecular and supramolecular structures [37, 38]. Fibrillar collagen type I and type III are the predominant components of the fibrillar structure of myocardial ECM. Collagen type I provides the myocardium with tensile strength, while collagen type III accounts for distensibility of the myocardium [4]. Myocardial collagen network exists at three levels: endomysium, epimysium and perimysium. The endomysial collagen surrounds individual muscle fiber, epimysial collagen surrounds a group of muscle fibers and the perimysium consists of thick, spiral shaped collagen bundles which connect the epimysial and the endomysial networks [34, 39].

Following injury to heart muscle, myocardium undergoes remodelling and fibro-blasts (and other cells types) transdifferentiate into myofibroblasts [40]. Myofibro-blasts are activated fibroblasts and are the principle cells involved in production of ECM proteins and fibrosis [41]. The proliferation, migration and fibrillar collagen synthesis and secretion capacity of myofibroblasts are closely regulated by mechan-ical stretch [42, 43], ischemic myocardial injury [44, 45], autocrine or paracrine effects of neurohormonal factors, vasoactive peptides (e.g. Ang II), growth factors (e.g. TGF-β_1, CTGF and PDGF), angiogenic factor (e.g. VEGF), transcriptional factors (e.g. Smad proteins), circulating hormones (e.g., aldosterone), and proin-flammatory cytokines (e.g., TNF-α and interleukin-1, -6 and 13) [23, 46–48]. These factors cause an imbalance in ECM turnover resulting in excessive deposition or aberrant degradation of ECM proteins.

Transforming growth factor-β (TGFβ) is a multifunctional cytokine that has been shown to be a potent mediator of myofibroblast transformation and collagen pro-duction. Among the three isoforms of TGFβ (TGFβ1, TGFβ2, and TGFβ3), TGFβ1 is most strongly linked to fibrosis in the myocardium. All TGFβ isoforms are tran-scribed with the latency-associated pro-protein (LAP). LAP is cleaved intracellular-ly but stays non-covalently associated with TGFβ, forming the small latent complex (SLC). SLC is secreted as the large latent complex (LLC) bound to LTBP (Latent TGFβ-Binding Protein) via a disulphide bond. After secretion, TGFβ is sequestered in the ECM through covalent binding to LTBPs which is a member of the fibrillin superfamily comprising the microfibrils. TGFβ activation requires the release of the 25 kDa homodimer from the latent complex through proteolytic cleavage of LTBP from ECM and subsequent release of TGFβ from LAP. This can be mediated by MMPs such as MMP2, MMP9 or MT1-MMP [49, 50] or through the integrin-mediated processes [51]. The integrin-mediated TGFβ activation is most relevant in response to mechanical stress such as in cardiac pressure-overload. In a rat model of myocardial infarction, increased expression of integrin-α3 enhanced its interaction with collagen VI and contributed to fibroblast transdifferentiation, collagen produc-tion and accumulation [42]. Therefore, MMPs can also contribute to synthesis of matrix proteins, and as such the traditional dogma on the role of MMPs only as matrix-degrading enzymes is no longer valid.

This step is required for release of TGFβ ligand and its binding to its receptors (TGFRI/II) and activation of the downstream pathways (Fig. 1). Other proteases such as plasmin and bone morphogenetic protein-1 (BMP-1) have also been shown to cleave the LTBP-ECM bond, but only inhibition of MMPs prevented release of TGFβ homodimer from LAP [52]. Upon binding of the active TGFβ to its recep-tor, TGFβRII, it dimerizes with and phosphorylates TGFβRI, inducing intracellular signaling. The main TGFβ-mediated signaling pathways are the canonical Smad pathway, and the non-canonical MAPK pathways (ERK, JNK and p38) [53, 54] (Fig. 1). In canonical TGFβ signaling, the TGFβRI/II complex serine phosphory-lates Smad2/3, which then form a heteromeric complex with Smad4, translocate into the nucleus and regulate transcription of target genes including collagen, elastin and -MMPs [55]. TGFβ also induces expression of inhibitory Smad6 and Smad7 which prevent activation of Smad2/3 thereby acting as an autoinhibitory feedback

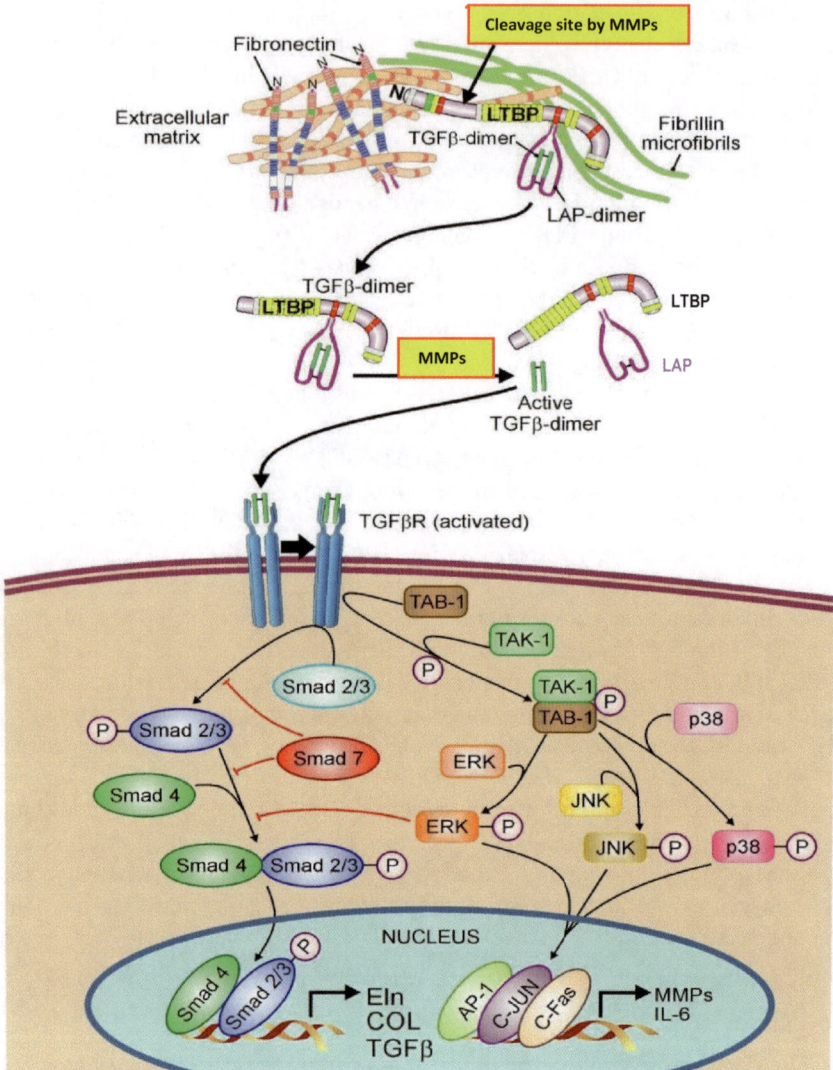

Fig. 1 Molecular pathway for TGFβ activation and its downstream signaling pathways. TGFβ dimer (25 kDa) is secreted from VSMC non-covalently associated with the latency-associated peptide (*LAP*), and bound to LTBP (Latent TGFβ Binding Protein). This complex is called the Large Latent Complex (*LLC*) which binds to components of ECM (fibronectin). LTBP can be proteolytically cleaved, small latency complex (TGFβ+LAP) is released, and subsequently TGFβ dimer is dissociated from LAP and binds to its receptor (TGFβRI/II) activating the downstream pathways. Phosphorylation of Smad2/3 mediates their association with Smad4 and their nuclear translocation and induction of a number of genes, mainly ECM proteins, elastin and collagen. In parallel, MAPK pathways (ERK, JNK and p38) can also be activated triggering synthesis of cytokines and MMPs

mechanism [56]. In non-canonical TGFβ signaling, upon binding of TGFβ, TGFβ RI is tyrosine phosphorylated, allowing it to recruit and directly phosphorylate SchcA proteins on tyrosine and serine, leading to phosphorylation of ERK MAP kinase [53]. This pathway is activated upon excess TGFβ bioavailability, and would account for the Smad-independent functions of TGFβ as well as inhibiting the Smad pathway [57]. TGFβ1 has been strongly linked to fibrosis since it is upregulated in the LV of patients with idiopathic hypertrophic cardiomyopathy [58], dilated cardiomyopathy [59], as well as in animal models of myocardial infarction [60, 61], pressure overload [50, 62].

Collagens are produced as triple helix pro-collagens consisting of three α-chains with the characteristic Gly-X-Y repeat sequence [63] and are secreted to the extracellular compartment for post-translational modifications. Multiple intracellular and extracellular steps are required for completion of collagen deposition in the interstitium after it is synthesized. Pro-collagen-α is produced in the ribosome of cFB and is imported to the endoplasmic reticulum (ER) where it undergoes hydroxylation of proline and lysine residues, N- and O- linked glycosylation, trimerization, disulphide bonding, isomerization and finally folding of the triple helix structure [38]. In the ER, stability of the triple helix collagen molecule is maintained by a chaperone molecular, Hsp47 [64]. The soluble pro-collagen triple helix is then transported to the Golgi apparatus and packed inside secretory COPII vesicles, and then secreted from the fibroblasts into the interstitial space [65]. Recently, the ubiquitin ligases CUL3-KLHL12 were identified as regulators of large COPII coat formation which is essential to accommodate the large size of the collagen triple helix prior to their secretion from the cFB [65]. In the extracellular space, the pro-collagen requires enzymatic removal of the N- and C-propeptides. The N- propeptide can be cleaved by **a d**isintegrin **a**nd **m**etalloproteinase with a thrombospondin motif (ADAMTS)-2, −3, −14, while removal of the C-propeptide can be mediated by BMP1 [35, 37], whose activity is enhanced by procollagen C-endopeptidase enhancer (PCPE 1 and 2) and the secreted frizzled-related protein (sFRP2) [66–69]. PCPE 1 and 2 show highest expression levels in the myocardium and can increase the efficiency of C-propeptidases by up to 20 folds [5, 68]. Metalloproteases meprin α and meprin β have been shown to function as both C- and N-propeptide proteinases [70]. The C-terminal and N-terminal propeptides of type I procollagen (PICP, PINP), and those of type III procollagen (PIIICP and PIIINP) are released during biosynthesis of these collagen fibrils in a stoichiometric manner, and are considered as biomarkers of collagen synthesis [12] discussed later in this chapter.

Following the removal of the C- and N-propeptides, lysyl hydroxylase (PLOD1) and lysyl oxidase (LOX) mediate hydroxylation and oxidative deamination of collagens, leading to cross-linking and stabilization of collagen fibres [71, 72]. Further post-translational regulation of collagen fibres is mediated by matricellular proteins, the non-structural ECM molecules that can mediate collagen stabilization and accumulation in the interstitium [73]. Among the matricellular proteins, SPARC (Secreted protein, acidic and rich in cysteine) and Osteopontin (OPN) have been linked to cardiac fibrosis [5, 73]. In normal physiology, SPARC regulates pro-collagen-α processing and interact with fibroblast cell surface, alters collagen formation in cardiac interstitium without altering the baseline systolic functions or blood pressure [74]. OPN similarly stabilizes collagen fibres, however OPN-null mice have increased heart rate, lower blood pressure and enhanced arterial compliance [75], suggesting that OPN may play additional roles in the cardiovascular system. Deletion of SPARC or OPN resulted in increased rate of LV rupture following myocardial infarction (MI) [76–78], indicating the critical role of these matricellular proteins in assembly of collagen fibres in generation of supportive matrix in healing the post-MI scar. On the other hand, following cardiac pressure overload, lack of OPN or SPARC led to attenuation of myocardial fibrosis, absence of myocardial hypertrophy and reduced diastolic stiffness [79, 80]. These studies clearly demonstrate the differential impact of replacement fibrosis post-MI versus reactive fibrosis following mechanical stress in the heart and important role of post-translational modification of collagen in these processes.

5   Collagen Degradation and Related Biomarkers in Heart Disease

ECM is a dynamic entity which undergoes constant turnover. In health, the ECM structural proteins, mainly collagens are degraded by Zn^{2+}-dependant proteases, matrix metalloproteinases (MMPs). To date, 28 MMPs have been identified and classified based on the substrate(s) that they were initially identified to process, although now it is well established that the function of MMPs is not limited to these substrates but each MMP can target multiple ECM and non-ECM proteins [81, 82]. The MMPs that have been shown to be involved in heart disease include collagenases MMP1, MMP13, and MMP8; gelatinases MMP2 and MMP9; stromelysins/matrilysins, MMP3 and MMP7; and the membrane type MMP, MT1-MMP [79, 83, 84]. Recently, MMP28 has been reported to be produced by macrophages and to contribute to post-MI remodeling [85]. MMPs can be produced by various cell types including cFBs, cardiomyocytes and inflammatory cells [8, 17].

MMPs are synthesized as inactive zymogens (pro-MMPs) and are activated by removal of the amino terminal pro-peptide domain and exposure of their catalytic domain. Once activated, activity of MMPs is regulated by their physiological inhibitors, the tissue inhibitor of metalloproteinases, TIMPs [8], which non-covalently bind to activated MMPs in 1:1 stoichiometric complexes and reversibly inhibit their proteolytic function. Other physiological inhibitors of MMPs have also been identified, such as alpha 2-macroglobulin and RECK (Reversion-inducing-cysteine-rich protein with kazal motifs) [86]. However, α2-macroglobulin can only block MMP activity in plasma and liquid compartments with minimal inhibitory function in the tissue [87], while baseline expression of RECK in the heart is minimal [88] and its role in cardiac function has not yet been identified. Therefore, TIMPs are considered the dominant MMP inhibitors in the myocardium.

MMP-mediated degradation of ECM proteins is a part of ECM turnover. However, aberrant activity MMPs in heart disease can result in excess degradation of collagen fibres leading to compromised integrity of the ECM network structure. During ECM turnover, collagen fibres are degraded and the telopeptides in the C- and/or N-terminal of type I collagen (CITP, NITP) and type III collagen (CIIITP and NIIITP) are cleaved (Fig. 2), which have been proposed to serve as biomarkers for collagen degradation in disease [89] as discussed later in this chapter. While MMP inhibition initially appeared to be an attractive therapy for heart disease patients [90], studies on MMP-deficient mice has shown that inhibiting MMPs will not invariably result in beneficial effects [85, 91, 92] suggesting their potential beneficial role in cardiac response to disease and that additional considerations need to be given in targeting MMPs in heart disease [44].

6 Clinical Markers of Collagen Synthesis and Degradation, and Their Prognostic Significance

Fibrillar collagen type I and III are the predominant components of cardiac ECM. These collagens are synthesized, transported in large COPII vesicles and secreted into the interstitial space as procollagens, which are then processed into mature collagen molecules upon cleavage of their propeptides by procollagen C- and N-proteinases (PCP and PNP, respectively), resulting in spontaneous assembly of collagen fibrils [5, 70, 93, 94]. The C-terminal and N-terminal propeptides of type I (PICP, PINP) and type III procollagens (PIIICP and PIIINP) are released during biosynthesis of these collagen fibrils in a stoichiometric manner, and hence have been considered as biomarkers of collagen synthesis [12]. During ECM turnover, degradation of collagen fibres, by MMPs or other proteases, results in cleavage and release of telopeptides from the C- or N-terminal of collagen type I (CITP, NITP) and type III (CIIITP and NIIITP) in the plasma (Fig. 2). As such, these telopeptides have been considered as biomarkers of collagen degradation [89].

Different reports have been made with respect to the use of plasma biomarkers for collagen turnover (synthesis and degradation) as a marker of LV remodeling and function in hypertensive patients. Reduced serum CITP, a marker of collagen type I degradation, was first reported to explain collagen deposition and fibrosis in hypertensive patients [95], but a later study showed increased CITP and PICP in hypertensive patients with LV fibrosis and diastolic dysfunction indicating enhanced collagen turnover in hypertensive patients [96]. A direct correlation was reported between serum PICP and collagen volume fraction in hypertensive patients using endomyocardial biopsies, as PICP levels were higher in patients with severe fibrosis compared to those with less severe fibrosis [97]. Similarly, serum levels of PINP, a marker of collagen synthesis, correlated with diastolic dysfunction in hypertensive patients [98]. These studies collectively suggest that serum markers of collagen synthesis (PICP or PINP) could serve as biomarkers for fibrosis in hypertensive patients. In patients with congestive heart failure, high serum levels of PIIINP and PICP, markers of collagen synthesis, were associated with poor outcome [99]. High plasma PICP and CITP levels were found in atrial fibrillation patients, however,

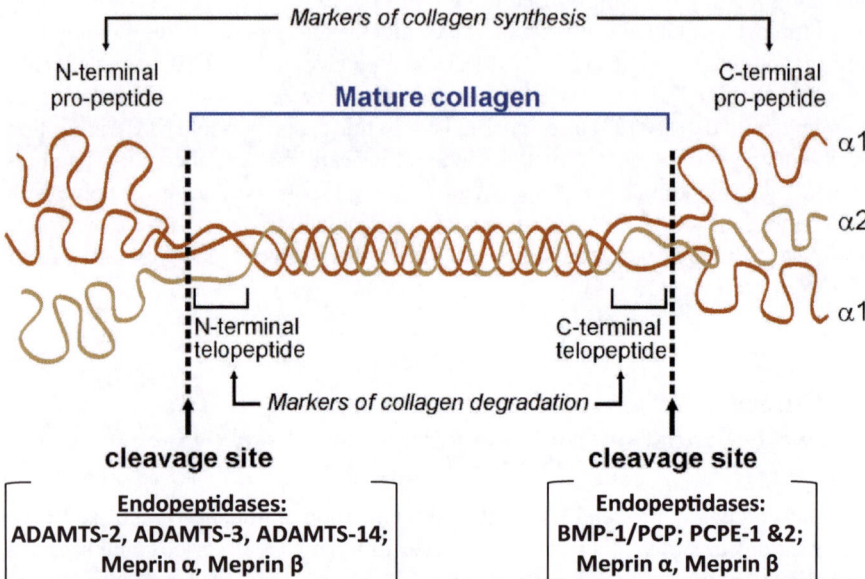

Fig. 2 Procollagen structure, cleaveage sites and enzymes mediating procollagen processing. Procollagen is comprised of two alpha-1 chains and one alpha-2 chain intertwined into a triple helix. Pro-peptide domain at the carboxy- and amino-terminals are cleaved by specific endopeptidases resulting in formation of mature collagen. When collagen is degraded, during physiological turnover or pathological adverse remodeling, telopeptides (from the amino- or carboxy-terminals) are cleaved and released into the plasma. *ADAM-TS* a distingetrin and metalloproteinase with a thrombospondin domain; *BMP*-1 Bone morphogenic protein-1; *PCPE* procollagen C-endopeptidase enhancer. Modified from Fan et al. [12]

persistent atrial fibrillation patients had high plasma PICP (synthesis), but not CITP (degradation) levels, suggesting that the degree of collagen type I synthesis and degradation may depend on the severity of fibrillation [100]. High PICP levels were also found in patients with repaired tetralogy of Fallot which could reflect RV fibrosis and link to adverse markers of clinical outcomes [101]. Although plasma CITP and PIIINP levels were associated with heart failure and LV dysfunction in aortic stenosis patients [102], they were found to be unreliable surrogate measures of myocardial fibrosis which did not affect the results of surgery or long-term survival [103].

In patients with coronary artery disease but no myocardial infarction, serum levels of PIIINP but not PINP correlated well with the number of diseased vessels and severity of coronary artery disease [104]. In acute MI patients, elevated serum PIIINP levels during the first few days post-MI were also associated with suppressed LV function, increased LV volume over 1-year follow up [105], poor overall prognosis and survival [106]. Early post-MI serum PIIINP levels have been suggested to serve as a marker of long term LV remodeling and prognosis. However, a more recent study reported that CITP, compared to PINP, PIIINP and TIMP1, served as the most suitable prognostic tool in patients with acute and chronic MI [107].

In using plasma biomarkers for ECM remodeling in patients with heart disease, it is important to keep in mind that collagen type I is the most abundant collagen in human body and is ubiquitously expressed in almost all organs and tissues. Although collagen type III shows a relatively more tissue-specific expression pattern, it is also highly expressed in the skin, lungs and the vasculature. In addition, once released into the plasma, the collagen propeptides and telopeptides can be cleared from circulation over time. Therefore, depending on the disease stage, these proteins may or may not be detectable in the plasma. These factors could likely explain the discrepancies in some of the reported studies.

7   Management of Cardiac Fibrosis

Cardiac fibrosis is not an undesired event in all cases. For instance, cardiac fibrosis post-MI plays a critical role in maintaining the integrity of the cardiac structure and prevents the infarcted heart from rupture. Adenoviral overexpression of SPARC in WT mice increased collagen deposition and improved maturation without alteration of the infarct size, and further prevented cardiac dilation and dysfunction post-MI [73, 76]. Therefore, management of fibrosis could include inducing normal fibrosis in specific conditions, as long as it is tightly regulated for the required duration.

However, when uncontrolled deposition of ECM structural proteins replace the functioning myocardium, or interrupt the connection and interaction between the cardiomyocytes, this leads to cardiac dysfunction and to heart failure if not managed. A number of potential therapies have been developed specifically targeting the pathogenesis of fibrosis. According to the mechanisms of cardiac fibrosis, four potential strategies have aimed to target fibrosis by interrupting collagen production and processing at the source cell, procollagen transportation and processing, and cross-linking and stabilization. (1) Inhibiting formation of myofibroblasts. TGFβ1 induces endothelial cells to undergo endothelial-mesenchymal transition (EndMT) and formation of myofibroblasts, whereas BMP-7 preserves the endothelial phenotype [108]. Systemic administration of recombinant human BMP-7 reduced EndMT and the progression of cardiac fibrosis in mouse models of pressure overload and chronic allograft rejection [108]. Targeting TGFβ1 has also been shown to be an effective strategy to control reactive fibrosis [16, 50]. Many elements of the innate and adaptive immune response participate in the differentiation and activation of fibroblasts [16] which could also serve as potential targets in limiting the population of myofibroblasts. (2) Suppression of procollagen transportation to the interstitium. Blocking CUL3 expression or its activity would reduce collagen secretion and its subsequent deposition in the interstitial space, thereby neutralizing an increase in collagen mRNA synthesis. Consistently, agonists of CUL3-KLHL12 were used to induce fibrosis in recessive syndromes such as cranio-lenticulo-sutural dysplasia in order to preserve intact structures [65]. (3) Interruption of procollagen processing.

Activity of BMP-1, a procollagen peptidase, can be specifically increased by PCPE-1 and PCPE-2 leading to enhanced collagen fibre formation and deposition [5, 69, 93]. Inhibiting the action of PCPEs could serve as a potential therapeutic approach against excess fibrogenesis [93]. (4) Inhibiting collagen cross-linking and stabilization. Inhibition of PLOD, LOX and their related enzymes would interrupt the cross-linking of collagen, rendering them more susceptible to degradation. An inhibitory monoclonal antibody against LOX-like-2 (LOXL2), an enzyme that catalyzes the cross-linking of collagen, reduced fibrosis and is being explored as a treatment for cardiac fibrosis in patients [16, 109]. SPARC binds to and stabilizes collagen fibres, and its suppression results in collagen association with fibroblast cell surface which is sensitive to degradation by collagenases [6, 73]. Stem cell transplantation has also been considered as a potential therapy for cardiac fibrosis which likely targets a number of the steps discussed above [110]. Cardiac fibrosis is often characterized by the activation of multiple profibrotic pathways, so a multipronged approach would be necessary to slow or stop the progression of fibrosis. However, a major obstacle slowing the development of antifibrotic drugs is the lack of disease-specific biomarkers which can be used to identify patients who might benefit from a specific therapy [16]. Several molecules involved in collagen synthesis and assembly, such as TGFβ1, CUL3, KLHL12, BMP-1, BMP-7, PCPE, PLOD1, LOX and LOXL2, which could be screened together with clinical fibrotic markers as discussed previously to decipher the mechanisms of cardiac fibrosis which would lead to an effective and specific therapy.

8   Conclusions

Cardiac fibrosis, deposition of ECM structural proteins in the myocardium, is not always an undesired process. In case of myocardial necrosis (e.g. MI), fibrosis is essential in preserving cardiac structure and preventing rupture of the myocardium. Therefore, attempts to prevent myocardial fibrosis should be limited to cases such as hypertensive cardiomyopathies where reactive fibrosis replaces existing cardiomyocytes. In addition, it is important to note that a number of steps are involved from the mRNA synthesis of collagens until collagen fibres are formed and deposited in the interstitium. Therefore, in assessing myocardial fibrosis, or in exploring the underlying mechanism of fibrosis, multitude of factors should be considered. While serum biomarkers of collagen synthesis and degradation could provide useful information about ECM turnover, it is critical to note that collagens type I and type III are present in many organs and tissues in the body, and as such change in the plasma levels could reflect information about other organs. Therefore, development of a biomarker to detect a protein specific to cardiac ECM will be highly valuable as a cardiac specific biomarker for myocardial fibrosis and ECM turnover.

References

1. Burchfield JS, Xie M, Hill JA (2013) Pathological ventricular remodeling: mechanisms: part 1 of 2. Circulation 128:388–400
2. Heidenreich PA, Trogdon JG, Khavjou OA et al (2011) Forecasting the future of cardiovascular disease in the United States: a policy statement from the American Heart Association. Circulation 123:933–944
3. Diez J, Querejeta R, Lopez B et al (2002) Losartan-dependent regression of myocardial fibrosis is associated with reduction of left ventricular chamber stiffness in hypertensive patients. Circulation 105:2512–2517
4. Segura AM, Frazier OH, Buja LM (2014) Fibrosis and heart failure. Heart Fail Rev 41(4):389–394
5. Goldsmith EC, Bradshaw AD, Spinale FG (2013) Cellular mechanisms of tissue fibrosis. 2. Contributory pathways leading to myocardial fibrosis: moving beyond collagen expression. Am J Physiol Cell Physiol 304:C393–402
6. Harris BS, Zhang Y, Card L et al (2011) SPARC regulates collagen interaction with cardiac fibroblast cell surfaces. Am J Physiol Heart Circ Physiol 301:H841–847
7. Leening MJ, Steyerberg EW (2013) Fibrosis and mortality in patients with dilated cardiomyopathy. JAMA 309:2547–2548
8. Moore L, Fan D, Basu R et al (2012) Tissue inhibitor of metalloproteinases (TIMPs) in heart failure. Heart Fail Rev 17:693–706
9. Spinale FG, Zile MR (2013) Integrating the myocardial matrix into heart failure recognition and management. Circ Res 113:725–738
10. Fan D, Creemers EE, Kassiri Z (2014) Matrix as an interstitial transport system. Circ Res 114:889–902
11. Camelliti P, Borg TK, Kohl P (2005) Structural and functional characterisation of cardiac fibroblasts. Cardiovasc Res 65:40–51
12. Fan D, Takawale A, Lee J, Kassiri Z (2012) Cardiac fibroblasts, fibrosis and extracellular matrix remodeling in heart disease. Fibrogenesis Tissue Repair 5:15
13. Petrov VV, Fagard RH, Lijnen PJ (2002) Stimulation of collagen production by transforming growth factor-beta1 during differentiation of cardiac fibroblasts to myofibroblasts. Hypertension 39:258–263
14. Houser SR, Margulies KB, Murphy AM et al (2012) Animal models of heart failure a scientific statement from the american heart association. Circ res 111:131–150
15. Roger VL, Go AS, Lloyd-Jones DM et al (2012) Heart disease and stroke statistics—2012 update a report from the american heart association. Circulation 125:e2–e220
16. Wynn TA, Ramalingam TR (2012) Mechanisms of fibrosis: therapeutic translation for fibrotic disease. Nat Med 18:1028–1040
17. Lambert JM, Lopez EF, Lindsey ML (2008) Macrophage roles following myocardial infarction. Int J Cardiol 130:147–58
18. Haudek SB, Xia Y, Huebener P et al (2006) Bone marrow-derived fibroblast precursors mediate ischemic cardiomyopathy in mice. Proc Natl Acad Sci U S A 103:18284–18289
19. van Amerongen MJ, Bou-Gharios G, Popa E et al (2008) Bone marrow-derived myofibroblasts contribute functionally to scar formation after myocardial infarction. J Pathol 214:377–386
20. van Amerongen MJ, Harmsen MC, van Rooijen N et al (2007) Macrophage depletion impairs wound healing and increases left ventricular remodeling after myocardial injury in mice. Am J Pathol 170:818–829
21. Widgerow AD (2011) Cellular/extracellular matrix cross-talk in scar evolution and control. Wound Repair and Regen 19:117–133
22. Weber KT, Sun Y, Bhattacharya SK et al (2013) Myofibroblast-mediated mechanisms of pathological remodelling of the heart. Nat Rev Cardiol 10:15–26
23. Wynn TA (2008) Cellular and molecular mechanisms of fibrosis. J Pathol 214:199–210

24. Robert V, Heymes C, Silvestre JS, et al (1999) Angiotensin AT1 receptor subtype as a cardiac target of aldosterone: role in aldosterone-salt-induced fibrosis. Hypertension 33:981–986

25. Schnee JM, Hsueh WA (2000) Angiotensin II, adhesion, and cardiac fibrosis. Cardiovasc Res 46:264–268

26. Lijnen P, Petrov V (2000) Induction of cardiac fibrosis by aldosterone. J Mol Cell Cardiol 32:865–879

27. Tsukamoto Y, Mano T, Sakata Y et al (2013) A novel heart failure mice model of hypertensive heart disease by angiotensin II infusion, nephrectomy, and salt loading. Am J Physiol Heart Circ Physiol 305:H1658–1667

28. Sharpe N, Smith H, Murphy J et al (1991) Early prevention of left ventricular dysfunction after myocardial infarction with angiotensin-converting-enzyme inhibition. Lancet 337:872–876

29. Granger CB, McMurray JJ, Yusuf S et al (2003) Effects of candesartan in patients with chronic heart failure and reduced left-ventricular systolic function intolerant to angiotensin-converting-enzyme inhibitors: the CHARM-Alternative trial. Lancet 362:772–776

30. Watanabe T, Barker TA, Berk BC (2005) Angiotensin II and the endothelium: diverse signals and effects. Hypertension 45:163–169

31. Bishop J, Laurent G (1995) Collagen turnover and its regulation in the normal and hypertrophying heart. Eur Heart J 16:38–44

32. Vasan RS, Levy D (2000) Defining diastolic heart failure: a call for standardized diagnostic criteria. Circulation 101:2118–21

33. Aljaroudi W, Alraies MC, Halley C et al (2012) Impact of progression of diastolic dysfunction on mortality in patients with normal ejection fraction. Circulation 125:782–788

34. Weber KT (1989) Cardiac interstitium in health and disease: the fibrillar collagen network. J Am Coll Cardiol 13:1637–1652

35. Prockop DJ, Kivirikko KI (1995) Collagens: molecular biology, diseases, and potentials for therapy. Annu Rev Biochem 64:403–434

36. Shamhart PE, Meszaros JG (2010) Non-fibrillar collagens: key mediators of post-infarction cardiac remodeling? J Mol Cell Cardiol 48:530–537

37. Kadler KE, Baldock C, Bella J, Boot-Handford RP (2007) Collagens at a glance. J Cell Sci 120:1955–1958

38. Vuorio E, de Crombrugghe B (1990) The family of collagen genes. Annu Rev Biochem 59:837–872

39. Borg TK, Sullivan T, Ivy J (1982) Functional arrangement of connective tissue in striated muscle with emphasis on cardiac muscle. Scan Electron Microsc 1775–1784 PMID: 7184150

40. Hinz B, Phan SH, Thannickal VJ et al (2007) The myofibroblast: one function, multiple origins. Am J Pathol 170:1807–1816

41. Baum J, Duffy HS (2011) Fibroblasts and myofibroblasts: what are we talking about? J Cardiovasc Pharmacol 57:376–379

42. Bryant JE, Shamhart PE, Luther DJ et al (2009) Cardiac myofibroblast differentiation is attenuated by alpha(3) integrin blockade: potential role in post-MI remodeling. J Mol Cell Cardiol 46:186–192

43. van den Borne SW, Diez J, Blankesteijn WM et al (2010) Myocardial remodeling after infarction: the role of myofibroblasts. Nat Rev Cardiol 7:30–37

44. Yabluchanskiy A, Li Y, Chilton RJ, Lindsey ML (2013) Matrix metalloproteinases: drug targets for myocardial infarction. Curr Drug Targets 14:276–286

45. Shinde AV, Frangogiannis NG (2013) Fibroblasts in myocardial infarction: a role in inflammation and repair. J Mol Cell Cardiol 70:74–82

46. Leask A (2010) Potential therapeutic targets for cardiac fibrosis: TGFbeta, angiotensin, endothelin, CCN2, and PDGF, partners in fibroblast activation. Circ Res 106:1675–1680

47. Dostal DE, Hunt RA, Kule CE et al (1997) Molecular mechanisms of angiotensin II in modulating cardiac function: intracardiac effects and signal transduction pathways. J Mol Cell Cardiol 29:2893–2902

48. Border WA, Noble NA (1994) Transforming growth factor beta in tissue fibrosis. N Engl J Med 331:1286–1292
49. Yu Q, Stamenkovic I (2000) Cell surface-localized matrix metalloproteinase-9 proteolytically activates TGF-beta and promotes tumor invasion and angiogenesis. Genes Dev 14:163–176
50. Kassiri Z, Defamie V, Hariri M et al (2009) Simultaneous transforming growth factor beta-tumor necrosis factor activation and cross-talk cause aberrant remodeling response and myocardial fibrosis in Timp3-deficient heart. J Biol Chem 284:29893–29904
51. Wipff PJ, Hinz B (2008) Integrins and the activation of latent transforming growth factor beta1– an intimate relationship. Eur J Cell Biol 87:601–615
52. Ge G, Greenspan DS (2006) BMP1 controls TGFbeta1 activation via cleavage of latent TGF-beta-binding protein. J Cell Biol 175:111–120
53. Lee MK, Pardoux C, Hall MC et al (2007) TGF-beta activates Erk MAP kinase signalling through direct phosphorylation of ShcA. EMBO J 26:3957–3967
54. Yamashita M, Fatyol K, Jin C et al (2008) TRAF6 mediates Smad-independent activation of JNK and p38 by TGF-beta. Mol Cell 31:918–924
55. Kang JS, Liu C, Derynck R (2009) New regulatory mechanisms of TGF-beta receptor function. Trends Cell Biol 19:385–394
56. Nakao A, Afrakhte M, Moren A et al (1997) Identification of Smad7, a TGFbeta-inducible antagonist of TGF-beta signalling. Nature 389:631–635
57. Holm TM, Habashi JP, Doyle JJ et al (2011) Noncanonical TGFbeta signaling contributes to aortic aneurysm progression in Marfan syndrome mice. Science 332:358–361
58. Khan R, Sheppard R (2006) Fibrosis in heart disease: understanding the role of transforming growth factor-beta in cardiomyopathy, valvular disease and arrhythmia. Immunology 118:10–24
59. Westermann D, Lindner D, Kasner M et al (2011) Cardiac inflammation contributes to changes in the extracellular matrix in patients with heart failure and normal ejection fraction. Circ Heart Fail 4:44–52
60. Frantz S, Hu K, Adamek A et al (2008) Transforming growth factor beta inhibition increases mortality and left ventricular dilatation after myocardial infarction. Basic Res Cardiol 103:485–492
61. Bujak M, Frangogiannis NG (2007) The role of TGF-beta signaling in myocardial infarction and cardiac remodeling. Cardiovasc Res 74:184–195
62. Kuwahara F, Kai H, Tokuda K et al (2002) Transforming growth factor-beta function blocking prevents myocardial fibrosis and diastolic dysfunction in pressure-overloaded rats. Circulation 106:130–135
63. Jugdutt BI (2003) Remodeling of the myocardium and potential targets in the collagen degradation and synthesis pathways. Curr Drug Targets Cardiovasc Haematol Disord 3:1–30
64. Lamande SR, Bateman JF (1999) Procollagen folding and assembly: the role of endoplasmic reticulum enzymes and molecular chaperones. Semin Cell Dev Biol 10:455–464
65. Jin L, Pahuja KB, Wickliffe KE et al (2012) Ubiquitin-dependent regulation of COPII coat size and function. Nature 482:495–500
66. Kobayashi K, Luo M, Zhang Y et al (2009) Secreted Frizzled-related protein 2 is a procollagen C proteinase enhancer with a role in fibrosis associated with myocardial infarction. Nat Cell Biol 11:46–55
67. Baicu CF, Zhang Y, Van Laer AO et al (2012) Effects of the absence of procollagen C-endopeptidase enhancer-2 on myocardial collagen accumulation in chronic pressure overload. Am J Physiol Heart Circ Physiol 303:H234–240
68. Steiglitz BM, Keene DR, Greenspan DS (2002) PCOLCE2 encodes a functional procollagen C-proteinase enhancer (PCPE2) that is a collagen-binding protein differing in distribution of expression and post-translational modification from the previously described PCPE1. J Biol Chem 277:49820–49830

69. Vadon-Le Goff S, Kronenberg D, Bourhis JM et al (2011) Procollagen C-proteinase enhancer stimulates procollagen processing by binding to the C-propeptide region only. J Biol Chem 286:38932–38938

70. Broder C, Arnold P, Vadon-Le Goff S et al (2013) Metalloproteases meprin alpha and meprin beta are C- and N-procollagen proteinases important for collagen assembly and tensile strength. Proc Natl Acad Sci U S A 110:14219–14224

71. Zuurmond AM, van der Slot-Verhoeven AJ, van Dura EA et al (2005) Minoxidil exerts different inhibitory effects on gene expression of lysyl hydroxylase 1, 2, and 3: implications for collagen cross-linking and treatment of fibrosis. Matrix Biol 24:261–270

72. Eyre D, Shao P, Weis MA, Steinmann B (2002) The kyphoscoliotic type of Ehlers-Danlos syndrome (type VI): differential effects on the hydroxylation of lysine in collagens I and II revealed by analysis of cross-linked telopeptides from urine. Mol Genet Metab 76:211–216

73. Frangogiannis NG (2012) Matricellular proteins in cardiac adaptation and disease. Physiol Rev 92:635–688

74. Fitzgerald MC, Schwarzbauer JE (1998) Importance of the basement membrane protein SPARC for viability and fertility in Caenorhabditis elegans. Curr Biol 8:1285–1288

75. Myers DL, Harmon KJ, Lindner V, Liaw L (2003) Alterations of arterial physiology in osteopontin-null mice. Arterioscler Thromb Vasc Biol 23:1021–1028

76. Schellings MW, Vanhoutte D, Swinnen M et al (2009) Absence of SPARC results in increased cardiac rupture and dysfunction after acute myocardial infarction. J Exp Med 206:113–123

77. Murry CE, Giachelli CM, Schwartz SM, Vracko R (1994) Macrophages express osteopontin during repair of myocardial necrosis. Am J Pathol 145:1450–1462

78. Trueblood NA, Xie Z, Communal C et al (2001) Exaggerated left ventricular dilation and reduced collagen deposition after myocardial infarction in mice lacking osteopontin. Circ Res 88:1080–1087

79. Bradshaw AD, Baicu CF, Rentz TJ et al (2009) Pressure overload-induced alterations in fibrillar collagen content and myocardial diastolic function: role of secreted protein acidic and rich in cysteine (SPARC) in post-synthetic procollagen processing. Circulation 119:269–280

80. Xie Z, Singh M, Singh K (2004) Osteopontin modulates myocardial hypertrophy in response to chronic pressure overload in mice. Hypertension 44:826–831

81. Morrison CJ, Butler GS, Rodriguez D, Overall CM (2009) Matrix metalloproteinase proteomics: substrates, targets, and therapy. Curr Opin Cell Biol 21:645–653

82. Rodriguez D, Morrison CJ, Overall CM (2010) Matrix metalloproteinases: what do they not do? New substrates and biological roles identified by murine models and proteomics. Biochim Biophys Acta 1803:39–54

83. Kassiri Z, Oudit GY, Sanchez O et al (2005) Combination of tumor necrosis factor-alpha ablation and matrix metalloproteinase inhibition prevents heart failure after pressure overload in tissue inhibitor of metalloproteinase-3 knock-out mice. Circ Res 97:380–390

84. Spinale FG, Janicki JS, Zile MR (2013) Membrane-associated matrix proteolysis and heart failure. Circ Res 112:195–208

85. Ma Y, Chiao YA, Zhang J et al (2012) Matrix metalloproteinase-28 deletion amplifies inflammatory and extracellular matrix responses to cardiac aging. Microsc Microanal 18:81–90

86. Oh J, Takahashi R, Kondo S et al (2001) The membrane-anchored MMP inhibitor RECK is a key regulator of extracellular matrix integrity and angiogenesis. Cell 107:789–800

87. Sottrup-Jensen L, Birkedal-Hansen H (1989) Human fibroblast collagenase-alpha-macroglobulin interactions. Localization of cleavage sites in the bait regions of five mammalian alpha-macroglobulins. J Biol Chem 264:393–401

88. Nuttall RK, Sampieri CL, Pennington CJ et al (2004) Expression analysis of the entire MMP and TIMP gene families during mouse tissue development. FEBS Lett 563:129–134

89. Zannad F, Rossignol P, Iraqi W (2010) Extracellular matrix fibrotic markers in heart failure. Heart Fail Rev 15:319–329

90. Creemers EE, Cleutjens JP, Smits JF, Daemen MJ (2001) Matrix metalloproteinase inhibition after myocardial infarction a new approach to prevent heart failure? Circ Res 89:201–210

91. Matsumura S, Iwanaga S, Mochizuki S et al (2005) Targeted deletion or pharmacological inhibition of MMP-2 prevents cardiac rupture after myocardial infarction in mice. J Clin Invest 115:599–609

92. Yabluchanskiy A, Ma Y, Iyer RP et al (2013) Matrix metalloproteinase-9: many shades of function in cardiovascular disease. Physiology (Bethesda) 28:391–403

93. Bourhis JM, Vadon-Le Goff S, Afrache H et al (2013) Procollagen C-proteinase enhancer grasps the stalk of the C-propeptide trimer to boost collagen precursor maturation. Proc Natl Acad Sci U S A 110:6394–6399

94. Sundstrom J, Vasan RS (2006) Circulating biomarkers of extracellular matrix remodeling and risk of atherosclerotic events. Curr Opin Lipidol 17:45–53

95. Laviades C, Varo N, Fernandez J, et al (1998) Abnormalities of the extracellular degradation of collagen type I in essential hypertension. Circulation 98:535–540

96. Lindsay MM, Maxwell P, Dunn FG (2002) TIMP-1: a marker of left ventricular diastolic dysfunction and fibrosis in hypertension. Hypertension 40:136–141

97. Querejeta R, Varo N, Lopez B et al (2000) Serum carboxy-terminal propeptide of procollagen type I is a marker of myocardial fibrosis in hypertensive heart disease. Circulation 101:1729–1735

98. Lin YH, Chiu YW, Shiau YC et al (2006) The relation between serum level of amioterminal propeptide of type I procollagen and diastolic dysfunction in hypertensive patients without diabetes mellitus: a pilot study. J Hum Hypertens 20:964–967

99. Zannad F, Alla F, Dousset B et al (2000) Limitation of excessive extracellular matrix turnover may contribute to survival benefit of spironolactone therapy in patients with congestive heart failure: insights from the randomized aldactone evaluation study (RALES). Rales Investigators. Circulation 102:2700–2706

100. Kallergis EM, Manios EG, Kanoupakis EM et al (2008) Extracellular matrix alterations in patients with paroxysmal and persistent atrial fibrillation: biochemical assessment of collagen type-I turnover. J Am Coll Cardiol 52:211–215

101. Chen CA, Tseng WY, Wang JK et al (2013) Circulating biomarkers of collagen type I metabolism mark the right ventricular fibrosis and adverse markers of clinical outcome in adults with repaired tetralogy of Fallot. Int J Cardiol 167:2963–2968

102. Piestrzeniewicz K, Luczak K, Maciejewski M et al (2014) Clinical outcome, echocardiographic assessment, neurohormonal and collagen turnover markers in "low flow" severe aortic stenosis with high transvalvular gradient. Pol Arch Med Wewn 124:1–2, 19–26

103. Kupari M, Laine M, Turto H et al (2013) Circulating collagen metabolites, myocardial fibrosis and heart failure in aortic valve stenosis. J Heart Valve Dis 22:166–176

104. Lin YH, Ho YL, Wang TD et al (2006) The relation of amino-terminal propeptide of type III procollagen and severity of coronary artery disease in patients without myocardial infarction or hibernation. Clin Biochem 39:861–866

105. Poulsen SH, Host NB, Jensen SE, Egstrup K (2000) Relationship between serum amino-terminal propeptide of type III procollagen and changes of left ventricular function after acute myocardial infarction. Circulation 101:1527–1532

106. Host NB, Jensen LT, Bendixen PM et al (1995) The aminoterminal propeptide of type III procollagen provides new information on prognosis after acute myocardial infarction. Am J Cardiol 76:869–873

107. Manhenke C, Orn S, Squire I et al (2011) The prognostic value of circulating markers of collagen turnover after acute myocardial infarction. Int J Cardiol 150:277–282

108. Zeisberg EM, Tarnavski O, Zeisberg M et al (2007) Endothelial-to-mesenchymal transition contributes to cardiac fibrosis. Nat Med 13:952–961

109. Barry-Hamilton V, Spangler R, Marshall D et al (2010) Allosteric inhibition of lysyl oxidase-like-2 impedes the development of a pathologic microenvironment. Nat Med 16:1009–1017

110. Elnakish MT, Kuppusamy P, Khan M (2013) Stem cell transplantation as a therapy for cardiac fibrosis. J Pathol 229:347–354

Matrix Metalloproteinase 9 (MMP-9)

The Middle-Man of Post-myocardial Infarction Extracellular Matrix Remodeling

Fouad A. Zouein, Ashley DeCoux, Yuan Tian, Jared A. White, Yu-Fang Jin and Merry L. Lindsey

Abstract The substantial involvement of matrix metalloproteinase-9 (MMP-9) in adverse cardiac extracellular matrix (ECM) remodeling makes it one of the most widely investigated MMPs. MMP-9 functions primarily by directly degrading and activating ECM structural and non-structural molecules to regulate cardiac tissue remodeling. This activity is opposed under physiological conditions by a set of endogenous inhibitors known as tissue inhibitors of metalloproteinases (TIMPS). Following myocardial infarction (MI), this constraint is diminished and MMP-9 tissue and plasma levels acutely increase concomitant with a decline in cardiac function. MMP-9 loss-of-function experiments in multiple animal models of MI demonstrate an overall beneficial effect and emphasize the importance of MMP-9 inhibition as a therapeutic intervention. This chapter summarizes MMP-9 structure, transcriptional regulation, post-translational modification, and downstream ECM substrates. We also explore the overall important role of MMP-9 in adverse cardiac remodeling post-MI and its potential utility as a pathophysiological biomarker. Finally, we highlight MMP-9 endogenous and pharmacological inhibitors and the challenges that must be overcome to achieve clinical translation. This is a comprehensive review of MMP-9, from its biochemical structure to its potential role in clinical trials, and can serve as an introduction to young researchers who just joined this research area.

M. L. Lindsey (✉) · F. A. Zouein · A. DeCoux · Y. Tian · Y-F Jin
San Antonio Cardiovascular Proteomics Center, San Antonio, USA
e-mail: MLLindsey@umc.edu

F. A. Zouein · A. DeCoux · Y. Tian · J. A. White · M. L. Lindsey
Mississippi Center for Heart Research, Department of Physiology and Biophysics, University of Mississippi Medical Center, Jackson, MS, USA

Y-F Jin
Department of Electrical and Computer Engineering, The University of Texas at San Antonio, San Antonio, TX, USA

M. L. Lindsey
Research Service, G.V. (Sonny) Montgomery Veterans Affairs Medical Center, Jackson, MS, USA

© Springer International Publishing Switzerland 2015   237
I.M.C. Dixon, J. T. Wigle (eds.), *Cardiac Fibrosis and Heart Failure: Cause or Effect?*,
Advances in Biochemistry in Health and Disease 13, DOI 10.1007/978-3-319-17437-2_13

Keywords ECM substrates · Inflammation · Tissue inhibitors of metalloproteinases ·
Myocardial infarction · Cardiac remodeling · Collagen · Scar formation · Proteomics

1   Introduction

The extracellular matrix (ECM) was historically perceived as a structural entity
providing mechanical support, protection, and a signaling platform for intercellular
interaction and communication [1]. Major ECM roles change under physiological
or pathophysiological conditions, reflecting growth, wound healing, and fibrosis
[2, 3]. Cardiac ECM components (e.g. collagens, elastin, fibronectin, laminin, and
proteoglycans) are synthesized and maintained mainly by fibroblasts, the most numerous cell type in the myocardium [4–6]. The ECM scaffold, outlined by collagen fibers, supports the bulky cardiomyocyte cells and offers a sturdy, yet resilient
framework for the left ventricle to accommodate a wide variation in chamber size
or wall stress, while preventing cardiomyocytes from overstretching and synchro-
nizing myocardial contraction and relaxation during each cardiac cycle [1]. As a
dynamic structure, the ECM also serves as a storage depot for growth factors and
matricellular peptides that influence cardiac development, homeostasis, and remodeling [3, 7, 8].

The matrix metalloproteinases (MMPs) are a key family of enzymes that regulate
ECM turnover. MMPs are a family of 25 members (22 in human), which combined
can cleave all structural elements of the ECM as well as non-ECM substrates [9].
MMP-9 is one of the most widely studied endopeptidases due to its key role in cardiac remodeling in cardiovascular diseases such as hypertension, atherosclerosis,
myocardial infarction (MI), and heart failure [10]. MMP-9 is expressed by a wide
range of cardiac and non-cardiac cells including cardiomyocytes, fibroblasts, vascular smooth muscle cells, endothelial cells, and innate immune cells. The focus of
this chapter will be the evaluation of MMP-9 roles in cardiac remodeling post-MI.

2   Regulation of Transcription and mRNA

The MMP-9 gene is a 2335 bp consisting of 13 exons sequence located on human
chromosome 20q11.2. MMP-9 is translated into a 707 amino acid protein [11]. The
MMP-9 promoter contains several *cis*-elements that contribute to the regulation
of gene expression by interacting with different trans-activators including activator protein-1 (AP-1), nuclear factor-κB (NF-κB), polyomavirus enhancer A-binding
protein-3 (PEA-3), specificity protein 1 (Sp-1), and the β-catenin/Tcf-4 complex
[12, 13]. Multiple growth factors (e.g. transforming growth factor-beta (TGF-β)) and
cytokines (e.g. interleukin-1 beta (IL-1β) and tumor necrosis factor-alpha (TNF-α))
regulate the expression, nuclear translocation and binding of these transcription factors to the MMP-9 promoter to control MMP-9 expression [10]. Transcriptional

regulation is also affected by polymorphisms in the promoter region of the MMP-9 gene. For instance, promoters with increased CA di-nucleotide repeats near the AP-1 binding site are associated with higher transcriptional activity [14]. In context of acute MI, patients had a significantly higher incidence of the-1562 C>T single nucleotide polymorphism (SNP) in the MMP-9 promoter compared to their control group. Higher levels of plasma MMP-9 were reported in the SNP patient group then in the group without [15].

Regulation of transcription is also mediated by transcription factor expression, promoter sequence integrity and accessibility, as well as epigenetic modifications [16]. The latter includes DNA methylation and alterations in chromatin structure. Genetic and pharmacologic studies demonstrate a negative relationship between MMP-9 promoter methylation and MMP-9 mRNA and protein expression [17]. MMP-9 expression is reduced by histone deacetylase recruitment to the MMP-9 promoter due to deacetylated chromatin remodeling and subsequent reduced DNA accessibility [18].

MMP-9 transcription often requires a coordinated effort by more than one transcription factor. Essential MMP-9 transcription factors such as NF-κB and AP-1 possess binding sites to the promoter region and coordinate MMP-9 upregulation in response to a number of stimuli, such as interleukins, reactive oxygen species, angiotensin II, and the glycoprotein, osteopontin [10, 14, 19–24]. Similarly, binding sites for SP-1 and PEA3 transcriptional factors both cooperates with the Ap-1 binding site to achieve a maximum level of transcriptional activation [25]. More recently, miRNAs have also been implicated in the up-regulation of MMP-9 expression [26, 27].

In contrast to genetic up-regulation, induction of MMP-9 transcriptional activity can be counterbalanced by many factors to decrease gene expression. Proteins that interfere with transcription factors binding and promoter activation, including kisspeptin, reversion-inducing-cysteine-rich protein with kazal motifs (RECK), early growth response protein 1 (EGR-1), CDK5 regulatory subunit associated protein 3 (LZAP), ataxin1 (ATXN1), and SP2 transcription factor/Kruppel-like factor 6 (SP2/KLF6) complex have all been shown to suppress MMP-9 expression [14].

3   MMP-9 Structure

Based on *in vitro* substrate specificity, MMPs are loosely divided into five groups: collagenases, stomelysins, matrilysins, gelatinases, and membrane type metalloproteinases [9, 10, 28]. Most MMPs are synthesized and released extracellularly as zymogens, where further processing by tissue and plasma proteinase generates fully active enzyme [29]. MMP-9 was first termed 92 kDa gelatinase (gelatinase B) due to its ability to degrade gelatin [30]. MMP-9 possesses four distinct domains: (1) 10 kDa NH_2-terminal pro-domain, (2) catalytic domain, (3) linker domain, and (4) hemopexin-like domain (Fig. 1).

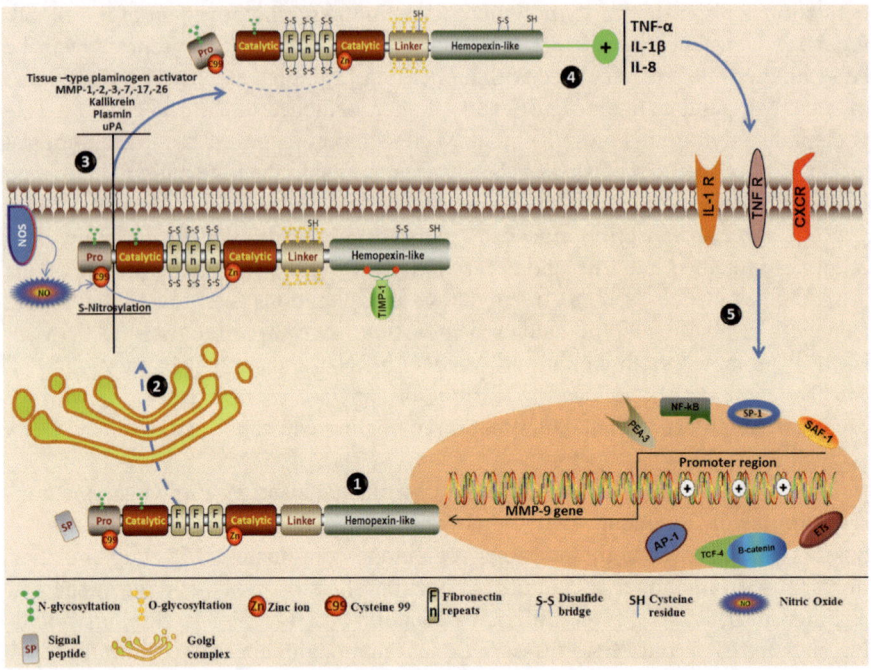

Fig. 1 Positive feedback loop of MMP-9 expression and synthesis. (*1*) ProMMP-9 is co-translationally N-glycosylated at the pro- and catalytic domains. Enzymatic activity is limited by the cysteine switch between C99 of the pro-domain and zinc ion of the catalytic domain. (*2*) Following synthesis, MMP-9 is processed through the Golgi complex where it acquires its tertiary structure characterized by seven disulfide bridges and heavy linker domain O-glycosylation. (*3*) In response to stress such as MI, MMP-9 can be activated by either intracellular components such as S-nitrosylation or released and extracellularly activated by other components such as MMPs, kallikrein, plasmin, and urokinase-type plasminogen activator. Extracellular components activate MMP-9 by disrupting the cysteine switch and releasing the pro-domain. (*4*) Upon activation, MMP-9 proteolytically processes a wide range of substrates, including multiple pro-inflammatory cytokines such as TNF-α, IL-1β, and IL-8, which bind and activate their receptors. (*5*) Receptors activation triggers an intracellular signaling cascade that recruits multiples trans-activators to which MMP-9 possess binding sites at the promoter region. This feeds back to stimulate further MMP-9 expression. Abbreviations: *AP-1* activator protein-1, *ETs* E-26 transcription factors, *IL-1β* interleukin-1 beta, *IL-8* interleukin-8, *NF-κB* nuclear factor-κB, *NOS* nitric oxide synthase, *PEA-3* polyomavirus enhancer A-binding protein-3, *SAF-1* serum amyloid A-activating factor-1, *Sp-1* specificity protein 1, *TIMP-1* tissue inhibitor of metalloproteinases-1, and *TNF-α* tumor necrosis factor alpha

3.1   Pro-, Active, and Inactive MMP-9

In humans, proMMP-9 is 92 kDa and the active form is 88 kDa. In mice, proMMP-9 is 105 kDa and the active form is 88 kDa [28]. MMP-9 activity is mediated by a ~ 170 amino acid catalytic domain, which binds two zinc ions and five calcium ions [31]. When one of the two catalytic zinc ions is structurally associated with the cysteine switch (cys99) of the NH_2-terminal 80 amino acid pro-domain, the enzyme is held in its inactive zymogenic form [31]. The fully active enzymatic form is released by cleavage of the propeptide by tissue kallilkrein, plasminogen activator, and certain MMPs including MMP-1, MMP-2, MMP-3, and MMP-7. However, MMP-9 possesses enzymatic activity even in the presence of its pro-domain [32].Using in situ zymography, Bannikov et al. showed that proMMP-9 of human placental tissue is enzymatically active with the presence of its NH2-terminal propeptide. The enzyme activity, however, is 10 fold lower than the proteolytically activated enzyme. Activity of proMMP-9 is further increased 600-fold higher when bound to a substrate than when free in solution [32]. This suggests that upon substrate binding, proMMP-9 acquires certain conformational changes that activate its catalytic domain, albeit to a lesser extent than the proteolytically processed enzyme.

In addition to the cysteine switch, the catalytic domain contains three repeats of fibronectin-type II domain (~58 amino acids) essential for binding gelatin and conserved among gelatinase MMP members (MMP-2 and MMP-9) [33, 34]. The C-terminal hemopexin-like domain consists of 210 amino acids and is linked to the catalytic domain via an elongated proline-rich 64 amino acid linker domain known as O-glycosylated (OG) domain [35, 36]. The OG domain increases MMP-9 accessibility to large gelatinous substrates. It also controls enzyme flexibility by allowing inter- and intra-domain conformational changes post substrate binding, a feature important for both substrate cleavage and unwinding [37–39].

The hemopexin-like domain plays an important role in modulating MMP-9 function. Structurally, this domain has four blades and one disulfide bridge that drive its interaction with substrates and binding to cell surface receptors. It also provides the optimum conformational positioning for full MMP-9 catalytic activity, which makes it an allosteric site for the endogenous inhibitors TIMPs (Fig. 1) [40].

4   Post-translational Modifications

During synthesis and processing, MMP-9 acquires numerous post-translational modifications resulting in structural and functional variation (Fig. 1). Disulfide bond formation between MMP-9 cysteine residues is the most well-known modification. The MMP-9 translated sequence possesses 19 cysteine residues, two of them located in the signal peptide [41]. The latter is cleaved during translation, leaving only 17 cysteine residues within the pro-MMP-9 form [42]. Fourteen of these residues contribute to the formation of seven disulfide inter-domain bridges that define and fix the folded structure. Six of these disulfide bonds are located within the fibronectin-type II repeats of the catalytic domain and are important for pro-MMP-9 secretion [43]. The remaining bond interconnects blade IV to blade I within the hemopexin domain [44]. Of the three residual cysteines, Cys99 of the pro-peptide domain interacts with the zinc ion of the catalytic domain and forms a switch that limits the enzyme activity. Post-translationally, this switch is subjected to interaction with nitric oxide and subsequent S-nitrosylation, resulting in full enzyme activation

[45]. The remaining Cys468 and Cys674 are located within the O-glycosylated and hemopexin domains, respectively. It is postulated that these residues are involved in intermolecular interaction and formation of MMP-9 multimers [46].

O- and N-Glycosylation are other potent forms of MMP-9 post-translational modifications with high impact on the enzyme catalytic efficiency and oligomerization, as well as intra- and interdomain flexibility. Asn38 of the propeptide domain and Asn120 of the catalytic domains are co-translationally N-glycosylated by oligosaccharyl transferase [47, 48]. O-glycosylation on the other hand, occurs heavily in the OG domain (over 14 potential sites) and during later post-translational stages in the Golgi apparatus [48].

Although most studies focus on the most abundant monomer form, MMP-9 is also found in oligomeric, complexed and truncated forms that are worth investigating. Oligomerization of MMP-9 monomers is assumed to be mediated essentially via O-glycans of the OG domain with potential contribution of Cys468 and Cys674 amino acids [46]. Most known heteromeric postranslational complexes involve MMP-9 covalent and/or non-covalent association with neutrophil gelatinase-associated lipocalin (NGAL) and/or TIMP-1 [49, 50]. Notably mouse MMP-9 lacks the Cys at amino acid 87 that binds NGAL. Chondroitin sulfate proteoglycan (CSPG) is shown to interact with MMP-9 forming a high molecular weight ($\sim$300 kDa) complex produced and secreted by macrophage cell-line THP-1 and postulated to have an important physiological function [51–54]. Truncated forms of MMP-9 are also produced by different proteases, including MMP-3, kallikrein-related peptidase 7 and meperin-α. Some forms lack both the aminoterminal propeptide and the carboxyterminal hemopexin domain, while others lack the carboxyterminal domains only [55–58]. In both cases, the resulting structures could not dimerize with TIMP-1, escaping its inhibitory effect, and may consequently play an important role in mediating the pathophysiological effects of MMP-9.

5   MMP-9 and Adverse Cardiac Remodeling Post-MI

In response to MI injury or stress, the cardiac ECM responds in a manner that is acutely appropriate, but often proves detrimental to the heart. By definition, adverse cardiac remodeling is the progressive series of histopathological events in the left ventricle that ultimately impairs cardiac structure and performance and can lead to heart failure [59]. These structural changes are mediated by two overlapping processes: inflammation and ECM scar formation [60, 61]. Elevated MMP-9 activity is conducive to pathophysiological immune response and fibrotic pathway stimulation, thereby exacerbating disease progression (Table 1). In both components, MMP-9 is the middle-man responsive factor connecting what is an appropriate reparative or reactive action to the initial assault to a detrimental dysregulated reaction.

Table 1 A selection of studies that evaluated MMP-9 as a pathophysiological biomarker

Patients	Sample type	Time of sample aquisition	MMP-9 concentration in MI	MMP-9	Reference
Selected patients from PINRISK study population Acute MI (n=120)	Serum[a]	When admitted to study (no more than 8 years after diagnosis of MI)	56±32.7 ng/ml	↑ in the MI group compared to controls	[145]
AMI (n=22)	Plasma	Upon admission within 3–9 h of chest pain onset	49±28 ng/mL	↑ in AMI compared to control patients	[146]
AMI (n=34)	Plasma	24 h and after 6 months post-MI	93.56±53.7 ng/ml	↑ 24 h after AMI but significantly 6 months post-MI	[147]
AMI (n=60)	Plasma	Within 6–12 of onset of symptoms, and after delivery of thrombolysis, when appropriate	49±11 ng/ml	↑ in AMI patients and in LV volume (LV dysfunction)	[148]
Anterior AMI (n=20)	Great cardiac vein plasma	Within 12 h of onset of AMI	37±5 in AMI patients vs. 15±2 in controls	Levels higher in patients with AMI vs controls	[149]
AMI (n=20)	Plasma	Patients with AMI were admitted within 6 h of symptom onset	195±63 ng/mL in patients with a large MI 78±19 ng/mL in patients with moderate MI	Levels were above baseline in both moderate and large MI patients on admission. Levels in those with large MI were higher than that of moderate MI	[150]
AMI (n=52) patients were recruited from participants in the OPTIMAAL Trial	Plasma	1 month, 1 year, and >4 years after AMI	–	↑ at all time points from acute phase until 4 years post-MI	[151]
Death from cardiac rupture (n=20)	LV	Autopsy of post-MI ruptures	–	↑ in ruptured LV	[152]

[a] Serum levels are artifactually high since clotting can releases MMP-9 from leukocytes

5.1   In Cardiac Inflammation

Following an MI, MMP-9 levels increase substantially [62, 63]. This increase is mediated by activated neutrophil degranulation, which stores pre-formed MMP-9 in gelatinase granules [64–66]. MMP-9 then proteolytically processes and activates several immune chemotactic molecules including CXCL5, CXCl6 and CXCL8 which attracts in a feed forward mechanism more MMP-9 producers into the injured area, such as macrophages and lymphocytes [67]. This pro-inflammatory environment is further enhanced by the ability of MMP-9 to cleave and activate potent inflammatory cytokines such as IL1-β, IL-8 and TNF-α [68, 69], which stimulate pro-inflammatory transcription factors in immune and cardiac cells, namely, NF-B, PEA3, AP-1, Sp-1, serum amyloid A-activating factor-1, and E-26 transcription factors. MMP-9 possesses several response elements to these transcription factors [70–77]. This creates a positive feedback loop [78], which explains the early and persistent elevation of MMP-9 levels in response to inflammatory stress (Fig. 1). However, the major sources of sustained production of MMP-9 shifts from innate immune cells in the inflammatory phase to fibroblasts during the reparative remodeling phase [60]. Although inflammation is initially critical for adequate cardiac repair and adaptation to stress, sustained inflammation can lead to unresolved wound healing [61, 79–86].

5.2   In Cardiac Fibrosis

Cardiac remodeling entails prominent alterations in the ECM [87]. In normal adult hearts, ECM turnover is minimal and highly regulated. Diseased hearts, however, are characterized by robust ECM remodeling of either an increase in collagen deposition or degradation depending upon the diseased state [88]. Following an MI, ECM remodeling is biphasic. In the early stages, ECM is infiltrated by immune cells and degraded to clear debris and necrotic tissue [81]. In the later phase, fibroblasts infiltrate and proliferate and initiate scar formation mainly via collagen deposition [4], where collagen type I deposition dominates [89]. MMP-9 is critically involved in both collagen deposition and degradation.

5.3   In Post Myocardial Infarction

At the onset of MI, the ischemic environment, cell death, and the production of danger-associated molecular pattern molecules (DAMPs) elicit an immediate innate immune response [61, 81, 90]. Neutrophil and macrophage infiltration dominates in the inflammatory phase (0–7 days) and constitute the primary source of sustained MMP-9 production [60, 90], where ECM degradation, phagocytosis and elimination of necrotic tissue and cellular debris occur. Both inflammation and ECM

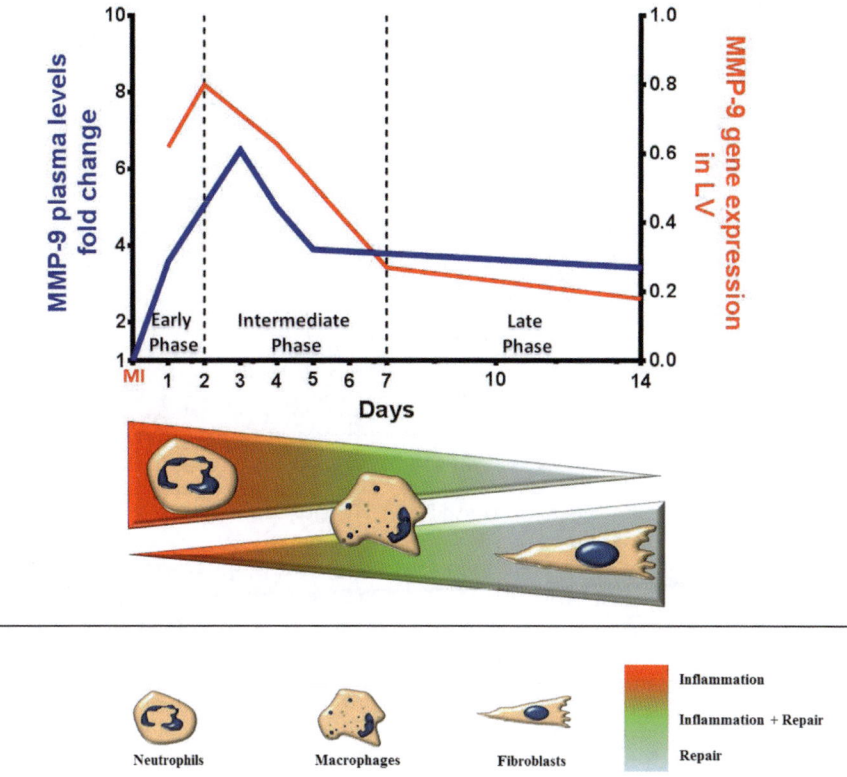

Fig. 2 MMP-9 plasma levels and LV gene expression time course post MI. MMP-9 plasma levels (*blue line*) peaked around 3 days post-MI followed by a significant decrease by days 5–7, where levels afterwards were maintained through day 14 post-MI. MMP-9 gene expression levels in the LV (*red line*) peaked around 2 days post-MI followed by a significant decrease by days 5–7, where levels afterwards were maintained through day 14 post-MI. In the early phase post-MI, inflammation dominates and neutrophils constitute the primary source of MMP-9. In the intermediate phase, inflammation and repairing mechanisms overlap and macrophages become the major source of MMP-9. In the late phase inflammation resolve and repairing and scar formation dominate. At this stage, fibroblasts constitute a source of MMP-9 production

degradation allow fibroblast migration and infiltration into the injured environment where they proliferate and initiate remodeling [4]. In the subsequent reparative phase, fibroblasts become the main source of MMP-9 production and control collagen synthesis and ECM deposition to produce a mature scar. Although, collagen type III deposition is observed as early as two days post-MI, mature scar formation is predominantly formed by collagen I [89] (Fig. 2). The transition from pro-inflammatory and pro-ECM degradation phase to reparative pro-ECM synthesis phase is associated with pathways that turn off inflammation and promote repairing machinery [3, 61].

Table 2 A selection of studies linking MMP-9 to post-MI cardiac remodeling

Animal model	Post-MI remodeling findings		Reference
MMP-9 null mice	Reduced ↑ in LV dimensions compared to WT	↓ LV rupture	[92, 94]
		↑ neovascularization	
	↑ ES pressure	↓ septal wall thickness (compensatory hypertrophy)	
	↑ dP/Dt$_{max}$		
	↓ collagen accumulation in infarct	↑ infarct-to septal wall thickness (less infarct thinning)	
Overexpression of MMP-9 in macrophages	MMP-9 overexpression did not affect macrophage infiltration or neutrophil numbers in the infarct region	↑ ejection fraction (improved function)	[142]
		↓ inflammation	
		↓ M1 activation markers	
	Reduced ↑ in end systolic volume	↓ ECM synthesis and fibrosis	
Aging C57Bl/6 WT mice	↑ MMP-9 activity and LV remodeling	↑ LV rupture in middle aged mice	[143]
WT mice with MMP-9 promoter fused to gelatinase B/LacZ promoter	Selective MMP-9 spatiotemporal induction	↑ MMP-9 promoter induction at day 3 and peaks at day 7	[144]
	MMP-9 promoter activity was highest at the border between the infarct and remote regions		

MMP-9 levels are increased and maintained during both the inflammatory and reparative phases. Notably, this biphasic profile of MMP-9 levels correlates with an early pro-inflammatory detrimental effect and a late beneficial pro-repairing and preservation of LV function effect [91]. Several studies show the beneficial effect of MMP-9 deletion early post-MI (Table 2). Ducharme et al. showed that MMP-9 deletion in the mouse reduces the pro-inflammatory response post-MI and attenuates LV enlargement and collagen deposition [92]. Lindsey et al. documented an overall improvement in LV remodeling mainly due to the stimulation of neovascularization in a mouse MI model of permanent occlusion with targeted MMP-9 deletion [93]. Others have reported the beneficial impact of MMP-9 deletion on post-MI cardiac rupture [94, 95]. Similar findings were also obtained by use of MMP-9 pharmacological inhibitors [96–99] or via indirect MMP-9 attenuation [100]. Lauer et al. reported the beneficial effect of compound 21, an angiotensin II type 2 receptor agonist, on cardiac function and remodeling in later stages post MI [100], mainly via MMP-9 attenuation. The impact of direct and indirect pharmacological inhibition of MMP-9 is detailed in the MMP-9 inhibitors paragraph.

6 MMP-9 Bioactive ECM Substrates

The interstitial matrix and basement membrane constituting the ECM are composed of a variety of highly organized structural proteins like fibronectin, laminin, fibrillin, and various types of collagen, as well as a diversity of non-structural cell-cell and cell-matrix regulatory matricellular proteins including tenascin, thrombospondin, decorin and others [6–8]. Many of these molecules constitute major bioactive substrates to MMP-9 and play important roles in cardiac remodeling [101]. Laminin levels for instance, are increased early post-MI during the ECM degradation phase and have an important role in modulating macrophage infiltration and decreasing cardiac rupture [102]. MMP-9-mediated laminin cleavage makes this positive effect inversely proportional to MMP-9 levels [103]. MMP-9-generated fibronectin fragments act as chemotactic molecules for many cell types involved in the infarct healing process [101, 104]. Type IV and XVIII collagen components of the basement membrane are cleaved by MMP-9 into active anti-angiogenic tumstatin and endostatin respectively [105, 106]. In addition, plasminogen, a matrix bound protein, is cleaved by MMP-9 to generate an NH_2 terminal fragment known as angiostatin and shown to inhibit endothelial cells proliferation and subsequently, angiogenesis [107]. The marked generation of anti-angiogenic fragments by MMP-9 is postulated to be one of the main reasons behind the progressive maladaptive transition to heart failure and likely explains why MMP-9 deletion increases post-MI neovascularization [93, 107].

Galectin-3, a carbohydrate-binding protein, is a powerful participant in ECM cell-cell and cell-matrix adhesion due to the capacity to bind to glycosylated ECM components, such as laminin, fibronectin, and tenascin [108, 109]. It possesses a collagen-like domain highly prone to cleavage by MMP-9 [110]. Galectin-3 release along with high MMP-9 activity is highly associated with inflammation, fibrosis and adverse cardiac remodeling making it a powerful biomarker for poor remodeling prognosis [111]. Thrombospondin-1 is also upregulated post MI but, unlike Galectin-3, acts to preserve LV function mainly by reducing MMP-9 activity, favoring matrix preservation, and stimulating angiogenesis [8, 112]. Tenascin-C is a marker of active remodeling and plays an important role during the proliferative phase of the healing infarct [8, 101, 113]. Higher plasma levels of tenascin-C are associated with poor survival outcome in heart failure patients [114]. Decorin is a collagen-associated proteoglycan that serves as a depot for the anti-inflammatory cytokine, TGF-β [115]. Once cleaved, decorin releases TGF-β, which is further processed by MMP-9 to generate its active form [115]. Last but not least, osteopontin, a phosphorylated acidic glycoprotein is markedly increased post cardiac injury and plays an important role in regulating collagen assembly and deposition [8]. In summary, MMP-9 bioactive ECM substrates possess both detrimental and beneficial effects, further highlighting the Janus-like role of MMP-9 in cardiac remodeling and the importance of spatiotemporal considerations in targeting MMP-9 therapeutically. Recent proteomics studies of the diseased cardiac ECM reveal many potential new substrates for MMP-9 with unknown pathophysiological activity [101, 116]. An understanding of these potential bioactive candidates may provide insight into achieving better and more efficient therapeutic outcomes.

7   A Pathophysiological Biomarker

Heart failure management can be broadly classified into two approaches: decelerating disease progression and relieving comorbid symptoms. There is no treatment that reverses heart failure and approximately 50% of patients with heart failure die within 5 years of diagnosis [117]. This emphasizes the need to stop or slow the progression to heart failure [118]. Several plasma or serum proteins have been categorized as disease biomarkers due to their powerful correlation with screening, diagnosis, prognosis, disease recurrence, or therapeutic monitoring [119, 120]. A better understanding of these molecules is shifting treatment for heart failure prevention or management toward effective personalized medicine. The goal would be that, instead of bundling a group of patients under one treatment protocol, based on their biomarkers profile one could categorize them as low, intermediate or high-risk groups and treat them accordingly. ECM remodeling occurs at very early stages of heart disease and persists throughout progression, making ECM markers a potentially powerful tool for disease assessment. MMP-9's spatial and temporal characteristics place it at the top of the list of the most prominent new biomarkers candidates for cardiac remodeling [60]. In the pre-clinical and clinical settings, increasing MMP-9 levels highly correlate with adverse cardiac remodeling and poor prognostic profile (Table 1) [60]. For example, higher levels of MMP-9 early post-MI associate with a substantial higher risk of LV rupture [95]. This adverse effect is either related to a direct action of MMP-9 on the ECM (i.e, degradation or via activation of bioactive matricellular proteins). In conclusion, there is a strong rationale supporting the usage of MMP-9 levels alone or in combination with other biomarkers to assess, characterize, and better treat cardiac remodeling diseases [60].

8   MMP-9 Inhibitors

Tight regulation of MMP-9 is crucial to maintain its normal physiological function. The temporal MMP-9 activity pattern makes it difficult to discern between its beneficial and detrimental pathophysiological effect during disease progression [91]. Nevertheless, there is general consensus that higher MMP-9 levels are associated with adverse cardiac remodeling and often poor survival outcome [60]. The utility of inhibiting MMP-9 is supported by the significant positive effects observed post-MI in MMP-9 null mice. Here, we present a summary of endogenous MMP-9 inhibitors as well as a selective list of the most prominent specific and non-specific

MMP-9 inhibitors shown to be efficacious pre-clinically in managing cardiac re-modeling post-MI. Their proven effectiveness in animal studies makes them strong candidates for clinical trials.

8.1 Endogenous MMP-9 Inhibitors

Excessive and continuous MMP-9 activity is limited and balanced by a set of endogenous inhibitors in both the circulation with α2-macroglobulin and in tissues with RECK and TIMPs. TIMPs have four members (TIMP 1–4) of size 20–30 kDa, all known to inhibit interstitial MMP-9 activity [121]. The N-terminal of both TIMP-1 and TIMP-2 inhibit MMP-9 by directly interacting with its catalytic domain. The C-terminal domain of TIMP-1 can form a tight complex with the hemopexin domain of pro-MMP-9 in the Golgi apparatus post-translationally and before secretion [122, 123]. Similarly, TIMP-3 is shown to directly interact with pro-MMP-9's hemopexin domain [124].

All TIMPs are highly involved in MI progression and their absence is proven detrimental to cardiac remodeling. TIMP-1 knockout mice have larger post-MI LV volumes [125]. Likewise, absence of TIMP-2, TIMP-3 and TIMP-4 leads to greater infarct expansion and worsened LV dysfunction and increased mortality due to high risk of cardiac rupture [126–128]. All these effects are reversed with MMP inhibition, emphasizing the importance of TIMP-MMP balance in reducing MI-induced cardiac dysfunction. However, although using TIMPs as a potential therapy could be promising, their therapeutic effectiveness is limited by their multifunctional properties and their widespread inhibition of all MMPs. Therefore, more selective MMP-9 inhibitors need to be developed.

8.2 Pharmacological MMP-9 Inhibitors

Many MMP pharmacological inhibitors have recently emerged as potential therapeutic candidates for inflammatory and cardiovascular diseases [129–131]. Tetracycline and their chemically modified derivatives are most widely used to target MMP-9 due to their ability to inhibit the enzyme and lower its levels [132, 133]. More selective MMP-9 inhibitors such as REGA-3G12, SB-3CT and salvianolic acid have also surfaced as promising drugs [134–136].

MMP-9 is also targeted by many drugs that are prominently used in cardiovascular diseases management. For instance, ACE inhibitors decrease MMP-9 activity in hamster MI model by directly binding to the enzyme active site [137]. In heart failure dogs, MMP-9 activity decreased in response to the aldosterone receptor blocker, eplerenone [138]. Propanolol and Carvedilol also exert inhibitory effect by reducing the enzyme levels as seen in rat MI model and patients with idiopathic cardiomyopathy respectively [139, 140].

8.3   MMP-9 Inhibition in Translational Medicine

Although promising, inhibiting MMP-9 activity is yet to reach its maturity as a primary beneficial therapeutic strategy in clinical settings of cardiovascular disease management. Two major problems support this limitation. First, MMP-9 is a member of a wide range of enzymes that share structural similarities but exert different functions. In this context, usage of non-selective or less selective inhibitors increases the risk of adverse effects, which could outweigh the predicted therapeutic outcome. Furthermore, selective MMP-9 targeting is still challenging at the moment and warrants further investigation. Understanding the full spectrum of MMP-9 substrates will help to better define inhibition strategies. Second, MMP-9 spatiotemporal characteristics and beneficial versus detrimental contradictory effect, limit the therapeutic intervention to a specific set of circumstances and disease conditions that are yet to be elucidated.

9   Conclusion and Future Directions

A significant amount of evidence supports a substantial role for MMP-9 in adverse cardiac ECM remodeling. MMP-9 expression and plasma levels are acutely increased in response to cardiac injury, which correlates with adverse cardiac effects and poor survival outcomes, making MMP-9 a potentially powerful biomarker. MMP-9 acts mostly as an intermediary protein that proteolytically processes crucial ECM bioactive molecules involved in cardiac matrix remodeling to activate or inactivate them. Although it is generally known for its extracellular activity, recent studies have shed light on interesting intracellular roles for MMP-9 [141]. Exploring this novel activity may provide better insight into unexplained or uncharted effects of MMP-9 and raise the possibility of new intracellular targeted therapies. In addition to the better studied monomeric form, post-translational modifications of MMP-9 result in oligomeric, complexed and truncated forms with undefined function that warrant further investigation. Also, a better understanding of MMP-9 spatiotemporal characteristics during disease progression is crucial to devise therapeutic strategies that boost beneficial over detrimental outcomes. To accomplish the goal of targeting MMP-9 therapeutically, several key questions await resolution: (1) can MMP-9 changes allow us to phenotype post-MI patients in a more precise and specific way; (2) can we sort out different sets of patients that may undergo post-MI remodeling by different selective mechanisms; and (3) how much of ECM remodeling is cause and how much is consequence? Targeting MMP-9 in a judicious manner to achieve constructive repair of the infarcted heart is challenging, but given the recent progress made in MMP-9 research, this goal is now within sight.

Acknowledgements We acknowledge support from National Institutes of Health (NIH) for T32HL105324, HL101430, HL075360, HL051971, GM104357, and HHSN 268201000036C (N01-HV-00244) for the San Antonio Cardiovascular Proteomics Center, and from the Biomedical Laboratory Research and Development Service of the Veterans Affairs Office of Research and Development Award 5I01BX000505.

References

1. Lockhart M1, Wirrig E, Phelps A, Wessels A (2011) Extracellular matrix and heart development. Birth Defects Res A Clin Mol Teratol 91:535–550
2. Ma Y, Halade GV, Lindsey ML (2012) Extracellular matrix and fibroblast communication following myocardial infarction. J Cardiovasc Transl Res 5:848–857
3. Dobaczewski M, Gonzalez-Quesada C, Frangogiannis NG (2010) The extracellular matrix as a modulator of the inflammatory and reparative response following myocardial infarction. J Mol Cell Cardiol 48:504–511
4. Brown RD, Ambler SK, Mitchell MD, Long CS (2005) The cardiac fibroblast: therapeutic target in myocardial remodeling and failure. Annu Rev Pharmacol Toxicol 45:657–687
5. Jugdutt BI (2003) Remodeling of the myocardium and potential targets in the collagen degradation and synthesis pathways. Curr Drug Targets Cardiovasc Haematol Disord 3:1–30
6. Bosman FT, Stamenkovic I (2003) Functional structure and composition of the extracellular matrix. J Pathol 200:423–428
7. Schellings MW, Pinto YM, Heymans S (2004) Matricellular proteins in the heart: possible role during stress and remodeling. Cardiovasc Res 64:24–31
8. Frangogiannis NG (2012) Matricellular proteins in cardiac adaptation and disease. Physiol Rev 92:635–688
9. Sternlicht MD, Werb Z (2001) How matrix metalloproteinases regulate cell behavior. Annu Rev Cell Dev Biol 17:463–516
10. Yabluchanskiy A, Ma Y, Iyer RP, Hall ME, Lindsey ML (2013) Matrix metalloproteinase-9: many shades of function in cardiovascular disease. Physiology (Bethesda) 28:391–403
11. Minond D, Lauer-Fields JL, Cudic M, Overall CM, Pei D, Brew K, Visse R, Nagase H, Fields GB (2006) The roles of substrate thermal stability and P2 and P1' subsite identity on matrix metalloproteinase triple-helical peptidase activity and collagen specificity. J Biol Chem 281(50):38302–38313
12. Yan C, Boyd DD (2007) Regulation of matrix metalloproteinase gene expression. J Cell Physiol 211:19–26
13. Sato H, Seiki M (1993) Regulatory mechanism of 92 kda type iv collagenase gene expression which is associated with invasiveness of tumor cells. Oncogene 8:395–405
14. Peters DG, Kassam A, St Jean PL, Yonas H, Ferrell RE (1999) Functional polymorphism in the matrix metalloproteinase-9 promoter as a potential risk factor for intracranial aneurysm. Stroke 30:2612–2616
15. Koh YS, Chang K, Kim PJ, Seung KB, Baek SH, Shin WS, Lim SH, Kim JH, Choi KB (2008) A close relationship between functional polymorphism in the promoter region of matrix metalloproteinase-9 and acute myocardial infarction. Int J Cardiol 127:430–432
16. Vandooren J, Van den Steen PE, Opdenakker G (2013) Biochemistry and molecular biology of gelatinase b or matrix metalloproteinase-9 (mmp-9): the next decade. Crit Rev Biochem Mol Biol 48:222–272
17. Chicoine E, Esteve PO, Robledo O, Van Themsche C, Potworowski EF, St-Pierre Y (2002) Evidence for the role of promoter methylation in the regulation of mmp-9 gene expression. Biochem Biophys Res Commun 297:765–772
18. Yan C, Wang H, Toh Y, Boyd DD (2003) Repression of 92-kda type iv collagenase expression by mta1 is mediated through direct interactions with the promoter via a mechanism, which is both dependent on and independent of histone deacetylation. J Biol Chem 278:2309–2316

19. Yang CM, Lee IT, Hsu RC, Chi PL, Hsiao LD (2013) Nadph oxidase/ros-dependent pyk2 activation is involved in tnf-alpha-induced matrix metalloproteinase-9 expression in rat heart-derived h9c2 cells. Toxicol Appl Pharmacol 272:431–442

20. Rouet-Benzineb P, Gontero B, Dreyfus P, Lafuma C (2000) Angiotensin ii induces nuclear factor-kappa b activation in cultured neonatal rat cardiomyocytes through protein kinase c signaling pathway. J Mol Cell Cardiol 32:1767–1778

21. Rangaswami H, Bulbule A, Kundu GC (2004) Nuclear factor-inducing kinase plays a crucial role in osteopontin-induced mapk/ikappabalpha kinase-dependent nuclear factor kappab-mediated promatrix metalloproteinase-9 activation. J Biol Chem 279:38921–38935

22. Clark IM, Swingler TE, Sampieri CL, Edwards DR (2008) The regulation of matrix metalloproteinases and their inhibitors. Int J Biochem Cell Biol 40:1362–1378

23. He C (1996) Molecular mechanism of transcriptional activation of human gelatinase b by proximal promoter. Cancer Lett 106:185–191

24. Yokoo T, Kitamura M (1996) Dual regulation of il-1 beta-mediated matrix metalloproteinase-9 expression in mesangial cells by nf-kappa b and ap-1. Am J Physiol 270:F123–130

25. Chandrasekar B, Mummidi S, Mahimainathan L, Patel DN, Bailey SR, Imam SZ, Greene WC, Valente AJ (2006) Interleukin-18-induced human coronary artery smooth muscle cell migration is dependent on nf-kappab- and ap-1-mediated matrix metalloproteinase-9 expression and is inhibited by atorvastatin. J Biol Chem 281:15099–15109

26. Lin HY, Chiang CH, Hung WC (2013) Stat3 upregulates mir-92a to inhibit reck expression and to promote invasiveness of lung cancer cells. Br J Cancer 109:731–738

27. Limana F, Esposito G, D'Arcangelo D, Di Carlo A, Romani S, Melillo G, Mangoni A, Bertolami C, Pompilio G, Germani A, Capogrossi MC (2011) Hmgb1 attenuates cardiac remodelling in the failing heart via enhanced cardiac regeneration and mir-206-mediated inhibition of timp-3. PLoS ONE 6:e19845

28. Papazafiropoulou A, Tentolouris N (2009) Matrix metalloproteinases and cardiovascular diseases. Hippokratia 13:76–82

29. Borkakoti N (2000) Structural studies of matrix metalloproteinases. J Mol Med (Berl) 78:261–268

30. Nagase H, Visse R, Murphy G (2006) Structure and function of matrix metalloproteinases and TIMPs. Cardiovasc Res 69:562–573

31. Rowsell S, Hawtin P, Minshull CA, Jepson H, Brockbank SM, Barratt DG, Slater AM, McPheat WL, Waterson D, Henney AM, Pauptit RA (2002) Crystal structure of human MMP9 in complex with a reverse hydroxamate inhibitor. J Mol Biol 319:173–181

32. Bannikov GA, Karelina TV, Collier IE, Marmer BL, Goldberg GI (2002) Substrate binding of gelatinase B induces its enzymatic activity in the presence of intact propeptide. J Biol Chem 277:16022–16027

33. Allan JA, Docherty AJ, Barker PJ, Huskisson NS, Reynolds JJ, Murphy G (1995) Binding of gelatinases A and B to type-I collagen and other matrix components. Biochem J 309:299–306

34. Steffensen B, Wallon UM, Overall CM (1995) Extracellular matrix binding properties of recombinant fibronectin type II-like modules of human 72-kDa gelatinase/type IV collagenase. High affinity binding to native type I collagen but not native type IV collagen. J Biol Chem 270:11555–11566

35. De Souza SJ, Pereira HM, Jacchieri S, Brentani RR (1996) Collagen/collagenase interaction: does the enzyme mimic the conformation of its own substrate? FASEB J 10:927–930

36. Gomis-Rüth FX, Gohlke U, Betz M, Knäuper V, Murphy G, López-Otín C, Bode W (1996) The helping hand of collagenase-3 (MMP-13): 2.7 A crystal structure of its C-terminal haemopexin-like domain. J Mol Biol 264:556–566

37. Vandooren J, Geurts N, Martens E, Van den Steen PE, Jonghe SD, Herdewijn P, Opdenakker G (2011) Gelatin degradation assay reveals MMP-9 inhibitors and function of O-glycosylated domain. World J Biol Chem 2:14–24

38. Rosenblum G, Van den Steen PE, Cohen SR, Grossmann JG, Frenkel J, Sertchook R, Slack N, Strange RW, Opdenakker G, Sagi I (2007) Insights into the structure and domain flexibility of full-length pro-matrix metalloproteinase-9/gelatinase B. Structure 15:1227–1236

39. Rosenblum G, Van den Steen PE, Cohen SR, Bitler A, Brand DD, Opdenakker G, Sagi I (2010) Direct visualization of protease action on collagen triple helical structure. PLoS ONE 5:e11043
40. Nagase H, Visse R, Murphy G (2006) Structure and function of matrix metalloproteinases and TIMPs. Cardiovasc Res 69:562–573
41. Wilhelm SM, Collier IE, Marmer BL, Eisen AZ, Grant GA, Goldberg GI (1989) SV40-transformed human lung fibroblasts secrete a 92-kDa type IV collagenase which is identical to that secreted by normal human macrophages. J Biol Chem 264:17213–17221
42. Hibbs MS, Hasty KA, Seyer JM, Kang AH, Mainardi CL (1985) Biochemical and immunological characterization of the secreted forms of human neutrophil gelatinase. J Biol Chem 260:2493–2500
43. Khan MM, Simizu S, Suzuki T, Masuda A, Kawatani M, Muroi M, Dohmae N, Osada H (2012) Protein disulfide isomerase-mediated disulfide bonds regulate the gelatinolytic activity and secretion of matrix metalloproteinase-9. Exp Cell Res 318:904–914
44. Cha H, Kopetzki E, Huber R, Lanzendörfer M, Brandstetter H (2002) Structural basis of the adaptive molecular recognition by MMP9. J Mol Biol 320:1065–1079
45. Gu Z, Kaul M, Yan B, Kridel SJ, Cui J, Strongin A, Smith JW, Liddington RC, Lipton SA (2002) S-nitrosylation of matrix metalloproteinases: signaling pathway to neuronal cell death. Science 297:1186–1190
46. Van den Steen PE, Van Aelst I, Hvidberg V, Piccard H, Fiten P, Jacobsen C, Moestrup SK, Fry S, Royle L, Wormald MR, Wallis R, Rudd PM, Dwek RA, Opdenakker G (2006) The hemopexin and O-glycosylated domains tune gelatinase B/MMP-9 bioavailability via inhibition and binding to cargo receptors. J Biol Chem 281:18626–18637
47. Kotra LP, Zhang L, Fridman R, Orlando R, Mobashery S (2002) N-Glycosylation pattern of the zymogenic form of human matrix metalloproteinase-9. Bioorg Chem 30:356–370
48. Van den Steen PE, Rudd PM, Dwek RA, Opdenakker G (1998) Concepts and principles of O-linked glycosylation. Crit Rev Biochem Mol Biol 33:151–208
49. Kiczak L, Tomaszek A, Bania J, Paslawska U, Zacharski M, Noszczyk-Nowak A, Janiszewski A, Skrzypczak P, Ardehali H, Jankowska EA, Ponikowski P (2013) Expression and complex formation of MMP9, MMP2, NGAL, and TIMP1 in porcine myocardium but not in skeletal muscles in male pigs with tachycardia-induced systolic heart failure. Biomed Res Int 2013:283856
50. Triebel S, Bläser J, Reinke H, Tschesche H (1992) A 25 kDa alpha 2-microglobulin-related protein is a component of the 125 kDa form of human gelatinase. FEBS Lett 314:386–388
51. Winberg JO, Kolset SO, Berg E, Uhlin-Hansen L (2000) Macrophages secrete matrix metalloproteinase 9 covalently linked to the core protein of chondroitin sulphate proteoglycans. J Mol Biol 304:669–680
52. Malla N, Berg E, Theocharis AD, Svineng G, Uhlin-Hansen L, Winberg JO (2013) In vitro reconstitution of complexes between pro-matrix metalloproteinase-9 and the proteoglycans serglycin and versican. FEBS J 280:2870–2887
53. Malla N, Sjøli S, Winberg JO, Hadler-Olsen E, Uhlin-Hansen L (2008) Biological and pathobiological functions of gelatinase dimers and complexes. Connect Tissue Res 49:180–184
54. Malla N, Berg E, Uhlin-Hansen L, Winberg JO (2008) Interaction of pro-matrix metalloproteinase-9/proteoglycan heteromer with gelatin and collagen. J Biol Chem 283:13652–13665
55. Okada Y, Gonoji Y, Naka K, Tomita K, Nakanishi I, Iwata K, Yamashita K, Hayakawa T (1992) Matrix metalloproteinase 9 (92-kDa gelatinase/type IV collagenase) from HT 1080 human fibrosarcoma cells. Purification and activation of the precursor and enzymic properties. J Biol Chem 267:21712–21719
56. Bellini T, Trentini A, Manfrinato MC, Tamborino C, Volta CA, Di Foggia V, Fainardi E, Dallocchio F, Castellazzi M (2012) Matrix metalloproteinase-9 activity detected in body fluids is the result of two different enzyme forms. J Biochem 151:493–499
57. Geurts N, Becker-Pauly C, Martens E, Proost P, Van den Steen PE, Stöcker W, Opdenakker G (2012) Meprins process matrix metalloproteinase-9 (MMP-9)/gelatinase B andenhance the activation kinetics by MMP-3. FEBS Lett 586:4264–4269

58. Ramani VC, Kaushal GP, Haun RS (2011) Proteolytic action of kallikrein-related peptidase 7 produces unique active matrix metalloproteinase-9 lacking the C-terminal hemopexin domains. Biochim Biophys Acta 1813:1525–1531

59. Cohn JN, Ferrari R, Sharpe N (2000) Cardiac remodeling concepts and clinical implications: a consensus paper from an international forum on cardiac remodeling. Behalf of an International Forum on Cardiac Remodeling. J Am Coll Cardiol 35:569–582

60. Halade GV, Jin YF, Lindsey ML (2013) Matrix metalloproteinase (MMP)-9: a proximal biomarker for cardiac remodeling and a distal biomarker for inflammation. Pharmacol Ther 139:32–40

61. Frantz S, Bauersachs J, Ertl G (2009) Post-infarct remodelling: contribution of wound healing and inflammation. Cardiovasc Res 81:474–481

62. Etoh T, Joffs C, Deschamps AM, Davis J, Dowdy K, Hendrick J, Baicu S, Mukherjee R, Manhaini M, Spinale FG (2001) Myocardial and interstitial matrix metalloproteinase activity after acute myocardial infarction in pigs. Am J Physiol Heart Circ Physiol 281:H987–94

63. Romanic AM, Burns-Kurtis CL, Gout B, Berrebi-Bertrand I, Ohlstein EH (2001) Matrix metalloproteinase expression in cardiac myocytes following myocardial infarction in the rabbit. Life Sci 68:799–814

64. Mukherjee R, Colbath GP, Justus CD, Bruce JA, Allen CM, Hewett KW, Saul JP, Gourdie RG, Spinale FG (2010) Spatiotemporal induction of matrix metalloproteinase-9 transcription after discrete myocardial injury. FASEB J 24:3819–3828

65. Faurschou M, Borregaard N (2003) Neutrophil granules and secretory vesicles in inflammation. Microbes Infect 5:1317–1327

66. Borregaard N, Cowland JB (1997) Granules of the human neutrophilic polymorphonuclear leukocyte. Blood 89:3503–3521

67. Van den Steen PE, Wuyts A, Husson SJ, Proost P, Van Damme J, Opdenakker G (2003) Gelatinase B/MMP-9 and neutrophil collagenase/MMP-8 process the chemokines human GCP-2/CXCL6, ENA-78/CXCL5 and mouse GCP-2/LIX and modulate their physiological activities. Eur J Biochem 270:3739–3749

68. Gearing AJ, Beckett P, Christodoulou M, Churchill M, Clements J, Davidson AH, Drummond AH, Galloway WA, Gilbert R, Gordon* JL, Leber TM, Mangan M, Miller K, Nayee P, Owen K, Patel S, Thomas W, Wells G, Wood andLM, Woolley K (1994) Processing of tumour necrosis factor-α precursor by metalloproteinases. Nature 370:555–557

69. Schönbeck U, Mach F, Libby P (1998) Generation of biologically active IL-1β by matrix metalloproteinases: a novel caspase-1-independent pathway of IL-1β processing. J Immunol 161:3340–3346

70. Benbow U, Brinckerhoff CE (1997) The AP-1 site and MMP gene regulation: what is all the fuss about? Matrix Biol 15:519–526

71. Lindsey ML, Zamilpa R (2012) Temporal and spatial expression of matrix metalloproteinases and tissue inhibitors of metalloproteinases following myocardial infarction. Cardiovasc Ther 30:31–41

72. Wasylyk C, Gutman A, Nicholson R, Wasylyk B (1991) The c-Ets oncoprotein activates the stromelysin promoter through the same elements as several non-nuclear oncoproteins. EMBO J 10:1127–1134

73. Wu CY, Hsieh HL, Sun CC, Yang CM (2009) IL-1β induces MMP-9 expression via a Ca^{2+}-dependent CaMKII/JNK/c-JUN cascade in rat brain astrocytes. Glia 57:1775–1789

74. Ray A, Bal BS, Ray BK (2005) Transcriptional induction of matrix metalloproteinase-9 in the chondrocyte and synoviocyte cells is regulated via a novel mechanism: evidence for functional cooperation between serum amyloid A-activating factor-1 and AP-1. J Immunol 175:4039–4048

75. Oda N, Abe M, Sato Y (1999) ETS-1 converts endothelial cells to the angiogenic phenotype by inducing the expression of matrix metalloproteinases and integrin $β_3$. J Cell Physiol 178:121–132

76. Murthy S, Ryan AJ, Carter AB (2012) SP-1 regulation of MMP-9 expression requires Ser586 in the PEST domain. Biochem J 445:229–236

77. Taheri F, Bazan HE (2007) Platelet-activating factor overturns the transcriptional repressor disposition of Sp1 in the expression of MMP-9 in human corneal epithelial cells. Invest Ophthalmol Vis Sci 48:1931–1941

78. Opdenakker G, Van den Steen PE, Dubois B, Nelissen I, Van Coillie E, Masure S, Proost P, Van Damme J (2001) Gelatinase B functions as regulator and effector in leukocyte biology. J Leukoc Biol 69:851–859

79. Liaudet L, Rosenblatt-Velin N (2013) Role of innate immunity in cardiac inflammation after myocardial infarction. Front Biosci (Schol Ed) 5:86–104

80. Abe Y, Kawakami M, Kuroki M, Yamamoto T, Fujii M, Kobayashi H, Yaginuma T, Kashii A, Saito M, Matsushima K (1993) Transient rise in serum interleukin-8 concentration during acute myocardial infarction. Br Heart J 70:132–134

81. Nian M, Lee P, Khaper N, Liu P (2004) Inflammatory cytokines and postmyocardial infarction remodeling. Circ Res 94:1543–1553

82. Kubota T, McTiernan CF, Frye CS, Slawson SE, Lemster BH, Koretsky AP, Demetris AJ, Feldman AM (1997) Dilated cardiomyopathy in transgenic mice with cardiac-specific overexpression of tumor necrosis factor-α. Circ Res 81:627–635

83. Bryant D, Becker L, Richardson J, Shelton J, Franco F, Peshock R, Thompson M, Giroir B (1998) Cardiac failure in transgenic mice with myocardial expression of tumor necrosis factor-alpha. Circulation 97:1375–1381

84. Jordan JE, Zhao ZQ, Vinten-Johansen J (1999) The role of neutrophils in myocardial ischemia-reperfusion injury. Cardiovasc Res 43:860–878

85. Vasan RS, Sullivan LM, Roubenoff R, Dinarello CA, Harris T, Benjamin EJ, Sawyer DB, Levy D, Wilson PW, D'Agostino RB, Framingham Heart Study (2003) Inflammatory markers and risk of heart failure in elderly subjects without prior myocardial infarction: the Framingham Heart Study. Circulation 107:1486–1491

86. Amsterdam EA, Stahl GL, Pan HL, Rendig SV, Fletcher MP, Longhurst JC (1995) Limitation of reperfusion injury by a monoclonal antibody to C5a during myocardial infarction in pigs. Am J Physiol 268:H448–57

87. Tyagi SC (1997) Proteinases and myocardial extracellular matrix turnover. Mol Cell Biochem 168:1–12

88. Kusachi SN (2003) Fibrogenesis, cellular and molecular basis. In: Razzaque MS (eds) Myocardial infarction and cardiac fibrogenesis. New York, Kluwer Academic, 77–96

89. Zimmerman SD, Thomas DP, Velleman SG, Li X, Hansen TR, McCormick RJ (2001) Time course of collagen and decorin changes in rat cardiac and skeletal muscle post-MI. Am J Physiol Heart Circ Physiol 281:H1816–22

90. Yabluchanskiy A, Chilton RJ, Lindsey ML (2013) Left ventricular remodeling: one small step for the extracellular matrix will translate to a giant leap for the myocardium. Congest Heart Fail 19:E5–E8

91. Kelly D, Cockerill G, Ng LL, Thompson M, Khan S, Samani NJ, Squire IB (2007) Plasma matrix metalloproteinase-9 and left ventricular remodelling after acute myocardial infarction in man: a prospective cohort study. Eur Heart J 28:711–718

92. Ducharme A, Frantz S, Aikawa M, Rabkin E, Lindsey M, Rohde LE, Schoen FJ, Kelly RA, Werb Z, Libby P, Lee RT (2000) Targeted deletion of matrix metalloproteinase-9 attenuates left ventricular enlargement and collagen accumulation after experimental myocardial infarction. J Clin Invest 106:55–62

93. Lindsey ML, Escobar GP, Dobrucki LW, Goshorn DK, Bouges S, Mingoia JT, McClister DM Jr, Su H, Gannon J, MacGillivray C, Lee RT, Sinusas AJ, Spinale FG (2006) Matrix metalloproteinase-9 gene deletion facilitates angiogenesis after myocardial infarction. Am J Physiol Heart Circ Physiol 290:H232–H239

94. Heymans S, Luttun A, Nuyens D, Theilmeier G, Creemers E, Moons L, Dyspersin GD, Cleutjens JP, Shipley M, Angellilo A, Levi M, Nübe O, Baker A, Keshet E, Lupu F, Herbert JM, Smits JF, Shapiro SD, Baes M, Borgers M, Collen D, Daemen MJ, Carmeliet P (1999) Inhibition of plasminogen activators or matrix metalloproteinases prevents cardiac rupture but impairs therapeutic angiogenesis and causes cardiac failure. Nat Med 5:1135–1142

95. van den Borne SW, Cleutjens JP, Hanemaaijer R, Creemers EE, Smits JF, Daemen MJ, Blankesteijn WM (2009) Increased matrix metalloproteinase-8 and -9 activity in patients with infarct rupture after myocardial infarction. Cardiovasc Pathol 18:37–43

96. Villarreal FJ, Griffin M, Omens J, Dillmann W, Nguyen J, Covell J (2003) Early short-term treatment with doxycycline modulates postinfarction left ventricular remodeling. Circulation 108:1487–1492

97. Gutierrez FR, Lalu MM, Mariano FS, Milanezi CM, Cena J, Gerlach RF, Santos JE, Torres-Dueñas D, Cunha FQ, Schulz R, Silva JS (2008) Increased activities of cardiac matrix metalloproteinases matrix metalloproteinase (MMP)-2 and MMP-9 are associated with mortality during the acute phase of experimental Trypanosoma cruzi infection. J Infect Dis 197:1468–1476

98. Mukherjee R, Brinsa TA, Dowdy KB, Scott AA, Baskin JM, Deschamps AM, Lowry AS, Escobar GP, Lucas DG, Yarbrough WM, Zile MR, Spinale FG (2003) Myocardial infarct expansion and matrix metalloproteinase inhibition. Circulation 107:618–25

99. Rohde LE, Ducharme A, Arroyo LH, Aikawa M, Sukhova GH, Lopez-Anaya A, McClure KF, Mitchell PG, Libby P, Lee RT (1999) Matrix metalloproteinase inhibition attenuates early left ventricular enlargement after experimental myocardial infarction in mice. Circulation 99:3063–3070

100. Lauer D, Slavic S, Sommerfeld M, Thöne-Reineke C, Sharkovska Y, Hallberg A, Dahlöf B, Kintscher U, Unger T, Steckelings UM, Kaschina E (2014) Angiotensin type 2 receptor stimulation ameliorates left ventricular fibrosis and dysfunction via regulation of tissue inhibitor of matrix metalloproteinase 1/matrix metalloproteinase 9 axis and transforming growth factor $\beta1$ in the rat heart. Hypertension 63:e60–e67

101. Zamilpa R, Lopez EF, Chiao YA, Dai Q, Escobar GP, Hakala K, Weintraub ST, Lindsey ML (2010) Proteomic analysis identifies in vivo candidate matrix metalloproteinase-9 substrates in the left ventricle post-myocardial infarction. Proteomics 10:2214–2223

102. Matsumura S, Iwanaga S, Mochizuki S, Okamoto H, Ogawa S, Okada Y (2005) Targeted deletion or pharmacological inhibition of MMP-2 prevents cardiac rupture after myocardial infarction in mice. J Clin Invest 115:599–609

103. Horstmann S, Kalb P, Koziol J, Gardner H, Wagner S (2003) Profiles of matrix metalloproteinases, their inhibitors, and laminin in stroke patients: influence of different therapies. Stroke 34:2165–2170

104. van Dijk A, Niessen HW, Ursem W, Twisk JW, Visser FC, van Milligen FJ (2008) Accumulation of fibronectin in the heart after myocardial infarction: a putative stimulator of adhesion and proliferation of adipose-derived stem cells. Cell Tissue Res 332:289–298

105. Bendrik C, Robertson J, Gauldie J, Dabrosin C (2008) Gene transfer of matrix metalloproteinase-9 induces tumor regression of breast cancer in vivo. Cancer Res 68:3405–3412

106. Panchal VR, Rehman J, Nguyen AT, Brown JW, Turrentine MW, Mahomed Y, March KL (2004) Reduced pericardial levels of endostatin correlate with collateral development in patients with ischemic heart disease. J Am Coll Cardiol 43:1383–1387

107. Cornelius LA, Nehring LC, Harding E, Bolanowski M, Welgus HG, Kobayashi DK, Pierce RA, Shapiro SD (1998) Matrix metalloproteinases generate angiostatin: effects on neovascularization. J Immunol 161:6845–6852

108. Rosenberg I, Cherayil BJ, Isselbacher KJ, Pillai S (1991) Mac-2-binding glycoproteins. Putative ligands for a cytosolic beta-galactoside lectin. J Biol Chem 266:18731–18736

109. Sato S, Hughes RC (1992) Binding specificity of a baby hamster kidney lectin for H type I and II chains, polylactosamine glycans, and appropriately glycosylated forms of laminin and fibronectin. J Biol Chem 267:6983–6990

110. Ochieng J, Fridman R, Nangia-Makker P, Kleiner DE, Liotta LA, Stetler-Stevenson WG, Raz A (1994) Galectin-3 is a novel substrate for human matrix metalloproteinases-2 and -9. BioChemistry 33:14109–14114

111. Ho JE, Liu C, Lyass A, Courchesne P, Pencina MJ, Vasan RS, Larson MG, Levy D (2012) Galectin-3, a marker of cardiac fibrosis, predicts incident heart failure in the community. J Am Coll Cardiol 60:1249–1256

112. Iruela-Arispe ML (2008) Regulation of thrombospondin1 by extracellular proteases. Curr Drug Targets 9:863–868

113. Sato A, Aonuma K, Imanaka-Yoshida K, Yoshida T, Isobe M, Kawase D, Kinoshita N, Yazaki Y, Hiroe M (2006) Serum tenascin-C might be a novel predictor of left ventricular remodeling and prognosis after acute myocardial infarction. J Am Coll Cardiol 47:2319–2325

114. Franz M, Berndt A, Neri D, Galler K, Grün K, Porrmann C, Reinbothe F, Mall G, Schlattmann P, Renner A, Figulla HR, Jung C, Küthe F (2013) Matrix metalloproteinase-9, tissue inhibitor of metalloproteinase-1, B^+ tenascin-C and ED-A$^+$ fibronectin in dilated cardiomyopathy: potential impact on disease progression and patients prognosis. Int J Cardiol 168:5344–5351

115. Imai K, Hiramatsu A, Fukushima D, Pierschbacher MD, Okada Y (1997) Degradation of decorin by matrix metalloproteinases: identification of the cleavage sites, kinetic analyses and transforming growth factor-beta1 release. Biochem J 322:809–814

116. Barallobre-Barreiro J, Didangelos A, Schoendube FA, Drozdov I, Yin X, Fernández-Caggiano M, Willeit P, Puntmann VO, Aldama-López G, Shah AM, Doménech N, Mayr M (2012) Proteomics analysis of cardiac extracellular matrix remodeling in a porcine model of ischemia/reperfusion injury. Circulation 125:789–802

117. Writing Group Members, Roger VL, Go AS, Lloyd-Jones DM, et al (2012) On behalf of the American Heart Association Statistics Committee and Stroke Statistics Subcommittee; On behalf of the American Heart Association Statistics Committee and Stroke Statistics Subcommittee. Heart disease and stroke statistics-2012 update: a report from the American Heart Association. Circulation 125:e2–e220

118. Zouein FA, Kurdi M, Booz GW (2013) LIF and the heart: just another brick in the wall? Eur Cytokine Netw 24:11–19

119. Marko-Varga G, Lindberg H, Löfdahl CG, Löfdahl CG, Jönsson P, Hansson L, Dahlbäck M, Lindquist E, Johansson L, Foster M, Fehniger TE (2005) Discovery of biomarker candidates within disease by protein profiling: principles and concepts. J Proteome Res 4:1200–1212

120. Vasan RS (2006) Biomarkers of cardiovascular disease: molecular basis and practical considerations. Circulation 113:2335–2362

121. Brew K, Nagase H (2010) The tissue inhibitors of metalloproteinases (TIMPs): an ancient family with structural and functional diversity. Biochim Biophys Acta 1803:55–71

122. Roderfeld M, Graf J, Giese B, Salguero-Palacios R, Tschuschner A, Müller-Newen G, Roeb E (2007) Latent MMP-9 is bound to TIMP-1 before secretion. Biol Chem 388:1227–1234

123. Goldberg GI, Strongin A, Collier IE, Genrich LT, Marmer BL (1992) Interaction of 92-kDa type IV collagenase with the tissue inhibitor of metalloproteinases prevents dimerization, complex formation with interstitial collagenase, and activation of the proenzyme with stromelysin. J Biol Chem 267:4583–4591

124. Baker AH, Edwards DR, Murphy G (2002) Metalloproteinase inhibitors: biologicalactions and therapeutic opportunities. J Cell Sci 115:3719–3727

125. Ikonomidis JS, Hendrick JW, Parkhurst AM, Herron AR, Escobar PG, Dowdy KB, Stroud RE, Hapke E, Zile MR, Spinale FG (2005) Accelerated LV remodeling after myocardial infarction in TIMP-1-deficient mice: effects of exogenous MMP inhibition. Am J Physiol Heart Circ Physiol 288:H149–158.

126. Kandalam V, Basu R, Abraham T, Wang X, Awad A, Wang W, Lopaschuk GD, Maeda N, Oudit GY, Kassiri Z (2010) Early activation of matrix metalloproteinases underlies the exacerbated systolic and diastolic dysfunction in mice lacking TIMP3 following myocardial infarction. Am J Physiol Heart Circ Physiol 299:H1012–1023.

127. Kandalam V, Basu R, Abraham T, Wang X, Soloway PD, Jaworski DM, Oudit GY, Kassiri Z (2010) TIMP2 deficiency accelerates adverse post-myocardial infarction remodeling because of enhanced MT1-MMP activity despite lack of MMP2 activation. Circ Res 106:796–808

128. Koskivirta I, Kassiri Z, Rahkonen O, Kiviranta R, Oudit GY, McKee TD, Kytö V, Saraste A, Jokinen E, Liu PP, Vuorio E, Khokha R (2010) Mice with tissue inhibitor of metalloproteinases 4 (Timp4) deletion succumb to induced myocardial infarction but not to cardiac pressure overload. J Biol Chem 285:24487–24493

129. Hu J, Van den Steen PE, Sang QX, Opdenakker G (2007) Matrix metalloproteinase inhibitors as therapy for inflammatory and vascular diseases. Nat Rev Drug Discov 6:480–498

130. Muroski ME, Roycik MD, Newcomer RG, Van den Steen PE, Opdenakker G, Monroe HR, Sahab ZJ, Sang QX (2008) Matrix metalloproteinase-9/gelatinase B is a putative therapeutic target of chronic obstructive pulmonary disease and multiple sclerosis. Curr Pharm Biotechnol 9:34–46

131. Bench TJ, Jeremias A, Brown DL (2011) Matrix metalloproteinase inhibition with tetracyclines for the treatment of coronary artery disease. Pharmacol Res 64:561–566

132. Romero-Perez D, Fricovsky E, Yamasaki KG, Griffin M, Barraza-Hidalgo M, Dillmann W, Villarreal F (2008) Cardiac uptake of minocycline and mechanisms for *in vivo* cardioprotection. J Am Coll Cardiol 52:1086–1094

133. Salo T, Soini Y, Oiva J, Kariylitalo, Nissinen A, Biancari F, Juvonen T, Satta J (2006) Chemically modified tetracyclines (CMT-3 and CMT-8) enable control of the pathologic remodellation of human aortic valve stenosis via MMP-9 and VEGF inhibition. Int J Cardiol 111:358–364

134. Martens E, Leyssen A, Van Aelst I, Fiten P, Piccard H, Hu J, Descamps FJ, Van den Steen PE, Proost P, Van Damme J, Liuzzi GM, Riccio P, Polverini E, Opdenakker G (2007) A monoclonal antibody inhibits gelatinase B/MMP-9 by selective binding to part of the catalytic domain and not to the fibronectin or zinc binding domains. Biochim Biophys Acta 1770:178–186

135. Hadass O, Tomlinson BN, Gooyit M, Chen S, Purdy JJ, Walker JM, Zhang C, Giritharan AB, Purnell W, Robinson CR 2nd, Shin D, Schroeder VA, Suckow MA, Simonyi A, Y Sun G, Mobashery S, Cui J, Chang M, Gu Z (2013) Selective inhibition of matrix metalloproteinase-9 attenuates secondary damage resulting from severe traumatic brain injury. PLoS ONE 8:e76904

136. Jiang B, Li D, Deng Y, Teng F, Chen J, Xue S, Kong X, Luo C, Shen X, Jiang H, Xu F, Yang W, Yin J, Wang Y, Chen H, Wu W, Liu X, Guo DA (2013) Salvianolic acid A, a novel matrix metalloproteinase-9 inhibitor, prevents cardiac remodeling in spontaneously hypertensive rats. PLoS ONE 8:e59621

137. Yamamoto D, Takai S, Jin D, Inagaki S, Tanaka K, Miyazaki M (2007) Molecular mechanism of imidapril for cardiovascular protection via inhibition of MMP-9. J Mol Cell Cardiol 43:670–676

138. Rastogi S, Mishra S, Zacà V, Alesh I, Gupta RC, Goldstein S, Sabbah HN (2007) Effect of long-term monotherapy with the aldosterone receptor blocker eplerenone on cytoskeletal proteins and matrix metalloproteinases in dogs with heart failure. Cardiovasc Drugs Ther 21:415–422

139. Deten A, Volz HC, Holzl A, Briest W, Zimmer HG (2003) Effect of propranolol on cardiac cytokine expression after myocardial infarction in rats. Mol Cell Biochem 251:127–137

140. Pauschinger M, Rutschow S, Chandrasekharan K, Westermann D, Weitz A, Peter Schwimmbeck L, Zeichhardt H, Poller W, Noutsias M, Li J, Schultheiss HP, Tschope C (2005) Carvedilol improves left ventricular function in murine coxsackievirus-induced acute myocarditis association with reduced myocardial interleukin-1beta and MMP-8 expression and a modulated immune response. Eur J Heart Fail 7:444–452

141. de Castro Brás LE, Cates CA, Deleon-Pennell KY, Ma Y, Iyer RP, Halade GV, Yabluchanskiy A, Fields GB, Weintraub ST, Lindsey ML (2014) Citrate synthase is a novel *in vivo* matrix metalloproteinase-9 substrate that regulates mitochondrial function in the post-myocardial infarction left ventricle. Antioxid Redox Signal 21:1974–1985

142. Zamilpa R, Ibarra J, de Castro Brás LE, Ramirez TA, Nguyen N, Halade GV, Zhang J, Dai Q, Dayah T, Chiao YA, Lowell W, Ahuja SS, D'Armiento J, Jin YF, Lindsey ML (2012) Transgenic overexpression of matrix metalloproteinase-9 in macrophages attenuates the

inflammatory response and improves left ventricular function post-myocardial infarction. J Mol Cell Cardiol 53:599–608

143. Yang Y, Ma Y, Han W, Li J, Xiang Y, Liu F, Ma X, Zhang J, Fu Z, Su YD, Du XJ, Gao XM (2008) Age-related differences in postinfarct left ventricular rupture and remodeling. Am J Physiol Heart Circ Physiol 294:H1815–22
144. Mukherjee R, Mingoia JT, Bruce JA, Austin JS, Stroud RE, Escobar GP, McClister DM Jr, Allen CM, Alfonso-Jaume MA, Fini ME, Lovett DH, Spinale FG (2006) Selective spatio-temporal induction of matrix metalloproteinase-2 and matrix metalloproteinase-9 transcription after myocardial infarction. Am J Physiol Heart Circ Physiol 291:H2216–28
145. Renko J, Kalela A, Jaakkola O, Laine S, Höyhtyä M, Alho H, Nikkari ST (2004) Serum matrix metalloproteinase-9 is elevated in men with a history of myocardial infarction. Scand J Clin Lab Invest 64:255–261
146. Kai H, Ikeda H, Yasukawa H, Kai M, Seki Y, Kuwahara F, Ueno T, Sugi K, Imaizumi T (1998) Peripheral blood levels of matrix metalloproteases-2 and -9 are elevated in patients with acute coronary syndromes. J Am Coll Cardiol 32:368–372
147. Símová J, Skvor J, Slovák D, Mazura I, Zvárová J (2013) Serum levels of matrix metalloproteinases 2 and 9 in patients with acute myocardial infarction. Folia Biol (Praha) 59:181–187
148. Squire IB, Evans J, Ng LL, Loftus IM, Thompson MM (2004) Plasma mmp-9 and mmp-2 following acute myocardial infarction in man: correlation with echocardiographic and neurohumoral parameters of left ventricular dysfunction. J Card Fail 10:328–333
149. Inokubo Y, Hanada H, Ishizaka H, Fukushi T, Kamada T, Okumura K (2001) Plasma levels of matrix metalloproteinase-9 and tissue inhibitor of metalloproteinase-1 are increased in the coronary circulation in patients with acute coronary syndrome. Am Heart J 141:211–217
150. Kaden JJ, Dempfle CE, Sueselbeck T, Brueckmann M, Poerner TC, Haghi D, Haase KK, Borggrefe M (2003) Time-dependent changes in the plasma concentration of matrix metalloproteinase 9 after acute myocardial infarction. Cardiology 99:140–144
151. Orn S, Manhenke C, Squire IB, Ng L, Anand I, Dickstein K (2007) Plasma mmp-2, mmp-9 and n-bnp in long-term survivors following complicated myocardial infarction: relation to cardiac magnetic resonance imaging measures of left ventricular structure and function. J Card Fail 13:843–849
152. van den Borne SW, Cleutjens JP, Hanemaaijer R (2009) Increased matrix metalloproteinase-8 and -9 activity in patients with infarct rupture after myocardial infarction. Cardiovasc Pathol 18:37–43

Collagen Processing and its Role in Fibrosis

Christopher A. McCulloch and Nuno M. Coelho

Abstract In several diseases involving the heart such as pressure overload, diabetic cardiomyopathy or myocardial infarction, fibrosis is a common disorder of myocardial extracellular matrix structure and function. The clinical significance of fibrosis is that accumulation of disorganized fibrillar collagen in the cardiac interstitium can inhibit diastolic and systolic function. Fibrosis is mediated by several different cellular and extracellular processes including disruptions of fibroblast differentiation, perturbations of post-translational processing and assembly of matrix molecules, and inappropriately organized matrix degradation by proteases and intracellular digestion. The enlargement of transformed fibroblast and myofibroblast populations in the diseased cardiac interstitium plays a critical role in the disorganized matrix remodeling that occurs after pressure overload or diabetes because these cells do not process and remodel interstitial collagen in a physiological fashion. New data that have examined the regulation of pro-collagen processing by molecules such as pro-collagen C-endopeptidase enhancer and modulation of collagen assembly by the secreted protein acidic and rich in cysteine, have suggested novel therapeutic targets for ameliorating cardiac fibrosis. Further, studies of transmembrane matrix metalloproteinases, such as MT-1, indicate the remarkable breadth of function and complexity of the matrix proteolytic family since MT-1 can break down the matrix and is also important in mediating collagen degradation by phagocytosis. Our growing recognition that the myocardial matrix is highly dynamic and comprises a wide range of matricellular and non-structural proteins and proteases in addition to well-defined structural proteins, suggests new approaches for myocardial fibrosis in a spectrum of cardiac diseases.

Keywords Collagen · Phagocytosis · Matrix · Turnover · Proteases · Assembly · Matrix metalloproteinases

C. A. McCulloch (✉) · N. M. Coelho
Matrix Dynamics Group, University of Toronto, Room 244, Fitzgerald Building, 150 College Street, M5S 3E2 Toronto, ON, Canada
e-mail: christopher.mcculloch@utoronto.ca

© Springer International Publishing Switzerland 2015   261
I.M.C. Dixon, J. T. Wigle (eds.), *Cardiac Fibrosis and Heart Failure: Cause or Effect?*,
Advances in Biochemistry in Health and Disease 13, DOI 10.1007/978-3-319-17437-2_14

1   Introduction and Clinical Considerations

Several cardiac diseases feature prominent alterations of the structure and function of the cardiac interstitium, which is seen after myocardial infarction, reperfusion injury, diabetes, dilated cardiomyopathy and in the responses to prolonged pressure overload associated with chronic hypertension. Here we will consider the cell phenotypes and alterations in collagen processing and digestion that are particularly relevant to hypertension and diabetes.

In the hypertensive patient, left ventricular hypertrophy manifests as disorganized extracellular matrix structure, increased collagen deposition, and the formation of pro-fibrotic cells in the cardiac interstitium (Fig. 1). These features of cardiac fibrosis markedly reduce cardiac function and increase the risk of heart failure and sudden death. In hypertensive patients, increased mechanical forces generated by cardiac muscle are thought to be critically important for the adverse cardiac remodeling of the ventricular wall.

While considerable progress has been made in understanding the development of cardiac fibrosis in response to humoral cues, little is known about how the mechanical forces associated with hypertension promote the fibrotic response and how collagen processing may contribute to fibrosis of the interstitium and alteration of normal cardiac function. Notably, hypertension is a major public health burden and is one of the leading medical diagnoses for adults in Western countries. For example, approximately 7.5 million Canadians (total population 33 million) currently suffer from hypertension. More than 50 % of cardiovascular diseases, including 6 in 10 strokes and 13 % of premature deaths, are linked to hypertension. These medical conditions cost the Canadian healthcare system over $ 7 billion/year and $ 13 billion/year in lost productivity, which is considerable for a country with a relatively small population.

In the context of hypertension, current therapies are aimed primarily at reducing blood pressure; but these approaches so far have been less effective in treating heart failure and myocardial fibrosis. New therapeutic approaches for cardiac fibrosis may need to consider the roles of fibroblasts and activated myofibroblasts in the pathological remodeling of the cardiac extracellular matrix and how inappropriate processing of prominent matrix molecules such as collagen may provide new therapeutic targets for reversal of fibrosis [1].

Diabetic cardiomyopathy, a specific condition of some individuals afflicted with diabetes, can be defined as the alterations induced by diabetes mellitus in cardiac structure and function in the absence of ischemic heart disease, hypertension or other co-morbidities [2]. Left ventricular hypertrophy, perivascular and interstitial fibrosis are fundamental histopathological features in the hearts of patients with diabetic cardiomyopathy, regardless of the presence or absence of ischemic heart disease [3–6]. Detailed studies of chamber compliance in diabetic patients with either impaired or preserved systolic function and heart failure, but normal renal function, demonstrate increased myocyte stiffness and interstitial fibrosis as potential mediators of cardiac stiffness [7]. Reduced chamber compliance leads to diastolic dysfunction, for which diabetics are particularly prone [8]. Indeed, current

Relationship between Cardiac Development and
Fibrotic Events

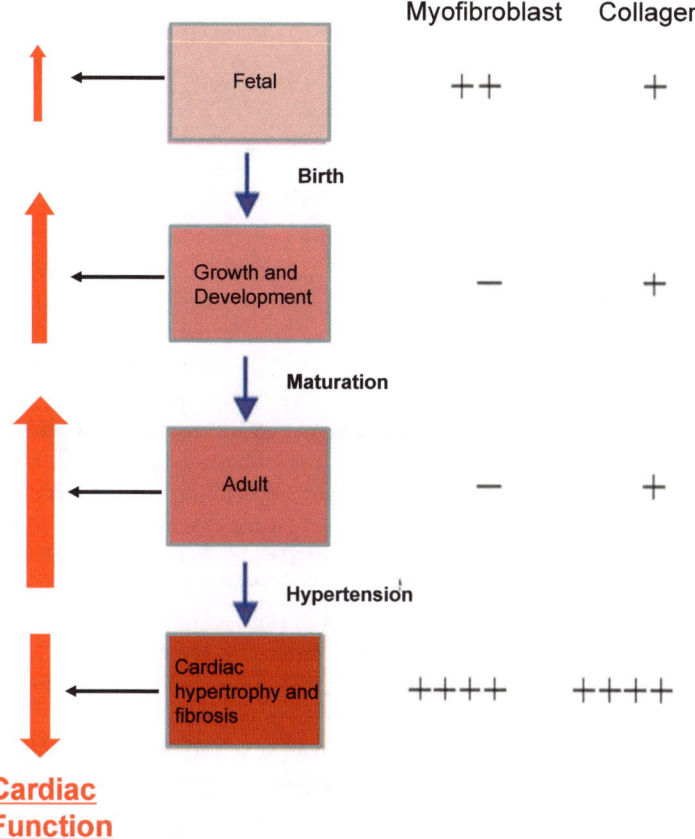

Fig. 1 Relationship between cardiac development and fibrotic events. During development, cardiac function increases and there is minimal fibrosis in the cardiac interstitium. During the development of fibrosis as a result of pressure overload or in diabetes, myofibroblasts appear and there is the formation of a collagen-rich, fibrotic matrix, which contributes to reduced cardiac output

non-invasive assessments of cardiac function by echocardiography have demonstrated that ~30–50 % of hospitalizations for heart failure occur in patients with preserved left ventricular systolic function but with diastolic dysfunction of the left ventricle ("heart failure with preserved ejection fraction" [9, 10]). Recent studies documenting the syndrome of heart failure with preserved ejection fraction demonstrate similar morbidity and mortality as subjects with systolic heart failure [9, 10]. Since diastolic dysfunction is an important cause of symptomatic heart failure [11], it is notable that in diabetic cardiomyopathy that there is adverse accumulation and structural remodeling of the cardiac fibrillar collagen matrix [7]. Despite these findings there are no evidence-based therapies to target diabetes-induced cardiac fibrosis, and treatment remains empirical [12]. Accordingly, it would seem that an improved understanding of pathophysiological processes in the cardiac matrix of the diabetic heart, including the metabolism and processing of prominent cardiac matrix molecules, is needed to develop new therapeutic strategies for clinical management of fibrosis.

2   Functional Importance of Cardiac Extracellular Matrix

The cardiac interstitium is composed of non-myocytic cells and a structural protein network that plays a dominant role in governing the architecture and mechanical behavior of the myocardium [13–15]. The cardiac extracellular matrix (ECM) is composed predominantly of collagen fibers and a variety of other ECM proteins including fibronectin, laminin, tenascin and a spectrum of matricellular molecules like osteopontin. Cardiac muscle contains about 6-fold more collagen than skeletal muscle. Collagen is by far the most abundant protein of mammals [16] and in the cardiac interstitium its physical and biological properties exert powerful effects on cellular and mechanical processes that affect cardiac function. Indeed, differences in the resting tension relationships in cardiac and skeletal muscle may result largely from differences in the connective tissue matrix [17]. The fibrillar elements form a stress-tolerant network that facilitates the distribution of forces generated in the heart and provide for appropriate alignment of cardiac myocytes [18].

The long-term performance of cardiac muscle is regulated by a complex but poorly understood group of feedback mechanisms in which mechanical loading controls the organization of myofibrils, the size of muscle fibers, the expression of muscle-specific genes and the synthesis and secretion of a wide variety of ECM products and trophic factors [19]. This phenomenon is tightly regulated during growth or adaptive responses. Increases of muscle mass occur because of hypertrophic enlargement of terminally differentiated cardiomyocytes, increased numbers of fibroblasts and increased ECM volume [20].

3   Responses of the Cardiac Extracellular Matrix in Disease

When afterload is increased in hypertension, the adult heart adapts by hypertrophy (Fig. 1). This compensatory response in adult hearts is associated with up to a 6-fold increase of type I and III collagens [21, 22] and an increase of the ratio of type III/I collagens [18]. Contemporaneous with increased collagen synthesis is reduced collagen degradation, possibly mediated by reduced collagenolytic activity [23]. There is a large, multi-member family of matrix metalloproteinases (Fig. 2), some of

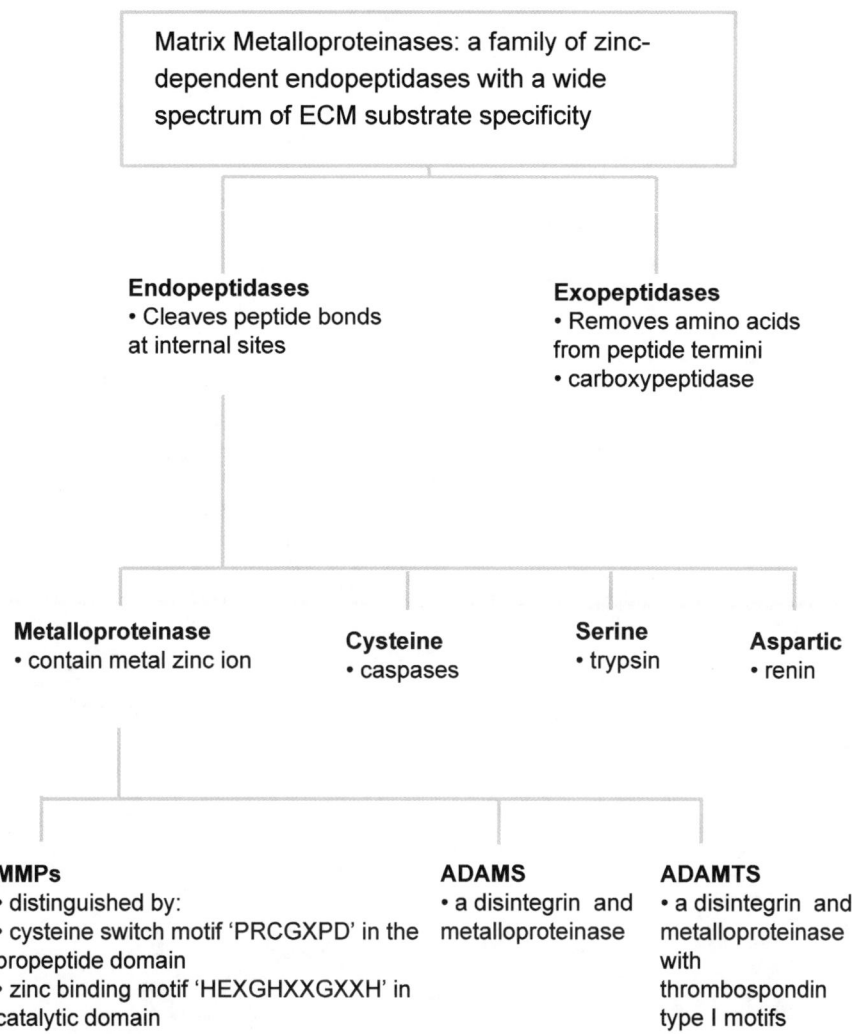

Fig. 2 Matrix Metalloproteinases: a family of zinc-dependent endopeptidases with a wide spectrum of ECM substrate specificity. Diagram to illustrate the structural organization and main members of the proteins that comprise the matrix metallproteinases

which are capable of degrading fibrillar collagen molecules in the adult heart. The collagenolytic activity of these molecules is tightly regulated in time and space and recent evidence indicates that extracellular collagenolysis may be strongly complemented by an intracellular degradation pathway (Fig. 3), which is important for the appropriate organization of the normal interstitium (see below).

Increased deposition of disorganized collagen in the interstitium is an important component of pathological hypertrophy since it may account for abnormal myocardial stiffness [24]. Over the long-term, this adaptive response can contribute to

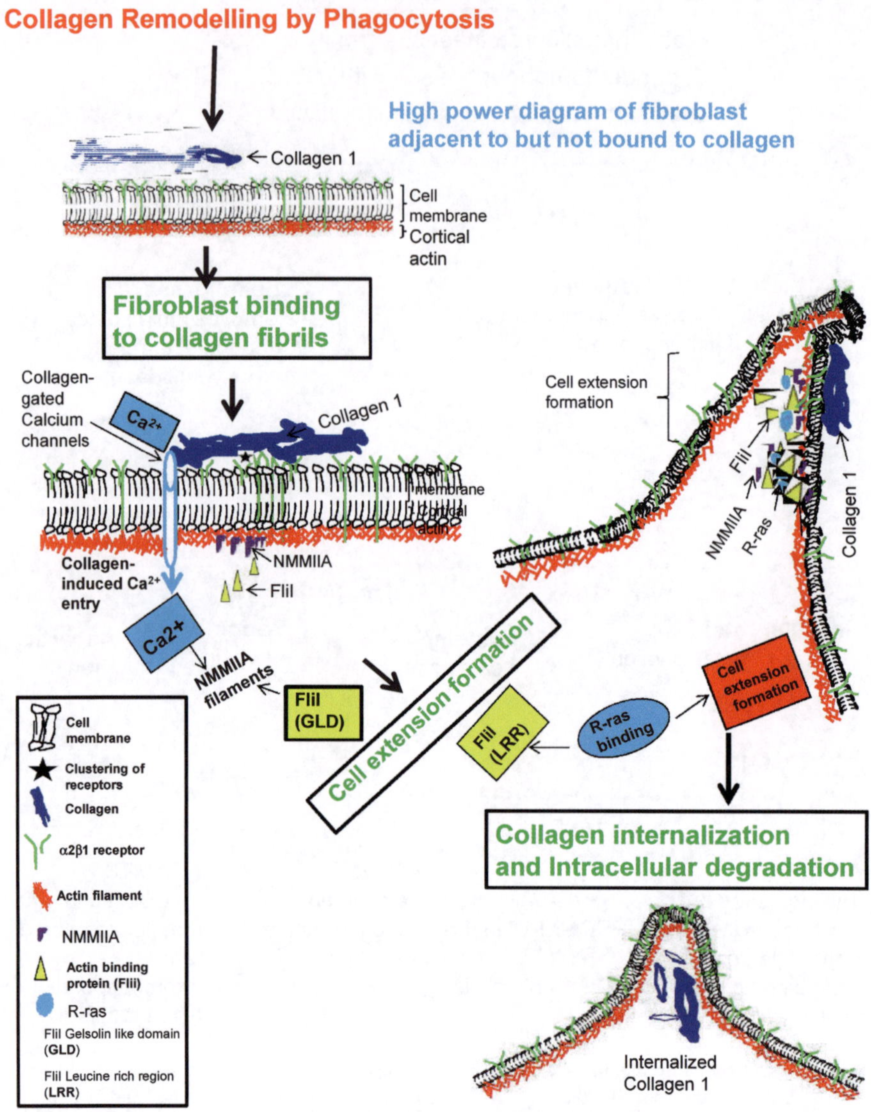

Fig. 3 Collagen Remodelling by Phagocytosis. Schematic diagram to show the main steps in temporal sequence by which fibroblasts remodel collagen by internalization and the critical role of the cytoskeleton in mediating these processes

impairment of cardiac function and heart failure [25]. Thus the regulatory mechanisms that are related to the fibrous tissue response in various cardiovascular diseases (e.g. hypertensive heart disease, dilated cardiomyopathy, post-myocardial infarction, aortic stenosis) are of primary clinical interest [13]. In diabetes, there may also be over-production of collagen [26], which can also contribute to diastolic dysfunction [27] and the formation of a disorganized matrix (Fig. 4). In the context of

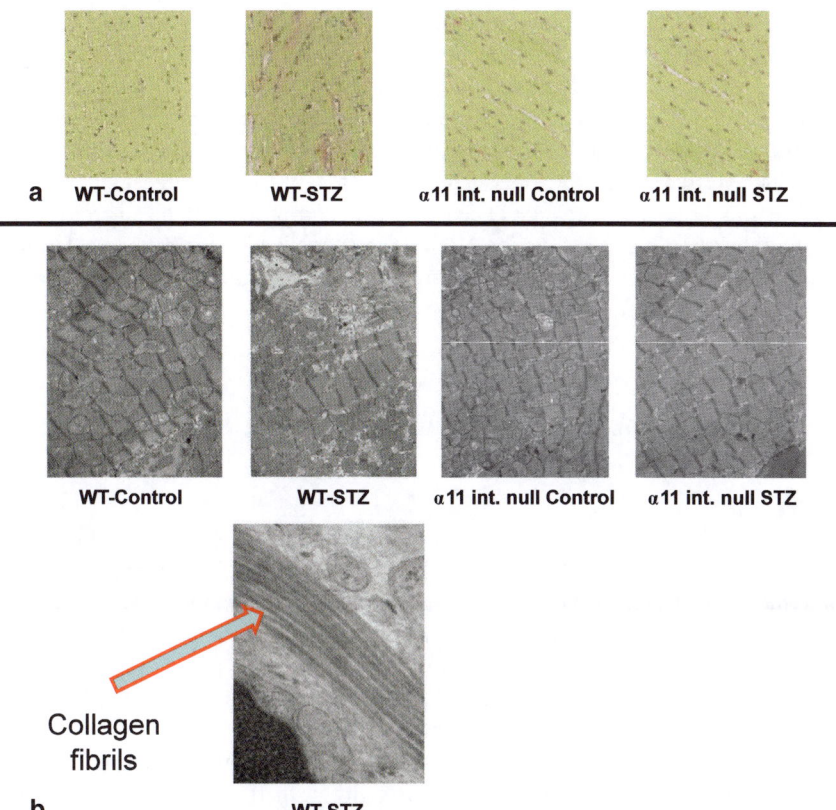

Fig. 4 Morphological analysis of extracellular matrix in streptozotocin-induced diabetes in mice. **a.** Picrosirius red stained paraffin sections of ventricles from mice treated with vehicle or streptozotocin (STZ) to induce diabetes. Mice were killed 8 weeks after STZ. Sections are representative of 5 different mice from each group. Note that interstitial collagen staining after STZ is not detectable in α11 integrin null mice treated with STZ but is readily seen in WT-STZ mice. **b.** Uranyl acetate-stained electron microscopy sections of ventricles from mice treated with vehicle or streptozotocin (STZ) to induce diabetes. Mice were killed 8 weeks after STZ. Sections are representative of 5 different mice from each group. Note that myocyte structure is similar in all groups but in discrete areas of WT-STZ group, there was abundant interstitial collagen, consistent with the picrosirius red staining

the fibrous tissue response in diabetic cardiomyopathy, these alterations are of primary clinical interest in the pathogenesis of the disease [28]. Alterations in collagen cross-linking attributable to diabetic metabolites have been shown to have a marked negative impact on cardiac function [29]. Consistent with these observations, increased serum levels of advanced glycation end products are associated with greatly reduced left ventricular diastolic function in patients with type 1 diabetes [30]. Diabetes, which is known to be an important risk factor for cardiac hypertrophy [31], increases the stiffness of the ECM, contributes to reduced cardiac compliance [32] and alters electro-mechanical connectivity in the myocardium [33]. Currently we do not understand the relative importance of glycated collagen cross-linking versus the synthesis of disorganized collagen by pro-fibrotic myofibroblasts, in contributing to diastolic dysfunction. An improved understanding of how post-translational collagen processing in diabetes and how fibroblasts mediate inappropriate remodeling of myocardial collagen could lead to improved prevention and identification of treatments that interfere with fibrosis. Such an approach has been suggested in animal models using drugs that break advanced glycation end-product-induced cross-links, thereby enhancing cardiac function [34]. Taken together, these data indicate that inappropriate post-translational modifications of collagen and collagen processing may be important determinants of cardiac health.

4 Cardiac Fibroblasts and Myofibroblasts in Response to Hypertension and Diabetes

While cardiac myocytes comprise the largest volume fraction of the adult heart, they represent <25% of cell number [35]. By far the most abundant non-myocyte cell in the myocardium is the fibroblast (30–50% of cell number) [36]. Fibroblasts are the principal cell type involved in the synthesis and remodeling of the ECM and therefore play a central role in cardiac responses to hypertension [37]. The conversion of the fibroblast to the myofibroblast is a critical step since myofibroblasts elaborate a poorly organized collagen matrix that impairs diastolic function [38]. The cardiac fibroblast also plays other, crucial roles in hypertensive responses: (1) cardiac hypertrophy induces expression of angiotensin II receptors in fibroblasts [21]; (2) angiotensin II stimulates fibroblast proliferation [39]; (3) angiotensin II stimulates the autocrine production of transforming growth factor-beta (a pro-fibrotic cytokine) in adult cardiac fibroblasts [40, 41]; (4) angiotensin II promotes the expression of trophic factors that act on myocytes [42]; (5) angiotensin II strongly increases integrin-mediated contractility [43].

When cardiac fibroblasts convert to myofibroblasts in hypertensive conditions, myofibroblasts further contribute to cardiac dysfunction as a result of conduction disturbances [44–47]. Collectively these findings indicate that cardiac fibroblasts and myofibroblasts are of central importance in elaborating and remodeling the ECM [48] in cardiac responses to hypertension. What it is not well understood now is how different fibroblast phenotypes remodel and process interstitial collagen in health compared with hypertensive disease or diabetes. But the opportunity to alter pathological remodeling to effect reversal of fibrosis, possibly by targeting molecules that affect collagen processing [49–51] could open up novel therapeutic approaches. Indeed, the role of matricellular proteins like the secreted protein acidic and rich in cysteine (SPARC) in collagen processing has recently been implicated in cardiac fibrosis [49] and suggests that matricellular molecules may provide productive research opportunities for future anti-fibrotic drugs.

Another prominent example of matricellular molecules that influence cardiac fibrosis and myofibroblast formation is osteopontin, a multifunctional protein that is strongly expressed in healing wounds and fibrotic lesions, both of which are characterized by the formation of myofibroblasts. We found that osteopontin is required for the differentiation and activity of myofibroblasts formed in response to the pro-fibrotic cytokine transforming growth factor-β1 [52]. Despite these encouraging findings, currently we do not understand how pathologically processed matrix molecules like collagen and osteopontin contribute to the alterations of the fibroblast phenotypes that are evidently important in pressure overload-induced cardiac fibrosis [53] but the reciprocal nature of matrix signaling and its impact on cell differentiation processes [54] suggests that this could be fertile area for clinical investigation.

5   Conversion of Fibroblasts to Pro-Fibrotic Myofibroblasts

In normal wound healing, myofibroblasts expressing type I and type III collagens, SMA and the ED-A fibronectin splice isoform transiently form in response to TGF-β and to cell-generated mechanical tension [55, 56]. These contractile cells help to close open wounds and, 4–7 days after wounding, they usually disappear by apoptosis [57]. However, in many pathological situations including the conversion of fibroblasts to myofibroblast in cardiac fibrosis, myofibroblasts persist and continue to synthesize and remodel the ECM, resulting in the formation of a disorganized interstitium [58]. While some of the cytokines that drive fibrosis are known (e.g. TGF-β, endothelin-1, CCN2 [59]), it is not known how the altered matrix in fibrosis is involved in myofibroblast formation and in the persistence of myofibroblast in the hypertensive or diabetic heart.

As noted above, a central feature of cardiac fibrosis and reduced diastolic function in hypertension is the conversion of the fibroblast to the myofibroblast [38, 53, 60]. But what is not currently understood is how the matrix and in particular processing of collagen, may contribute to myofibroblast formation. A number of *in vitro* and *in vivo* studies have suggested critical roles for specific types of cell-ECM adhesions [61, 62] and increased mechanical stiffness of the ECM [56] in determining myofibroblast formation. Myofibroblasts are critical for the formation of disorganized collagen matrix in the myocardium in hypertension [53] and after myocardial infarction [58] and it is conceivable that inappropriately processed matrix collagen molecules that are elaborated by myofibroblasts may enable these cells to persist long after the initial pro-fibrotic stimulus is removed. Therefore identification of post-translational modifications in collagen that may enable pathological prolongation of myofibroblast in fibrotic lesions may suggest new avenues for treatment of critical cell populations that contribute to fibrosis.

6   Remodeling of Collagen

Degradation of collagen, which occurs during growth and development and in response to functional demands in adult connective tissues, is begun within the pericellular environment of fibroblasts and other cells that reside in the matrix. There is little currently known about the role of pericellular collagen remodelling in the adult heart but in view of the important role that the pericellular matrix plays in cardiac remodelling, this would seem to be a potentially productive research field. The pericellular matrix is an important conductor of information transfer to the cell and is replete with dynamic interchange of biologically active molecules, growth factors, and cytokines that regulate cell function. When cells bind collagen and other extracellular matrix components, they also affect the structural integrity of the matrix. Accordingly, homeostasis of the pericellular cardiac matrix is likely a key factor in modulating both fibroblastic and myocytic responses in cardiac function.

Physiological remodeling of the cardiac matrix is critically dependent on the time-appropriate degradation of existing collagen matrices. While a great deal of work has been devoted to understanding the role of matrix metalloproteinases in the extracellular degradation pathway of collagen, many elements of collagen turnover in the cardiac interstitium are not well defined. Recent developments in optical imaging and the development of assays to assess collagen turnover in situ have identified cell types and molecules that are critical in this process in skin [63]; conceivably, these processes may be important in the cardiac interstitium as well. Madsen and colleagues injected collagen into the dermis of mice and the exogenous collagen was rapidly phagocytosed and then degraded in lysosomes. They found that collagen phagocytosis was mediated by receptor-mediated uptake by M2-like macrophages and fibroblasts [63]. These new approaches support a long-standing interest in the phagocytosis of collagen as an important mechanism for regulating the structure and function of the extracellular matrix [64]. Consideration of the phagocytic pathway of collagen degradation and how post-translational modifications to collagen in this process are considered below.

7   Collagen Degradation by Phagocytosis

Collagen receptor-dependent phagocytosis uniquely engages markedly different cortical structures and membrane protrusions than phagocytosis that is initiated by complement or immunoglobulin receptors [65]. In contrast to "professional" phagocytic cells such as macrophages in which cells first extend processes and then bind their targets [66], the initial, rate-limiting step in collagen phagocytosis by fibroblasts is controlled by cell adhesion to collagen, which in turn is dependent on β1 integrin activation (Fig. 5) [67, 68]. The small GTPase Rap1 and NMMIIA [67] are important components of this activation step. Electron microscopic analysis of fibroblast-mediated remodeling of collagen *in vivo* show that following adhesion to

Sequential steps in collagen degradation by phagocytosis

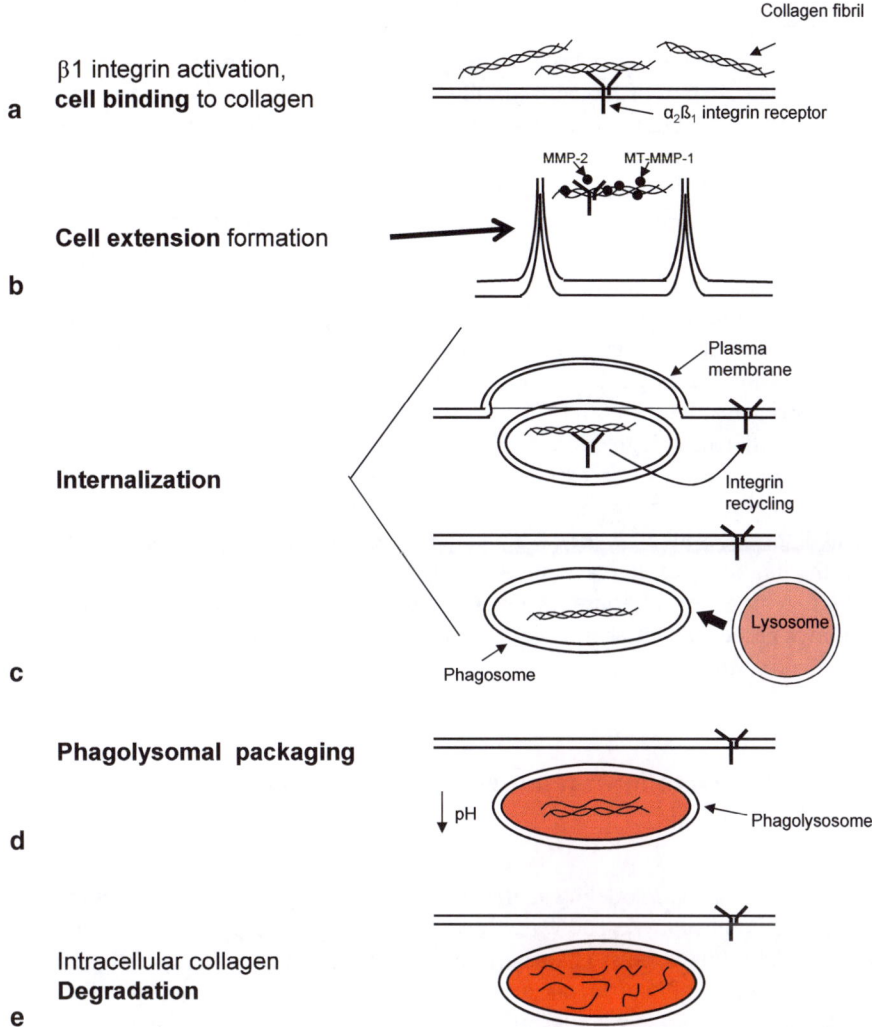

Fig. 5 Sequential steps in collagen degradation by the phagocytic pathway

collagen, actin-rich pseudopods are formed, which pull and reshape collagen fibrils [69]. *In vitro* studies have indicated that these critical steps in collagen phagocytosis are dependent on NMMII [70] and on actin filaments. Accordingly, the actin cytoskeleton, in concert with attachment to integrins (which is mediated by actin binding proteins), provides a dynamic system [71] that enables fibroblasts to interact dynamically and reciprocally with extracellular matrix proteins like collagen and underscores the importance of the cytoskeleton in controlling collagen remodeling.

Removal of collagen is necessary for the reorganization of collagen fibers as tissues grow. In contrast, in mature tissues, existing collagen fibers are replaced by new fibers that are frequently aligned according to altered tensional forces. In physiological remodeling fibroblasts are responsible for, and control both the formation and degradation of the collagen fibers, which are very resistant to proteolytic fragmentation. Degradation of collagen can occur through either extracellular or intracellular pathways [64]. Extracellular degradation of collagen fibers is mediated by matrix metalloproteinases (MMPs) that include MMP-1, MMP-13, and MMP-14 (MT1-MMP), some of the few extracellular proteases capable of degrading native collagen [72]. These enzymes cleave iso-leucine-glycine and leucine-glycine bonds in the helical region of collagen type I collagen. As a result, 3/4- and 1/4 collagen fragments are produced, which unfold at physiological temperature and pH. Although the kinetics of collagen cleavage are relatively slow because of the challenges in obtaining access to scissile bonds [73], in cooperation with gelatinases that cleave denatured 3/4- and 1/4-fragments, collagenolytic matrix metalloproteinases can efficiently degrade the pericellular collagenous matrix [72]. Notably, the degradation of collagen in extracellular environments is often associated with developmental processes, when then there is rapid growth of tissues. These developmental processes involve, for example, the secretion and activation of pro-enzymes and the inhibition of activated enzymes by tissue inhibitors of metalloproteinases [74]. In remodeling adult tissues, collagen degradation occurs primarily via a largely underappreciated intracellular pathway in which fragments of collagen fibers are engulfed by fibroblasts [64].

8 Matrix Metalloproteinases and Processing in Collagen Phagocytosis

It is thought that in remodeling adult tissues, degradation of collagen occurs primarily through a phagocytic pathway. However, although various steps in the phagocytic pathway have been characterized, the enzyme that is required to initially fragment collagen fibrils for subsequent phagocytosis is not completely defined. In work undertaken by Jaro Sodek and co-workers, with the use of confocal microscopy, transmission electron microscopy, and biochemical assays, it was found that human fibroblasts initiated degradation of collagen because of the collagenolytic activity of the membrane-bound metalloproteinase MT1-MMP [75]. Degradation of natural and reconstituted collagen substrates correlated with the expression of MT1-MMP, which was restricted to sites of collagen cleavage at the surface of the cells. Collagen cleavage was also observed inside of cells. The degradation of collagen was blocked with the use of small interfering RNA treatment specific for MT1-MMP. Notably, the gelatinolytic activity of MMP-2 was not needed for collagen phagocytosis. Taken together, these studies showed a critical role of catalytically active MT1-MMP for enabling collagen fibril degradation by the phagocytic pathways [75].

Arising from these earlier studies, post-translational modifications of collagen that occur in diabetes may be important for mediating the pro-fibrotic changes that occur in the diabetic heart, particularly because modifications to collagen structure can affect cell binding to collagen, the first step in collagen degradation by phagocytosis [76]. Notably, in diabetes, glycation-induced modifications of critical amino acids in collagen binding are an important process by which post-translational modifications to collagen affect collagen remodeling by phagocytosis [77]. Treatment with aminoguanidine, a nucleophilic hydrazine compound that blocks the formation of glucose-derived collagen cross-links by derivatizing methylglyoxal, a potent glucose metabolite, prevents glycation-induced inhibition of collagen binding and enables normal collagen phagocytosis to occur [78]. To identify the possible mechanisms involved in inhibition of collagen adhesion, the structural changes caused by methylglyoxal were examined, in particular the cross-linking of lysine residues and modifications of arginine residues of collagen that are mediated by methylglyoxal. In particular, the sequence in collagen that mediates binding through integrins (glycine-phenylalanine-hydroxyproline-glycine-glutamic acid-arginine), is the principal fibrillar collagen binding site for the α2β1 integrin in fibrillar collagens I and II and therefore appears to be important for the initial binding step of collagen phagocytosis. Notably, triple helical collagen peptides (36 amino acids) that contain the α2β1 integrin binding site of type I collagen were used for more detailed molecular analysis of the interference of cell binding that is mediated by collagen glycation. By mass spectrometry, it was found that arginine residues in the collagen binding sequence were modified by methylglyoxal, indicating that modification of arginine residues within the critical α2β1 integrin binding site of collagen may alter binding to cells and as a result, the initial step of collagen phagocytosis. From the standpoint of cardiac matrix biology and diabetes, these findings indicate that methylglyoxal-induced modification of arginine residues in the α2β1 integrin binding region of type I collagen molecules strongly inhibits collagen binding that is required for phagocytosis of collagen fibers. Since phagocytosis is an important step for collagen degradation and remodeling in mature cardiac tissues, this inhibition could be an important post-translational modification in collagen that is involved in the fibrotic processes in the diabetic cardiac matrix.

9 Future Prospects

The fibrotic cardiac interstitium is a challenge both to the research scientist and to the practicing clinician. Fibrosis is an accumulation of disorganized fibrillar collagen in the cardiac interstitium that can have an important and seemingly irreversible effect on cardiac function. Here we have considered the role of collagen phagocytosis in maintenance of connective tissue homeostasis and in particular, how post-translational modifications in diabetes can affect the regulation of this process. Because there are very limited data on how internalization and intracellular processing of collagen contribute to fibrosis, it would seem that these are fruitful areas for examining how perturbations of post-translational processing and digestion of collagen may contribute to the functional disorders in hypertension and diabetes-associated cardiac disease. We have emphasized above the role of the the enlargement of transformed fibroblast and myofibroblast populations in the diseased cardiac interstitium and how they may contribute to the disorganized matrix remodeling observed in pressure overload or diabetes. These cells do not seem to process or remodel interstitial collagen in a physiological manner. The dynamic nature of the myocardial matrix and the poorly understood nature of how matricellular and non-structural proteins and proteases contribute to fibrosis, suggest many different opportunities to address cardiac fibrosis and inappropriate collagen processing in new and exciting ways.

References

1. Goldsmith EC, Bradshaw AD, Spinale FG (2013) Cellular mechanisms of tissue fibrosis. 2. Contributory pathways leading to myocardial fibrosis: moving beyond collagen expression. Am J Physiol Cell Physiol 304:C393–C402
2. Maya L, Villarreal FJ (2009) Diagnostic approaches for diabetic cardiomyopathy and myocardial fibrosis. J Mol Cell Cardiol 48:524–529
3. Srivastava PM, Calafiore P, Macisaac RJ, Patel SK, Thomas MC, Jerums G, Burrell LM (2008) Prevalence and predictors of cardiac hypertrophy and dysfunction in patients with Type 2 diabetes. Clin Sci (Lond) 114:313–320
4. Heerebeek L van, Borbely A, Niessen HW, Bronzwaer JG, Velden J van der, Stienen GJ, Linke WA, Laarman GJ, Paulus WJ (2006) Myocardial structure and function differ in systolic and diastolic heart failure. Circulation 113:1966–1973
5. Heerebeek L van, Hamdani N, Handoko ML, Falcao-Pires I, Musters RJ, Kupreishvili K, Ijsselmuiden AJ, Schalkwijk CG, Bronzwaer JG, Diamant M, Borbely A, Velden J van der, Stienen GJ, Laarman GJ, Niessen HW, Paulus WJ (2008) Diastolic stiffness of the failing diabetic heart: importance of fibrosis, advanced glycation end products, and myocyte resting tension. Circulation 117:43–51
6. Frustaci A, Kajstura J, Chimenti C, Jakoniuk I, Leri A, Maseri A, Nadal-Ginard B, Anversa P (2000) Myocardial cell death in human diabetes. Circ Res 87:1123–1132
7. Asbun J, Villarreal FJ (2006) The pathogenesis of myocardial fibrosis in the setting of diabetic cardiomyopathy. J Am Coll Cardiol 47:693–700
8. Poirier P, Garneau C, Bogaty P, Nadeau A, Marois L, Brochu C, Gingras C, Fortin C, Jobin J, Dumesnil JG (2000) Impact of left ventricular diastolic dysfunction on maximal treadmill performance in normotensive subjects with well-controlled type 2 diabetes mellitus. Am J Cardiol 85:473–477
9. Bhatia RS, Tu JV, Lee DS, Austin PC, Fang J, Haouzi A, Gong Y, Liu PP (2006) Outcome of heart failure with preserved ejection fraction in a population-based study. N Engl J Med 355:260–269
10. Owan TE, Hodge DO, Herges RM, Jacobsen SJ, Roger VL, Redfield MM (2006) Trends in prevalence and outcome of heart failure with preserved ejection fraction. N Engl J Med 355:251–259
11. Burlew BS, Weber KT (2002) Cardiac fibrosis as a cause of diastolic dysfunction. Herz 27:92–98
12. Howlett JG, McKelvie RS, Arnold JM, Costigan J, Dorian P, Ducharme A, Estrella-Holder E, Ezekowitz JA, Giannetti N, Haddad H, Heckman GA, Herd AM, Isaac D, Jong P, Kouz S, Liu P, Mann E, Moe GW, Tsuyuki RT, Ross HJ, White M (2009) Canadian Cardiovascu-

lar Society Consensus Conference guidelines on heart failure, update 2009: diagnosis and management of right-sided heart failure, myocarditis, device therapy and recent important clinical trials. Can J Cardiol 25:85–105

13. Brilla CG, Maisch B, Rupp H, Funck R, Zhou G, Weber KT (1995) Pharmacological modulation of cardiac fibroblast function. Herz 20:127–134

14. Brilla CG, Maisch B, Weber KT (1992) Myocardial collagen matrix remodelling in arterial hypertension. Eur Heart J 13(Suppl D):24–32

15. Weber KT, Sun Y, Tyagi SC, Cleutjens JP (1994) Collagen network of the myocardium: function, structural remodeling and regulatory mechanisms. J Mol Cell Cardiol 26:279–292

16. Perez-Tamayo R (1978) Pathology of collagen degradation. A review. Am J Pathol 92:508–566

17. Covell JW (1990) Cardiac myocyte connective tissue interactions in health and disease, vol 13. pp. 99–112

18. Carver W, Nagpal ML, Nachtigal M, Borg TK, Terracio L (1991) Collagen expression in mechanically stimulated cardiac fibroblasts. Circ Res 69:116–122

19. Yamazaki T, Komuro I, Yazaki Y (1998) Signalling pathways for cardiac hypertrophy. Cell Signal 10:693–698

20. Olson EN, Srivastava D (1996) Molecular pathways controlling heart development. Science 272:671–676

21. Sun Y, Weber KT (1996) Cells expressing angiotensin II receptors in fibrous tissue of rat heart. Cardiovasc Res 31:518–525

22. Butt RP, Laurent GJ, Bishop JE (1995) Mechanical load and polypeptide growth factors stimulate cardiac fibroblast activity. Ann N Y Acad Sci 752:387–393

23. Gonzalez A, Lopez B, Ravassa S, Beaumont J, Arias T, Hermida N, Zudaire A, Diez J (2009) Biochemical markers of myocardial remodelling in hypertensive heart disease. Cardiovasc Res 81:509–518

24. Wilke A, Funck R, Rupp H, Brilla CG (1996) Effect of the renin-angiotensin-aldosterone system on the cardiac interstitium in heart failure. Basic Res Cardiol 91(Suppl 2):79–84

25. Keating MT, Sanguinetti MC (1996) Molecular genetic insights into cardiovascular disease. Science 272:681–685

26. Singh VP, Baker KM, Kumar R (2008) Activation of the intracellular renin-angiotensin system in cardiac fibroblasts by high glucose: role in extracellular matrix production. Am J Physiol Heart Circ Physiol 294:H1675–H1684

27. Aragno M, Mastrocola R, Alloatti G, Vercellinatto I, Bardini P, Geuna S, Catalano MG, Danni O, Boccuzzi G (2008) Oxidative stress triggers cardiac fibrosis in the heart of diabetic rats. Endocrinology 149:380–388

28. Tsujino T, Kawasaki D, Masuyama T (2006) Left ventricular diastolic dysfunction in diabetic patients: pathophysiology and therapeutic implications. Am J Cardiovasc Drugs 6:219–230

29. Herrmann KL, McCulloch AD, Omens JH (2003) Glycated collagen cross-linking alters cardiac mechanics in volume-overload hypertrophy. Am J Physiol Heart Circ Physiol 284:H1277–H1284

30. Berg TJ, Snorgaard O, Faber J, Torjesen PA, Hildebrandt P, Mehlsen J, Hanssen KF (1999) Serum levels of advanced glycation end products are associated with left ventricular diastolic function in patients with type 1 diabetes. Diabetes Care 22:1186–1190

31. Adeghate E (2004) Molecular and cellular basis of the aetiology and management of diabetic cardiomyopathy: a short review. Mol Cell Biochem 261:187–191

32. Zieman S, Kass D (2004) Advanced glycation end product cross-linking: pathophysiologic role and therapeutic target in cardiovascular disease. Congest Heart Fail 10:144–149; quiz 150–141 (Greenwich, Conn)

33. Casis O, Echevarria E (2004) Diabetic cardiomyopathy: electromechanical cellular alterations. Curr Vasc Pharmacol 2:237–248

34. Asif M, Egan J, Vasan S, Jyothirmayi GN, Masurekar MR, Lopez S, Williams C, Torres RL, Wagle D, Ulrich P, Cerami A, Brines M, Regan TJ (2000) An advanced glycation endproduct

cross-link breaker can reverse age-related increases in myocardial stiffness. Proc Natl Acad Sci U S A 97:2809–2813

35. Grove D, Zak R, Nair KG, Aschenbrenner V (1969) Biochemical correlates of cardiac hypertrophy. IV. Observations on the cellular organization of growth during myocardial hypertrophy in the rat. Circ Res 25:473–485

36. Eghbali M (1992) Cardiac fibroblasts: function, regulation of gene expression, and phenotypic modulation. Basic Res Cardiol 87(Suppl 2):183–189

37. Khan R, Sheppard R (2006) Fibrosis in heart disease: understanding the role of transforming growth factor-beta in cardiomyopathy, valvular disease and arrhythmia. Immunology 118:10–24

38. Kuwahara F, Kai H, Tokuda K, Kai M, Takeshita A, Egashira K, Imaizumi T (2002) Transforming growth factor-beta function blocking prevents myocardial fibrosis and diastolic dysfunction in pressure-overloaded rats. Circulation 106:130–135

39. Sadoshima J, Izumo S (1997) The cellular and molecular response of cardiac myocytes to mechanical stress. Annu Rev Physiol 59:551–571

40. Campbell SE, Katwa LC (1997) Angiotensin II stimulated expression of transforming growth factor-beta1 in cardiac fibroblasts and myofibroblasts. J Mol Cell Cardiol 29:1947–1958

41. Lee AA, Dillmann WH, McCulloch AD, Villarreal FJ (1995) Angiotensin II stimulates the autocrine production of transforming growth factor-beta 1 in adult rat cardiac fibroblasts. J Mol Cell Cardiol 27:2347–2357

42. Kim NN, Villarreal FJ, Printz MP, Lee AA, Dillmann WH (1995) Trophic effects of angiotensin II on neonatal rat cardiac myocytes are mediated by cardiac fibroblasts. Am J Physiol 269:E426–E437

43. Burgess ML, Carver WE, Terracio L, Wilson SP, Wilson MA, Borg TK (1994) Integrin-mediated collagen gel contraction by cardiac fibroblasts. Effects of angiotensin II. Circ Res 74:291–298

44. Miragoli M, Gaudesius G, Rohr S (2006) Electrotonic modulation of cardiac impulse conduction by myofibroblasts. Circ Res 98:801–810

45. Miragoli M, Salvarani N, Rohr S (2007) Myofibroblasts induce ectopic activity in cardiac tissue. Circ Res 101:755–758

46. Zlochiver S, Munoz V, Vikstrom KL, Taffet SM, Berenfeld O, Jalife J (2008) Electrotonic myofibroblast-to-myocyte coupling increases propensity to reentrant arrhythmias in two-dimensional cardiac monolayers. Biophys J 95:4469–4480

47. Askar SF, Ramkisoensing AA, Schalij MJ, Bingen BO, Swildens J, Laarse A van der, Atsma DE, Vries AA de, Ypey DL, Pijnappels DA (2011) Antiproliferative treatment of myofibroblasts prevents arrhythmias in vitro by limiting myofibroblast-induced depolarization. Cardiovasc Res 90:295–304

48. Porter KE, Turner NA (2009) Cardiac fibroblasts: at the heart of myocardial remodeling. Pharmacol Ther 123:255–278

49. Bradshaw AD, Baicu CF, Rentz TJ, Laer AO Van, Bonnema DD, Zile MR (2010) Age-dependent alterations in fibrillar collagen content and myocardial diastolic function: role of SPARC in post-synthetic procollagen processing. Am J Physiol Heart Circ Physiol 298:H614–H622

50. Bradshaw AD (2009) The role of SPARC in extracellular matrix assembly. J Cell Commun Signal 3:239–246

51. Bradshaw AD, Baicu CF, Rentz TJ, Laer AO Van, Boggs J, Lacy JM, Zile MR (2009) Pressure overload-induced alterations in fibrillar collagen content and myocardial diastolic function: role of secreted protein acidic and rich in cysteine (SPARC) in post-synthetic procollagen processing. Circulation 119:269–280

52. Lenga Y, Koh A, Perera AS, McCulloch CA, Sodek J, Zohar R (2008) Osteopontin expression is required for myofibroblast differentiation. Circ Res 102:319–327

53. Leslie KO, Taatjes DJ, Schwarz J, vonTurkovich M, Low RB (1991) Cardiac myofibroblasts express alpha smooth muscle actin during right ventricular pressure overload in the rabbit. Am J Pathol 139:207–216

54. Goldsmith EC, Bradshaw AD, Zile MR, Spinale FG (2014) Myocardial fibroblast-matrix interactions and potential therapeutic targets. J Mol Cell Cardiol 70C:92–99
55. Desmouliere A, Geinoz A, Gabbiani F, Gabbiani G (1993) Transforming growth factor-beta 1 induces alpha-smooth muscle actin expression in granulation tissue myofibroblasts and in quiescent and growing cultured fibroblasts. J Cell Biol 122:103–111
56. Arora PD, Narani N, McCulloch CA (1999) The compliance of collagen gels regulates transforming growth factor-beta induction of alpha-smooth muscle actin in fibroblasts. Am J Pathol 154:871–882
57. Gabbiani G (2003) The myofibroblast in wound healing and fibrocontractive diseases. J Pathol 200:500–503
58. Tomasek JJ, Gabbiani G, Hinz B, Chaponnier C, Brown RA (2002) Myofibroblasts and mechano-regulation of connective tissue remodelling. Nat Rev Mol Cell Biol 3:349–363
59. Shi-Wen X, Renzoni EA, Kennedy L, Howat S, Chen Y, Pearson JD, Bou-Gharios G, Dashwood MR, Bois RM du, Black CM, Denton CP, Abraham DJ, Leask A (2007) Endogenous endothelin-1 signaling contributes to type I collagen and CCN2 overexpression in fibrotic fibroblasts. Matrix Biol 26:625–632
60. Kitamura M, Shimizu M, Ino H, Okeie K, Yamaguchi M, Funjno N, Mabuchi H, Nakanishi I (2001) Collagen remodeling and cardiac dysfunction in patients with hypertrophic cardiomyopathy: the significance of type III and VI collagens. Clin Cardiol 24:325–329
61. Pittet P, Lee K, Kulik AJ, Meister JJ, Hinz B (2008) Fibrogenic fibroblasts increase intercellular adhesion strength by reinforcing individual OB-cadherin bonds. J Cell Sci 121:877–886
62. Hinz B, Pittet P, Smith-Clerc J, Chaponnier C, Meister JJ (2004) Myofibroblast development is characterized by specific cell-cell adherens junctions. Mol Biol Cell 15:4310–4320
63. Madsen DH, Leonard D, Masedunskas A, Moyer A, Jurgensen HJ, Peters DE, Amornphimoltham P, Selvaraj A, Yamada SS, Brenner DA, Burgdorf S, Engelholm LH, Behrendt N, Holmbeck K, Weigert R, Bugge TH (2013) M2-like macrophages are responsible for collagen degradation through a mannose receptor-mediated pathway. J Cell Biol 202:951–966
64. Everts V, Zee E van der, Creemers L, Beertsen W (1996) Phagocytosis and intracellular digestion of collagen, its role in turnover and remodeling. Histochem J 28:229–245
65. Olazabal IM, Caron E, May RC, Schilling K, Knecht DA, Machesky LM (2002) Rho-kinase and myosin-II control phagocytic cup formation during CR, but not FcgammaR, phagocytosis. Curr Biol 12:1413–1418
66. Rougerie P, Miskolci V, Cox D. Generation of membrane structures during phagocytosis and chemotaxis of macrophages: role and regulation of the actin cytoskeleton. Immunol Rev 256:222–239
67. Arora PD, Conti MA, Ravid S, Sacks DB, Kapus A, Adelstein RS, Bresnick AR, McCulloch CA (2008) Rap1 activation in collagen phagocytosis is dependent on nonmuscle myosin II-A. Mol Biol Cell 19:5032–5046
68. Arora PD, Glogauer M, Kapus A, Kwiatkowski DJ, McCulloch CA (2004) Gelsolin mediates collagen phagocytosis through a rac-dependent step. Mol Biol Cell 15:588–599
69. Melcher AH, Chan J (1981) Phagocytosis and digestion of collagen by gingival fibroblasts *in vivo*: a study of serial sections. J Ultrastruct Res 77:1–36
70. Meshel AS, Wei Q, Adelstein RS, Sheetz MP (2005) Basic mechanism of three-dimensional collagen fibre transport by fibroblasts. Nat Cell Biol 7:157–164
71. Hay ED (1981) Extracellular matrix. J Cell Biol 91:205s–223s
72. Visse R, Nagase H (2003) Matrix metalloproteinases and tissue inhibitors of metalloproteinases: structure, function, and biochemistry. Circ Res 92:827–839
73. Tam EM, Moore TR, Butler GS, Overall CM (2004) Characterization of the distinct collagen binding, helicase and cleavage mechanisms of matrix metalloproteinase 2 and 14 (gelatinase A and MT1-MMP): the differential roles of the MMP hemopexin c domains and the MMP-2 fibronectin type II modules in collagen triple helicase activities. J Biol Chem 279:43336–43344
74. Egeblad M, Werb Z (2002) New functions for the matrix metalloproteinases in cancer progression. Nat Rev Cancer 2:161–174

75. Lee H, Overall CM, McCulloch CA, Sodek J (2006) A critical role for the membrane-type 1 matrix metalloproteinase in collagen phagocytosis. Mol Biol Cell 17:4812–4826
76. Arora PD, Manolson MF, Downey GP, Sodek J, McCulloch CA (2000) A novel model system for characterization of phagosomal maturation, acidification, and intracellular collagen degradation in fibroblasts. J Biol Chem 275:35432–35441
77. Yuen A, Laschinger C, Talior I, Lee W, Chan M, Birek J, Young EW, Sivagurunathan K, Won E, Simmons CA, McCulloch CA Methylglyoxal-modified collagen promotes myofibroblast differentiation. Matrix Biol 29:537–548
78. Chong SA, Lee W, Arora PD, Laschinger C, Young EW, Simmons CA, Manolson M, Sodek J, McCulloch CA (2007) Methylglyoxal inhibits the binding step of collagen phagocytosis. J Biol Chem 282:8510–8520

Mechanisms of Cardiac Fibrosis and Heart Failure

Keith Dadson, Vera Kovacevic and Gary Sweeney

Abstract The cardiac extracellular matrix (ECM) is the dynamic interstitial scaffolding environment that plays an important role in optimal cardiac function. The strength of the cardiac ECM is known to confer significant protection against myocardial rupture, and the elasticity supports cardiomyocyte contractile function. Upon haemodynamic or ischemic stress, ECM remodeling occurs and is now well established to play a role in the progression of heart disease. Increased turnover of ECM is promoted when resident fibroblasts differentiate into the active myofibroblast phenotype. Although ECM remodeling can be initially beneficial under some circumstances, prolonged and extensive fibrosis is typically associated with decreased contractility, diastolic dysfunction and poor clinical outcome. In this chapter we review the structure and function of the cardiac ECM from development to various pathological states. We will also discuss the role of fibroblasts and the activation of myofibroblasts as well as highlight new findings in the study of reverse cardiac remodelling following unloading of the left ventricle.

Keywords Fibrosis · Cardiac remodeling · Cardiac fibroblast · Ischemia · Pressure overload · Volume overload · Myocardial infarction

1 Structure and Function of the Extracellular Matrix

The cardiac extracellular matrix (ECM) is the highly organized interstitial network of structural proteins that form a scaffold to anchor and tether adjacent cardiac myocytes [1–3]. In the normal heart, the ECM undergoes slow turnover and remains largely quiescent. In the pathology of cardiovascular disease, the cardiac ECM undergoes important changes; such as to provide vital stress bearing support during overload, and to reinforce areas weakened by myocyte apoptosis and necrosis in order to maintain myocardial integrity and prevent rupture in ischemic heart disease [4, 5]. The dynamic re-organization of the cardiac ECM following myocardial stress

G. Sweeney (✉) · K. Dadson · V. Kovacevic
Department of Biology, York University, 4700 Keele Street,
M3J1P3 Toronto, ON M3J1P3, Canada
e-mail: gsweeney@yorku.ca

© Springer International Publishing Switzerland 2015
I.M.C. Dixon, J. T. Wigle (eds.), *Cardiac Fibrosis and Heart Failure: Cause or Effect?*,
Advances in Biochemistry in Health and Disease 13, DOI 10.1007/978-3-319-17437-2_15

or injury is commonly accepted as an early support mechanism to counteract tissue overload and injury, but when left unchecked excessive remodelling of the ECM leads to deleterious matrix accumulation termed cardiac fibrosis. This is associated with myocardial stiffness, diminished cardiac compliance, and poor clinical outcomes, especially in the later stages of heart failure [3, 6, 7].

The cardiac ECM is separated into three layers: the outer layer, the *epimysium,* surrounds the myocardium and lies below the endothelial layer of the epi- and endo-cardium. The *perimysium* is a sheath of connective tissue consisting of tendon-like extensions of the epimysium. The perimysium aggregates myocytes into muscle fibres, bears shearing forces during contraction and relaxation, and minimizes the dissipation of force generated by cardiomyocytes. The *endomysium* surrounds and connects neighbouring myocytes and the intramuscular vasculature [1]. It is proposed that myocardial forces are transmitted through the endomysial ECM to the internal cytoskeleton of individual myocytes via ECM-integrin interactions at the cell surface [8, 9]. All three layers of the ECM are composed mostly of collagen type I which has a tensile strength greater than that of steel [1], and allows for the mechanical co-ordination of forces generated by the myocytes. The ECM also contributes to LV expansion during diastole as coiled fibres release potential energy stored through compression in systole [10]. With a half-life of approximately 80–120 days in the healthy myocardium, the ECM undergoes a slow rate of turnover at approximately 0.6 % per day [1, 11]. The cardiac ECM is composed primarily of fibrillar collagens: collagen type I (~85 %), type III (~10 %), and type IV (~5 %), although the endomysial and perimysial layers also have significant amounts of elastin, allowing more flexibility during each cardiac cycle [2, 3, 11].

2   Cardiac Fibroblasts and Myofibroblasts

Cardiac ECM regulation is principally conducted by cardiac fibroblasts (CFs) which constitute 60–70 % of the total cellular population of the myocardium [11–13]. CFs play a central role in myocardial organization beginning in the developing heart, the earliest organ to form in the incipient embryo, and later in life during the pathogenesis of heart disease. Developmentally, CFs are derived from a migratory cell population that originate in the proepicardial organ. These cells then migrate and infiltrate throughout the forming embryonic heart while responding to a number of growth factors including fibroblast growth factors (FGFs), transforming growth factors (TGFs) and platelet derived growth factors (PDGFs), to undergo epithelial-to-mesenchymal transition (EMT) and then differentiation into CFs [14] (Fig. 1). There is also evidence that endocardial cells may undergo endocardial-to-mesenchymal transition (EndMT) to form CFs resident in atrioventricular valve leaflets [15]. Embryonic CFs throughout the developing heart are responsible for constructing a comprehensive three-dimensional myocardial matrix framework, and for stimulating cardiomyocyte (CM) growth and proliferation during ventricular compaction until birth [16, 17]. In fact, embryonic CF expression of fibronectin, collagen,

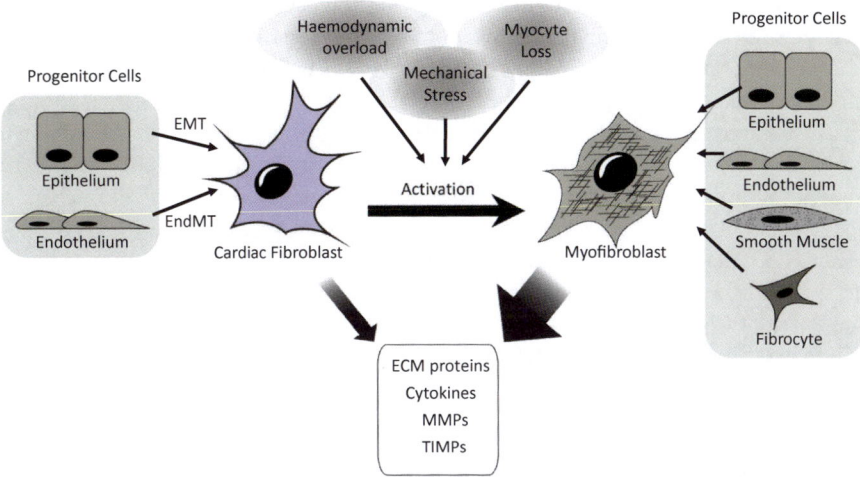

Fig. 1 Schematic summary of various inputs mediating the activation of cardiac fibroblasts to myofibroblasts

and heparin-binding EGF-like growth factor was shown to induce CM proliferation through β1-integrin signalling, while adult CF instead induced CM hypertrophy [17]. These findings implicate CFs as facilitators of the well-established post-natal shift in cardiac growth from embryonic hyperplasia to adult hypertrophy [17].

As the final step in the developmental maturation process, CFs play a principal role in the compensatory mechanism sustaining heart function against the substantial increase in systolic pressure following birth through modification of the ECM to efficiently distribute mechanical stress to the ventricles undergoing severe hypertrophy [14]. This period occurs for a short time after birth, after which mature CFs lie in a more quiescent state in the normal heart. While the mature CF population throughout the adult heart is considered heterogeneous, exhibiting different characteristics based on chamber (atria vs ventricle [18]), and region (infarct vs remote region [19]), they nevertheless collectively contribute to regulation of basal ECM turnover, maintain signalling interactions with cardiomyocytes (CM) and retain their migratory ability to elicit a response to stressors [20, 21].

Throughout development and in the mature heart the constant close proximity of CMs and CFs facilitates bi-directional paracrine and direct cell-cell cross-talk that plays a central role in regulating the structural, mechanical and electrical profile of the heart under both normal and stressed conditions [22]. CFs express several important endocrine factors and are now appreciated as key players contributing to the cardiac-secretome [13]. *In vitro* co-culture of CFs and CM have shown that CM contractile capacity [21] or natriuretic peptide expression [20] are altered following exposure to CFs. Currently, endocrine factors (transforming growth factor-β (TGFβ), fibroblast growth factor-2 (FGF-2), interleukin-6 and -33), ECM proteins (collagen I and III) and direct cell-cell interactions (gap junctions) are all established as important players in fibroblast-myocyte crosstalk [23].

The importance of CF-CM cross-talk *in vivo* was demonstrated in an elegant study by Takeda et al. [24]. Kruppel-like factor-5 (Klf5), a crucial transcription factor for normal embryonic development, is a significant player in the adaptive response to cardiovascular remodelling [25]. Haplosufficiency of Klf5 in CF was shown to minimize fibrosis and hypertrophy in response to moderate PO, whereas Klf5 deletion in CM had no effect. Moreover, following severe overload, CF specific Klf5 deletion severely accentuated adverse remodeling. When transcriptional activation of Klf5 was enhanced in CF, this conferred cardiac protection through induction of CM hypertrophy via insulin-like growth factor-1 (IGF-1) activity [24]. These data suggest that CFs, independent of their actions on the ECM, are important paracrine players in the adaptive response to pressure overload.

In the progression of heart disease, CFs become active, and along with other precursor cells, differentiate into myofibroblasts (myoCF) [2, 13, 14, 24, 26–28]. These fibroblast-smooth muscle hybrid cells become a key source of cardio-active agents, and develop contractile apparatus to aid in preventing LV rupture upon the loss of cardiomyocytes to apoptosis or necrosis [28–30]. It has been proposed that oxidative stress-induced cardiomyocyte necrosis is a mechanism preceding and stimulating myoCF differentiation [2, 31]. The current understanding is that myoCF progenitor cells include not only fibroblasts but also endothelial or epithelial cells (following EndMT and EMT, respectively), smooth muscle cells and bone-marrow derived cells (e.g. fibrocytes, monocytes) [26, 30]. Owing to the diversity of parental cell lineages, myoCF differentiation is a complex process that varies by cell type and stimulus, although AngII, TGFβ and mechanical force are three well established mechanisms to induce increased myoCF expression in the heart [30].

AngII is significantly upregulated in many cases of heart disease, functions through GPCR signalling, and stimulates ERK and p38 signalling [32] and PARP-1 activity [33]. Inhibition of p38 attenuates AngII induced myoCF differentiation [34], while AngII stimulation induces TGFβ and Collagen I expression from CFs [35, 36]. TGFβ is a primary and potent mediator of myoCF differentiation that is upregulated and secreted into the ECM by mesenchymal cells, macrophages, monocytes, and resident CFs [30]. TGFβ signalling is mediated by the TGFβ receptors TBFβRI/RII which initiate SMAD2/3 phosphorylation which, upon activation, complexes with SMAD4. TGFβ-SMAD2/3 signalling leads to myoCF differentiation, and αSMA, Collagen I, Collagen VI, and MMP expression. TGFβ also initiates myoCF differentiation through p38 signalling via TGFβ/RII [30]. Mechanical stretch can induce angiotensinogen or TGFβ expression and increase p38 phosphorylation [37–40]. When retained in the ECM, TGFβ1 forms a complex with latent TGFβ binding proteins and fibrillin. In the progression of myocardial fibrosis, αSMA mediated myoCF contraction of the increasingly rigid ECM liberates TGFβ from this intracellular pool of signalling proteins, inducing a positive feedback loop which further stimulates myoCF differentiation, and secretion of the ECM and myoCF derived cytokines [30].

In summary, the impact of myoCFs occurs via secreting numerous factors involved in cardiac remodelling: TNFα, IL-1β, IL-6, TGFβ, Ang II, ET-1, the NFF (ANP, and BNP), and VEGF [2, 13, 14, 27, 28]. Activation of CFs and subsequent upregulation of the myoCF secretome is largely accepted as an early adaptive event to maintain homeostatic cardiac function; however, prolonged presence of myoCF and chronic deposition of collagen in the ECM is considered a detrimental, maladaptive mechanism leading to poor clinical outcomes [27].

3   Collagen Synthesis and Regulation

The collagen family comprises over 18 distinct isoforms of collagen with diverse physical characteristics whose relative abundance varies from tissue to tissue [41]. Interstitial collagen consists of 3 polypeptide α-chains, composed of large helical domains, which contain a high proportion of hydroxyproline [41]. Collagen type I (Collagen I) and collage type III (Collagen III) are the major structural components in the cardiac extracellular matrix [1]. Collagen I and III are secreted by CFs and myoCFs as pro-collagen precursors, with N-terminal and C-terminal propeptides that are cleaved leaving short non-helical regions at either end suitable for inter-molecular crosslinks [42]. Pro-collagen molecules are released into the extracelluar space where the enzyme lysyl oxidase mediates covalent crosslinking of collagen molecules into rigid, mature fibrils [43].

Approximately 2–4% of the myocardium is collagen in the normal adult heart, but because collagen has a high tensile strength, small changes in collagen concentration can impact the structural properties of the heart [44]. While Collagen I accounts for approximately 85% of all collagen in the myocardium [3], the much less rigid Collagen III comprises the majority of the remaining collagen content and helps confer elasticity to the myocardium. There are numerous methods to quantify the relative ratio of collagen isoforms within the heart (quantitative PCR, Western blot, ELISA, Masson's trichrome staining). However, two of the most exemplary techniques to accurately image the structure of the collagen matrix are (i) picrosirius red staining coupled with polarized light microscopy (picrosirius red detects the high hydroxyproline content in collagen, while polarized light microscopy permits the distinction of Collagen I and Collagen III with high precision [45]), and (ii) direct imaging using scanning electron microscopy (SEM). In SEM micrographs, Collagen I appears thick and rigid, while Collagen III appears much finer, and is present in significant quantity in the healthy myocardium [5]. It is believed that changes in the relative ratios of Collagen I:Collagen III, as well as alterations in the organization and cross-linking of the collagen matrix, influence myocardial compliance. Specifically, late-stage LV remodelling in pressure overload (PO) hypertrophy is associated with elevated Collagen I which promotes myocardial stiffness, resulting in impaired contractile function [3, 5]. Numerous pathologies including diabetic cardiomyopathy and aortic valve stenosis are associated with increased LV fibrosis, decreased LV compliance, and cardiac dysfunction. Microscopic imaging analysis

Table 1 Structural proteins in the ECM. Adapted from [41, 49, 51]

Structural protein	Location	Role
Collagen I	Interstitium	Structural
Collagen III	Interstitium	Structural
Collagen IV	Basement membrane	Cell attachment, substrate filter
Collagen V	Interstitium	Structural, substrate filter
Collagen VI	Interstitium	Structural
Laminin	Basement membrane	Cell attachment
Elastin	Interstitium, connective tissue	Elastic recoil and resilience
Fibrillin	Interstitium, connective tissue	Elastic recoil and resiience

of dilated (idiopathic) cardiomyopathic hearts showed a significant increase in Collagen I:Collagen III ratio compared to normal samples, localized to the endomysial layer surrounding the cardiomyocyte fibrils [46, 47]. Similarly, Collagen I expression [48] is significantly increased in patients with essential hypertension.

In addition to Collagen I and Collagen III, the ECM hosts a number of non-fibrillar collagens such as collagen IV, V, and VI, as well as other structural proteins such as laminin, fibrillin, and elastin (Table 1) that form the *in situ* adhesion substrate for cardiomyocytes and cardiac fibroblasts [41, 49]. Cellular behavior is in part dictated by the relative ratio of these structural proteins in the substrate: Collagen I and III increase CF proliferation and migration, Collagen VI increases myoCF differentiation and decreases migration, Collagen V imparts anti-proliferative effects on smooth muscle cells [50], Collagen IV induces smooth muscle cell differentiation [49], and the highly distensible fibre elastin influences smooth muscle cell proliferation and phenotype [51]. It is hypothesized that the increased expression of the basement membrane collagens (Collagen IV, V) serves to slow the migration of several cell types, and to aid in maintaining structural integrity in the post-ischemic heart [49].

Cardiac fibrosis occurs with high incidence in arrhythmic myopathies [28]. Recent data has recognized the importance of non-myocyte reorganization in the pathology of electrophysiological remodelling leading to cardiac arrhythmias [28]. Although CFs are typically considered non-excitable cells, they in fact have a depolarized resting membrane potential albeit lower than that of neighbouring CMs, implying that CFs then may affect stimulus spread by inducing a minor impedance as they convey electrical signals between CMs [28]. Unfortunately, CF heterogeneity across different areas of the heart makes definitive conclusions regarding the participation of CFs in electrical signalling elusive [28]. However, expansion of the collagen matrix following PO and myocardial infarction (MI) has been shown to play a detrimental role in myocardial electrical coupling [52]. Various patterns of fibrosis (compact, patchy, interstitial, diffuse) have differential impacts on the propagation of electrical signals between myocytes. Patchy or severe fibrosis, where myocytes are separated by a cellular regions of collagen, poses the greatest risk for arrhythmias, as electrical propagation is forced to take a circuitous route through the tissue slowing conduction velocities [52]. Fibrosis also promotes uncoordinated triggers originating in aberrant myocytes, which then propagate to and depolarize the few neighbouring myocytes leading to fibrillation [52].

Clearly, the collagen ECM plays a significant, initially adaptive and subsequently detrimental role in the diseased myocardium. In the next section, regulation of the ECM will be discussed.

4 Matrix Metalloproteinases and Tissue Inhibitors of MMPs

The rate of ECM turnover in the normal heart is mediated by the concerted effects of (i) matrix metalloproteinases (MMPs) which have the ability to degrade structural proteins, (ii) tissue inhibitors of metalloproteinases (TIMPs) and (iii) the continual expression of structural proteins principally by cardiac fibroblasts, with smaller contributions from cardiac myocytes and other cell types (Table 2) [3].

The expression of MMP isoforms is upregulated in response to both acute and chronic stresses such as MI and hypertension [53, 54]. Patients suffering from dilated cardiomyopathy were shown to have increased ventricular MMP-2 and MMP-9 activity [55], while spontaneously hypertensive rats with heart failure were found to have elevated MMP-2 activity when compared against normotensive controls [56]. By contrast mice lacking MMP-2 or MMP-9 showed a decrease in MI-induced left ventricular hypertrophy [57, 58], while exogenous MMP-2 treatment was found to dilate and decrease the cardiac tensile strength of ventricular preparations from spontaneously hypertensive rats [59].

Table 2 ECM substrates of MMPs, and activity change with LVH or CHF. (Adapted from [11])

Enzyme	Alternate name	ECM substrate	Δ in LVH	Δ in CHF
MMP1	Collagenase-1	Collagens (I, II, III, VII, VIII, & X), gelatin, proteoglycan link protein	–	↑
MMP2	Gelatinase A	Collagens (I, IV, V, VII, X & XIV), gelatin, elastin, fibronectin, laminin	↑	↑↑
MMP8	Collagenase-2	Collagens (I, II, III, V, VII, VIII, & X), gelatin, aggrecan	–	–
MMP9	Gelatinase B	Collagens (IV, V, VII, X, & XIV), gelatin, aggrecan, elastin, fibronectin	↑	↑↑
MMP14	MT1-MMP	Collagens (I, II and III), casein, elastin, fibronectin, gelatin, laminin, proteoglycans	–	↑

LVH LV hypertrophy, defined as increased LVPWd with normal EF, *CHF* congestive heart failure

MMPs are under transcriptional regulation by a number of factors, although two major *cis*-acting elements are found in the majority of the MMP promoters: polyoma enhancer A binding protein-3 (PEA-3) which interacts with the Ets family of transcription factors, and activator protein-1 (AP-1) which interacts with the Fos and Jun family of transcription factors [5]. The MMP-2 promoter, however, lacks both of these elements possibly indicating more selective regulation of expression of the MMP-2 protein, although MMP-2 expression may be increased in the failing heart after MI [60].

To protect the cell from unregulated damage, MMP's are typically synthesized as inactive zymogens, and must undergo post-translational modification to become proteolytically active [53]. Membrane type-1 MMP (MT1-MMP, also called MMP-14) is a zinc dependent MMP that plays an important role in peri-cellular ECM digestion leading to cell migration and tissue invasion [61]. MT1-MMP is itself synthesized as a pro-peptide, activated through removal of the inhibitory pro-domain in a two-step degradation/cleavage mechanism mediated by the protease furin [62]. Mature, membrane bound MT1-MMP's activity may be regulated through control of its trafficking, internalization and degradation via ubiquitination [63]. In addition to direct regulation of the peri-cellular ECM, MT1-MMP activity facilitates activation of MMP-2 and MMP-13 leading to processing of collagen substrates (e.g. collagen IV) that MT1-MMP is incapable of degrading [61] (Table 2). Cell surface activation of MMP2 requires the homo-dimerization of MT1-MMP which then binds an MMP-2/TIMP-2 complex leading to cleavage of the MMP-2 pro-domain and the liberation of active MMP2 [64, 65]. In obesity and diabetes, adiponectin and leptin may be important regulators of cardiac ECM. Adiponectin stimulates MT1-MMP cell surface localization, MMP-2 activation and cardiac fibroblast migration through the phosphorylation of AMPK [66]. Leptin also has been shown to increase MMP2 activity, and cell surface expression of MT1-MMP, and also to stimulate Collagen-I secretion [67]. Upstream activators of MMP expression include many of the common cytokines involved in the inflammatory response, cardiac dysfunction and metabolic disorders: TNF-α, IL-1β, ET-1, oxidative stress, sympathetic activation, TGF-β and mechanical stretch [68]. Accordingly, MMP activity is upregulated in many pathological conditions including diabetic cardiomyopathy, PO, volume overload (VO) and also in animal models such as the spontaneously hypertensive rat (SHR) [11].

The regulatory function of TIMPs is primarily via acting as endogenous inhibitors of MMPs [5, 69, 70]. However, as described above, TIMP2 is also involved in the cleavage and activation of MMP-2 by MT1-MMP at the cell surface [5]. Nevertheless, it is generally believed that an increase in TIMP expression may serve to limit MMP activity and induce collagen accumulation in the myocardium. Accordingly, in the SHR model TIMP-4 expression is increased in compensated LVH, while the progression of CHF is characterized by an increase in MMP-2 activity and TIMP4 expression is decreased [11]. Similarly, TIMP expression is reduced in end-stage DCM which is associated with increased ECM degradation and LV dilation [11]. Indeed, evidence suggests that TIMP2 and 4 may have opposing roles in heart failure. TIMP2 was found to have a much higher specificity than the other TIMPs for MMP-2 and play a role in its activation [64], while TIMP-4 has a broad general affinity as an inhibitor for various MMPs [71], suggesting that increased TIMP-2 expression may facilitate increased MMP-2 activation, while increased TIMP-4 may lead to an increase in fibrosis.

An intracellular role for MMPs may be of underappreciated significance as it has been shown that active forms of intracellular MMP-2 and MMP-9 have been found in cardiomyocytes of patients with dilated cardiomyopathy which may lead to degradative loss of sarcomeres and impaired contractile function [60].

5 Cardiac Remodelling Following Myocardial Infarction

Necrosis and apoptosis of cardiac myocytes due to nutrient deprivation following an MI quickly initiates a reparative cascade to support and maintain ventricular wall integrity and prevent myocardial rupture. This process is characterized by the infiltration of inflammatory cells and fibroblasts, as well as the initiation of angiogenesis to supply nutrients to the newly forming infarct scar [72]. The infarct scar matures 6–8 weeks post MI and is generally considered a relatively inert expanse of collagen populated principally by the contractile myoCFs which may persist in the myocardium for many years [73]. However, cardiac fibrosis in the post-MI myocardium is now appreciated as an ongoing, dynamic restructuring not limited to the initial site of MI, termed 'reactive fibrosis', which has a significant impact on cardiac function and long term prognosis [74, 75].

Chronic upregulation of ECM secretion is considered maladaptive so, with respect to dermal wound healing, myofibroblasts do not persist in the healing scar but are instead stimulated to undergo apoptosis to allow for ECM degradation, scar regression, and tissue regeneration [74]. By contrast, in the myocardium, cardiomyocytes are considered terminally differentiated and, outside of small numbers [76, 77], are incapable of repopulating the heart to restore contractile function. Persistent mechanical stress in the infarct scar is thought to inhibit fibroblast apoptosis in the heart, such that interventions to reduce stress promote myoCF apoptosis [78]. It has also been proposed that reduction of substrate rigidity may induce myoCF dedifferentiation into more quiescent CFs [79]. Furthermore, CF apoptosis is known to be regulated by Fas/Fas ligand interaction following MI [80], while TGF-β signalling can in some circumstances mediate reversion of myoCF to CF [81].

Indeed, collagen upregulation has also been found in the non-infarcted/remote myocardium [72]. In contrast to the activities of MMPs in the overloaded myocardium in which cardiac remodelling supports the myocytes to mitigate diastolic or systolic stress, ECM accumulation following MI is a necessary reparative mechanism replacing lost myocytes and preventing LV rupture. This in turn necessitates an acute switch of ECM regulation towards the rapid accumulation of collagen, of which collagen type I and type III are the predominant collagen isoforms in the infarct scar [72]. Indeed, a wide range of studies have demonstrated that attenuation of MMP activity confers cardiac protection following MI in mice [5]. Numerous genetically modified mouse models have been employed to tease out the role of the various MMPs and TIMPs in the restructuring of the myocardium post-MI. MMP-7 deletion in mice subjected to MI for 7 days significantly improved survival despite similar indices of infarct scar size, and LV dilation between MMP7-KO and WT mice [82]. However, the absence of MMP-7 mediated degradation of the gap junction protein connexin-43 in MMP7-KO mice which significantly improved myocardial conduction velocity compared to WT mice and protected against the development of arrhythmias [82]. Similarly, post-MI survival is improved and LV rupture reduced in MMP2-KO mice as well as in WT mice administered an MMP2 selective inhibitor, despite similar levels of LV dilation and infarct size between groups [57]. Interestingly, macrophage infiltration was slower in mice with reduced MMP2 activity [57]. Clearly, attenuating ECM degradation through reduction of MMP activity promotes survival and confers secondary beneficial effects in the myocardium. Accordingly, overexpression of the MMP2 activator MT1-MMP is associated with increased fibrosis, decreased ejection fraction, and lower survival [83].

Modification of TIMP expression also affects post-MI cardiac structure. TIMP-1, −2, and −3 deficiency were independently shown to increase LV dilation, while TIMP-4 KO exhibit increased LV rupture following MI [26]. It has been proposed that TIMP-1, -2, and -3 affect overall cardiac structure following MI, while TIMP-4 functions primarily in the infarct zone [26].

6   Cardiac Remodelling, Pressure, and Volume Overload

PO and VO exhibit distinct hypertrophic remodeling patterns. PO is commonly associated with increased LV wall thickness, decreased chamber size, and significant accumulation of matrix proteins, especially the rigid collagen I isoform, which is associated with increased myocardial stiffness [5]. Comparatively, VO hypertrophy is associated with a decreased expression of the ECM and dissolution of existing collagen by the matrix metalloproteinases [5] (Fig. 2).

Changes in collagen-I and -III levels are the principle features of ECM remodeling following the induction of PO [5]. Procollagen-III mRNA is elevated in the acute phase following the induction of PO using minimally invasive transverse aortic banding (MTAB) surgery, while accumulation of fibrillar collagen occurs at a later timepoint, perhaps due to the decreased rate of collagen degradation. [84]. Pressure overload in MMP2-KO mice results in reduced myocardial hypertrophy and fibrosis, and also reduced heart weight [85]. MMP-9 deletion however only induced partial resolution of hypertrophy and fibrosis following MTAB [86]. Recently, TIMP2-KO mice were shown to exhibit greater LV dilation and dysfunction due to increased stabilization of collagen fibres following the induction of pressure overload [87]. TIMP3-KO mice subjected to MTAB display increased LV remodeling and dysfunction, and also severe fibrosis [88, 89].

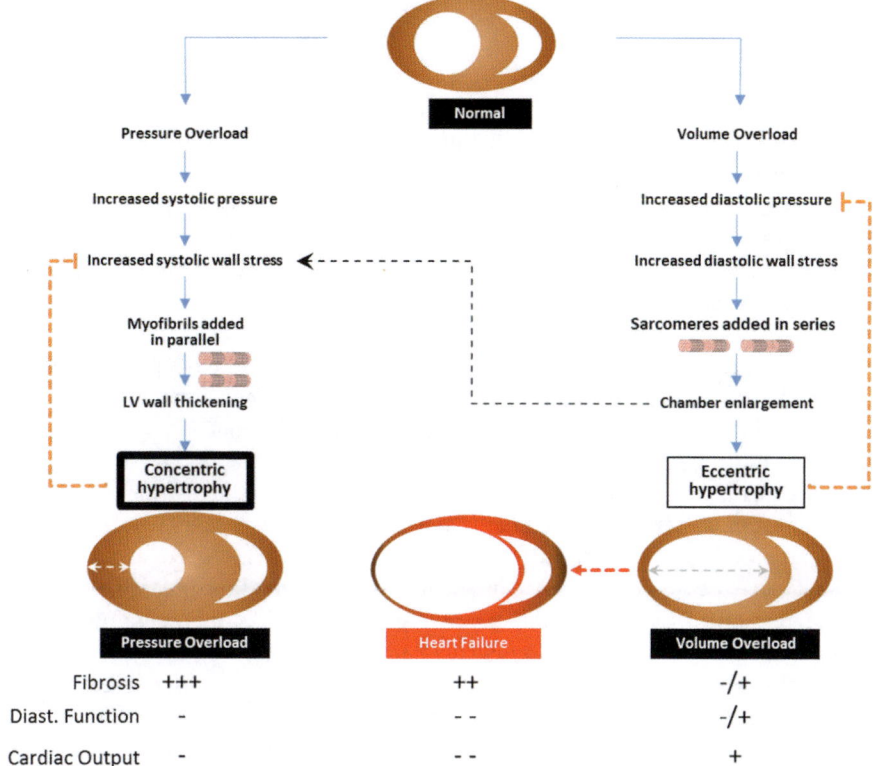

Fig. 2 Characteristics of heart failure with comparison of pressure overload (*PO*) and volume overload (*VO*)

The temporal onset of fibrosis and cardiac hypertrophy are closely linked, and there is growing appreciation of the direct role the ECM plays in activating pro-hypertrophic signalling. ECM-integrin interaction may play a crucial role in transducing hemodynamic/mechanical load into biochemical events through 'outside-in' cascades: mechanical stress first exerts force upon the cardiomyocytes and the ECM which then stresses the focal adhesion complex and the integrins embedded in the cell membrane, thereby initiating intracellular signalling events such as activation of the pro-hypertrophic Ras-Raf-ERK1/2 pathway [6, 9]. The α3β1 integrin heterodimer expressed in cardiomyocytes can bind numerous ECM ligands including collagen I, fibronectin, and laminin, and indeed the integrin subunits α1, α5, β1 and β3 are upregulated and/or activated by hypertrophy and PO [8, 29, 90, 91]. VO however is associated with a progressive decrease in integrin expression [92]. Clearly, mechanisms mediating cardiac hypertrophy and cardiac fibrosis are intertwined in the overloaded myocardium. In the following section the effects of ventricular unloading on the remodelled/hypertrophic heart will be discussed.

7   Cardiac Fibrosis in Reverse Remodelling

While remodelling events induced by excessive haemodynamic load support the myocardium to preserve function, numerous *in vivo* studies have demonstrated that when cardiac remodelling is inhibited through genetic modification or pharmacological treatment, normal cardiac geometry and function are retained even when subjected to increased myocardial load. Echocardiography based findings from the Framingham Heart Study identified left ventricular hypertrophy (LVH) as a significant risk factor for cardiovascular morbidity and death, even after normalization for other significant risk factors such as blood pressure, cigarette use and cholesterol profile [93]. This suggested that although hypertrophy and ECM remodeling are viewed as initially beneficial, these changes may in fact be playing a permissive role in the progression towards heart failure. Accordingly, anti-hypertensive therapies (beta-blockers, ACE inhibitors and angiotensin II receptor antagonists) have been enormously successful in delaying the onset of maladaptive remodelling. In severe cases of congestive heart failure, patients have benefitted from aggressive surgical techniques such as valve replacement and left ventricular assist device implantation, both of which directly alleviate LV PO [94, 95]. Indeed, accumulating clinical and experimental evidence has demonstrated restoration of cardiac function and regression of LV remodelling following LV unloading, a process now commonly termed *reverse remodelling* [96, 97].

There has been significant recent interest in developing *in vivo* models mimicking unloading of the left ventricle. Particularly, LV unloading of PO induced by MTAB in rodents through surgical removal of the aortic band (debanding), has been shown to initiate reverse hypertrophic remodelling, restoration of normal cardiac function [98–100], activation of a unique gene expression profile [101] and increased autophagic flux [102]. However, if debanding surgery is performed after the heart enters decompensated failure, an inability of the heart to recover was noted [103]. Other novel LV unloading methods have yielded similar findings. Heterotopic transplantation was shown to elicit atrophic remodelling of the rat heart [104]. The injection of mesenchymal stem cells was shown to significantly improve hemodynamic performance, reverse LV remodelling and improve exercise tolerance following aortic banding [105], while increasing EF and capillary density in a model of doxorubicin-induced heart failure [106].

Detailed examination of the ECM following LV unloading presents a valuable opportunity to understand the dynamic potential of the myocardial collagen network in greater detail as there is little opportunity for accrued fibrosis to rescind in models of heart disease. Visualization of interstitial collagen through picrosirius red staining of heart sections showed a clear reduction in fibrosis following debanding, although collagen levels remained elevated compared to sham-treated mice [100]. Temporal evaluation of the principle collagen isoforms in the recovering myocardium showed an early (3 day) increase in collagen I, followed by an increase in collagen III and VIII 7 days after debanding [107]. LV unloading also increased MMP2 and TIMP1 gene expression [107]. These data suggest that the cardiac ECM is a dynamically regulated structure which can undergo bidirectional remodeling to fulfill the mechanical requirements of the heart. Specific factors initiating reverse remodeling, and the role of myoCFs in this process remain to be fully elucidated.

8 Regulation of the ECM in Diabetes and the Metabolic Syndrome

Diabetic patients are known to have a two to four fold higher chance of developing heart disease, while approximately 65% of diabetes related deaths are due to CVD [108, 109]. Obesity is also a significant risk factor for the development of heart disease and the onset of cardiac remodeling [110]. Hyperglycaemia, hyperinsulinemia and dyslipidemia are all known to exert direct and potent effects on cardiac function and structure, although it is elevated glucose levels that are believed to be primarily responsible for changes in the structure of the ECM. Hypertension is also closely associated with diabetes and obesity in the definition of the metabolic syndrome, and is known to induce a switch in cardiac energy substrate to glucose [111]. Hyperglycaemia is poorly handled by mitochondria in the heart, leading to increased ROS production, activation of poly (ADP-ribose) polymerase-1 (PARP-1), and the apoptosis cascade [112]. PARP-1 activity itself suppresses GAPDH activity, resulting in multiple pathways leading to cellular damage: advanced glycation end product (AGE) accumulation, the polyol pathway, protein kinase C activation and overactivity of the hexosamine pathway [113]. AGE accumulation in the myocardium is closely associated with collagen cross-linking, and HG induced oxidative stress also leads to increased collagen deposition [114, 115], myocardial fibrosis [116] and diastolic stiffness [117, 118]. Ultimately, the diabetic heart exhibits severe cardiac dysfunction [119] and widespread cellular necrosis [120], while cardiac fibroblasts undergo phenotypic change to directly mediate fibrosis [117].

Angiotensin II, a strong marker of cardiac dysfunction and member of the renin-angiotensin-aldosterone system (RAAS), is closely associated with many features of diabetic cardiomyopathy [121]. Angiotensin receptor density is increased in the diabetic heart [119] and was shown to induce insulin resistance, and stimulate the switch in cardiac substrate utilization through increased PDK expression [122]. Importantly, activation of the local RAAS in diabetic cardiomyopathy may induce oxidative damage, cardiomyocyte apoptosis, and cardiac fibrosis [120]. However, cardiomyocyte derived IGF-I can downregulate local AngII expression and cardiomyocyte apoptosis [121]. Therapeutically, it has been shown that AngII stimulated cardiac hypertrophy and fibrosis was attenuated by pioglitazone, a PPARγ targeting anti-diabetic therapy that has been extensively used in DBCM patients [123].

9 Conclusions

The extracellular matrix is now accepted as a major player in the progression of heart disease. As detailed in this chapter, the ECM is a dynamic environment composed of structural proteins, a vasculature and numerous cell types. The characteristics of chronic (PO) and acute (MI) cardiovascular stressors vary widely in nature, and so the myocardium must be able to robustly adapt in order to maintain cardiac function. The ECM is a principal player in this adaptive response and significant research has uncovered many of the underlying cellular and molecular players. Newer data proposes that changes to the collagen ECM are not finite and that upon resolution of stress the cardiac ECM can regress, limiting myocardial fibrosis and further promoting optimal cardiac function.

References

1. Weber KT (1989) Cardiac interstitium in health and disease: the fibrillar collagen network. J Am Coll Cardiol 13(7):1637–1652
2. Weber KT et al (2013) Myofibroblast-mediated mechanisms of pathological remodelling of the heart. Nat Rev Cardiol 10(1):15–26
3. Fedak PW et al (2005) Cardiac remodeling and failure from molecules to man (part II). Cardiovasc Pathol 14(2):49–60
4. Bishop JE et al (1995) The regulation of collagen deposition in the hypertrophying heart. Ann N Y Acad Sci 752:236–239
5. Spinale FG (2007) Myocardial matrix remodeling and the matrix metalloproteinases: influence on cardiac form and function. Physiol Rev 87(4):1285–1342
6. Lorell BH, Carabello BA (2000) Left ventricular hypertrophy: pathogenesis, detection, and prognosis. Circulation 102(4):470–479
7. Segura AM, Frazier OH, Buja LM (2012) Fibrosis and heart failure. Heart Fail Rev 19(2):173–185
8. Harston RK, Kuppuswamy D (2011) Integrins are the necessary links to hypertrophic growth in cardiomyocytes. J Signal Transduct 2011:521742
9. Ross RS, Borg TK (2001) Integrins and the myocardium. Circ Res 88(11):1112–1119
10. Miner EC, Miller WL (2006) A look between the cardiomyocytes: the extracellular matrix in heart failure. Mayo Clin Proc 81(1):71–76
11. Graham HK, Horn M, Trafford AW (2008) Extracellular matrix profiles in the progression to heart failure. European young physiologists symposium keynote lecture-Bratislava 2007. Acta Physiol (Oxf) 194(1):3–21
12. Banerjee I et al (2007) Determination of cell types and numbers during cardiac development in the neonatal and adult rat and mouse. Am J Physiol Heart Circ Physiol 293(3):H1883–H1891
13. Porter KE, Turner NA (2009) Cardiac fibroblasts: at the heart of myocardial remodeling. Pharmacol Ther 123(2):255–278
14. Lajiness JD, Conway SJ (2013) Origin, development, and differentiation of cardiac fibroblasts. J Mol Cell Cardiol 70:2–8
15. de Lange FJ et al (2004) Lineage and morphogenetic analysis of the cardiac valves. Circ Res 95(6):645–654
16. Goldsmith EC et al (2004) Organization of fibroblasts in the heart. Dev Dyn 230(4):787–794

17. Ieda M et al (2009) Cardiac fibroblasts regulate myocardial proliferation through beta1 integrin signaling. Dev Cell 16(2):233–244
18. Burstein B et al (2008) Differential behaviors of atrial versus ventricular fibroblasts: a potential role for platelet-derived growth factor in atrial-ventricular remodeling differences. Circulation 117(13):1630–1641
19. Zhang Y, Kanter EM, Yamada KA (2010) Remodeling of cardiac fibroblasts following myocardial infarction results in increased gap junction intercellular communication. Cardiovasc Pathol 19(6):e233–e240
20. Ikeda K et al (2008) Cellular physiology of rat cardiac myocytes in cardiac fibrosis: *in vitro* simulation using the cardiac myocyte/cardiac non-myocyte co-culture system. Hypertens Res 31(4):693–706
21. LaFramboise WA et al (2007) Cardiac fibroblasts influence cardiomyocyte phenotype *in vitro*. Am J Physiol Cell Physiol 292(5):C1799–C1808
22. Zhang P, Su J, Mende U (2012) Cross talk between cardiac myocytes and fibroblasts: from multiscale investigative approaches to mechanisms and functional consequences. Am J Physiol Heart Circ Physiol 303(12):H1385–H1396
23. Kakkar R, Lee RT (2010) Intramyocardial fibroblast myocyte communication. Circ Res 106(1):47–57
24. Takeda N et al (2010) Cardiac fibroblasts are essential for the adaptive response of the murine heart to pressure overload. J Clin Invest 120(1):254–265
25. Nagai R et al (2005) Significance of the transcription factor KLF5 in cardiovascular remodeling. J Thromb Haemost 3(8):1569–1576
26. Fan D et al (2012) Cardiac fibroblasts, fibrosis and extracellular matrix remodeling in heart disease. Fibrogenesis Tissue Repair 5(1):15
27. Turner NA, Porter KE (2013) Function and fate of myofibroblasts after myocardial infarction. Fibrogenesis Tissue Repair 6(1):5
28. Vasquez C, Benamer N, Morley GE (2011) The cardiac fibroblast: functional and electrophysiological considerations in healthy and diseased hearts. J Cardiovasc Pharmacol 57(4):380–388
29. Stewart JA Jr et al (2010) Temporal alterations in cardiac fibroblast function following induction of pressure overload. Cell Tissue Res 340(1):117–126
30. Davis J, Molkentin JD (2013) Myofibroblasts: Trust your heart and let fate decide. J Mol Cell Cardiol
31. Zhang X et al (2001) Differential vulnerability to oxidative stress in rat cardiac myocytes versus fibroblasts. J Am Coll Cardiol 38(7):2055–2062
32. Gao X et al (2009) Angiotensin II increases collagen I expression via transforming growth factor-beta1 and extracellular signal-regulated kinase in cardiac fibroblasts. Eur J Pharmacol 606(1–3):115–120
33. Huang D et al (2009) Angiotensin II promotes poly(ADP-ribosyl)ation of c-Jun/c-Fos in cardiac fibroblasts. J Mol Cell Cardiol 46(1):25–32
34. Davis J et al (2012) A TRPC6-dependent pathway for myofibroblast transdifferentiation and wound healing *in vivo*. Dev Cell 23(4):705–715
35. Campbell SE, Katwa LC (1997) Angiotensin II stimulated expression of transforming growth factor-beta1 in cardiac fibroblasts and myofibroblasts. J Mol Cell Cardiol 29(7):1947–1958
36. Lee AA et al (1995) Angiotensin II stimulates the autocrine production of transforming growth factor-beta 1 in adult rat cardiac fibroblasts. J Mol Cell Cardiol 27(10):2347–2357
37. Lal H et al (2008) Stretch-induced regulation of angiotensinogen gene expression in cardiac myocytes and fibroblasts: opposing roles of JNK1/2 and p38alpha MAP kinases. J Mol Cell Cardiol 45(6):770–778
38. Blaauboer ME et al (2011) Cyclic mechanical stretch reduces myofibroblast differentiation of primary lung fibroblasts. Biochem Biophys Res Commun 404(1):23–27
39. Lee AA et al (1999) Differential responses of adult cardiac fibroblasts to *in vitro* biaxial strain patterns. J Mol Cell Cardiol 31(10):1833–1843

40. van Wamel AJ et al (2001) The role of angiotensin II, endothelin-1 and transforming growth factor-beta as autocrine/paracrine mediators of stretch-induced cardiomyocyte hypertrophy. Mol Cell Biochem 218(1–2):113–124
41. Bishop JE, Laurent GJ, (1995) Collagen turnover and its regulation in the normal and hypertrophying heart. Eur Heart J 16(Suppl C):38–44
42. Trackman PC (2005) Diverse biological functions of extracellular collagen processing enzymes. J Cell Biochem 96(5):927–937
43. Gonzalez A et al (2011) New targets to treat the structural remodeling of the myocardium. J Am Coll Cardiol 58(18):1833–1843
44. Robinson TF et al (1988) Coiled perimysial fibers of papillary muscle in rat heart: morphology, distribution, and changes in configuration. Circ Res 63(3):577–592
45. Whittaker P et al (1994) Quantitative assessment of myocardial collagen with picrosirius red staining and circularly polarized light. Basic Res Cardiol 89(5):397–410
46. Pauschinger M et al (1999) Dilated cardiomyopathy is associated with significant changes in collagen type I/III ratio. Circulation 99(21):2750–2756
47. Marijianowski MM et al (1995) Dilated cardiomyopathy is associated with an increase in the type I/type III collagen ratio: a quantitative assessment. J Am Coll Cardiol 25(6):1263–1272
48. Diez J et al (1995) Increased serum concentrations of procollagen peptides in essential hypertension. Relation to cardiac alterations. Circulation 91(5):1450–1456
49. Shamhart PE, Meszaros JG (2010) Non-fibrillar collagens: key mediators of post-infarction cardiac remodeling? J Mol Cell Cardiol 48(3):530–537
50. Naugle JE et al (2006) Type VI collagen induces cardiac myofibroblast differentiation: implications for postinfarction remodeling. Am J Physiol Heart Circ Physiol 290(1):H323–H330
51. Kielty CM (2006) Elastic fibres in health and disease. Expert Rev Mol Med 8(19):1–23
52. Nguyen TP, Qu Z, Weiss JN (2013) Cardiac fibrosis and arrhythmogenesis: the road to repair is paved with perils. J Mol Cell Cardiol
53. Spinale FG (2002) Matrix metalloproteinases: regulation and dysregulation in the failing heart. Circ Res 90(5):520–530
54. Galis ZS, Khatri JJ (2002) Matrix metalloproteinases in vascular remodeling and atherogenesis: the good, the bad, and the ugly. Circ Res 90(3):251–262
55. Spinale FG et al (2000) A matrix metalloproteinase induction/activation system exists in the human left ventricular myocardium and is upregulated in heart failure. Circulation 102(16):1944–1949
56. Peterson JT et al (2001) Matrix metalloproteinase inhibition attenuates left ventricular remodeling and dysfunction in a rat model of progressive heart failure. Circulation 103(18):2303–2309
57. Matsumura S et al (2005) Targeted deletion or pharmacological inhibition of MMP-2 prevents cardiac rupture after myocardial infarction in mice. J Clin Invest 115(3):599–609
58. Ducharme A et al (2000) Targeted deletion of matrix metalloproteinase-9 attenuates left ventricular enlargement and collagen accumulation after experimental myocardial infarction. J Clin Invest 106(1):55–62
59. Mujumdar VS, Smiley LM, Tyagi SC (2001) Activation of matrix metalloproteinase dilates and decreases cardiac tensile strength. Int J Cardiol 79(2–3):277–286
60. Lalu MM et al (2005) Ischaemia-reperfusion injury activates matrix metalloproteinases in the human heart. Eur Heart J 26(1):27–35
61. Itoh Y (2006) MT1-MMP: a key regulator of cell migration in tissue. IUBMB Life 58(10):589–596
62. Golubkov VS et al (2007) Proteolysis of the membrane type-1 matrix metalloproteinase prodomain: implications for a two-step proteolytic processing and activation. J Biol Chem 282(50):36283–36291
63. Eisenach PA et al (2012) Membrane type 1 matrix metalloproteinase (MT1-MMP) ubiquitination at Lys581 increases cellular invasion through type I collagen. J Biol Chem 287(14):11533–11545

64. Morgunova E et al (2002) Structural insight into the complex formation of latent matrix metalloproteinase 2 with tissue inhibitor of metalloproteinase 2. Proc Natl Acad Sci U S A 99(11):7414–7419
65. Strongin AY et al (1995) Mechanism of cell surface activation of 72-kDa type IV collagenase. Isolation of the activated form of the membrane metalloprotease. J Biol Chem 270(10):5331–5338
66. Dadson K et al. (2013) Adiponectin mediated APPL1-AMPK signaling induces cell migration, MMP activation, and collagen remodeling in cardiac fibroblasts. J Cell Biochem
67. Schram K et al (2008) Increased expression and cell surface localization of MT1-MMP plays a role in stimulation of MMP-2 activity by leptin in neonatal rat cardiac myofibroblasts. J Mol Cell Cardiol 44(5):874–881
68. Deschamps AM, Spinale FG (2006) Pathways of matrix metalloproteinase induction in heart failure: bioactive molecules and transcriptional regulation. Cardiovasc Res 69(3):666–676
69. Vianello A et al (2009) Role of matrix metalloproteinases and their tissue inhibitors as potential biomarkers of left ventricular remodelling in the athlete's heart. Clin Sci (Lond) 117(4):157–164
70. Spinale FG, Janicki JS, Zile MR (2013) Membrane-associated matrix proteolysis and heart failure. Circ Res 112(1):195–208
71. Miragoli M, Salvarani N, Rohr S (2007) Myofibroblasts induce ectopic activity in cardiac tissue. Circ Res 101(8):755–758
72. Sun Y, Weber KT (2000) Infarct scar: a dynamic tissue. Cardiovasc Res 46(2):250–256
73. Willems IE et al (1994) The alpha-smooth muscle actin-positive cells in healing human myocardial scars. Am J Pathol 145(4):868–875
74. Czubryt MP (2012) Common threads in cardiac fibrosis, infarct scar formation, and wound healing. Fibrogenesis Tissue Repair 5(1):19
75. Pfeffer MA, Braunwald E (1990) Ventricular remodeling after myocardial infarction. Experimental observations and clinical implications. Circulation 81(4):1161–1172
76. Al Darazi F et al. (2014) Small dedifferentiated cardiomyocytes bordering on microdomains of fibrosis: evidence for reverse remodeling with assisted recovery. J Cardiovasc Pharmacol
77. Takeuchi T (2014) Regulation of cardiomyocyte proliferation during development and regeneration. Dev Growth Differ 56(5):402–409
78. Dobaczewski M, Gonzalez-Quesada C, Frangogiannis NG (2010) The extracellular matrix as a modulator of the inflammatory and reparative response following myocardial infarction. J Mol Cell Cardiol 48(3):504–511
79. Wang H et al (2012) Redirecting valvular myofibroblasts into dormant fibroblasts through light-mediated reduction in substrate modulus. PLoS ONE 7(7):e39969
80. Li Y et al (2004) Critical roles for the Fas/Fas ligand system in postinfarction ventricular remodeling and heart failure. Circ Res 95(6):627–636
81. Rosenkranz S (2004) TGF-beta1 and angiotensin networking in cardiac remodeling. Cardiovasc Res 63(3):423–432
82. Lindsey ML et al (2006) Matrix metalloproteinase-7 affects connexin-43 levels, electrical conduction, and survival after myocardial infarction. Circulation 113(25):2919–2928
83. Zavadzkas JA et al (2011) Direct regulation of membrane type 1 matrix metalloproteinase following myocardial infarction causes changes in survival, cardiac function, and remodeling. Am J Physiol Heart Circ Physiol 301(4):H1656–H1666
84. Eleftheriades EG et al (1993) Regulation of procollagen metabolism in the pressure-overloaded rat heart. J Clin Invest 91(3):1113–1122
85. Matsusaka H et al (2006) Targeted deletion of matrix metalloproteinase 2 ameliorates myocardial remodeling in mice with chronic pressure overload. Hypertension 47(4):711–717
86. Heymans S et al (2005) Loss or inhibition of uPA or MMP-9 attenuates LV remodeling and dysfunction after acute pressure overload in mice. Am J Pathol 166(1):15–25
87. Kandalam V et al (2011) Lack of tissue inhibitor of metalloproteinases 2 leads to exacerbated left ventricular dysfunction and adverse extracellular matrix remodeling in response to biomechanical stress. Circulation 124(19):2094–2105

88. Kassiri Z et al (2005) Combination of tumor necrosis factor-alpha ablation and matrix me-talloproteinase inhibition prevents heart failure after pressure overload in tissue inhibitor of metalloproteinase-3 knock-out mice. Circ Res 97(4):380–390
89. Kassiri Z et al (2009) Simultaneous transforming growth factor beta-tumor necrosis factor activation and cross-talk cause aberrant remodeling response and myocardial fibrosis in Timp3-deficient heart. J Biol Chem 284(43):29893–29904
90. Suryakumar G et al (2010) Lack of beta3 integrin signaling contributes to calpain-me-diated myocardial cell loss in pressure-overloaded myocardium. J Cardiovasc Pharmacol 55(6):567–573
91. Willey CD et al (2008) STAT3 activation in pressure-overloaded feline myocardium: role for integrins and the tyrosine kinase BMX. Int J Biol Sci 4(3):184–199
92. Stewart JA Jr et al. (2013) Temporal changes in integrin-mediated cardiomyocyte adhesion secondary to chronic cardiac volume overload in rats. Am J Physiol Heart Circ Physiol
93. Levy D et al (1990) Prognostic implications of echocardiographically determined left ven-tricular mass in the Framingham Heart Study. N Engl J Med 322(22):1561–1566
94. Hellawell JL, Margulies KB (2012) Myocardial reverse remodeling. Cardiovasc Ther 30(3):172–181
95. Kirkpatrick JN, John Sutton MS (2012) Assessment of ventricular remodeling in heart fail-ure clinical trials. Curr Heart Fail Rep 9(4):328–336
96. Baba HA, Wohlschlaeger J (2008) Morphological and molecular changes of the myocar-dium after left ventricular mechanical support. Curr Cardiol Rev 4(3):157–169
97. Zerkowski HR et al. (2000) Reverse remodeling by surgery–fact or fiction? Z Kardiol 89(Suppl 7):76–84
98. Bjornstad JL et al (2012) A mouse model of reverse cardiac remodelling following banding-debanding of the ascending aorta. Acta Physiol (Oxf) 205(1):92–102
99. Stansfield WE et al (2007) Characterization of a model to independently study regression of ventricular hypertrophy. J Surg Res 142(2):387–393
100. Gao XM et al (2005) Regression of pressure overload-induced left ventricular hypertrophy in mice. Am J Physiol Heart Circ Physiol 288(6):H2702–H2707
101. Stansfield WE et al. (2009) Regression of pressure-induced left ventricular hypertrophy is characterized by a distinct gene expression profile. J Thorac Cardiovasc Surg 137(1):232–238, 238e1–238e8
102. Hariharan N et al (2013) Autophagy plays an essential role in mediating regression of hy-pertrophy during unloading of the heart. PLoS ONE 8(1):e51632
103. Andersen NM et al (2012) Recovery from decompensated heart failure is associated with a distinct, phase-dependent gene expression profile. J Surg Res 178(1):72–80
104. Sharma S et al (2006) Atrophic remodeling of the transplanted rat heart. Cardiology 105(2):128–136
105. Molina EJ et al. (2008) Improvement in hemodynamic performance, exercise capacity, in-flammatory profile, and left ventricular reverse remodeling after intracoronary delivery of mesenchymal stem cells in an experimental model of pressure overload hypertrophy. J Thorac Cardiovasc Surg 135(2):292–299, 299e1
106. Arsalan M et al. (2013) The reverse remodeling effect of mesenchymal stem cells is inde-pendent from the site of epimyocardial cell transplantation. Innovations (Phila) 8(6):433–439
107. Bjornstad JL et al (2011) Collagen isoform shift during the early phase of reverse left ven-tricular remodelling after relief of pressure overload. Eur Heart J 32(2):236–245
108. Grundy SM et al (1999) Diabetes and cardiovascular disease: a statement for healthcare professionals from the American Heart Association. Circulation 100(10):1134–1146
109. Grundy SM et al (1998) Primary prevention of coronary heart disease: guidance from Framingham: a statement for healthcare professionals from the AHA task force on risk reduction. American Heart Association. Circulation 97(18):1876–1887
110. Abel ED, Litwin SE, Sweeney G (2008) Cardiac remodeling in obesity. Physiol Rev 88(2):389–419

111. Taegtmeyer H et al (2004) Linking gene expression to function: metabolic flexibility in the normal and diseased heart. Ann N Y Acad Sci 1015:202–213

112. Battiprolu PK et al (2010) Diabetic cardiomyopathy: mechanisms and therapeutic targets. Drug Discov Today Dis Mech 7(2):e135–e143

113. Giacco F, Brownlee M (2010) Oxidative stress and diabetic complications. Circ Res 107(9):1058–1070

114. Siwik DA, Pagano PJ, Colucci WS (2001) Oxidative stress regulates collagen synthesis and matrix metalloproteinase activity in cardiac fibroblasts. Am J Physiol Cell Physiol 280(1):C53–C60

115. Tang M et al (2007) High glucose promotes the production of collagen types I and III by cardiac fibroblasts through a pathway dependent on extracellular-signal-regulated kinase 1/2. Mol Cell Biochem 301(1–2):109–114

116. Aragno M et al (2008) Oxidative stress triggers cardiac fibrosis in the heart of diabetic rats. Endocrinology 149(1):380–388

117. Hutchinson KR et al (2013) Cardiac fibroblast-dependent extracellular matrix accumulation is associated with diastolic stiffness in type 2 diabetes. PLoS ONE 8(8):e72080

118. van Heerebeek L, Somsen A, Paulus WJ (2009) The failing diabetic heart: focus on diastolic left ventricular dysfunction. Curr Diab Rep 9(1):79–86

119. Boudina S, Abel ED (2007) Diabetic cardiomyopathy revisited. Circulation 115(25):3213–3223

120. Frustaci A et al (2000) Myocardial cell death in human diabetes. Circ Res 87(12):1123–1132

121. Fang ZY, Prins JB, Marwick TH (2004) Diabetic cardiomyopathy: evidence, mechanisms, and therapeutic implications. Endocr Rev 25(4):543–567

122. Mori J et al (2013) ANG II causes insulin resistance and induces cardiac metabolic switch and inefficiency: a critical role of PDK4. Am J Physiol Heart Circ Physiol 304(8):H1103–H1113

123. Li P et al (2010) Evidence for the importance of adiponectin in the cardioprotective effects of pioglitazone. Hypertension 55(1):69–75

Mathematical Simulations of Sphingosine-1-Phosphate Actions on Mammalian Ventricular Myofibroblasts and Myocytes

K. A. MacCannell, L. Chilton, G. L. Smith and W. R. Giles

Abstract Mathematical modeling has been used to explore the consequences of the actions of sphingosine-1-phosphate (S-1-P) within the ventricular myocardium. Electrophysiological data obtained from rabbit cultured myofibroblasts [1] provided the basis for formulation of our model of electrotonic coupling between ventricular myocytes and fibroblasts [2]. Specifically, our in silico fibroblast/myocyte hybrid model was modified to account for the electrophysiological properties that are characteristic of the myofibroblast (the wound healing phenotype of the fibroblast). In addition, equations describing an S-1-P-induced current that can be activated in the myofibroblast were added.

The sets of simulations that constitute this paper demonstrate that S-1-P can cause a significant depolarization of the resting membrane potential in both the myofibroblast and myocyte. When the myocyte to fibroblast coupling ratio is 1:1, this concentration-dependent effect is due to ligand-gated current in the myofibroblast depolarizing the myocyte through heterotypic connexin-mediated intercellular junctions. In addition to changing the resting potential in the myocyte, the S-1-P induced current resulted in significant changes in action potential waveform.

A second set of simulations was done for the purpose of exploring the effects of S-1-P on myocytes that have some of the main electrophysiological properties of those from the failing heart. In these computations, the ten Tusscher model of the human ventricular myocyte was modified by reducing parameters as follows: cell capacitance, inward rectifier K^+ current, delayed-rectifier K^+ currents (I_{Ks} and I_{Kr}), and transient outward K^+ current. In combination, these changes (each of which

W. R. Giles (✉)
Faculty of Kinesiology, University of Calgary, 2500 University Drive NW, Calgary, AB T2N 1N4, Canada
e-mail: wgiles@ucalgary.ca

K. A. MacCannell
Worldwide Research and Development Pfizer Inc., Cambridge, MA, USA

L. Chilton
College of Public Health, Medical & Veterinary Science, James Cook University, Townsville, Australia

G. L. Smith
Institute of Biomedical and Life Sciences, University of Glasgow, Glasgow, UK

© Springer International Publishing Switzerland 2015
I.M.C. Dixon, J. T. Wigle (eds.), *Cardiac Fibrosis and Heart Failure: Cause or Effect?*, Advances in Biochemistry in Health and Disease 13, DOI 10.1007/978-3-319-17437-2_16

is associated with heart failure), resulted in prolongation of action potential duration. Simulations of electrotonic coupling between this model 'failing' myocyte and myofibroblasts demonstrated that the resting potential and APD in the failing myocyte is more susceptible to modulation by electrotonic influences from S-1-P-stimulated myofibroblasts when a 'failing' electrophysiological phenotype in the ventricular myocyte is introduced.

In summary, our simulations draw attention to important effects of S-1-P on the ventricular myocardium even when this paracrine substance acts only on the fibroblast cell population. These cell-specific S-1-P effects alter the myocyte action potential via electrotonic coupling. It is apparent that myofibroblasts can have significant effects on myocyte action potentials; and that these effects would be expected to be more pronounced in the presence of ligand-gated effects on the myofibroblast. The general setting that we have attempted to replicate with this first order model has some similarities to acute or sterile inflammation in the myocardium wherein S-1-P concentrations in the interstitium are relatively high.

Keywords Sphingosine-1-phosphate · Electrotonic coupling · Myocyte · Fibroblast · Myofibroblast · Mathematical simulations

1 Introduction

Sphingosine-1-phosphate (S-1-P) is a naturally occurring sphingolipid. It may be produced locally, and can act as an autocrine/paracrine factor. However, S-1-P is also stored in high concentrations in platelets [3]. S-1-P is known to have both extracellular and intracellular signaling roles, including regulation of cell proliferation, differentiation, survival and motility in a number of different tissues. These responses are initiated following binding of S-1-P to the so-called EDG family of surface membrane receptors. This ligand/receptor interaction can activate a number of different ion channels, in some cases this effect is mediated through well characterized G protein-dependent mechanisms [4].

In the heart, S-1-P can influence fundamental aspects of myofibroblast physiology e.g., wound healing and cell migration [3, 4]. Cardiac myofibroblasts are the wound-healing phenotype of native fibroblasts as is described in detail elsewhere in this series of papers. Myofibroblasts are consistently detected in relatively large numbers in the infarct zone soon after a myocardial infarct [5], perhaps as a result of an acute inflammatory response within the injured myocardium.

The main objective of the present theoretical study was to develop a novel approach for defining cell specific functional electrophysiological effects of S-1-P in the mammalian myocardium. This chapter illustrates some of the ways in which mathematical modeling can contribute integrative capabilities and conceptual insights regarding this complex response within a heterogeneous syncytium: the mammalian ventricular myocardium. Where possible, our mathematical approaches have been guided by our own experimental work in which the electrophysiological

effects of S-1-P were studied. A published mathematical model of the action potential in the human ventricular myocyte [6] was chosen since we have previously
utilized it in conjunction with our model of the mammalian ventricular fibroblast/
myofibroblast [2].

Our experimental work has provided three new insights into the mechanisms of
action of S-1-P in the rabbit ventricle. First, S-1-P, when applied to myofibroblasts
at nanomolar concentrations, can activate an approximately linear ligand-gated current having a reversal potential near -20 mV. Second, S-1-P, at nanomolar levels
has no effect on ventricular myocyte electrophysiology as baseline [1]. Third, S-
1-P application (short term superfusion) onto a co-culture preparation resulted in
a reversible loss of excitability, and in some cases a significant depolarization of
the myocyte resting potential in these myocyte/myofibroblast co-cultures. These
electrophysiological effects were absent when the gap junction blocker heptanol
was applied previously [1]. In combination, these findings were the basis for the
prediction that changes in ventricular electrophysiology may arise through indirect
(or electrotonic) modulation of myocyte excitability or action potential waveform.
In fact, direct tests of this hypothesis remain a significant technical challenge since
the required double and simultaneous microelectrode impalements of an identified
myocyte/myofibroblast are required. However, evaluation of the plausible consequences of this cell-to-cell communication can be approached by developing a suitable mathematical model.

It is now well established that there exists electrotonic coupling between cardiac
myocyte and fibroblasts or myofibroblasts. In the important first studies, Kohl et al.
(1994) simultaneously recorded action potentials in myocytes and related electrotonic changes in membrane potential in fibroblasts in rat atrial preparations *in vitro* [7]. Later, the Kohl group provided further evidence (based on fluorescent dye
transfer) for functional coupling between rabbit sinoatrial node myocytes and fibroblasts [8]. They also demonstrated expression of connexins in the scarred regions
resulting from myocardial infarction of sheep hearts [9]. In co-cultures of neonatal
cardiac myocytes and fibroblasts, electrotonic changes in membrane potential between myocytes and fibroblasts have been studied in detail [10–14]. Somewhat
similar findings have been presented from adult rabbit myocyte and fibroblasts
preparations in co-culture [15].

In contrast to this evidence for electrotonic coupling between cardiac myocytes
and fibroblasts, or myofibroblasts, there is little information concerning how such
coupling may alter the electrophysiological responses in either of these two cell
types. In both the neonatal and the adult cardiac syncytium, electrotonic coupling
is very complex in part because it depends upon many factors, including intercellular resistance, relative numbers of myocytes vs. fibroblasts and the heterogeneous
prevalent paracrine and autocrine environment. In addition, it is likely that the functional details of this new modality within cardiac electrophysiology is altered significantly when the patterns of ion channel expression and/or their densities change
in the myocyte. In fact, such changes are known to occur in heart failure [34]. In
this setting, the capacitance of the myocyte is also reduced substantially due to the
'loss' of significant amounts of the transverse tubule system [16]. In addition and

simultaneously, the density of the expression of a number of different K^+ channels decreases substantially. Specifically, the inward rectifier K^+ (I_{ks}), delayed rectifiers (I_{ks} and I_{kr}) and transient outward K^+ current densities (I_{to}) are all reduced [17–22]. It is apparent that these changes will result in the failing myocyte having a higher input impedance/resistance, and a reduced repolarization reserve, thus rendering it more susceptible to electrotonic influences.

In principle, mathematical models can provide a means to characterize complex cell-cell interactions. We [2] and others [7, 23, 24–26] have developed models of the electrophysiological behaviour of cardiac myocytes coupled with fibroblasts. Using this approach, we have illustrated how coupling of a adult human myocyte to one or more fibroblasts (having basal K^+ currents) may alter the myocyte action potential waveform [2], while also resulting in significant electrotonic changes in membrane potential in the fibroblast. Subsequent reports of similar simulations have documented these findings [23, 24, 26]. In the sets of computations which form the basis of this paper we have modified; (i) the ten Tusscher model of the human ventricular action potential, and (ii) our previously published mathematical model of the ventricular fibroblast/myofibroblast so that sets of simulations could begin to evaluate:

1. if the intrinsic biophysical/electrophysiological properties of the myofibroblast, compared to the fibroblast, represents an important difference as judged by simulated cell-to-cell electrotonic interactions;
2. whether a S-1-P ligand-gated current in myofibroblasts can significiantly change well known electrophysiological characteristics of the ventricular myocyte to which it is electrotonically coupled;
3. the consequences of reducing myocyte capacitance and also the K^+ currents, I_{K1}, I_{KS}, I_{Kr}, and I_{to}, as a first attempt to mimic known electrophysiological changes the conditions in the failing heart);
4. the effects of activating the S-1-P-gated current in myofibroblasts coupled to ventricular myocytes in which these selected electrophysiological changes that are known to occur in the failing ventricle have been incorporated.

In this chapter, the main features of these sets of computations are presented. Our results are discussed with respect to the utility of such first order mathematical models for study of multidisciplinary problems in cardiac electrophysiology and pathophysiology, including electrophysiological abnormalities associated with the syndrome often descried as 'heart failure'.

2 Model Development

We have previously applied the ten Tusscher model of the human ventricular epicardial action potential [6] to study the effects of coupling between ventricular myocytes and cardiac fibroblasts [2]. In this Chapter this approach was utilized again as a basis for in silico Simulation of co-cultured myofibroblast-myocyte interactions.

2.1 Mathematical Model Abbreviations

I_{S-1-P}	S-1-P-induced Myofibroblast Current
g_{S-1-P}	S-1-P-induced Myofibroblast Conductance
E_{mMyoFb}	Myofibroblast Membrane Potential
C_{mMyoFb}	Myofibroblast Membrane Capacitance
I_{gap}	Gap Junction Current
G_{gap}	Gap Junction Conductance
E_{mMyo}	Myocyte Membrane Potential
t	Time
I_{net}	Net Current
I_{Kv}	Time- and Voltage-dependent K^+ Current (in the myocyte and myofibroblasts)
I_{K1}	Inwardly Rectifying K^+ Current (in the myocyte and myofibroblasts)
I_{NaK}	Na^+/K^+ Electrogenic Pump Current

2.2 Myocyte Equations

The ventricular myocyte was modeled as described previously [2]. However, when a heart failure setting was considered, the parent ten Tusscher model was modified to (1) mimic the loss of I_{K1}, I_{KS}, I_{Kr} and I_{Kto} observed in myocytes of the failing heart, (2) mimic the lower capacitance associated with detubulation observed in myocytes of the failing heart, and (3) include coupling with a given number of ventricular myofibroblasts. Myocyte capacitance was set at 183 pF under control conditions, and decreased to 80 pF in the myocyte from the failing heart. Myocyte currents, resting potential and action potentials were simulated as described previously [2, 6].

2.3 Myofibroblast Equations

Our previously published model [2] was used as a basis for simulating the currents recorded in ventricular myofibroblasts. Myofibroblasts exhibit somewhat similar K^+ currents, that is I_{to}, K_V and I_{K1} currents to those in fibroblasts from the same [1, 27]. However, myofibroblasts have a much larger capacitance than fibroblasts (average capacitance in rabbit ventricular myofibroblasts, 21 pF, [1] vs. 6.3 pF in rat ventricular fibroblasts, [27]. Our experiment have revealed an S-1-P-induced current in rabbit ventricular myofibroblasts [1] it was introduced into the myofibroblast model. A number of key features of this ligand-gated conductance in the mammalian myofibroblast have not been characterized completely. For the purposes of this mathematical model, a linear S-1-P current was added to the myofibroblast. Its reversal potential was set at −30 mV. Thus this current can be described by the following expression:

$$I_{S-1-P} = g_{S-1-P}\left(E_{MyoFb} + 30mV\right) \tag{1}$$

Where g_{S-1-P} is the linear conductance for this S-1-P current. We have assumed that I_{S-1-P} magnitude is directly related to S-1-P concentration. Our experimental data describing the current-concentration relationship [1] was normalized and fitted to a sigmoid equation of the form:

$$g_{s-1-p} = g_{max}\cfrac{1}{1+\exp\left(a\left(\cfrac{[S-1-P]-[S-1-P]_o}{b}\right)\right)} \tag{2}$$

Where g_{max} is the maximum steady-state conductance of this current in the presence of S-1-P and has been evaluated to be 1.182 nS/pF. The reversal potential E_R was calculated to be −18.33 mV. The value [S-1-P] represents the variable concentration of S-1-P. A three-parameter regression was performed to determine the values of the constants a, b and c.

2.4 Electrotonic Coupling

Each myofibroblast was coupled through a linear fixed resistance to a single myocyte. This coupling conductance was varied between 0.3 and 3 nS, as specified in Results. Coupling conductance was assumed to be constant, that is it did not vary as a function of either time or voltage. Accordingly the current through the gap junction (I_{gap}) was the product of the gap junction conductance (g_{gap}), and the net driving force through the gap junction:

$$I_{gap} = g_{gap}\left(E_{mMyo} - E_{mMyoFb}\right) \tag{3}$$

as was described previously for coupling of myocytes and fibroblasts [2].

2.5 Myofibroblast Membrane Potential

Myofibroblast membrane potential (E_{mMyoFb}) is given by the expression:

$$\frac{\partial E_{mMyoFb}}{\partial t} = -\frac{I_{net}}{C_{mMyoFb}} \tag{4}$$

Where

$$I_{net} = I\left(E_{mMyoFb}, t\right) - I_{gap} \tag{5}$$

We defined I_{gap} to be opposite in direction, but identical in magnitude to that calculated in the myocyte. This signifies that an outward current through the gap junction in the myocyte was equivalent to an inward current of identical magnitude in the myofibroblast; and *vice-versa*. The function $I\left(E_{mMyoFb}, t\right)$ was given by the expression:

$$I\left(E_{mMyoFb}, t\right) = I_{K1} + I_{Kv} + I_{NaK} + I_{S-1-P} \tag{6}$$

2.6   *In silico Syncytium*

In this study, one isolated ventricular myocyte was directly coupled to 3 homogeneous myofibroblasts through a fixed coupling resistance. This heterogeneous group of cells was stimulated at 1 Hz. At a selected time (t=3.5 s), the S-1-P-induced conductance was increased. Thus S-1-P conductance was varied from 0 to 500 pS, in an attempt to simulate this ligand gated conductance in myofibroblast. In each simulation, at time t=75 s, S-1-P conductance was set to 0 pS, to mimic the washout of S-1-P.

3   Results

3.1   *Baseline Simulations of the Effects of S-1-P*

In the challenged/injured myocardium, or during inflammation, fibroblasts differentiate into the wound healing phenotype. These cells are denoted myofibroblasts [5, 28]. In the adult rat heart, ventricular myofibroblasts express both background or time-independent and time- and voltage-dependent K^+ currents that are very similar to those in fibroblasts [27]. However, the transformation from fibroblast to myofibroblast results in an approximate 2—3-fold increase in cell capacitance.

In this study, our main objective was to utilize mathematical modeling to gain further insights into the electrotonic effects of cell-to-cell coupling among ventricular myocytes and myofibroblasts. With this information, we can explore the consequences of S-1-P-induced currents in the myofibroblast on the ventricular substrate/syncytium. A diagram of the equivalent circuit which forms the basis for our mathematical modeling is shown in Fig. 1. Panel A illustrates the interacting circuit elements; and Panel B shows the principal components of the equivalent circuit for each cell type and for the connexin-mediated intercellular resistance.

In the initial set of simulations, a control epicardial human ventricular myocyte (as formulated in the ten Tusscher model) was coupled through a fixed intercellular

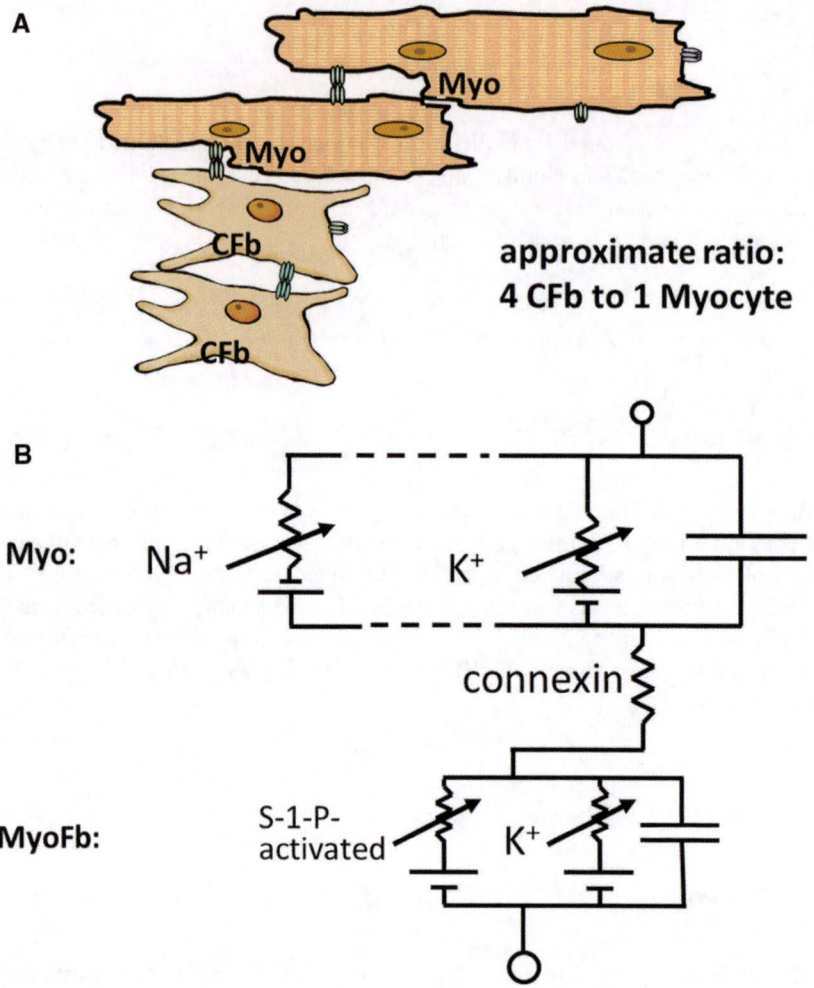

Fig. 1 Diagram (Panel **a**) and equivalent circuit (Panel **b**) of the chosen myocyte/myofilament intercellular coupling paradigm, consisting of an epicardial human ventricular myocyte and a fixed number of ventricular myofibroblasts. Panel A shows two myocytes coupled to two myofibroblasts. Panel B depicts the main elements of the equivalent circuit for a single myocyte (Myo) and one myofibroblast (CFb) communicating through a linear resistance corresponding to a connexin. Two of the time- and voltage-dependent conductances (Na^+ and K^+) in the ten Tusscher model of the human ventricular myocyte action potential are shown. The myofibroblast component shows that a number of K^+ selective conductances have been identified in these cells and illustrates the finding that S-1-P activates a ligand gated current in the myofibroblast only. See text for further details

conductance (3 nS) to a selected number of ventricular myofibroblasts (1 to 5 in progressively increasing numbers) in each successive simulation.

As shown in Fig. 2a, electrotonic coupling of 1 (solid blue line), 3 (dashed blue line) and 5 (dotted blue line) myofibroblasts to a single ventricular myocyte reduced

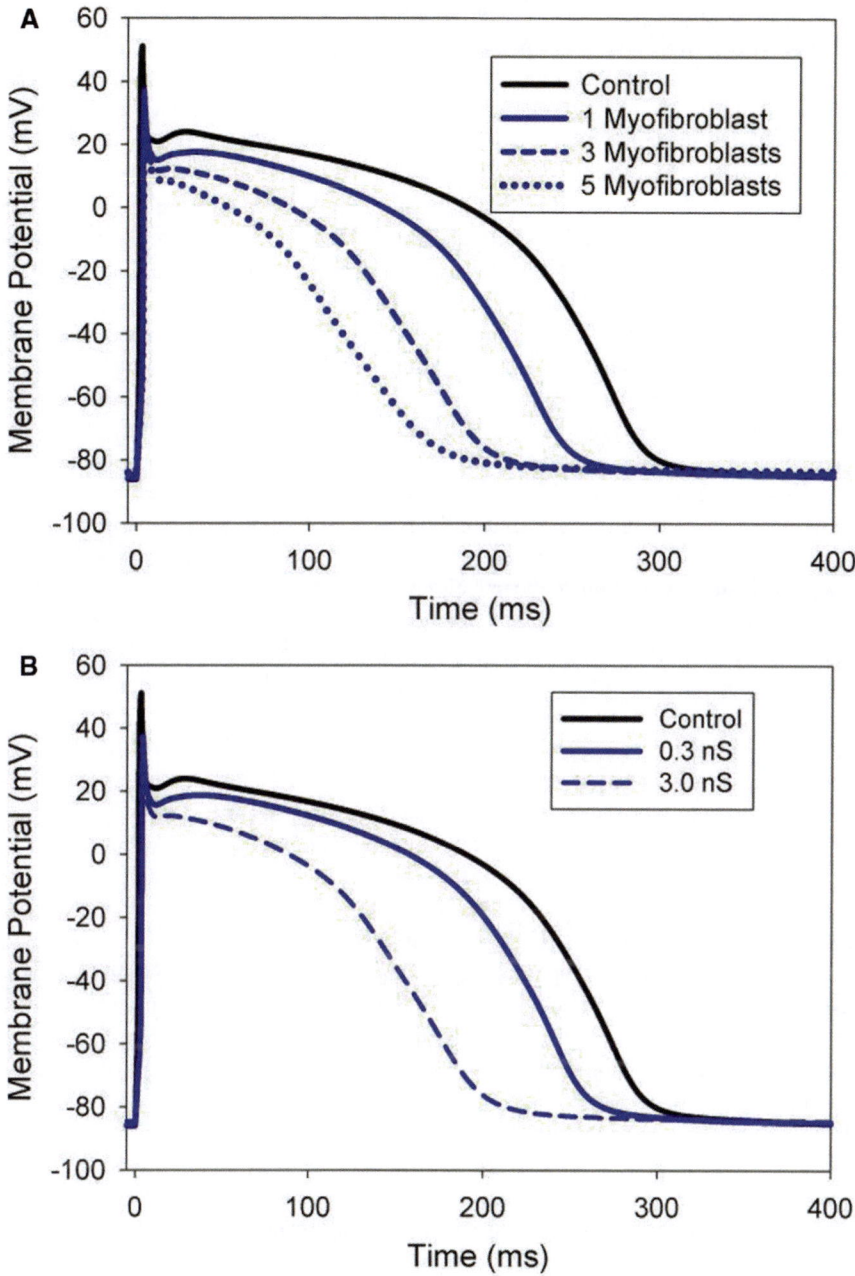

Fig. 2 The superimposed action potential waveforms from a human epicardial ventricular cardiac myocyte demonstrate the effects on action potential morphology of electrotonic coupling with ventricular myofibroblasts. In each trace shown in Panel **A**, a fixed number (0, 1, 3 or 5) of myofibroblasts was coupled to a myocyte through a coupling conductance of 3 nS. The action potential waveform was activated by a depolarizing current (1.9 nA for 3 ms). The superimposed action

APD relative to the APD in the uncoupled myocyte at baseline (black line). At 1 Hz the APD_{90} was 285 ± 1 ms in the uncoupled myocyte, and this value decreased to 245 ± 1 ms when this myocyte was coupled to one myofibroblast. Subsequent simulations in which the same control ventricular myocyte was coupled to either 3 or 5 myofibroblasts further reduced APD_{90} to 192 ± 1 ms and 163 ± 1 ms, respectively. Panel B of Fig. 2 summarizes the output of sets of computations that were done to illustrate the effects of altering intercellular conductance while holding the myocyte/myofibroblast ratio constant at 1:3. Under control conditions, APD_{90} was 285 ± 1 ms (black line). Note that progressively increasing the coupling conductance from 0.3 (solid blue line) to 3 nS (dashed blue line) in discrete steps caused significant reductions in APD_{90}: 253 ± 1 ms (at 0.3 nS) and 192 ± 1 ms (at 3 nS), with accompanying progressive changes in action potential morphology.

The simulations in Fig. 2 were done under conditions in which the myofibroblast was assumed to express only three different types of ion transfer mechanisms: K^+ conductances, a background Na^+ current, and a Na^+/K^+ ATPase electrogenic current. This is in accordance with our original model of the adult mammalian ventricular fibroblast [2]. A primary goal of the present work was to simulate the effects of the ligand-gated conductance that is activated by S-1-P. This paracrine substance is present in relatively high concentrations during an immune response and during acute inflammation [29, 30]. Our previous electrophysiological studies showed that S-1-P can induce a quasi-linear current in rabbit ventricular myofibroblasts at concentrations that are much too low to directly affect rabbit ventricular myocytes [1]. This quasi-linear current has a reversal potential between -35 and -40 mV. Panel A of Fig. 3 shows the steady-state current voltage relationship which depicts the original model (black). The blue line shows the S-1-P-induced current in a ventricular fibroblast. Note that S-1-P activates a substantial inward current at membrane potentials negative to -40 mV. In addition, the positive or outward current is approximately doubled. Panel B of Fig. 3 shows the effect of these S-1-P-induced currents in a myofibroblast, illustrated as an isochronal I-V relationship obtained at a 250 ms time point (i.e., during repolarization) in the ten Tusscher model of the human ventricular action potential. The black trace shows the I-V curve generated by the currents that are present *only* in the myocyte. The blue curve illustrates the marked changes in *net current* that result from electrotonic coupling (3 nS) of 3 myofibroblasts in which both the background K^+ conductances and the S-1-P-induced current are active.

Inspection of Fig. 3b leads to the prediction that electrotonic interactions between S-1-P-activated myofibroblasts and human ventricular myocytes may cause a significant depolarization of the resting potential in the ventricular myocyte. This extent of depolarization would markedly alter myocyte excitability.

---

potentials in Panel **B** demonstrate the effects of altering intercellular conductance while utilizing a fixed number of myofibroblasts (3) coupled to a single myocyte. Three different intercellular conductances (0, 0.3 or 3 nS) were evaluated. Note that alteration of either the number of myofibroblasts or the intercellular conductance (within the range that is relevant to the known microanatomy) can result in significant changes in action potential waveform (*height and duration*)

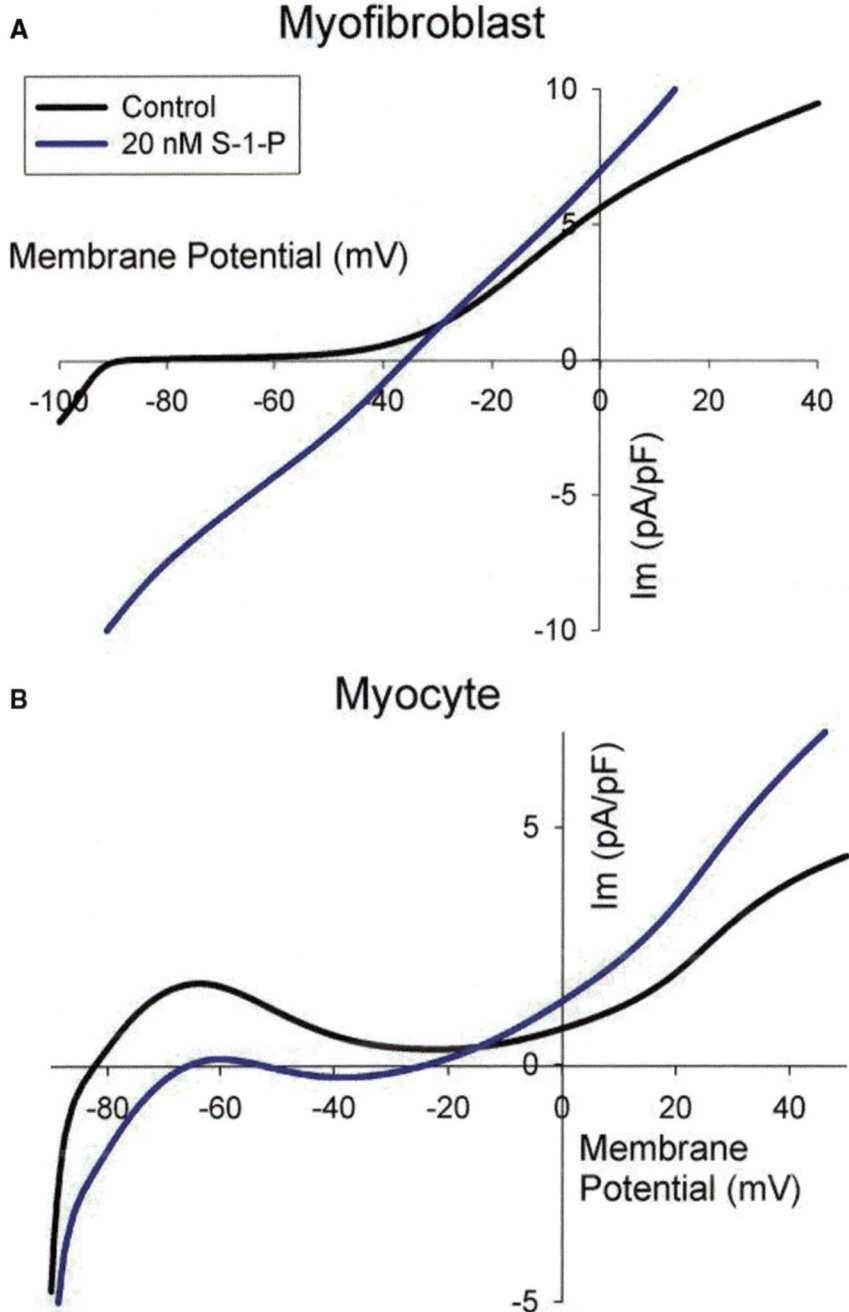

Fig. 3 Evaluation of the effects of S-1-P on selected membrane currents in ventricular myofi-broblasts (Panel **A**) and myocytes (Panel **B**), assuming an intercellular conductance of 3 nS. The two steady-state I-V curves in Panel A show the control background I-V relation (*black*) together

These possibilities were evaluated in the simulations shown in Fig. 4. Panel A shows 14 action potentials elicited at 1 Hz in the presence and absence of electrotonic coupling (3 nS), where it is also assumed that S-1-P activates the ligand-gated conductance in each of the three coupled myofibroblasts. Note that when S-1-P is present, the diastolic potential depolarizes significantly and action potential duration (as judged by the contour of final repolarization) shortens. The sigmoidal trace in Fig. 4b shows the dose response relationship of this S-1-P-induced effect on the myocyte resting potential. It is interesting that even very small concentrations of S-1-P activate the myofibroblast as originally shown in our experimental work [1]. In contrast, our results and those of others consistently demonstrate that much higher concentrations are needed to cause direct effects on the myocyte itself [31, 32]. Thus S-1-P may have its most significant electrophysiological effects in the human ventricle by acting in a selective, that is, cell selective manner.

This pattern of *in silico* results is consistent with the effects of S-1-P observed when studying myocytes and myofibroblasts in co-culture. Here, S-1-P caused a reversible diastolic depolarization [1]. The experimentally-determined effect of the S-1-P-induced current in the myofibroblast was concentration-dependent, with maximal effect produced by approximately 100 nM concentrations [1]. The simulations in Figs. 3 and 4 suggest that while electrotonic coupling of a single myocytes to 3 myofibroblasts with a coupling conductance of 3 nS may result in a reversible diastolic depolarization in the presence of e.g., 1–10 nM S-1-P, the large I_{K1} of the healthy ventricular myocyte prevents this depolarization from being great enough to inactivate Na^+ channels and block myocyte excitability.

3.2   Simulations of Heart Failure in Ventricular Myocytes

The second part of this study was done to address the fact that S-1-P has its actions predominantly in the compromised, as opposed to healthy ventricular myocardium. As an initial approach, a decision was made to attempt to alter the *in silico* substrate (the ventricular myocyte) so that its electrophysiological responses approximated the main features of the electrophysiological remodeling which occurs during heart failure. This remodeling has been a topic of a number of comprehensive electrophysiological studies in a variety of mammalian preparations. It is now known that many, if not all, of the repolarizing K^+ currents which are expressed in ventricular myocytes are reduced in heart failure, and that myocyte capacitance is also decreased, mainly due to a progressive loss of the transverse tubule system [17, 22]. Although other, and very significant changes in transmembrane currents have been described [33, 34] in heart failure models; these changes in K^+ currents and cell

---

with the I-V relation for the S-1-P-induced current in each of the 3 myofibroblasts. In Panel **B**, an isochronal I-V curve obtained at 250 ms is shown (*black trace*). The blue trace demonstrates the effect of the S-1-P-induced current (shown in Panel A) on this isochronal background I-V curve for the myocyte. The marked and nonlinear changes in the currents shown in Panel B are responsible for the changes in action potential morphology and resting potential in the following figures

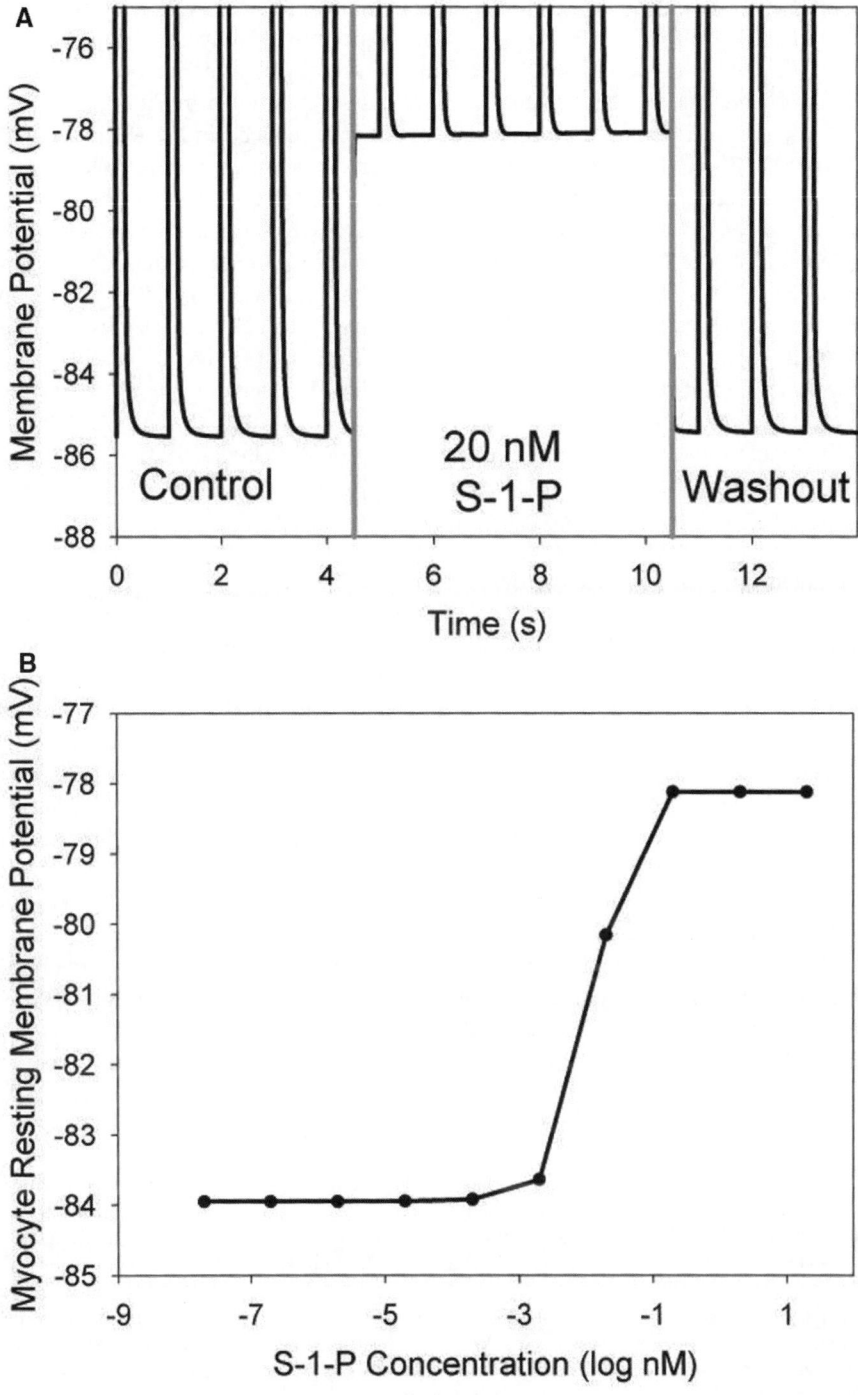

Fig. 4 Mathematical simulations of the changes in resting potential in a 'coupled' ventricular myocyte due to S-1-P (20 nM) acting only on the ventricular myofibroblasts. One ventricular myocyte was electrotonically coupled to 3 myofibroblasts through a junctional conductance of

capacitance were made as an initial approach in this project. Specifically, we modeled the effects of reduced capacitance and of progressive reduction of I_{K1}, I_{KS}, I_{Kr} and I_{to} in the ten Tusscher model of human epicardial ventricular myocyte [6]. The resulting changes on action potential morphology are shown in Fig. 5, Panel A. In these simulations myocyte capacitance was first reduced from baseline values (180 pF) to 80 pF. This resulted in a small increase in action potential duration. The additional 4 superimposed action potential waveforms demonstrate the progressive, but nonlinear, effects of simultaneously reducing each of the K^+ currents, I_{K1}, I_{KS}, I_{Kr} and I_{to} by 20, 50, 70, and then 90%. Note that action potential duration is lengthened markedly and also that the initial repolarization phase or notch of the action potential disappears. Given the role of I_{K1}, I_{KS}, and I_{Kr} in late repolarization (during phase 3) of the ventricular action potential [33, 35, 36], the reduction in APD_{90} and depolarization of myocyte RMP with progressive K^+ channel block were entirely predictable consequences.

The superimposed isochronal current I-V relationships in Fig. 5B provide further insight into the consequences of these changes in time- and voltage dependent and background K^+ currents. Specifically, this Figure shows the effects of progressive reductions in the selected K^+ currents plotted as an isochronal I-V relationship at 250 ms in an 80 pF 'failing' ventricular myocyte. The control I-V is shown in black. Reducing I_{K1}, I_{KS}, I_{Kr} and I_{to} by 20% (solid blue line), 50% (dashed blue line), 70% (dotted blue line) and 90% (dash+dotted blue line) resulted in a progressive loss of the outward limb of the I-V curve. As a consequence, the net current at this point in time (which would correspond to the repolarization of the action potential) is reduced and may even become a net inward current (due mainly to the remaining, slowly inactivating L-type Ca^{2+} current). We note also that in combination these changes have the consequence that the *in silico* heart failure myocyte has an increased resistance during much of the action potential duty cycle. For this reason, the failing ventricular myocyte would be expected to be more susceptible to electrotonic influences from within the syncytium (see Discussion). This prediction is tested in the simulations which are shown in Figs. 6–8.

The isochronal I-V relationships (at 250 ms) in Fig. 6 illustrate two important aspects of our findings. In these simulations, at the myocyte level, heart failure was simulated by reducing capacitance to 80 pF, and setting the magnitude of I_{K1}, I_{KS}, I_{Kr} and I_{to} to 50% of their control or baseline values. In addition, each failing myocyte was coupled to three myofibroblasts through a conductance of 3 nS. As was done

---

3 nS. In panel A (*left*), the effect of electrotonic coupling to myofibroblasts having only baseline currents is shown. Note that this model output is modified significantly by addition of 20 nM S-1-P: there is a small depolarization (to approx.–78 mV) but this does not significantly reduce myocyte excitability and action potential generation. The right side of this panel demonstrates a return to control conditions after the S-1-P-induced conductance was turned off. Panel **B** shows the changes in resting potential caused by three selected concentrations of S-1-P, (again based on previously published experimental data). In these simulations, the *in silico* model consisted of 1 myocyte coupled to 3 myofibroblasts through a conductance of 3 nS. Note that in all cases, addition of S-1-P resulted in significant depolarization of the myocyte due to electrotonic effects of the S-1-P-induced current activated only in the myofibroblasts

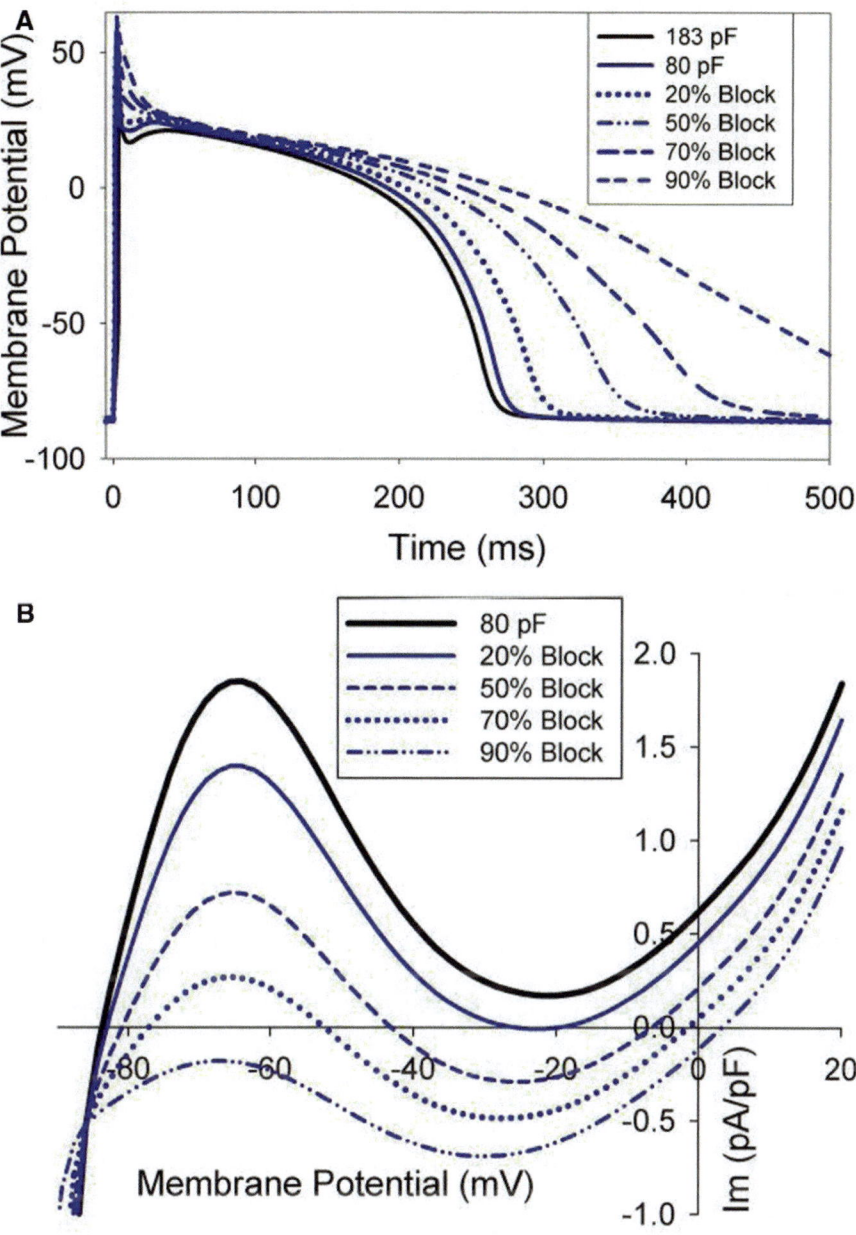

Fig. 5 Mathematical model of a human ventricular action potential waveform under conditions that are typical of a myocyte in the failing myocardium (Panel **a**). Each of the superimposed action potentials was computed based on the modifications of the ten Tusscher model of the epicardial ventricular myocyte. Specifically, the myocyte capacitance was reduced from 183 pF (*blue*) to 80 pF (*red*) consistent with the loss of T-tubules in the failing myocyte. Additional action potential waveforms demonstrate changes when the K^+ conductances I_{K1}, I_{KS}, I_{Kr} and I_{to} were decreased in

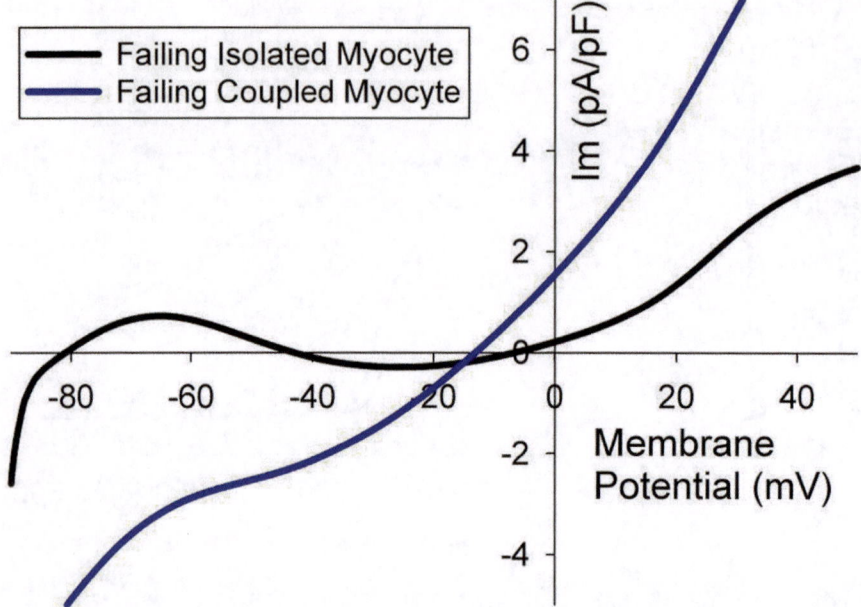

Fig. 6 Mathematical simulations that illustrate the effects of electrotonic coupling among 3 myo-fibroblasts and 1 'failing' myocyte in the absence (*black*) and in the presence (*blue*) of 20 nM S-1-P. Note that addition of the S-1-P-induced current results in significant curvilinear changes in the net current. These changes would be expected to result in a shortened action potential and depolarized resting potential (see text for further explanation)

in the computations shown in Figs. 3 and 4, S-1-P-induced current was activated by 'application' of 20 nM S-1-P. In Fig. 6, the black I-V relationship represents the isochronal data in the absence of S-1-P. The I-V relationship shown in blue depicts the S-1-P (or ligand-gated) current that is superimposed on the base line 'failing heart' electrophysiological parameters. Note that S-1-P tends to depolarize the ventricular myocyte (based on the significant additional inward current at all membrane potentials negative to −20 mV) and to shorten action potential duration (due to the additional outward current positive to −20 mV).

Figure 7 (in which the layout is analogous to Fig. 4) illustrates the effect of S-1-P on the resting potential in the 'failing' ventricular myocyte. Note that in this *in silico* experiment in which action potentials were elicited at 1 Hz, S-1-P produced a marked (approx. 30 mV) depolarization of the resting potential, and also a tendency for an after-depolarization. Panel B of Fig. 7 shows the concentration-effect relationships for S-1-P-induced changes in the myocyte resting potential under control

---

a step-wise fashion by 20, 50, 70, and finally 90%. Note that these changes in K^+ currents altered the early repolarization (*notch*), and significantly prolonged the action potential duration (APD) in the failing myocyte. The superimposed isochronal I-V relationships (250 ms) in Panel B illustrate the combined effects of the progressive reductions in I_{K1}, I_{KS}, I_{Kr} and I_{to}. These computations, cell capacitance was reduced to (80 pF)

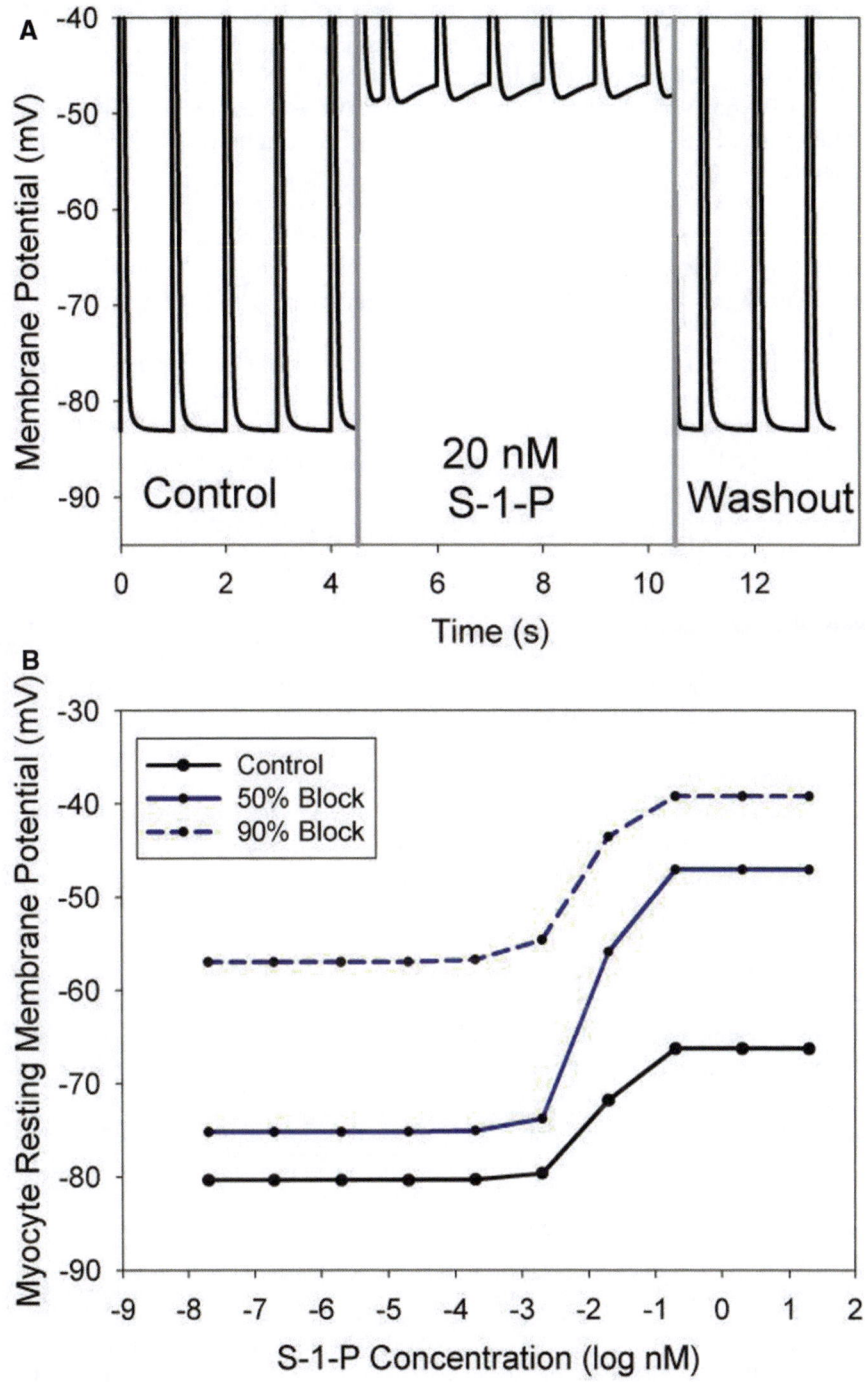

Fig. 7 Simulations of the effects of coupling a 'failing' ventricular myocyte to 3 myofibroblasts through a gap junction conductance of 3 nS in response to a 1 Hz stimulus. In Panel A (*left*), action potentials from the failing myocyte (80 pF; I_{K1}, I_{KS}, I_{Kr} and I_{to} reduced by 50%) coupled with 3

conditions (black), following 50% reduction of the selected K^+ currents (dark blue) and following 90% reduction of these currents (light blue). As expected, there is a progressive depolarization of the resting potential due to the reduction in I_{K1}; and the response to S-1-P results in a depolarization to a more depolarized 'resting' potential.

As shown in Fig. 7, when a single heart failure myocyte was paced at 3 Hz, and then coupled with 3 myofibroblasts through a conductance of 3 nS, 20 nM S-1-P-induced an effect on myocyte excitability that was more pronounced than in the same paradigm when a healthy myocyte was the S-1-P target (compare Fig. 7 A with Fig. 4 A).

The final set of computations was done to illustrate the consequences of these changes on action potential morphology. As shown in Fig. 8 coupling a ventricular myocyte from a failing heart, in which I_{K1}, I_{KS}, I_{Kr} and I_{to} were reduced by 50% and in addition cell capacitance was reduced to 80 pF, to three myofibroblasts and then activating the 20 nM S-1-P-induced current changed the myocyte action potential waveform dramatically. In Fig. 8a, the two superimposed action potential waveforms illustrate the differences resulting from the failing myocyte assumptions. The action potential shown in blue was computed on the basis of the ten Tusscher model baselilne conditions, with myocyte capacitance decreased from approximately 180–80 pF. The superimposed action potential shown in black illustrates the changes resulting from the reduction in K^+ currents by 50%. This change in action potential duration is similar to that shown in one of the traces in Fig. 5a.

In Fig. 8b, the effect of coupling a human ventricular myocyte in which capacitance is 80 pF and I_{K1}, I_{KS}, I_{Kr} and I_{to} were decreased by 50% to three myofibroblasts at 0.3 nS conductance is shown. In this Panel, the starting condition in which only basal currents are active in the myofibroblasts is shown in blue. The action potential waveform shown in black exhibits significant lengthening as expected from the reduced net outward K^+ currents which underlie our model of the 'failing' myocyte. These simulations further reinforce the very significant effect of coupling a failing myocyte with myofibroblasts, in the setting of inflammation, where S-1-P concentrations are likely to be in the high nM range.

The two superimposed action potential waveforms in Panel C of Fig. 8 illustrate the effects of 20 nM S-1-P 'applied' in the setting of 'heart failure'. The action potential waveform for the failing myocyte is shown in black; addition of S-1-P results in a small decrease in plateau height, a depolarization of the resting potential and a significant shortening of the entire repolarization phase of the action potential. Each of these effects are expected, based upon the interacting S-1-P or ligand-gated currents and intrinsic time- and voltage-dependent currents shown in Fig. 6.

---

myofibroblasts. At the 4.5 s time-point, the S-1-P-sensitive current was activated in the myofibroblasts (middle, '20 nM S-1-P'). This caused the resting membrane potential in the failing myocyte to depolarize. The 3 traces in Panel B illustrate this S-1-P-induced depolarization of the resting potential in the failing myocyte. The lower trace (*black*) shows the effect of coupling a failing (80 pF) myocyte, assuming baseline expression levels of all K^+ currents. The middle and upper curves in dark and light blue respectively illustrate the effects on resting membrane potential of 20 nM S-1-P when myocyte I_{K1}, I_{KS}, I_{Kr} and I_{Kto} currents were reduced by 50% (*continuous blue*) and then by 90% (*interrupted blue*). This pattern of results is expected from the net current change illustrated in Fig. 6 (see text for further explanation)

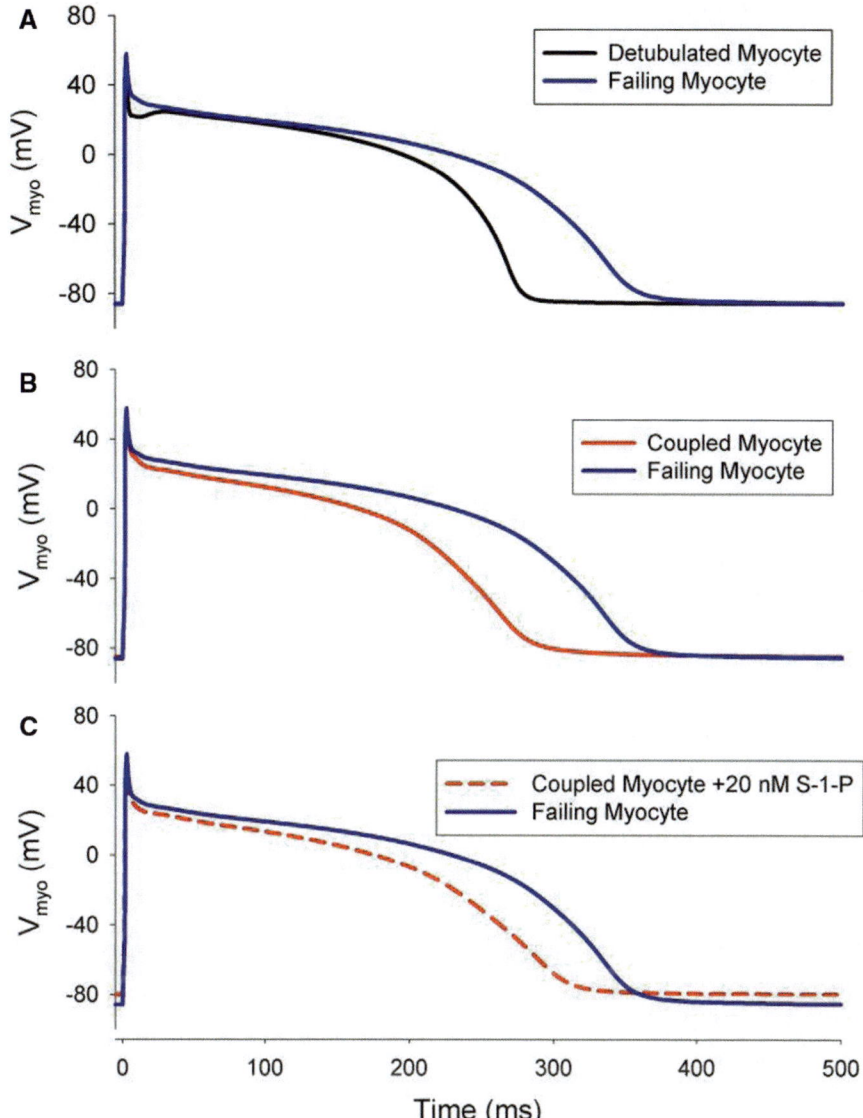

Fig. 8 Mathematical simulations illustrating the effects of electrotonic interactions between a fixed number of myofibroblasts (3) and one 'failing' ventricular myocyte in the absence (Panels **a** and **b**) and in the presence (Panel **c**) of S-1-P. In Panel A, the effects on action potential waveform of reducing I_{K1}, I_{KS}, I_{Kr} and I_{Kto} in a failing (80 pF) myocyte are shown. The two superimposed action potentials illustrate the starting condition, a failing myocyte (*black trace*); and the lengthened action potential (*shown in blue*) from a failing myocyte in which the K$^+$ conductances have been reduced by 50%. In Panel **b**, the action potential from a failing myocyte (80 pF, I_{K1}, I_{KS}, I_{Kr} and I_{to} reduced by 50%) is shown in blue; together with the action potential computed after this myocyte was coupled with 3 myofibroblasts at a coupling conductance of 0.3 nS (*red trace*). In Panel **c**, the effects of S-1-P are shown. The action potential waveform shown in blue is the same as the one shown in Panel B above. Note that activation of the S-1-P - induced current causes a small depolarization and shortens the action potential in this in silico failing ventricular substrate (*dashed red trace*). See text for further details

4   Discussion

The combination of previously published mathematical models for the human ventricular myocyte action potential and ventricular fibroblast electrophysiology provided the basis for addressing the main goal of this study. Since the principal biological compound of interest was sphingosine-1-phosphate (S-1-P), a ligand that is present in relatively high concentrations (approximately 0.1 μM) in the setting of inflammation, it was considered necessary to also attempt to simulate some electro-physiological properties of the compromised or failing ventricular myocardium. We therefore modified our mathematical model of the fibroblast so that it could repre-sent same of the electrophysiological/biophysical features of the ventricular myofi-broblast. In addition, the known changes during heart failure in ventricular myocyte capacitance and K^+ current expression were included by introducing appropriate decreases in these parameters. We recognize that the resulting hybrid first order model remains incomplete. Nevertheless, some salient and potentially functionally important features of S-1-P actions in the mammalian ventricular myocardium were able to be replicated, visualized and integrated. Perhaps the most significant in-sights are the following:

1. Although very low concentrations of S-1-P affect only the myofibroblast, the existing electrotonic coupling between the myofibroblast and myocyte cell pop-ulations in both the healthy and the failing ventricular myocardium can result in significant changes in action potential morphology and in small, but important changes in the resting membrane potential.
2. The consequences of S-1-P binding to Edge receptor(s) and then activating ion channels in the myofibroblast are very straightforward, the resulting changes in action potential waveform are complex. That is, S-1-P activates an approxi-mately linear non-specific cation conductance that has a reversal potential near -30 mV. This ligand-gated conductance appears to scale in a monotonic fash-ion as a function of ligand concentrations in the 0.1 to 50 nanomolar range. However, this conductance does *not* result in the expected depolarization of the myofibroblast because the electrotonic influence of the much larger myocyte is dominant. Thus, the apparent simplicity and hence predictability of the S-1-P action in the myofibroblast is attenuated dramatically when the indirect S-1-P-mediated effects are observed or recorded in the myocyte. The main reason for this is that the marked changes in complex impedance in the myocyte, brought about by the inwardly rectifying K^+ conductances, I_{K1} and I_{Kr}, serve to attenuate the otherwise straightforward properties of the current generator which is the S-1-P-sensitive channels in the myofibroblast population.
3. The principles that have been revealed and are summarized above and illustrated in Figs. 7 through 9, lead directly to some additional functional insights. The known properties of the failing myocardium, reduced cell capacitance due to progressive loss of the anatomical transverse tubule system, and commensurate (in fact, perhaps for this reason) reductions in both background and time- and voltage-dependent conductances, result in this population of myocytes exhibit-

ing a significantly altered impedance, or membrane resistance profile. Since the dominant characteristic is the expected increase in membrane resistance; *a priori* it would be expected that the S-1-P-induced current source in the subjacent myofibroblast population is more effective. Our computations illustrate patterns of results consistent with these starting principles and assertions. More definitive insights would require independent experimental verification of the possibility that S-1-P production or actions are heterogeneous. These data would also need to be complemented with information concerning the possibility that intercellular resistance may be altered in either of the predominant settings known to give rise to increased S-1-P liberation: acute inflammation and/or a significant ischaemic insult; e.g., in the setting of what may be referred to as heart failure or infarction scaring.

Inspection of the sets of computational findings that we have been produced identifies at least one other phenomenon of significant interest. It is noteworthy that perhaps the most important consequence of electrotonic interactions between myofibroblasts and fibroblasts is the consistent and significant depolarization of the *myofibroblast* which is phase locked with each ventricular action potential. In both the fibroblast population and in myofibroblasts, this depolarization is large enough that it would be expected to significantly modulate excitation-secretion coupling in the myofibroblast. The paracrine actions of S-1-P therefore need to be considered in the context of interactive, perhaps closed loop control mechanisms. It is very likely that the indirect actions of S-1-P on the ventricular myocyte action potential feed back to alter myofibroblast electrophysiological responses, including but not limited to significant components of known regulated excitation-secretion coupling mechanisms.

A final consideration that arises from this computational work concerns the ways in which a heterogeneous cell population, the ventricular myocardium, interacts/communicates on a cell-by-cell basis in both health and disease. It has been known for 30 years that the inward rectification of the predominant K^+ conductances in most cardiac myocytes, in smooth muscle cells, in both endothelial cells and pericytes, and in bone cells, effectively results in an ideal current generator. That is, even very small depolarizations from the resting potential result in marked increases in the resistance of these cells and this increased resistance results in what is often termed a lengthened 'space constant'. In a simplistic way, our computational work reveals a cautionary note which should be applied to this classical electrophysiological principle in the context of the cardiac syncytium or the endothelium. If it is the case that a ligand-gated current (induced by S-1-P) can effectively shunt the intrinsic properties of an inwardly rectifying background K^+ conductance, then the effective 'point spread function' of such a paracrine compound may be quite complex. With available electrophysiological techniques this complexity, may be difficult to reveal convincingly. In practice, this would require at least three independent electrophysiological measurements from identified myocytes and myofibroblasts in the same preparation. This is sufficiently difficult to make it likely that obtaining a reliable experimental data set will not be successful. The most plausible alternate approach, obtaining this data using voltage-sensitive dye methods, is also challenging since consistent, reliable recordings from two different cell types in the same syncytium are difficult; and since in the mammalian myocardium, motion artifacts remain a limiting technical factor in recording and accurate interpretation of electrotonic changes in membrane potential, or the expected small changes in action potential waveform. These considerations provide a basis for continuing to develop tools and platforms which allow mathematical simulations of some aspects of physiological and pathophysiological phenomena in the heart. It is although that this approach will be useful to provide the type of straight-forward, and semiquantitative, output which is the basis of this paper, more conclusive work must always arise from and be verified by directly relevant experimental data.

Acknowledgments This work was supported by grants from the Canadian Institutes of Health Research (WRG), the Heart and Stroke Foundation of Canada (WRG, LC), the Alberta Heritage Foundation for Medical Research, now Alberta Innovates - Health Solutions (WRG, LC) and the British Heart Foundation (GLS).

References

1. Chilton L, Giles WR, Smith GL (2007) Evidence of intercellular coupling between co-cultured adult rabbit ventricular myocytes and myofibroblasts. J Physiol 583(Pt 1):225–236.
2. MacCannell KA, Bazzazi H, Chilton L, Shibukawa Y, Clark RB, Giles WR (2007) A mathematical model of electrotonic interactions between ventricular myocytes and fibroblasts. Biophys J 92(11):4121–4132.
3. Pyne S, Pyne NJ (2000) Sphingosine 1-phosphate signalling in mammalian cells. Biochem J 349(Pt 2):385–402
4. Peters SL, Alewijnse AE (2007) Sphingosine-1-phosphate signaling in the cardiovascular system. Curr Opin Pharmacol 7(2):186–192.
5. Sun Y, Kiani MF, Postlethwaite AE, Weber KT (2002) Infarct scar as living tissue. Basic Res Cardiol 97(5):343–347.
6. ten Tusscher KH, Noble D, Noble PJ, Panfilov AV (2004) A model for human ventricular tissue. Am J Physiol Heart Circ Physiol 286(4):H1573–1589.
7. Kohl P, Kamkin AG, Kiseleva IS, Noble D (1994) Mechanosensitive fibroblasts in the sinoatrial node region of rat heart: interaction with cardiomyocytes and possible role. Exp Physiol 79(6):943–956
8. Camelliti P, Green CR, LeGrice I, Kohl P (2004) Fibroblast network in rabbit sinoatrial node: structural and functional identification of homogeneous and heterogeneous cell coupling. Circ Res 94(6):828–835.
9. Camelliti P, Devlin GP, Matthews KG, Kohl P, Green CR (2004) Spatially and temporally distinct expression of fibroblast connexins after sheep ventricular infarction. Cardiovasc Res 62(2):415–425.
10. Gaudesius G, Miragoli M, Thomas SP, Rohr S (2003) Coupling of cardiac electrical activity over extended distances by fibroblasts of cardiac origin. Circ Res 93(5):421–428.
11. Hachiro T, Kawahara K, Sato R, Yamauchi Y, Matsuyama D (2007) Changes in the fluctuation of the contraction rhythm of spontaneously beating cardiac myocytes in cultures with and without cardiac fibroblasts. Bio Syst 90(3):707–715.
12. Kizana E, Ginn SL, Smyth CM, Boyd A, Thomas SP, Allen DG, Ross DL, Alexander IE (2006) Fibroblasts modulate cardiomyocyte excitability: implications for cardiac gene therapy. Gene Ther 13(22):1611–1615.

13. Miragoli M, Gaudesius G, Rohr S (2006) Electrotonic modulation of cardiac impulse conduction by myofibroblasts. Circ Res 98(6):801–810.
14. Rook MB, van Ginneken AC, de Jonge B, el Aoumari A, Gros D, Jongsma HJ (1992) Differences in gap junction channels between cardiac myocytes, fibroblasts, and heterologous pairs. Am J Physiol 263(5 Pt 1):C959–977
15. Driesen RB, Dispersyn GD, Verheyen FK, van den Eijnde SM, Hofstra L, Thone F, Dijkstra P, Debie W, Borgers M, Ramaekers FC (2005) Partial cell fusion: a newly recognized type of communication between dedifferentiating cardiomyocytes and fibroblasts. Cardiovasc Res 68(1):37–46.
16. He J, Conklin MW, Foell JD, Wolff MR, Haworth RA, Coronado R, Kamp TJ (2001) Reduction in density of transverse tubules and L-type Ca(2+) channels in canine tachycardia-induced heart failure. Cardiovasc Res 49(2):298–307
17. Armoundas AA, Wu R, Juang G, Marban E, Tomaselli GF (2001) Electrical and structural remodeling of the failing ventricle. Pharmacol Ther 92(2–3):213–230
18. Beuckelmann DJ, Nabauer M, Erdmann E (1993) Alterations of K + currents in isolated human ventricular myocytes from patients with terminal heart failure. Circ Res 73(2):379–385
19. Janse MJ (2004) Electrophysiological changes in heart failure and their relationship to arrhythmogenesis. Cardiovasc Res 61(2):208–217
20. Kaab S, Nuss HB, Chiamvimonvat N, O'Rourke B, Pak PH, Kass DA, Marban E, Tomaselli GF (1996) Ionic mechanism of action potential prolongation in ventricular myocytes from dogs with pacing-induced heart failure. Circ Res 78(2):262–273
21. Kaye DM, Hoshijima M, Chien KR (2008) Reversing advanced heart failure by targeting Ca2 + cycling. Annu Rev Med 59:13–28.
22. Li GR, Lau CP, Ducharme A, Tardif JC, Nattel S (2002) Transmural action potential and ionic current remodeling in ventricles of failing canine hearts. Am J Physiol Heart Circ Physiol 283(3):H1031–1041.
23. Jacquemet V, Henriquez CS (2007) Modelling cardiac fibroblasts: interactions with myocytes and their impact on impulse propagation. Europace: European pacing, arrhythmias, and cardiac electrophysiology: journal of the working groups on cardiac pacing, arrhythmias, and cardiac cellular electrophysiology of the European Society of Cardiology 9(Suppl 6):vi29–37.
24. Sachse FB, Moreno AP, Abildskov JA (2008) Electrophysiological modeling of fibroblasts and their interaction with myocytes. Ann Biomed Eng 36(1):41–56.
25. Vasquez C, Siddiqui RA, Moreno AP, Berbari EJ (2002) A fibroblast-myocyte model which accounts for slow conduction and fractionated electrograms in infarct border zones. Comp Cardiol 29:245–248
26. Wang YJ, Sung RJ, Lin MW, Wu SN (2006) Contribution of BK(Ca)-channel activity in human cardiac fibroblasts to electrical coupling of cardiomyocytes-fibroblasts. J Membr Biol 213(3):175–185.
27. Chilton L, Ohya S, Freed D, George E, Drobic V, Shibukawa Y, Maccannell KA, Imaizumi Y, Clark RB, Dixon IM, Giles WR (2005) K + currents regulate the resting membrane potential, proliferation, and contractile responses in ventricular fibroblasts and myofibroblasts. Am J Physiol Heart Circ Physiol 288(6):H2931–2939.
28. Tomasek JJ, Gabbiani G, Hinz B, Chaponnier C, Brown RA (2002) Myofibroblasts and mechano-regulation of connective tissue remodelling. Nat Rev Mol Cell Biol 3(5):349–363.
29 Alvarez SE, Milstien S, Spiegel S (2007) Autocrine and paracrine roles of sphingosine-1-phosphate. Trends Endocrinol Metab: TEM 18(8):300–307
30. Rosen H, Goetzl EJ (2005) Sphingosine 1-phosphate and its receptors: an autocrine and paracrine network. Nature Rev Immunol 5(7):560–570.
31. MacDonell KL, Severson DL, Giles WR (1998) Depression of excitability by sphingosine 1-phosphate in rat ventricular myocytes. Am J Physiol 275(6 Pt 2):H2291–2299
32. McDonough PM, Yasui K, Betto R, Salviati G, Glembotski CC, Palade PT, Sabbadini RA (1994) Control of cardiac Ca2 + levels. Inhibitory actions of sphingosine on Ca2 + transients and L-type Ca2 + channel conductance. Circ Res 75(6):981–989

33. Nattel S, Maguy A, Le Bouter S, Yeh YH (2007) Arrhythmogenic ion-channel remodeling in the heart: heart failure, myocardial infarction, and atrial fibrillation. Physiol Rev 87(2):425–456.
34. Tomaselli GF, Marban E (1999) Electrophysiological remodeling in hypertrophy and heart failure. Cardiovasc Res 42(2):270–283
35. Bassani RA (2006) Transient outward potassium current and Ca2 + homeostasis in the heart: beyond the action potential. Braz J Med Biol Res = Revista brasileira de pesquisas medicas e biologicas/Sociedade Brasileira de Biofisica 39(3):393–403.
36. Roden DM, Balser JR, George AL, Jr., Anderson ME (2002) Cardiac ion channels. Ann Rev Physiol 64:431–475.

Extracellular Matrix and Cardiac Disease: Surgical and Scientific Perspectives

Holly E. M. Mewhort and Paul W. M. Fedak

Abstract Scientists and surgeons can each benefit from a clear understanding of cardiac ECM and its' influence on cardiac structure and function in both health and disease. This chapter highlights key concepts with respect to the influence of cardiac ECM in traditional surgical repairs and introduces emerging "biosurgical" strategies designed to leverage ECM biology and further enhance innovative surgical repairs to optimize cardiac performance.

Keywords Surgical repair · ECM architecture · Aortic disease · Valvular disease

1 Introduction

The primary role of the cardiac surgeon is to restore structure and improve function by reshaping and remodeling the heart. Cardiac structure and function has customarily been appreciated only in the context of its contractile muscle cells, the cardiomyocytes. The extracellular matrix (ECM), or interstitium, was recognized only for its supportive structural role in the heart and vasculature. Increasingly, clinicians now appreciate that ECM deficiencies and alterations are both a cause and consequence of cardiac disease. In parallel, scientists are beginning to recognize the critical regulatory role that ECM plays in providing a dynamic microenvironment for heart cells. ECM regulates critical cellular behaviors in response to tissue injury and plays an important role in both cardiac health and disease. Importantly, it is now accepted that ECM can be surgically modified to restore heart function. Scientists and surgeons can each benefit from a clear understanding of cardiac ECM and its' influence on cardiac structure and function in both health and disease. This chapter highlights key concepts with respect to the influence of cardiac ECM in traditional surgical repairs and introduces emerging "biosurgical" strategies designed to leverage ECM biology and further enhance innovative surgical repairs to optimize cardiac performance.

Paul W. M. Fedak (✉) · Holly E. M. Mewhort
Section of Cardiac Surgery, Department of Cardiac Sciences, Foothills Medical Centre,
C880G, 1403-29th Street NW, Calgary, AB T2N 2T9, Canada
e-mail: paul.fedak@gmail.com

© Springer International Publishing Switzerland 2015
I.M.C. Dixon, J. T. Wigle (eds.), *Cardiac Fibrosis and Heart Failure: Cause or Effect?*,
Advances in Biochemistry in Health and Disease 13, DOI 10.1007/978-3-319-17437-2_17

2   The ECM of the Heart

2.1   The Fibrous Skeleton

All surgical repairs that remedy damage from pathologic disease processes are performed on the structural foundation of the fibrous skeleton. The fibrous skeleton
is pivotal to cardiac function and therefore of great importance to the success of
surgical procedures. The cardiac fibrous skeleton incorporates the extracellular matrix of the myocardium, perivascular matrix, chordae tendineae and cardiac valves
(Fig. 1). It is composed of a complex and dynamic balance of fibrillar collagen,
elastin, microfibrillar proteins, proteoglycans, laminin and fibronectin. Fibrillar
collagen, the dominant protein in cardiac ECM, provides tensile strength to the
3-dimensional scaffold that houses the contractile components of the organ. Evidence suggests that collagen composition (quantity, quality and type) significantly
influences ventricular size, shape and function [1–3]. Collagens type I and III are
the most abundant subtypes found in cardiac ECM. By comparison type I collagen
is mature and heavily cross-linked giving it greater tensile strength then immature,
less cross-linked type III collagen, which demonstrates greater elasticity. The proportion of type I to type III collagen depends on the rate of collagen turnover, which
is tightly regulated. In essence, the fibrous skeleton is the macrorepresentation of
the continuity and connection of these extracellular matrix components throughout
the heart (Fig. 1). The fibrous skeleton provides the framework for the cardiac tissues to assemble into a functional and complex dynamic mechanical organ.

2.2   ECM Architecture and Cardiac Function

ECM architecture influences cell and muscle fiber orientation and in so doing,
regulates tissue function. The ECM of the myocardium can be classified into three
components: the epimysium, perimysium and endomysium (Fig. 1). The epimysium surrounds cardiomyocyte bundles and forms a structural continuum with the
fibrous skeleton of the heart connecting its passive structural and active functional
components. The perimysium surrounds and interconnects bundles of cardiomyocytes, dictating their specific orientation. The endomysium surrounds individual
cardiomyocytes within each bundle. The epimysium and perimysium are rich in
fibrillar collagen and thought to dictate the shape and distensibility of the cardiac
chambers, while the endomysium is composed of a combination of collagen and
elastin, thought to influence individual cell organization and attachment.

   In the healthy heart the precise 3-dimensional pattern by which fibrillar collagen networks of the perimysium are connected results in amplification of the 15 %
linear shortening of each individual myocyte to produce a coordinated contraction
of the left ventricle such that it ejects upwards of 60 % of its blood volume under normal conditions [4]. Spinale and co-workers showed that disruption of this
ECM network can decrease systolic performance without alterations in myocyte

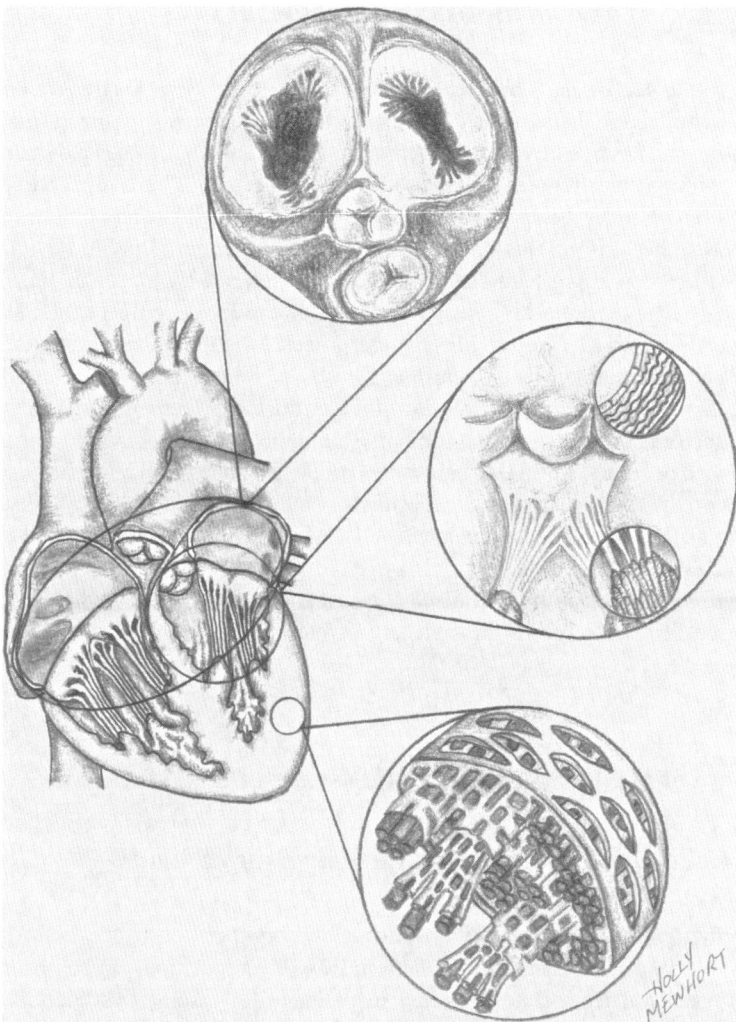

Fig. 1 The fibrous skeleton of the heart is depicted. The *top panel* highlights what is traditionally thought of as the fibrous skeleton. The *middle panel* highlights the continuity between traditional components of the fibrous skeleton such as the aortic valve and chordae tendoneae and the extracellular matrix components of the left ventricle and aortic wall that connect structure and functionality. The *bottom panel* highlights the epimysial, perimysial and endomysial levels of ECM organization within the left ventricular myocardium responsible for the translation of individual cellular contractions into organized LV ejection

contractility [5], demonstrating the significance by which the ECM contributes to the contractile function of the heart. Precise muscle fiber orientation also mitigates excessive stretch in diastole and contributes to elastic recoil in systole, enhancing the efficiency of myocardial contraction [6]. These relationships are important to respect when developing novel treatments for patients with heart failure.

2.3   ECM as a Dynamic Microenvironment

In addition to serving as a physical scaffold that provides tensile strength and support to tissue; the ECM serves as a framework for the precise organization of cells, determines the passive physical properties of tissue, and contributes a dynamic microenvironment that regulates cell proliferation, signaling, behavior and survival [7–11]. For example, the spatial and temporal expression of matrix components and cell-ECM adhesion proteins has been proven vital to normal organ development [7, 12]. In pressure overload states, the loss of myocyte-ECM attachment results in myocyte apoptosis marking the early stages of cardiac decompensation and systolic heart failure [13]. Key to all of these processes is the precise composition of the ECM is maintained by the cardiac fibroblast. Under steady-state conditions, cardiac fibroblasts are responsible for collagen turnover, and secrete a number of growth factors and cell signaling molecules that dictate cell behavior and survival. In response to injury, normally quiescent cardiac fibroblasts undergo a phenotypic conversion into their activated myofibroblast state. When activated, myofibroblasts alter the homeostatic balance to increase ECM deposition, leading to scar formation and eventual tissue healing. However, if myofibroblasts remain chronically activated, as they do in several disease states such as after myocardial infarction, they drive fibrosis resulting in a variety of pathologies that can require surgical management.

3   Role of the ECM in Cardiac Disease and Dysfunction

3.1   ECM Dysregulation and Cardiac Dysfunction

Just as the amount, type, stability, turnover and organization of the ECM plays a crucial role in cardiac health and function, ECM dysregulation can be observed in several cardiac disease states, and are responsible for cardiac dysfunction. Acute perturbations in ECM homeostasis can be adaptive, as is the case immediately following myocardial infarction (MI). Following ischemic insult, activated cardiac myofibroblasts increase ECM turnover promoting clearance of necrotic tissue and the formation of scar. Any delay in the clearance of necrotic tissue results in delayed healing [14] and prevents scar formation. Patients that fail to develop adequate scar tissue within the infarcted myocardium are at risk of developing mechanical complications of MI, such as free wall rupture, ischemic VSD or papillary muscle rupture and ischemic mitral regurgitation, because the structural integrity of the regional myocardium is lost. Mechanical complications of MI are often lethal and require emergent surgical correction when recognized.

Though acute changes in ECM turnover are in some situations adaptive, chronic perturbations in ECM homeostasis of a variety of etiologies inevitably result in fibrosis and subsequent heart failure. For example, regardless of etiology, degradation and disruption of cardiac ECM architecture is common to all patients with end stage heart failure (ESHF; Fig. 2) [15]. Evidence from animal models of ESHF suggests that the degradation and disruption of the cardiac ECM observed is the direct result of aberrant ECM homeostasis [16, 17].

In pressure overload situations, such as chronic hypertension, severe aortic stenosis (AS), or hypertrophic obstructive cardiomyopathy (HOCM) there is a gradual accumulation of excess myocardial collagen [18]. We have demonstrated that in patients with HOCM this is also likely the result of aberrant ECM homeostasis [19] favoring a net increase in collagen deposition as opposed to degradation. Aber-

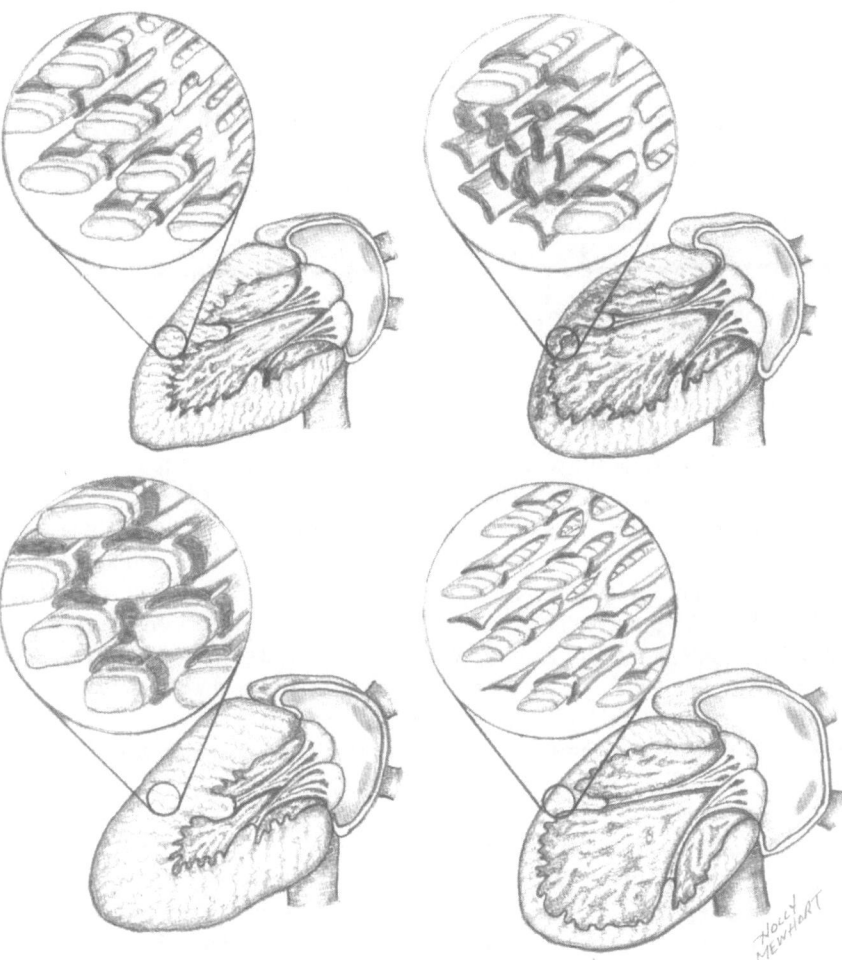

Fig. 2 Left ventricular geometry and ECM quality and composition in a normal heart (**a**) and the changes in LV geometry and ECM quality and composition that occur following ischemic injury (**b**), chronic pressure overload states (**c**) and chronic volume overload states (**d**; which also define ESHF)

rant ECM homeostasis and subsequent collagen accumulation results in decreased chamber compliance, complimenting the concentric hypertrophy that occurs in response to pressure overload. Following ischemic injury, an acute increase in collagen is initially adaptive. However, in many patients, the acute increase in collagen is followed by a chronic diffuse increase in myocardial collagen remote to the site of injury as a response to a compensatory increase in workload within the remaining healthy myocardium. In diabetic patients, diffuse fibrosis can be observed in the absence of pressure overload or ischemic injury [20]. Both situations may be the result of sustained fibroblast activation independent of mechanical stimuli.

In volume overload situations, such as severe aortic or mitral regurgitation, collagen accumulation is not initially observed, however an increase in the proportion of type III collagen has been observed [2]. The increased elasticity of type III collagen may be responsible for the LV dilatation seen in response to pressure overload. Progressive ventricular dilatation, a hallmark of heart failure, may also be mediated through ongoing ECM degradation by MMP activity [4, 21]. Though not observed initially, an increase in collagen accumulation seen in patients with dilated cardiomyopathy and ESHF [4], suggests collagen accumulation may define the final stages of cardiac dysfunction resulting from volume overload.

3.2 ECM and Aortic Disease

ECM dysregulation also plays a role in aneurysmal diseases of the aorta, which commonly require surgical management. Much of our knowledge about the role of the ECM in thoracic aortic aneurysm formation stems from the phenotypic characteristics of patients with genetically inherited defects in ECM components, such as those with Marfan's, Ehlers-Danlos, and Loeys-Dietz syndrome. Patients with Marfan's syndrome possess a mutation in the fibrillin-1 gene [22] that results in poor elastin filament alignment and lamellar unit disorganization resulting in aortic wall instability [23] that leads to an increased risk of aortic dissection and aneurysm formation [24]. Over 70 different mutation of the COL3A1 gene have bee identified in contribution to Ehlers-Danlos syndrome type IV. The genetic defect leads to dysfunctional type III collagen formation leaving patients at risk of aortic dilatation and dissection [25]. Patients with Loeys-Dietz syndrome possess a mutation of the gene that encodes a component of TGF-β-receptors, resulting in aberrant TGF-β signaling, a growth factor that significantly impacts ECM homeostasis by mediating cardiac fibroblast phenotype [26].

Though the specific genetic defect that effects patients with congenitally bicuspid aortic valves (BAV) has not been identified, ECM integrity has been implicated as a causal factor in the increased risk of ascending aortic dissection and thoracic aneurysm formation displayed by these patients. Robicsek and colleagues studied flow dynamics through BAVs vs. tricuspid aortic valves (TAVs) modeling their potential consequences on aortic wall structure concluding that aberrant flow dynamics resulting from BAV geometry was the likely cause of the ascending aortic dilatation more frequently observed in patients with BAVs [27]. We have shown that the distinct regional wall shear stress patterns measured by 4-dimensional magnetic resonance imaging are generated by the three different aortic valve cusp fusion patterns, and further, these distinct patterns are associated with the expression of specific BAV related aortopathies [28]. However evidence also exists to suggest that the increased incidence of ascending aortic dilatation and dissection is due to inherent abnormalities in the structural integrity of the aorta [29–31]. Hahn and colleagues demonstrated that patients with BAVs displayed a higher incidence of aortic root enlargement irrespective of age or abnormal valve hemodynamics [30]. Keane and colleagues demonstrated that patients with BAVs consistently display larger ascending aortic dimension that patients with TAVs irrespective of defects in valve related hemodynamics [29]. Yasuda and colleagues demonstrated that patients with BAVs experienced ascending aortic dilatation irrespective of whether or not aberrant valve hemodynamics were corrected by aortic valve replacement [31]. We demonstrated that the vasculature of BAV patients has prominent ECM dysregulation, including deficient fibrillin-1 and increased matrix metalloprotease (MMP) activity, possibly contributing to disruption of the aortic wall and aneurysm formation [32]. These data suggest that the increased incidence of aortic pathology displayed by patients with BAVs is related to an inherent structural defect of the aortic wall likely related to abnormalities of ECM homeostasis. It is possible that both inherited structural abnormalities of the aortic ECM in combination with aberrant flow dynamics due to valve cusp fusion contribute to aortic dilatation and a propensity to dissection.

Emerging research suggests that ECM dysregulation may be central to progressive aortic dilatation. An improved understanding of the mechanisms involved in BAV aortopathy is needed as there is a wide variability in surgical approaches to this disorder despite established guidelines [33]. Some of this variability appears to be secondary to beliefs about the mechanisms and causes of BAV aortopathy that may profoundly influence surgical practices. An improved understanding of the impact of ECM dysregulation may help identify patients at risk and guide surgical intervention. We are examining changes in ECM structure, composition and regulation in regional areas of the aortic wall subjected to high shear stress forces resulting from aberrant flow dynamics through BAVs as compared to normal areas and aortas of tricuspid aortic valve patients. This research may clarify controversies in the BAV literature with regards to whether aortic pathology is a result of aberrant flow dynamics and/or a genetically inherited aortopathy (Fig. 3), and lead to the establishment of more appropriate clinical guidelines identifying patients at risk and in need of surgical intervention.

Aortic dilatation related to ECM dysregulation and cardiac myofibroblast activity also occurs in patients without known congenital or inherited disorders. Vascular smooth muscle cells and aortic fibroblasts have been shown to respond to a number of environmental factors and acquired conditions contributing to the development of abdominal aortic aneurysm formation, including cigarette smoke, chronic obstructive pulmonary disease, hypertension, atherosclerosis and elevated body mass [34]. Elevated levels of MMPs responsible for the enzymatic degradation of ECM in response to oxidative stress have also been implicated in both thoracic aortic

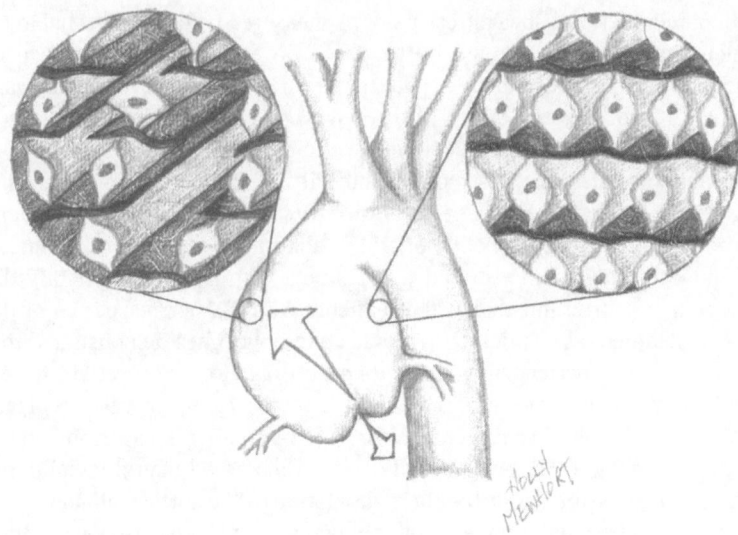

Fig. 3 Aortic root dilatation resulting in aortic valve insufficiency (*small arrow*). In patients with BAV disease aberrant flow patterns cause regional increases in aortic wall sheer stress (*large arrow*). Changes to the aortic wall ECM, specifically fragmentation of the elastic lamina, can be found in these regions (*left*) when compared to regions that are not exposed to aberrant wall shear stress (*right*)

aneurysm formation and dissection [35]. Recognition of the impact of these modifiable risk factors on ECM homeostasis may aid in the identification of new therapeutic targets and alter how we approach thoracic aortic diseases with respect to surgical approaches.

3.3   ECM and Valvular Heart Disease

Pathologic alterations in ECM composition and homeostasis also contributes to valvular diseases of the heart, particularly valvular insufficiencies. Genetic deficiencies in components of the ECM, such as Marfan's, Ehlers-Danlos, and Loeys-Dietz syndrome, result in remodeling and subsequent dilatation of the aortic root. Dilation of the sinotubular junction retracts the aortic valve cusps and reduces coaptation resulting in aortic valve insufficiency (Fig. 3).

Mitral regurgitation (MR) can also be a consequence of pathologic ECM composition and dysregulation. The functional anatomy of the mitral valve involves all components of the fibrous skeleton of the heart including the ECM of the left ventricle, the subvalvular apparatus (including the papillary muscles, and chordae tendineae), the mitral annulus, the mitral valve leaflets, and the ECM of the left atrium [36] (Fig. 1). MR can be classified according to pathologic changes observed in the functional anatomy (Carpentier's classification of MR) of the mitral valve; all of which are related to pathologic changes in the underlying ECM. MR occurs in the

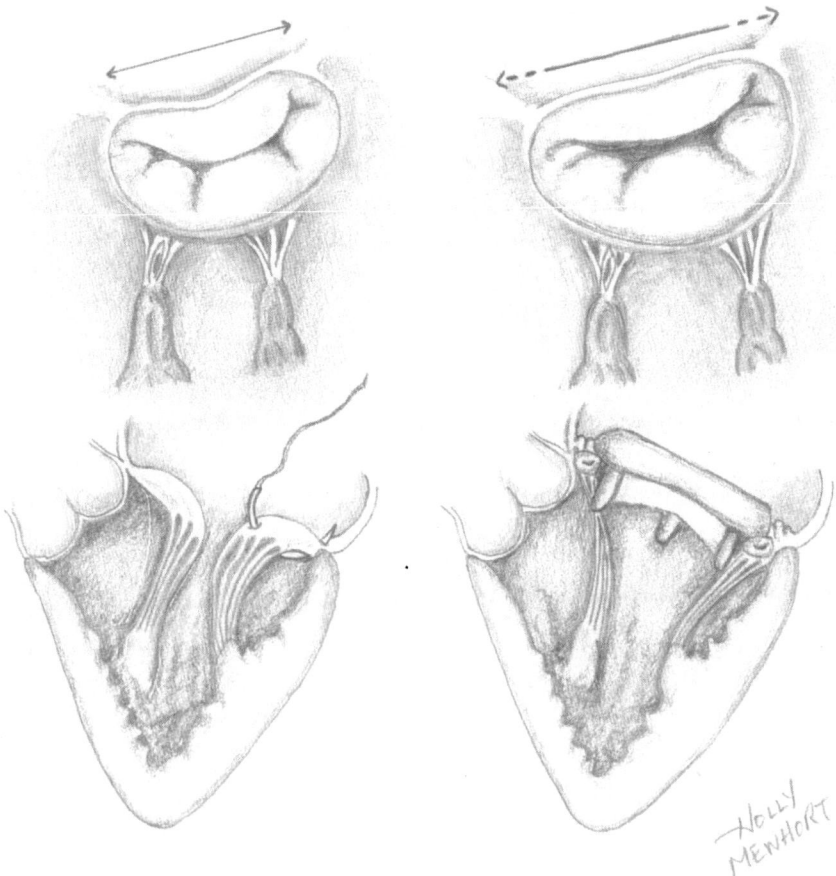

Fig. 4 Dilatation of the fibrous portion of the mitral valve annulus resulting in an increase in the intertrigonal distance previously thought to be fixed (*arrows*; *top panel*). Preservation of the mitral subvalvular apparatus (furrowing of the mitral valve leaflets to the mitral annulus and valve sewing ring) during mitral valve replacement surgery (*bottom panel*)

setting of normal leaflet motion (Carpentier's type I) when the mitral valve annulus dilates, pulling the leaflets apart and preventing complete coaptation (Fig. 4). The intertrigonal distance, a component of the mitral valve annulus belonging to the fibrous skeleton of the heart, was previously thought to be fixed, such that all annular dilatation observed in type I MR was thought to be the result of changes in the muscular component of the posterior mitral valve annulus [37]. However, the incidence of recurrent MR following successful mitral annuloplasty with an incomplete ring or flexible band that excluded the fibrous component of the mitral annulus from the repair has demonstrated that the intertrigonal distance is not fixed [36]. Perturbations in ECM composition and organization within the fibrous component of the mitral valve annulus can occur, resulting in annular dilatation and an increase in the intertrigonal distance [38]. This example highlights that cardiac surgeons have observed and documented alterations in the cardiac ECM that result in pathology. Surgeons then used such knowledge to develop novel surgical therapies to target these changes resulting in stable and predictable repairs.

MR can also occur as a result of abnormal mitral leaflet motion. Thickening and redundancy of the mitral valve leaflet(s) and elongation of even rupture of the chordae tendineae, which occurs with myxomatous disease, leads to prolapse of the mitral leaflet into the left atrium or in some cases a flail leaflet the moves freely into the left atrium (Carpentier's type II) preventing coaptation of the leaflets resulting in MR. The specific cellular and molecular mechanisms involved in the pathophysiology of myxomatous disease are incompletely understood. However, Rabkin and colleagues have demonstrated that myxomatous disease is likely the consequence of excessive ECM degradation [39], which has been further supported by Caira and colleagues [40]. Perturbations in ECM composition and homeostasis has been implicated in the weakening of the chordae leading to elongation or rupture [41]. Further, papillary muscle rupture secondary to MI is well known to reflect inadequate scar tissue formation following MI, as previously discussed.

Abnormal mitral leaflet motion can also occur as a result of restriction of the mitral valve leaflet(s) resulting from subvalvular fibrosis, typically caused by rheumatic disease (Carpentier's type IIIa), or papillary muscle displacement as a result of posterior-lateral ischemia (Carpentier's type IIIb). In both cases alterations in ECM composition and homeostasis are responsible for the fibrotic changes that contribute to type III MR. Though the etiology of pulmonary and tricuspid insufficiency is not as well characterized, they can also likely be explained by underlying changes in ECM composition and dysregulation of ECM homeostasis.

4 ECM and Surgical Treatment of Heart Disease

An understanding of the role of the ECM in cardiac function and how changes in ECM composition and homeostasis result in cardiac pathology can be leveraged toward development of surgical strategies to treat cardiac pathology exacerbated by ECM dysregulation. A surgical approach can be employed to treat several of the aforementioned cardiovascular pathologies, effecting changes in the ECM of heart and its subsequent function at the macro level of the fibrous skeleton down to the micro level of cells (fibroblasts) and molecules (MMPs, TIMPs, collagens). In the following paragraphs, we outline a number of key examples of how knowledge of ECM and cardiac function has influenced surgical strategies to treat cardiac pathology with cardiac surgery.

4.1   Reconstruction of the Fibrous Skeleton of the Heart

The fibrous skeleton can become destabilized by infection or congenital anomalies requiring highly complex surgical reconstruction procedures [42]. Disruption of the fibrous skeleton presents a significant challenge to the cardiac surgeon and mortality and morbidity is often high. Such reconstructions are typically performed using biologically inactive scaffolds such as glutaraldehyde-fixed bovine pericardium [43] or synthetic materials to reconstruct the biologically active fibrous skeleton of the heart. We believe these strategies have limited effects on promoting adaptive tissue remodeling but they remain the gold standard for contemporary surgical practice. As our knowledge of the ECM and its biologically active components expands, the use of novel biomaterials comprised of dynamic biologic endogenous ECM or its bioactive components will likely be used for more effective future repairs.

4.2   ECM and Mitral Valve Repair

The evolution of mitral valve annuloplasty is an example of how surgical intervention has contributed to our knowledge of the role of ECM in cardiovascular pathophysiology (Fig. 4). Mitral valve annuloplasty is a surgical approach that targets the pathology of the cardiac ECM. Restrictive mitral annuloplasty, a technique originally popularized by Bolling and colleagues [44], and later optimized by Miller and colleagues [45], was designed to surgically correct mitral annular dilatation. Late failure, characterized by the development of recurrent MR, resulting from the use of incomplete annuloplasty rings or flexible bands occurs because the repair did not address the entire ECM of the valve [36]. The intertrigonal distance (which is derived from the underlying ECM structure) is not "fixed" as once believed. The assumption that the intertrigonal distance was fixed is based on the traditional belief that the fibrous skeleton and its ECM is a passive structural support system and it is not a biologically active entity that itself can be remodeled. Subsequently, we have learned that the fibrous component of the mitral valve annulus can remodel and dilate, and that fixing the septal-lateral dimension with a rigid ring may prevent further dilatation of the intertrigonal fibrous component of the mitral annulus decreasing the incidence of recurrent MR and improving surgical outcomes [46]. These observations support the importance of understanding and respecting the dynamic nature of the ECM when developing surgical approaches to cardiac pathologies.

4.3   ECM and Mitral Valve Replacement

The importance of preserving components of the fibrous skeleton of the heart at the time of surgery and resultant effects on cardiac function was appreciated in early surgical interventions for the mitral valve. The significance of the subvalvular apparatus was first appreciated in the 1960s by C. Walton Lillehei, who suggested that preservation of the subvalvular apparatus at the time of mitral valve surgery would enhance functional recovery of the LV and preserve both chamber shape and size [37]. However, this practice was not adopted until the 1980's after Tirone David demonstrated that removal of the mitral chordeae at the time of mitral valve replacement in patients with chronic MR results in dilatation and dysfunction of the LV [47]. Today it is a standard of care and routine practice to preserve the mitral valve apparatus at the time of mitral valve replacement whenever feasible (Fig. 4). This surgical approach shows an appreciation that the components of the fibrous skeleton of the heart are not limited to the mitral valve leaflets and annulus, but extend to the subvalvular apparatus and the ECM surrounding the cardiac myocytes of the left ventricle.

4.4 Surgical Repair of the ECM at the Cellular and Molecular Level

Surgical interventions on the heart can greatly influence the ECM at the cellular and molecular levels. Reverse structural remodeling of the heart after surgical repairs such as aortic valve replacement (AVR) for severe AS or left ventricular assist device (LVAD) therapy for ESHF have been associated with restoration of ECM components and maintenance of ECM homeostasis.

Regression of left ventricular hypertrophy following AVR for AS is well documented [48] and presents an example how surgical interventions can influence ECM remodeling at the cellular level. Reductions in the transvalvular gradient resulting from AVR, whether a porcine or bovine tissue prosthesis [49], mechanical prosthesis, stented or stentless prosthesis [48], pulmonary valve (in the case of the Ross procedure) [50], or transcatheter aortic valve replacement [51] is performed, alters Renin-Angiotensin-Aldosterone System (RAAS) gene expression and favorably alters ECM homeostasis to reverse LV remodeling and myocardial fibrosis [52] (Fig. 5).

Mechanical unloading of the left ventricle with use of ventricular assist devices can also elicit changes at the cellular level with respect to ECM composition and recovery of heart size and function (Fig. 5). Patients with idiopathic dilated cardiomyopathy treated for cardiogenic shock with LVAD implantation demonstrated functional recovery secondary to ventricular unloading and the device was successfully explanted [53]. The idea that ESHF could be reversed is exciting and efforts to understand the underlying mechanism are important. Accumulating evidence suggests that changes in ECM architecture and homeostasis may contribute to reverse structural cardiac remodeling. Several studies have documented alterations in total myocardial collagen concentration and composition in response to LVAD therapy [53–58]. In addition, ECM regulation is significantly impacted by LVAD unloading including changes in MMPs, tissue inhibitors of MMPs [59], and disintegrin metalloproteinases [19].

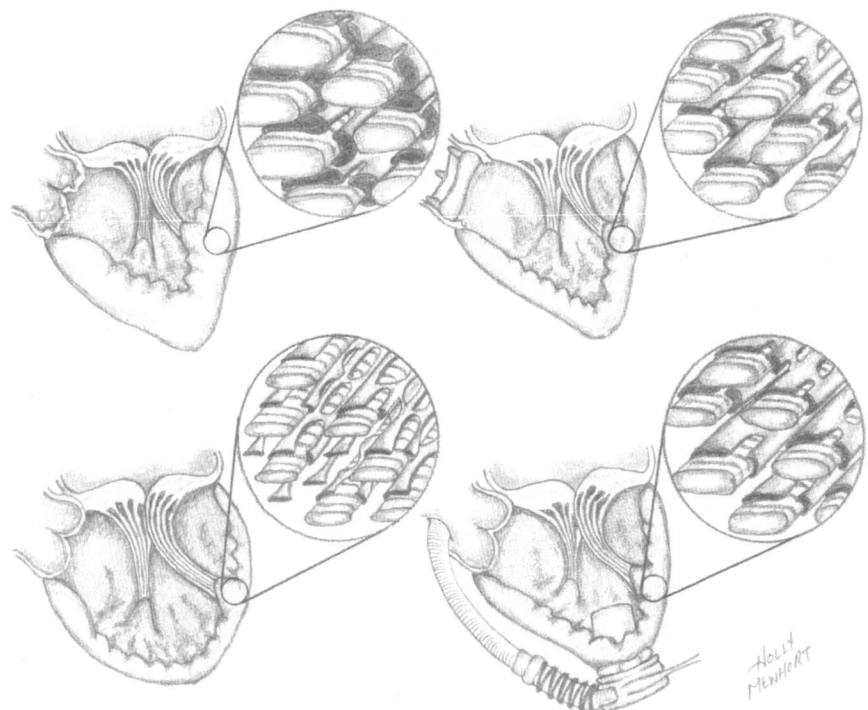

Fig. 5 Changes to LV geometry and ECM architecture that occur as a result of aortic valve stenosis (*top left*) reversed by aortic valve replacement (*top right*). Changes to LV geometry and ECM architecture that occur in ESHF (*bottom left*) reversed by LVAD therapy (*bottom right*)

5   Using ECM Therapeutically to Restore Heart Function

Recognizing the role of ECM in cardiac structure and function, an exciting prospect is the use of ECM and its components as a therapy itself. ECM plays an important role in dynamic cell signaling by proving a microenvironment for adjacent cells and tissues. ECM is increasingly implicated as a key regulator of endogenous repair mechanisms, including stem cells and regeneration of cardiac muscle. We believe that the properties of biologic ECM on endogenous repair can be leveraged to develop novel "biosurgical" strategies to improve cardiac structure and function. Using ECM as a targeted surgical therapy may provide surgeons with new tools to improve current repair strategies and perhaps even innovate novel therapies leading to cardiac regeneration and restoration of function in clinical scenarios previously intractable. An exciting and important proof-of-concept was shown by Taylor and colleagues who added stem cells to a cardiac ECM skeleton and were capable of restoring cardiac function [60] (Fig. 6). These data provide proof that cardiac ECM provides the needed signals and regenerative capacity to restore organ function if successfully repopulated with appropriate cells. The ECM may be capable of inducing profound effects on endogenous repair and cardiac restoration in the setting of disease.

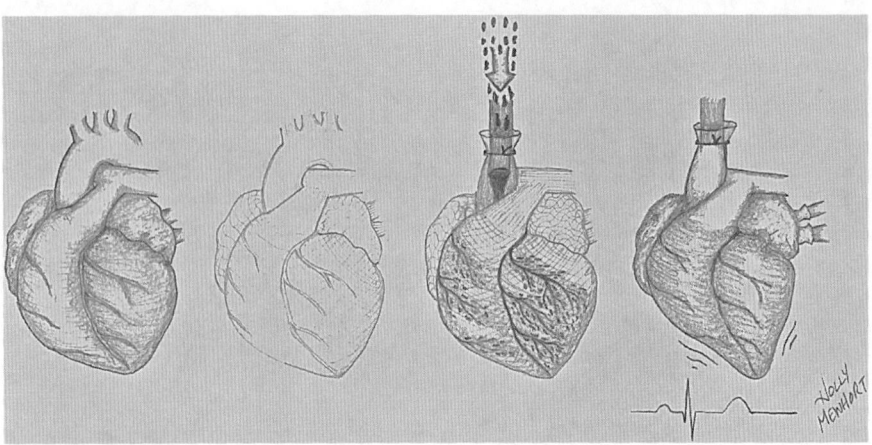

Fig. 6 (From *left to right*) an explanted rat heart *1* that has been completely decellularized leaving only the ECM fibrous skeleton of the heart intact *2* which serves as a scaffold for cellular repopulation using a cocktail of stem cells *3* to create *de novo* functional myocardial tissue capable of organized myocardial contraction *4*

5.1   Leveraging Cellular Therapies to Restore ECM Homeostasis

The importance of restoring ECM architecture and homeostasis has been appreciated as it was identified as one of the key mechanisms underlying the benefits of cell-based therapies [61]. Muscle cells injected into the failing heart have consistently demonstrated benefit with regards to limiting LV dilatation and improving cardiac function [62–64]. Menasche and colleagues demonstrated that skeletal myoblasts are capable of contraction after implantation into ischemic heart tissue [65]. Mc-Murray and colleagues demonstrated that cultured fetal and neonatal rat cardiomyocytes injected into the subcutaneous connective tissue of adult rat hindlimbs spontaneously form contractile cardiac-like muscle tissue [66]. In addition, several studies have demonstrated that transplanted cells integrate and communicate with host heart cells [8, 9, 67, 68]. These data suggest that cell-based therapies may result in functional improvement through an active process whereby the engrafted cells replace lost contractile elements and contract synchronously with the host myocardium. However, no study has demonstrated sustained presence of transplanted cells beyond the initials weeks following injection, even though the beneficial effects are sustained months after therapy. Some cell types have no regenerative capacity but still provide benefits on cardiac remodeling after injury [21]. This prompted our group to question if cell-based therapies benefit cardiac structure and function by influencing ECM [21]. We have shown that functional improvements observed with cell therapy are at least in part, due to a paracrine effect exerted by the engrafted cells on the ECM [1, 69].

As myocardial matrix disruption can decrease systolic performance without altering myocyte contractility [5], cell-based therapies may improve cardiac function without contributing themselves to contractile function, but instead restoring deficient interstitial matrix component in the failing heart. Several studies demonstrate that without physical adhesion to components of the ECM, cells will not proliferate, grow, or survive [7, 10, 11, 70]. Thus, in order to survive within the injured myocardium where matrix elements are deficient, engrafted cells must be able to compensate by secreting and incorporating new matrix elements to preserve cell survival in an autocrine fashion. These autocrine factors are likely to influence adjacent host myocardium in a paracrine fashion promoting native cell proliferation, growth and survival. We have demonstrated that smooth muscle cell transplantation, which has no capability to improve cardiac contraction directly, improves cardiac function by attenuating ECM degradation and restoring ECM architecture through paracrine mechanisms [21]. We have also demonstrated that vascular smooth muscle cells injected into rat hearts following MI improved ECM homeostasis and prevented maladaptive LV remodeling by attenuating myofibroblast activation through paracrine mechanisms involving fibroblast growth factor-2 (FGF-2) and tissue inhibitor of matrix metalloprotease-2 [62]. These data suggest that the mechanisms driving the improvement in cardiac function observed as a result of cell-based therapies are more likely paracrine in nature, influencing ECM homeostasis and restoring ECM architecture.

Evidence also indicates that cell-based therapies may also influence cardiac remodeling and heart function by stimulating angiogenesis through the release of paracrine factors such as vascular endothelial growth factor and FGF-2 [62]. The implanted cells themselves may also be capable of incorporating into newly forming blood vessels. Though angiogenesis is unlikely to result in myogenesis, increased perfusion may salvage hibernating myocardium, restore reversibly damaged cells and more importantly contribute to the repair of damaged ECM, facilitating donor cell incorporation, cell-cell communication, and structural support and tissue functionality.

5.2 Using ECM to Enhance Cardiac Tissue Engineering

Cell therapy for cardiovascular disease has had only a modest benefit in clinical applications. This blunted response has promoted efforts to create engineered functional tissue constructs or even engineered functional organs to improve results. The ECM may be a critical tool for successful tissue engineering. Cell seeded decellularized biologic ECM constructs have also been explored for tissue replacement post-MI in small animal models demonstrating that the use of decellularized biologic ECM seeded with human mesenchymal stromal cells improves cardiac function when the infarcted myocardium is replaced with the tissue engineered construct [71]. Early tissue engineering strategies utilized synthetic scaffolds, such as polyglycolic acid scaffolds [72] or ε-caprolactactone-co-L-lactide reinforced with knitted poly-L-lactide fabric [73], onto which the desired cell types can be seeded to create a functional construct. The later construct demonstrated improvement in cardiac function post-MI when the synthetic scaffold was seeded with vascular smooth muscle cells [73]. However, synthetic scaffolds lack ECM components we know are vital to cell communication, function and survival and in many cases elicit a detrimental immune response.

Reconstituted matrix components, such as collagen or gelatin based constructs, have been created into which targeted cell types can be seeded. For example, Akhyari and colleagues created beating cardiac tissue constructs by seeding human cardiac myocytes onto a gelatin scaffold [74]. Such substrates offer several advantages over synthetic constructs. They are biodegradable and capable of resorption after implantation by host MMPs and subsequently incorporated into the native tissue [75]. They can also be molded into a variety of shapes and sizes mimicking the tissue characteristics of the organ system they are designed to replace. From a surgical perspective however, these materials are challenging to work with, as they lack the tensile strength to withstand hemodynamic pressures with little to no hemostatic capabilities. Though biologic in nature, they still lack the many components that comprise a healthy balanced ECM environment capable of regulating cell and tissue function.

Decellularized biologic ECM offers several advantages over synthetic or gelatin based constructs. From a surgical perspective, decellularized biologic ECM demonstrates excellent tensile strength and hemostatic potential, and has been used for numerous cardiac surgical applications. Some of these include patch material for the repair of congenital abnormalities [76]. The use of decellularized biologic ECM constructs alone may also prove to be of benefit, mobilizing native cell populations driving *in vivo* cell seeding and tissue generation.

5.3   Using ECM Itself as a Novel Biosurgical Therapy

Given that dysregulation of the cardiac ECM alone can result in systolic dysfunction in the absence of decreased myocyte contractility [5], and the profound effect ECM homeostasis can have on cell behavior, survival, endogenous repair and tissue remodeling [7, 10, 11, 70], we believe that targeting ECM dysregulation by surgically introducing healthy biologic ECM may prove to be an effective therapeutic strategy. The application of biologic ECM constructs containing a healthy milieu of growth factors, cytokines and matricellular proteins may restore ECM homeostasis preserving ECM architecture and promoting cell proliferation, growth and survival in an otherwise dysregulated environment. To that end, we have demonstrated that epicardial implantation of a healthy biologic ECM construct, known as SIS-ECM (CorMatrix ECM, CorMatrix Cardiovascular Inc.), following MI restores tissue architecture within the infarcted myocardium, preventing LV dilatation, and improving myocardial contractility (Fig. 7) [77]. Further, we demonstrated that enhancing biologic SIS-ECM with paracrine factors known to have a positive influence on matrix remodeling and angiogenesis offered additional functional benefits. These data provide important evidence that ECM itself can be an effective therapy to restore cardiac function after injury.

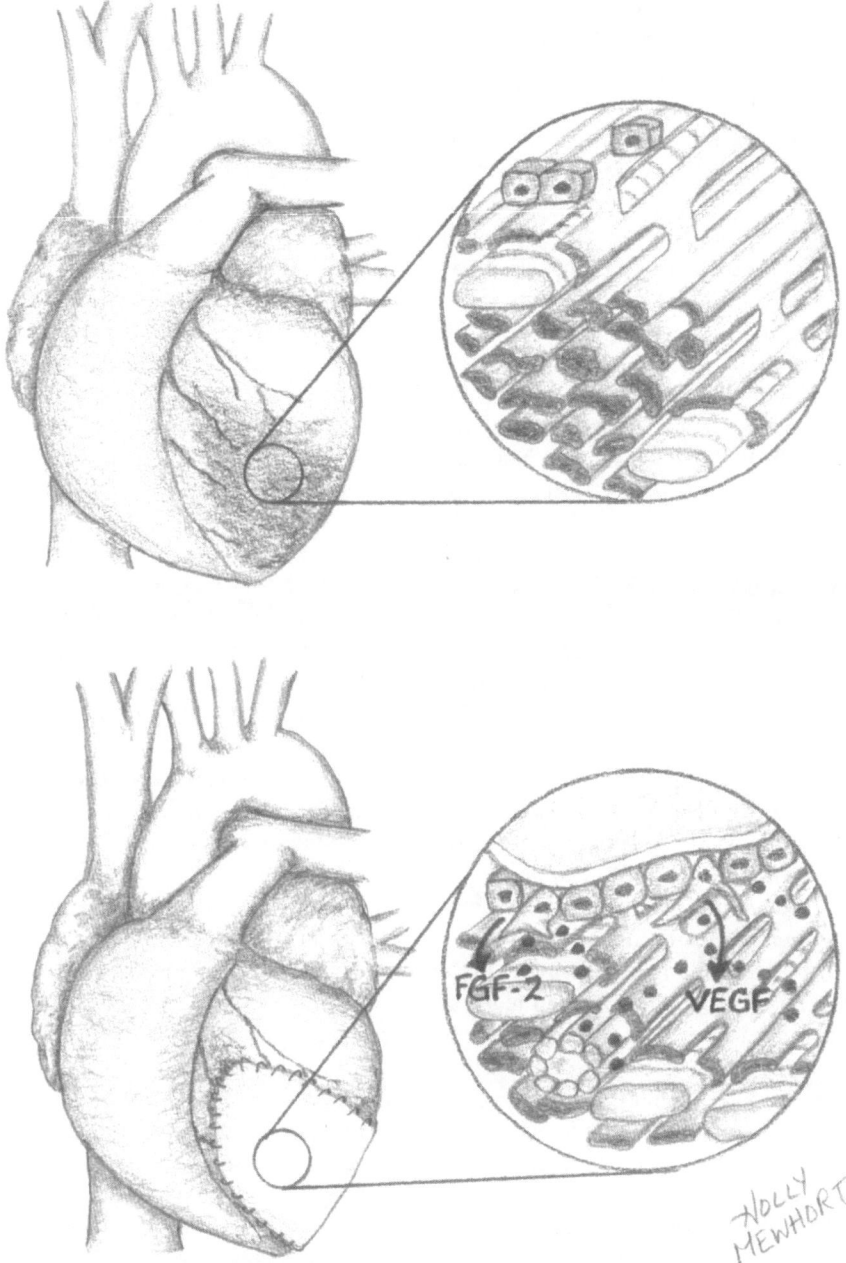

Fig. 7 Epicardial and myocardial damage resulting from ischemic injury and replacement with collagen-rich scar (*top panel*). Epicardial infarct repair with healthy biologic SIS-ECM improves myocardial functional recovery possibly by stimulating epithelial-mesenchymal transition, facilitating the release of paracrine factors, such as FGF-2 and VEGF, restoring ECM homeostasis, and stimulating vasculogenesis (*bottom panel*)

Porcine small intestinal submucosal extracellular matrix (SIS-ECM) can be extracted with retention of its native 3-D collagen architecture and adjacent matricellular proteins [78–80]. For example, SIS-ECM has been shown to retain endogenous FGF-2 in addition to VEGF, glycosaminoglycans, fibronectin, laminin, and other key ECM-based matricellular signaling biopeptides [78, 79, 81]. By employing specific methods of ECM extraction and processing to retain its native bioactive properties, SIS-ECM may facilitate adaptive and functional tissue formation when in contact with host cells after transplantation. SIS-ECM has been shown to support cell growth and promote tissue regeneration for diverse clinical applications [79]. Importantly, SIS-ECM is completely biodegradable with bioinductive consequences attributable to the release of ECM constituent molecules upon degradation [82]. Despite its non-cross-linked and xenogenic nature, *in vivo* studies suggest a good immune tolerance of SIS-ECM [83, 84]. Badylak and co-workers repaired the ventricular free wall of animals with SIS-ECM and noted adaptive tissue formation [85]. SIS-ECM is increasingly being used in clinical practice by surgeons for repair of intracardiac defects with good outcomes [86, 87].

Based on our work to date, we believe there are a number of putative mechanisms (Fig. 7). It is well established that ECM plays a critical role in endogenous repair mechanisms after ischemic injury, particularly in the epicardial compartment [88–90]. ECM influences injured myocardium through both direct and indirect pathways. Indirectly, through its biomechanical properties, ECM provides a lateral boundary (passive scaffold) to facilitate cellular mechanotransduction. Altered myocardial strain has been implicated in progression of maladaptive post-MI cardiac remodeling [91, 92]. It is possible that the biomechanical properties of the fixed ECM patch over the infarct area may have a *passive effect* by regional myocardial restraint that decreases myocardial strain. Reducing regional myocardial strain by passive myocardial restraint has been shown to provide significant benefits on post-MI remodeling [93–100].

Through more direct and active mechanisms, ECM presents a dynamic microenvironment for local cells by way of growth factors, matricellular proteins, and cell-ECM receptors. Increasingly, ECM is recognized as an essential player in endogenous tissue repair and cell regeneration mechanisms [101, 102]. Interestingly, a number of key stem cell niches are located in the epicardium [89]. Activation of endogenous epicardial-based repair mechanisms by epicardial infarct repair with SIS-ECM is thus an intriguing possibility. *Epithelial-mesenchymal transition* (EMT) of epicardial cells is believed to occur in response to ischemic injury and acts to enhance myocardial repair [87, 88, 90]. Artificial stimulation of EMT has been shown to improve post-MI cardiac remodeling [103–105], not unlike the results of our epicardial infarct repair technique. In addition, epicardial thickening in response to ischemic injury has been shown to act as a source of paracrine factors which condition the underlying myocardium for repair [104]. Importantly, FGF-2 has been implicated as a key paracrine factor in these epicardial repair mechanisms [89]. We believe that restoration of the damaged epicardium after ischemic injury with biologic ECM induces active endogenous repair mechanisms toward *ECM homeostasis* (by MMP/TIMP regulation and inhibition of myofibroblast activation), and activation of *epicardial EMT* promoting *angiogenesis and vasculogenesis*. The molecular and cellular effects of epicardial infarct repair with SIS-ECM should be further defined. Future studies will be required to address the mechanisms of benefit underlying epicardial infarct repair and further optimize the innovation for effective clinical translation.

6 Conclusion

This chapter has established that cardiac ECM contributes to the structure of the heart and in so doing, influences both myocardial tissue and cardiac function. Surgical strategies to effectively restore cardiac structure and function are optimal when they respect the underlying role of the ECM. This chapter also highlights cardiac ECM as a dynamic microenvironment that regulates tissue remodeling and repair in the face of injury. An emerging concept is that ECM itself can be leveraged to enhance cardiac repair. Future innovations in "biosurgical" approaches by cardiac surgeons with ECM-based platforms may improve outcomes and facilitate enhanced solutions to difficult cardiac diseases.

References

1. Fedak PWM, Bai L, Turnbull J et al (2012) Cell therapy limits myofibroblast differentiation and structural cardiac remodeling: basic fibroblast growth factor-mediated paracrine mechanism. Circ Heart Fail 5:349–356
2. Badenhorst D, Maseko M, Tsotetsi OJ et al (2003) Cross-linking influences the impact of quantitative changes in myocardial collagen on cardiac stiffness and remodelling in hypertension in rats. Cardiovasc Res 57:632–641
3. Mann DL, Spinale FG (1998) Activation of matrix metalloproteinases in the failing human heart: breaking the tie that binds. Circulation 98:1699–1702
4. Fedak PWM, Verma S, Weisel RD, Li R-K (2005) Cardiac remodeling and failure From molecules to man (Part II). Cardiovasc Pathol 14:49–60
5. Baicu CF, Stroud JD, Livesay VA et al (2003) Changes in extracellular collagen matrix alter myocardial systolic performance. Am J Physiol Heart Circ Physiol 284:H122–H132
6. Janicki JS, Brower GL (2002) The role of myocardial fibrillar collagen in ventricular remodeling and function. J Card Fail 8:S319–S325
7. Ross RS, Borg TK (2001) Integrins and the myocardium Circ Res 88:1112–1119
8. Koh GY, Soonpaa MH, Klug MG et al (1995) Stable fetal cardiomyocyte grafts in the hearts of dystrophic mice and dogs. J Clin Invest 96:2034–2042
9. Roell W, Lu ZJ, Bloch W et al (2002) Cellular cardiomyoplasty improves survival after myocardial injury. Circulation 105:2435–2441
10. Lukashev ME, Werb Z (1998) ECM signalling: orchestrating cell behaviour and misbehaviour. Trends Cell Biol 8:437–441
11. Lundgren E, Terracio L, Mårdh S, Borg TK (1985) Extracellular matrix components influence the survival of adult cardiac myocytes *in vitro*. Exp Cell Res 158:371–381
12. Carver W, Terracio L, Borg TK (1993) Expression and accumulation of interstitial collagen in the neonatal rat heart. Anat Rec 236:511–520

13. Ding B, Price RL, Goldsmith EC et al (2000) Left ventricular hypertrophy in ascending aortic stenosis mice: anoikis and the progression to early failure. Circulation 101:2854–2862

14. Nahrendorf M, Swirski FK, Aikawa E et al (2007) The healing myocardium sequentially mobilizes two monocyte subsets with divergent and complementary functions. J Exp Med 204:3037–3047

15. Weber KT, Pick R, Janicki JS et al (1988) Inadequate collagen tethers in dilated cardiopathy. Am Heart J 116:1641–1646

16. Fedak PWM, Altamentova SM, Weisel RD et al (2003) Matrix remodeling in experimental and human heart failure: a possible regulatory role for TIMP-3. Am J Physiol Heart Circ Physiol 284:H626–H634

17. Spinale FG (2002) Matrix metalloproteinases: regulation and dysregulation in the failing heart. Circ Res 90:520–530

18. Weber KT, Jalil JE, Janicki JS, Pick R (1989) Myocardial collagen remodeling in pressure overload hypertrophy. A case for interstitial heart disease. Am J Hypertens 2:931–940

19. Fedak PWM (2006) Altered expression of disintegrin metalloproteinases and their inhibitor in human dilated cardiomyopathy. Circulation 113:238–245

20. Weber KT, Brilla CG, Janicki JS (1993) Myocardial fibrosis: functional significance and regulatory factors. Cardiovasc Res 27:341–348

21. Fedak PWM, Szmitko PE, Weisel RD et al (2005) Cell transplantation preserves matrix homeostasis: a novel paracrine mechanism. J Thorac Cardiovasc Surg 130:1430–1439

22. Dietz HC, Loeys B, Carta L, Ramirez F (2005) Recent progress towards a molecular understanding of Marfan syndrome. Am J Med Genet C Semin Med Genet 139:4–9

23. Bunton TE, Biery NJ, Myers L et al (2001) Phenotypic alteration of vascular smooth muscle cells precedes elastolysis in a mouse model of Marfan syndrome. Circ Res 88:37–43

24. van Karnebeek CD, Naeff MS, Mulder BJ et al (2001) Natural history of cardiovascular manifestations in Marfan syndrome. Arch Dis Child 84:129–137

25. Germain DP (2007) Ehlers-Danlos syndrome type IV. Orphanet J Rare Dis 2:32

26. Loeys BL, Chen J, Neptune ER et al (2005) A syndrome of altered cardiovascular, craniofacial, neurocognitive and skeletal development caused by mutations in TGFBR1 or TGFBR2. Nat Genet 37:275–281

27. Robicsek F, Thubrikar MJ, Cook JW, Fowler B (2004) The congenitally bicuspid aortic valve: how does it function? Why does it fail? Ann Thorac Surg 77:177–185

28. Mahadevia R, Barker AJ, Schnell S et al (2014) Bicuspid aortic cusp fusion morphology alters aortic three-dimensional outflow patterns, wall shear stress, and expression of aortopathy. Circulation 129:673–682

29. Keane MG, Wiegers SE, Plappert T et al (2000) Bicuspid aortic valves are associated with aortic dilatation out of proportion to coexistent valvular lesions. Circulation 102:III35–III39

30. Hahn RT, Roman MJ, Mogtader AH, Devereux RB (1992) Association of aortic dilation with regurgitant, stenotic and functionally normal bicuspid aortic valves. JAC 19:283–288

31. Yasuda H, Nakatani S, Stugaard M et al (2003) Failure to prevent progressive dilation of ascending aorta by aortic valve replacement in patients with bicuspid aortic valve: comparison with tricuspid aortic valve. Circulation 108(Suppl 1):II291–4

32. Fedak PWM, de Sa MPL, Verma S et al (2003) Vascular matrix remodeling in patients with bicuspid aortic valve malformations: implications for aortic dilatation. J Thorac Cardiovasc Surg 126:797–806

33. Verma S, Yanagawa B, Kalra S et al (2013) Knowledge, attitudes, and practice patterns in surgical management of bicuspid aortopathy: a survey of 100 cardiac surgeons. J Thorac Cardiovasc Surg 146:1033–1040.e4

34. Ruddy JM, Jones JA, Ikonomidis JS (2013) Pathophysiology of thoracic aortic aneurysm (TAA): is it not one uniform aorta? Role of embryologic origin. Prog Cardiovasc Dis 56:68–73

35. Steed MM, Tyagi N, Sen U et al (2010) Functional consequences of the collagen/elastin switch in vascular remodeling in hyperhomocysteinemic wild-type, eNOS-/-, and iNOS-/- mice. Am J Physiol Lung Cell Mol Physiol 299:L301–L311

36. Fedak PWM, McCarthy PM, Bonow RO (2008) Evolving concepts and technologies in mitral valve repair. Circulation 117:963–974
37. Lillehei CW, Levy MJ, Bonnabeau RC (1964) Mitral valve replacement with preservation of papillary muscles and chordae tendineae. J Thorac Cardiovasc Surg 47:532–543
38. McCarthy PM (2002) Does the intertrigonal distance dilate? Never say never. J Thorac Cardiovasc Surg 124:1078–1079
39. Rabkin E, Aikawa M, Stone JR et al (2001) Activated interstitial myofibroblasts express catabolic enzymes and mediate matrix remodeling in myxomatous heart valves. Circulation 104:2525–2532
40. Caira FC, Stock SR, Gleason TG et al (2006) Human degenerative valve disease is associated with up-regulation of low-density lipoprotein receptor-related protein 5 receptor-mediated bone formation. J Am Coll Cardiol 47:1707–1712
41. Grande-Allen KJ, Griffin BP, Ratliff NB et al (2003) Glycosaminoglycan profiles of myxomatous mitral leaflets and chordae parallel the severity of mechanical alterations. JAC 42:271–277
42. David TE, Kuo J, Armstrong S (1997) Aortic and mitral valve replacement with reconstruction of the intervalvular fibrous body. J Thorac Cardiovasc Surg 114:766–771. Discussion 771–772
43. David TE (1998) The use of pericardium in acquired heart disease: a review article. J Heart Valve Dis 7:13–18
44. Bolling SF, Pagani FD, Deeb GM, Bach DS (1998) Intermediate-term outcome of mitral reconstruction in cardiomyopathy. J Thorac Cardiovasc Surg 115:381–638. Discussion 387–388
45. Tibayan FA, Rodriguez F, Langer F et al (2003) Annular remodeling in chronic ischemic mitral regurgitation: ring selection implications. Ann Thorac Surg 76:1549–1554. Discussion 1554–1555
46. Spoor MT, Geltz A, Bolling SF (2006) Flexible versus nonflexible mitral valve rings for congestive heart failure: differential durability of repair. Circulation 114:I67–I71
47. David TE, Burns RJ, Bacchus CM, Druck MN (1984) Mitral valve replacement for mitral regurgitation with and without preservation of chordae tendineae. J Thorac Cardiovasc Surg 88:718–725
48. Pibarot P, Dumesnil JG (2000) Hemodynamic and clinical impact of prosthesis-patient mismatch in the aortic valve position and its prevention. J Am Coll Cardiol 36:1131–1141
49. Suri RM, Zehr KJ, Sundt TM et al (2009) Left ventricular mass regression after porcine versus bovine aortic valve replacement: a randomized comparison. Ann Thorac Surg 88:1232–1237
50. Duebener LF, Stierle U, Erasmi A et al (2005) Ross procedure and left ventricular mass regression. Circulation 112:I415–I422
51. Tzikas A, Geleijnse ML, Van Mieghem NM et al (2011) Left ventricular mass regression one year after transcatheter aortic valve implantation. Ann Thorac Surg 91:685–691
52. Walther T, Schubert A, Falk V et al (2002) Left ventricular reverse remodeling after surgical therapy for aortic stenosis: correlation to Renin-Angiotensin system gene expression. Circulation 106:I23–I26
53. Müller J, Wallukat G, Weng YG et al (1997) Weaning from mechanical cardiac support in patients with idiopathic dilated cardiomyopathy. Circulation 96:542–549
54. Maybaum S, Mancini D, Xydas S et al (2007) Cardiac improvement during mechanical circulatory support: a prospective multicenter study of the LVAD Working Group. Circulation 115:2497–2505
55. Bruckner BA, Stetson SJ, Perez-Verdia A et al (2001) Regression of fibrosis and hypertrophy in failing myocardium following mechanical circulatory support. J Heart Lung Transplant 20:457–464
56. Klotz S, Foronjy RF, Dickstein ML et al (2005) Mechanical unloading during left ventricular assist device support increases left ventricular collagen cross-linking and myocardial stiffness. Circulation 112:364–374

57. Barbone A, Holmes JW, Heerdt PM et al (2001) Comparison of right and left ventricular responses to left ventricular assist device support in patients with severe heart failure: a primary role of mechanical unloading underlying reverse remodeling. Circulation 104:670–675
58. Butler CR, Jugdutt BI (2012) The paradox of left ventricular assist device unloading and myocardial recovery in end-stage dilated cardiomyopathy: implications for heart failure in the elderly. Heart Fail Rev 17:615–633
59. Li YY, Feng Y, McTiernan CF et al (2001) Downregulation of matrix metalloproteinases and reduction in collagen damage in the failing human heart after support with left ventricular assist devices. Circulation 104:1147–1152
60. Ott HC, Matthiesen TS, Goh S-K et al (2008) Perfusion-decellularized matrix: using nature's platform to engineer a bioartificial heart. Nat Med 14:213–221
61. Fedak PWM, Weisel RD, Verma S et al (2003) Restoration and regeneration of failing myocardium with cell transplantation and tissue engineering. Semin Thorac Cardiovasc Surg 15:277–286
62. Fedak PWM, Bai L, Turnbull J et al (2012) Cell therapy limits myofibroblast differentiation and structural cardiac remodeling: basic fibroblast growth factor-mediated paracrine mechanism. Circ Heart Fail 5:349–356
63. Yoo KJ, Li RK, Weisel RD et al (2000) Autologous smooth muscle cell transplantation improved heart function in dilated cardiomyopathy. Ann Thorac Surg 70:859–865
64. Yoo KJ, Li RK, Weisel RD et al (2000) Heart cell transplantation improves heart function in dilated cardiomyopathic hamsters. Circulation 102:III204–III209
65. Leobon B, Garcin I, Vilquin JT et al (2002) Do engrafted skeletal myoblasts contract in infarcted myocardium? Circulation 106:549–549
66. McMurray J, Pfeffer MA (2002) New therapeutic options in congestive heart failure: part II. Circulation 105:2223–2228
67. Soonpaa MH, Koh GY, Klug MG, Field LJ (1994) Formation of nascent intercalated disks between grafted fetal cardiomyocytes and host myocardium. Science 264:98–101
68. Ruhparwar A, Tebbenjohanns J, Niehaus M et al (2002) Transplanted fetal cardiomyocytes as cardiac pacemaker. Eur J Cardiothorac Surg 21:853–857
69. Fedak PWM (2008) Paracrine effects of cell transplantation: modifying ventricular remodeling in the failing heart. Semin Thorac Cardiovasc Surg 20:87–93
70. Hornberger LK, Singhroy S, Cavalle-Garrido T (2000) Synthesis of extracellular matrix and adhesion through β1 integrins are critical for fetal ventricular myocyte proliferation. Circ Res 87:508–515
71. Kang K, Sun L, Xiao Y et al (2012) Aged human cells rejuvenated by cytokine enhancement of biomaterials for surgical ventricular restoration. J Am Coll Cardiol 60:2237–2249
72. Langer R, Vacanti JP (1993) Tissue engineering. Science 260:920–926
73. Matsubayashi K, Fedak PWM, Mickle DAG et al (2003) Improved left ventricular aneurysm repair with bioengineered vascular smooth muscle grafts. Circulation 108(Suppl 1):II219–25
74. Akhyari P, Fedak PWM, Weisel RD et al (2002) Mechanical stretch regimen enhances the formation of bioengineered autologous cardiac muscle grafts. Circulation 106:I137–I142
75. Zimmermann W-H, Eschenhagen T (2003) Cardiac tissue engineering for replacement therapy. Heart Fail Rev 8:259–269
76. Quarti A, Nardone S, Colaneri M et al (2011) Preliminary experience in the use of an extracellular matrix to repair congenital heart diseases. Interact Cardiovasc Thorac Surg 13:569–572
77. Mewhort HEM, Turnbull JD, Meijndert HC et al (2014) Epicardial infarct repair with basic fibroblast growth factor-enhanced CorMatrix-ECM biomaterial attenuates postischemic cardiac remodeling. J Thorac Cardiovasc Surg 147:1650–1659
78. Voytik-Harbin SL, Brightman AO, Kraine MR et al (1997) Identification of extractable growth factors from small intestinal submucosa. J Cell Biochem 67:478–491
79. Badylak S, Freytes D, Gilbert T (2009) Extracellular matrix as a biological scaffold material: structure and function. Acta Biomaterialia 5:1–13

80. Brown BN, Barnes CA, Kasick RT et al (2010) Surface characterization of extracellular matrix scaffolds. Biomaterials 31:428–437
81. Brown B, Lindberg K, Reing J et al (2006) The basement membrane component of biologic scaffolds derived from extracellular matrix. Tissue Eng 12:519–526
82. Badylak SF (2007) The extracellular matrix as a biologic scaffold material. Biomaterials 28:3587–3593
83. Badylak SF (2004) Xenogeneic extracellular matrix as a scaffold for tissue reconstruction. Transpl Immunol 12:367–377
84. Daly KA, Stewart-Akers AM, Hara H et al (2009) Effect of the alphaGal epitope on the response to small intestinal submucosa extracellular matrix in a nonhuman primate model. Tissue Eng Part A 15:3877–3888
85. Badylak S, Obermiller J, Geddes L, Matheny R (2003) Extracellular matrix for myocardial repair. Heart Surg Forum 6:E20–E26
86. Witt RG, Raff G, Van Gundy J et al (2013) Short-term experience of porcine small intestinal submucosa patches in paediatric cardiovascular surgery. Eur J Cardiothorac Surg 44:72–76
87. Scholl FG, Boucek MM, Chan K-C et al (2010) Preliminary experience with cardiac reconstruction using decellularized porcine extracellular matrix scaffold: human applications in congenital heart disease. World J Pediatr Congenit Heart Surg 1:132–136
88. Balmer GM, Bollini S, Dubé KN et al (2014) Dynamic haematopoietic cell contribution to the developing and adult epicardium. Nat Commun 5:1–12
89. Smart N, Riley PR (2012) The epicardium as a candidate for heart regeneration. Future Cardiol 8:53–69
90. van Wijk B, Gunst QD, Moorman AFM, van den Hoff MJB (2012) Cardiac regeneration from activated epicardium. PLoS ONE 7:e44692
91. Cheng A, Langer F, Nguyen TC et al (2006) Transmural left ventricular shear strain alterations adjacent to and remote from infarcted myocardium. J Heart Valve Dis 15:209–218. Discussion 218
92. Joyce E, Hoogslag GE, Leong DP et al (2014) Association between left ventricular global longitudinal strain and adverse left ventricular dilatation after ST-segment-elevation myocardial infarction. Circ Cardiovasc Imaging 7:74–81
93. Shah PK (2005) Preservation of cardiac extracellular matrix by passive myocardial restraint: an emerging new therapeutic paradigm in the prevention of adverse remodeling and progressive heart failure. Circulation 112:1245–1247
94. Pilla JJ, Blom AS, Gorman JH et al (2005) Early postinfarction ventricular restraint improves borderzone wall thickening dynamics during remodeling. Ann Thorac Surg 80:2257–2262
95. Magovern JA, Teekell-Taylor L, Mankad S et al (2006) Effect of a flexible ventricular restraint device on cardiac remodeling after acute myocardial infarction. ASAIO J 52:196–200
96. Goldstein S (1999) Passive ventricular restraint. Cardiovasc Res 44:468–469
97. Ghanta RK, Rangaraj A, Umakanthan R et al (2007) Adjustable, physiological ventricular restraint improves left ventricular mechanics and reduces dilatation in an ovine model of chronic heart failure. Circulation 115:1201–1210
98. Dixon JA, Goodman AM, Gaillard WF et al (2011) Hemodynamics and myocardial blood flow patterns after placement of a cardiac passive restraint device in a model of dilated cardiomyopathy. J Thorac Cardiovasc Surg 142:1038–1045
99. Blom AS, Pilla JJ, Arkles J et al (2007) Ventricular restraint prevents infarct expansion and improves borderzone function after myocardial infarction: a study using magnetic resonance imaging, three-dimensional surface modeling, and myocardial tagging. Ann Thorac Surg 84:2004–2010
100. Acker MA, Bolling S, Shemin R et al (2006) Mitral valve surgery in heart failure: insights from the Acorn Clinical Trial. J Thorac Cardiovasc Surg 132:568–577 (e1–e4)

101. Bayomy AF, Bauer M, Qiu Y, Liao R (2012) Regeneration in heart disease-Is ECM the key? Life Sci 91:823–827
102. Mercer SE, Odelberg SJ, Simon H-G (2013) A dynamic spatiotemporal extracellular matrix facilitates epicardial-mediated vertebrate heart regeneration. Dev Biol 382:457–469
103. Winter EM, Grauss RW, Hogers B et al (2007) Preservation of left ventricular function and attenuation of remodeling after transplantation of human epicardium-derived cells into the infarcted mouse heart. Circulation 116:917–927
104. Zhou B, Honor LB, He H et al (2011) Adult mouse epicardium modulates myocardial injury by secreting paracrine factors. J Clin Invest 121:1894–1904
105. Smart N, Risebro CA, Melville AAD et al (2007) Thymosin beta4 induces adult epicardial progenitor mobilization and neovascularization. Nature 445:177–182

The Role of Neurohumoral Activation in Cardiac Fibrosis and Heart Failure

Nirmal Parajuli, Tharmarajan Ramprasath, Pavel Zhabyeyev, Vaibhav B. Patel and Gavin Y. Oudit

Abstract Heart failure is an emerging epidemic with an enormous economic burden and a high morbidity and mortality, thereby characterizing a public health problem. Heart failure is the most frequent cause of hospitalization in persons 65 years of age or older and heart failure continues to impose a substantial healthcare burden, despite recent treatment advances. From both an economical and clinical perspective, there is clearly an urgent need for advanced therapeutic strategies for heart failure. The key pathophysiological process that ultimately leads to chronic heart failure is cardiac remodeling, including cardiac fibrosis, in response to chronic injury. Cardiac fibrosis contributes to both the mechanical failure and the electrical disturbances in the failing heart. Several lines of experimental and clinical evidence implicate a key role for the neurohumoral activation in the pathophysiology of a number of cardiovascular diseases, such as myocardial infarction, hypertension, and heart failure. In this chapter, we will focus on the role of neurohumoral activation in cardiac fibrosis and heart failure. We will also review the pathophysiology of cardiac fibrosis and the development in therapies for heart failure utilizing the major neurohumoral systems.

Keywords Heart failure · Myocardial fibrosis · Natriuretic peptide · Sympathetic nervous system · Vasopressin · Rennin-angiotensin system

1 Introduction

Heart failure (HF) is a global epidemic that affects 14 M Europeans, 6 M in the US and over half a million Canadians and is the most common cause for hospitalization in elderly patients [1, 2]. There were 750,000 new cases of HF in North America in 2011 [2]. After diagnosis, the 1-year mortality is 25–40%. More than 50% of the cost associated with HF is hospitalization, resulting in total spending over $35B in the US and $4B in Canada and total costs are projected to double in the next 10–20

G. Y. Oudit (✉) · N. Parajuli · T. Ramprasath · P. Zhabyeyev · V. B. Patel
Division of Cardiology, Department of Medicine, Mazankowski Alberta Heart Institute,
University of Alberta, Edmonton, AB T6G 2S2 Canada
e-mail: gavin.oudit@ualberta.ca

© Springer International Publishing Switzerland 2015   347
I.M.C. Dixon, J. T. Wigle (eds.), *Cardiac Fibrosis and Heart Failure: Cause or Effect?*,
Advances in Biochemistry in Health and Disease 13, DOI 10.1007/978-3-319-17437-2_18

years. From both an economical and clinical perspective, there is clearly an unmet need for better therapies for these disorders. However, there is a paucity of novel therapies entering clinical trials [3, 4].

The CONSENSUS investigations enrolled patients with New York Heart Association (NYHA) class IV symptoms (at rest) that persisted despite contemporary optimal medical therapy. The trial was terminated early, after 253 patients had been randomized and an average follow-up of 188 days. One hundred and ninety-four patients were followed for 6 months (6-month mortality was the pre-defined endpoint) and 102 for 12 months. Six-month mortality in the enalapril group was 26% compared with 44% in the placebo group, giving a relative risk-reduction of 40% [5]. At 1 year these proportions were 36 and 52%, the placebo group mortality highlighting the dreadful prognosis in patients with severely symptomatic heart failure before the advent of modern disease-modifying therapies [6]. These seminal results were followed by multiple other clinical studies and trials examining the pharmacological antagonism of multiple other arms of the neurohumoral pathways and the resultant impact on heart failure [6]. In this book chapter, we will describe the role of the renin-angiotensin system (RAS), natriuretic peptide system (NPS), sympathetic nervous system (SNS), and vasopressin (ADH) in the progression of cardiac fibrosis and heart failure.

2 Molecular Pathogenesis of Cardiac Fibrosis and Heart Failure

The most common cause of heart failure is coronary artery disease, with hypertension also playing a major pathophysiological role. Occlusion of the coronary arteries leads to myocardial infarction (MI), which triggers major pathophysiological processes such as necrosis of an area of the myocardium, pathological remodeling (cardiac hypertrophy, cell death and fibrosis) and cardiac dysfunction [7, 8]. One of the most common histological features of the failing heart is myocardial fibrosis. Replacement fibrosis, often present in the terminal stages of heart failure, has been confirmed by histopathological autopsy studies. Pathophysiological mechanisms that lead to this fibrosis are various, with some being acute, as in MI and others being progressive and potentially reversible, as in hypertensive cardiomyopathy [9].

After myocardial injury, multiple neurohumoral factors and cytokines drive changes in cardiomyocytes and fibroblasts that collectively called as cardiac remodeling. When cardiomyocytes are subjected to neurohumoral stimulation, it causes cardiac hypertrophy which is one of the main ways in which cardiomyocytes respond to mechanical and neurohumoral stimuli [10, 11]. Two key neurohumoral systems activated in HF are the renin-angiotensin system (RAS) and sympathetic nervous system (SNS). In addition to causing further myocardial injury, these systemic responses have detrimental effects on many other tissues and account for clinical features of the HF syndrome. Interruption of these two key processes is the basis of much of the effective treatment of HF [8, 12]. The renin-angiotensin system (RAS) exists not only in the circulation where it is driven by renal renin, but also locally activated in many tissues and cells [13]. The RAS is often activated in human subjects and animal models with metabolic syndrome as well. Activation of the RAS and the subsequent generation of angiotensin (Ang) II are important mediators of myocardial fibrosis, pathological hypertrophy, and heart failure. Pathological hypertrophy and increased myocardial interstitial fibrosis contribute to increased ventricular wall stiffness, thereby impairing cardiac diastolic function, and represent an important risk factor for heart failure in experimental models and patients [14, 15]. Thus, the prevention and reversal of cardiac fibrosis are essential in the management of patients with heart failure. Understanding the regulation and pathophysiology of fibrosis is important to further our understanding and to enhance therapeutic strategies aimed at reducing cardiac fibrosis.

2.1 Renin-Angiotensin system

The renin-angiotensin system (RAS) is a peptide-based system that has been recognized for decades as a principal determinant of arterial blood pressure, fluid and electrolyte balance and as a mediator of adverse myocardial remodeling. Angiotensin converting enzyme (ACE) plays a key role in the production of Ang II and in catabolism of bradykinin, peptides involved in the modulation of vascular tone and in the proliferation of smooth muscle cells. In the heart, Ang II is also produced by another enzyme, named chymase [16]. Chymase, a human mast cell protease, has been observed to play role in the regulation of blood pressure via its Ang II forming activity. Chymase contributes to the regulation of blood pressure and plasma volume via aldosterone-regulated sodium excretion, sympathetic nervous system activity, and is involved in such diverse effects as proliferation, differentiation, regeneration, and apoptosis [17]. The classical view of the RAS focuses on the processing of angiotensinogen, the single obligate precursor to the active peptide Ang II and the interactions of Ang II with its receptors, primarily the Ang II type 1 receptor (AT1R). The contemporary view of the RAS is increasingly complex, involving a balance between multiple processing pathways for angiotensin generation and degradation [18] (Fig. 1a).

The multiple actions of Ang II are mediated via specific, highly complex intracellular signaling pathways that are stimulated following an initial binding of the peptide to its cell-surface receptors. The major actions of Ang II are mediated by two subtypes of G protein-coupled receptors, the AT1 and AT2 receptors [17]. Angiotensin-converting enzyme 2 (ACE2), a monocarboxypeptidase, cleaves one amino acid from either Ang I or Ang II, decreasing Ang II levels and increasing the level of Ang 1–7, which possesses vasodilator properties. Thus, the balance between ACE and ACE2 is an important factor controlling Ang II levels. Even though ACE is the primary enzyme leading to Ang II generation, in the heart the majority of Ang II is generated by chymase [19] (Fig. 1a). Renin is known to be a rate limiting factor for Ang II formation, and is a major determinant of tissue Ang II concentrations [20]. Renin is also present in many cardiovascular relevant organs (heart, kidneys, blood vessels, lungs, and brain) and exerts important actions in maintaining cardio-

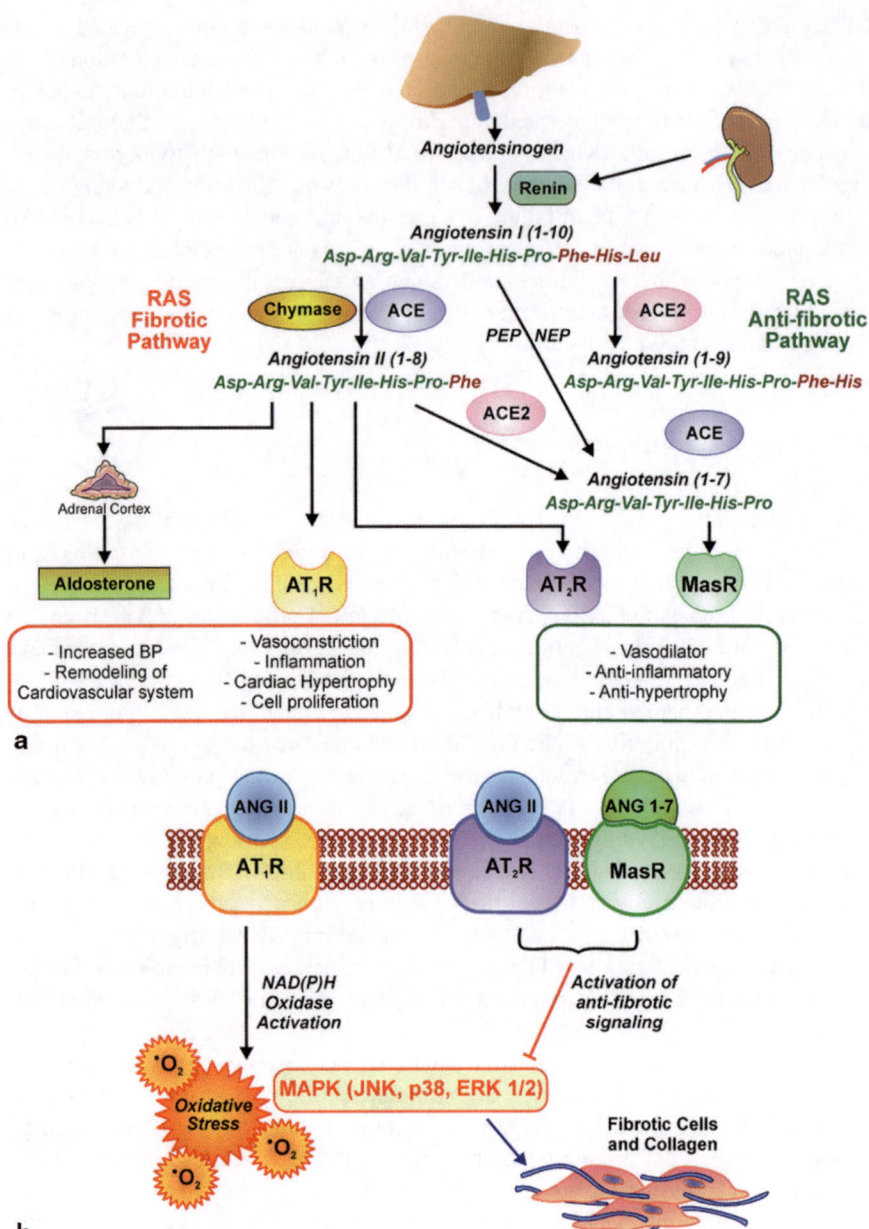

Fig. 1 RAS fibrotic and RAS antifibrotic pathways. a Angiotensinogen synthesized by liver, is converted into Ang I (1–10) which is again converted into either Ang II, by the action of ACE or Ang (1–9) by ACE2. Later, Ang (1–9) is converted into Ang (1–7) by ACE and Ang (1–8) is converted into Ang (1–7) by ACE2. Alternatively, Ang I (1–10) is directly converted into Ang (1–7) by PEP or NEP. By the action of ACE, the Ang (1–9) is converted into Ang (1–7). The Ang (1–7) which is a ligand for the Mas receptor, that participates in vasodilation, cardiac anti-hypertrophy and anti-inflammatory pathways. Ang II which is a ligand for the AT1 receptor participates in

vascular homeostasis. Altered expression and function of this enzyme is associated with cardiac and vascular disorders [21].

Aldosterone is another important family member of the RAS, is synthesized from cholesterol in the zona glomerulosa (ZG) of the adrenal cortex by a series of locus- and orientation-specific enzymatic reactions. A number of recent studies show that (i) aldosterone has marked effects in a wide range of non-epithelial tissues (eg. heart), (ii) it can be synthesized and regulated in a number of extra-adrenal tissues and (iii) it may act through alternative receptors in epithelial and nonepithelial tissue in a rapid non-genomic manner which is independent of gene transcription and translation [22]. Like Ang II, aldosterone activates NADPH oxidases in rat VSMCs. Moreover, aldosterone increases expression of the NADPH oxidase subunit p22^{phox} in human monocytes and promotes apoptosis in human renal proximal tubular cells. Ang II stimulates collagen production and degradation, resulting in net accumulation of interstitial collagen in cardiovascular tissues. Similarly, aldosterone may induce fibrosis by increasing collagen deposition [23] (Fig. 1a). Ang II acts on vascular smooth muscle cells to cause vasoconstriction and on the adrenal ZG to stimulate aldosterone production. The adrenal response to Ang II occurs within minutes, a time course that implies that no new protein synthesis is required. This acute Ang II-mediated release of aldosterone may involve in rapid synthesis [24]. Systemic administration of aldosterone increases oxidative stress in heart, vasculature and kidney, along with an increase in macrophage NADPH oxidase [25]. Aldosterone, increases tissue ACE activity and upregulates angiotensin receptors. This suggests the potential for a vicious cycle in which Ang II, through the AT1 receptor, stimulates the production of aldosterone, which, in turn, leads to an increase in tissue ACE activity, an additional increase in Ang II, and, therefore, an additional elevation in aldosterone levels [26]. Increased deposition of basement membrane collagen is a hallmark of the remodeling process, and it results in an increase in cardiac tissue stiffness which predisposes the patient to an increased risk of adverse cardiac events [27]. In humans, myocardial hypertrophy due to hemodynamic overload is characterized by increased deposition of extracellular matrix (ECM) constituents, proliferation of cardiac fibroblasts, and hypertrophic growth of cardiac myocytes [28]. In situations such as hypertension or chronic MI, heart responds to increased afterload by initiating adaptive remodeling processes. These include cardiomyocyte hypertrophy, fibrosis, ECM deposition and alterations of cardiac gene expression. Although these structural alterations represent the heart's efforts to maintain systolic function, they are deleterious over time and ultimately result in progressive heart failure. On the molecular level, cardiac remodeling is mediated by activation of several neurohumoral systems including the RAS, TGF-β1 and the β-adrenergic system [28]. The features of myocardial fibrosis are associated with

vasoconstriction, inflammation and cardiac hypertrophy pathways. The Ang II also triggers the synthesis of aldosterone by adrenal cortex. b When Ang II interacts with its receptor AT1R, it increases the oxidative stress myocardial tissue through the activation of NAD(P)H oxidase. This oxidative stress further activates the pathological signaling (eg. phosphorylation of JNK, p38, ERK 1/2) which leads to accumulation of collagen and myocardial fibrosis

hypertension and cardiac hypertrophy. Circulating Ang II and aldosterone are involved in the increase in fibrosis and resultant heterogeneity in tissue structure. The progressive interstitial fibrosis and perivascular fibrosis contribute to an increase in cardiac muscle stiffness and development of diastolic dysfunction. Many studies have demonstrated important roles of Ang II in the development of cardiomyocyte hypertrophy and cardiac fibrosis and modulation of cardiac fibroblast growth and collagen synthesis in humans as well as in animal models [29].

2.2 RAS-Mediated Redox Signaling in Heart Failure

Ang II mediates many of its cellular actions by stimulating the formation of reactive oxygen species (ROS), which play important roles in modulating inflammatory reactions. Evidences suggest that increased oxidative stress plays a pathological role in cardiovascular disease, including atherosclerosis, hypertension, and heart failure [30]. NADPH oxidase is a major source of reactive oxygen species (ROS) in vascular and cardiac tissues, and Ang II stimulates NADPH oxidase. It has been postulated that the increase in ROS is an important mechanism by which Ang II contributes to the pathogenesis of vascular disease, and perhaps cardiac remodeling [31]. Growing evidence highlights oxidative stress as an important mechanism for this maladaptation. Although ROS generation is a normal component of oxidative phosphorylation and plays a role in normal redox control of physiological signaling pathways [32], generation of excessive amount of ROS leads to oxidative stress that cannot be adequately countered by intrinsic antioxidant systems [33, 34]. Superoxide anion $(O_2^-.)$ can further combine with nitric oxide (NO), forming reactive compounds such as peroxynitrite, generating nitroso-redox imbalance. Altogether, Ang II affects vascular remodeling, ECM deposition with the possible role of oxidative stress [30]. The experiment conducted by Briones et al. (2009) showed that Ang II increased the expression of collagen type I, NF-κB, and AP-1 and decreased the expression and activity of matrix metalloproteinase in Wistar rats. Ang II-mediated collagen type I synthesis in cardiac fibroblasts was attenuated by the supplementation of statins. Moreover, downregulation of Ang II-induced NADPH oxidase subunit and Nox1 expression was also observed by the statin treatment which showed the role played by Ang-II in the induction of oxidative stress–induced vascular fibrosis (Fig. 1a). The importance of NADPH oxidase in processes associated with Ang II–mediated hypertension and cardiac hypertrophy has also been demonstrated in Ang II–infused animal models. Inhibition of NADPH oxidase activity with apocynin or gp91ds-tat, a chimeric peptide inhibitor that interferes with assembly of vascular NAD(P)H oxidase components, attenuates blood pressure increase, regresses cardiac and vascular remodeling, and improves endothelial function in Ang II-infused rodents [35–37].

2.3   Pro-hypertrophic and Pro-fibrotic Effects of RAS Signaling

Ang II, the main effector hormone of the RAS, modulates cardiac remodeling by causing myocyte hypertrophy and myocardial fibrosis. Myocardial interstitial changes, characterized by increases in total fibrillar collagen (types I and III), changes in the ratio of types I/III collagen, and alterations in collagen cross-linking may adversely affect cardiac diastolic and systolic function. Various models of cardiac remodeling and failure have demonstrated that Ang II is a powerful mediator of myocardial fibrosis. Several *in vitro* studies have shown that Ang II increases collagen production by cultured cardiac fibroblasts [38]. In addition to these earlier studies, demonstrating a direct effect of Ang II on cardiac fibroblast function, recent data also indicate an indirect effect of Ang II on cardiac fibroblasts mediated through cardiomyocytes [31]. Ang II mediates cardiac or renal fibrosis by binding and activating AT1 receptors [39]. This induces the production of reactive species and increases the oxidative stress in myocardial tissue through the activation of NAD(P)H oxidase. This oxidative stress further activates the pathological signaling (eg. phosphorylation of JNK, p38, ERK 1/2 etc.) which leads to the accumulation of collagen and fibrosis (Fig. 1b).

2.4   RAS Inhibitors for Cardiac Disease Management

With consideration of the significant role of RAS in cardiovascular disease, interrupting the RAS has become a leading therapeutic strategy for the alleviation of cardiac complications. However, a wide variety of potential renin inhibitors have been developed, due to low potency, poor bioavailability, and short duration of action after oral administration in humans, these compounds are not clinically used [40]. The ACE inhibitors (ACEI; eg. enalapril, ramipril, etc.) are used to lower Ang II, elevate levels of bradykinin, and reduce cardiovascular disease risk in high-risk individuals, resulting in increased survival and prevention of MI, strokes and HF. These inhibitors are also suggested for the prevention of diabetes mellitus, dementia, and atrial fibrillation [41, 42]. Similarly, angiotensin receptor blockers (ARBs, eg. telmisartan, valsartan, etc.) are also well established cornerstones in the prevention and treatment of hypertension and cardiovascular disease, alone or in combination with ARBs, as demonstrated by numerous clinical trials and worldwide clinical practice [43]. Aldosterone receptor antagonists, in addition to background ACE inhibitor and β-blocker therapy, have been found to be beneficial across all severities of systolic heart failure [44]. As the chymase inhibitors, unlike ACE inhibitors and ARBs, only play a small role in the regulation of the systemic RAS [45] and the possible applications of chymase inhibitors to prevent cardiovascular diseases are also poorly studied.

2.4.1 Angiotensin Converting Enzyme Inhibitors

ACE inhibitors have an established role as the first-line treatment for a number of cardiovascular and renal diseases. Their role in the management of hypertension is proven, and they have been shown to reduce mortality associated with HF. Furthermore, ACE inhibitors have been shown to reduce the rate of stroke, MI, and death in high-risk individuals without known HF [46]. The reports from the Heart Outcomes Prevention Evaluation (HOPE) and the European Trial on Reduction of Cardiac Events with Perindopril in Stable Coronary Artery Disease (EUROPA), state that patients with coronary or other vascular disease or with diabetes and another cardiovascular risk factors had reduced rates of death from cardiovascular causes or acute MI when they are assigned to ACE inhibitors [47].

2.4.2 Angiotensin II Receptor Antagonist

ARBs are used as a first-choice anti-hypertensive for the prevention of hypertension induced multiple organ injury. As ARBs are selective for the AT1 receptors they attenuate the deleterious effects of Ang II. The American College of Cardiology/ American Heart Association (ACC/AHA) recommend ARBs as an alternative to an ACEI for those have clinical signs of HF and acute MI. Similarly, the Heart Failure Society of America (HFSA) also recommends using ARBs as an alternative to an ACEI in ACEI intolerant and tolerant patients with HF with or without MI [48].

3 Natriuretic Peptides (NP)

The natriuretic peptide (NP) family a neurohormonal system localized in the heart, brain and other organs. Different cardiovascular diseases like chronic heart failure, systemic hypertension, coronary artery disease, endothelial dysfunction and others are responsible for increased NPS secretion [49]. The natriuretic peptide family in mammals consists of three homologous peptides: atrial natriuretic peptide (ANP), brain natriuretic peptide (BNP), and C-type natriuretic peptide (CNP) with unique physiological and biochemical properties [50–52]. These peptides share 17-amino-acid peptide ring with a cysteine bridge that is well preserved and binds to the specific receptors (Fig. 2b)[51, 52]. ANP is synthesized and secreted from the atria and BNP from the ventricles [50, 51, 53]. The released ANP and BNP primarily from the heart circulate as hormones in various tissues to induce vasodilation, natriuresis, and diuresis [54–56]. CNP is from endothelial cells that are synthesized in myocardial tissue to protect against cardiac remodeling. These neurohormones are involved in the long-term regulation of sodium and water balance, blood volume and arterial pressure [49, 57]. The major consequences of natriuretic peptide actions are the vasodilator effects and renal effects leading to natriuresis and diuresis [58]. Natriuretic peptides increases venous compliance and decreases central venous

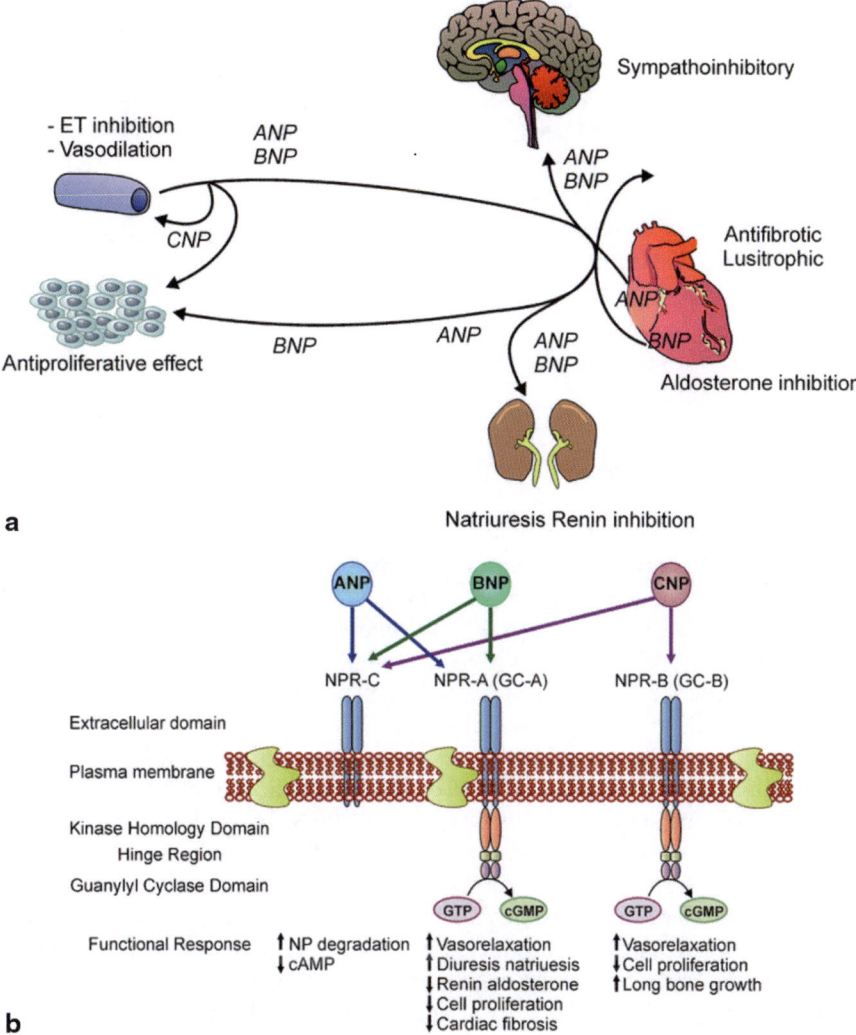

Fig. 2 Multisystem interaction and signaling in the natriuretic peptide system. a Interaction of the heart, kidney, vasculature and brain with the natriuretic peptide system. *ANP* A-type natriuretic peptide; *BNP* B-type natriuretic peptide; *CNP* C-type natriuretic peptide. **b** Signaling mediated by natriuretic peptide receptor A, natriuretic peptide receptor B and natriuretic peptide receptor C (NPR-A, NPR-B, NPR-C). NPR-A binds ANP and BNP, but not for CNP. NPR-B has weaker affinities for ANP and BNP, as it mainly acts as a receptor for CNP. NPR-A and NPR-B contain a transmembraine domain linked to an intracellular kinase-like domain linked to particulate guanylyl cyclase. NPs causes a conformational change to the guanylyl cyclase domain and allow GTP dephosphorylation to cGMP. NPR-C is not cGMP-linked, binds to all NPs with equal affinity, and works as a clearance receptor

pressure which ultimately reduce cardiac output by decreasing ventricular preload [54, 59]. Natriuretic peptides (i) decrease systemic vascular resistance and systemic arterial pressure by dilating arteries and (ii) increase glomerular filtration rate and filtration fraction in the kidneys. These effects ultimately cause increased sodium excretion (natriuresis) and increased fluid excretion (diuresis) [55, 56]. Natriuretic peptides also decrease renin release and decrease the circulating levels of Ang II and aldosterone [60, 61]. Thus, the natriuretic peptides control the homeostatic balance of blood volume, arterial pressure, central venous pressure, pulmonary capillary wedge pressure, and cardiac output [58, 62, 63] (Fig. 2a).

3.1   Atrial Natriuretic Peptide (ANP)

Atrial natriuretic peptide (ANP) is mainly produced by the cardiac ventricles in prenatal development, and in the atria of adult hearts. In adults, ANP, a 28-amino acid peptide, is mainly produced from atrial myocytes and in small fraction from the ventricles [64]. The ANP gene is located on human chromosome 1p36.2 and is released as a 151-amino-acid pre-propeptide precursor. After this, the precursor is spliced to a 126-amino-acid propeptide (proANP), localized in membrane bound secretary granules in the atria and released into the blood stream [50]. This ANP has vasodilatory, natriuretic, diuretic, and/or kaliuretic properties. Upon release, the propeptide is cleaved at specific sites by a transmembrane serine protease corin into a 28-amino acid C-terminal active form (ANP 99-126) and a 98-amino-acid N-terminal inactive form (proANP 1-98). Active peptides consisting of amino acids 1–30 (proANP 1-30; long-acting natriuretic peptide), amino acids 31–67 (proANP 31-67; vessel dilator), and amino acids 79–98 (proANP 79-98; kaliuretic peptide) are proteolytically cleaved from the N-terminus proANP 1-98 by proteases [50, 63–65].

Despite the short half-life of ANP, this peptide has potent direct and indirect physiological effects like natriuresis, diuresis, vasodilation, vasorelaxation, and negative inotropy. Further ANP increases venous capacity and reduces sympathetic tone in peripheral vasculature, and suppresses the RAS [65–67]. ANP also inhibit the cardiac fibroblasts growth and the collagen deposition to reduce the adverse cardiac remodeling. Further, ANP arrests cell cycle of cardiac myocyte during development and reduces the hypertrophic response in pathological conditions [65, 67].

3.2   B-type Natriuretic Peptide (BNP)

B-type natriuretic peptide was initially isolated from porcine brain tissue. However, the highest expression was found in cardiac ventricles [52, 56, 63, 68]. Like ANP, BNP also has a 17 peptide ring structure. BNP is initially synthesized as a preprohormone of 134 residues containing a signal sequence that is cleaved to yield a 108-amino-acid prohormone (proBNP). In humans, the BNP gene locus is on

chromosome 1p36.2. BNP is comprised of a biologically active C-terminal fragment the mature form of 32 amino acids (BNP 77-108) and an inactive 76 amino
acids as N-terminal fragment [63, 65]. It is very similar to ANP for overall function,
but it differs only with the mode of synthesis and secretion. The ventricular BNP
production is transcriptionally regulated by cardiac wall stretch resulting from volume overload. In contrast to the regulated release of ANP, BNP is secreted into the
circulation along a constitutive pathway. In the atria, ANP is expressed and stored
in granules and in the ventricle; BNP is produced and constitutively released. BNP
mRNA levels increase with chronic blood volume overload and cause hemodynamic effects similar to ANP. BNP has been shown to inhibit collagen secretion by
fibroblasts, thereby limiting cardiac remodeling [54, 59, 68].

3.3   C-type Natriuretic Peptide (CNP)

C-type natriuretic peptide, or CNP, was also initially discovered in porcine brain,
however, it was later determined to be mainly produced in the vascular endothelium. In humans CNP is a 22-amino acid peptide, which has gene locus on chromosome 2. Like ANP and BNP, CNP is made up of a 17 amino-acid peptide ring
structure which is essential for its biological activity [63, 65]. This peptide does
not have a C-terminal extension. CNP peptides and its receptors have a much wid-
er distribution and primarily localized in cytokine-exposed endothelial cells and
chondrocytes. CNP is the most conserved NP. Two mature forms of peptides are
produced by cleaving a 103-amino acid prohomone (proCNP 1-103) that is derived
from a 126-amino acid preproCNP. CNP is usually cleaved into either a 22 or 53
amino acid peptides as biologically active fragments. CNP-53 (CNP 51-103) pre-
dominates in tissues, whereas CNP-22 (CPN 82-103) is found mainly in plasma and
cerebrospinal fluid. Similar to ANP and BNP, CNP is a powerful vasorelaxant of
vascular smooth muscle; however, it causes only mild diuresis and natriuresis [50,
51, 61]. Despite low levels, CNP acts to control fluid movement across capillaries, CNP also plays a critical role in normal skeletal development and regulate the
growth of bone and cartilage. Further, CNP has been reported to be an antifibrotic
and antihypertrophic agent in cardiac dysfunction [56, 68].

3.4   NP Receptors

The circulating NPs interact with different NP receptors (NPRs); NPR-A, NPR-B
and NPR-C (Fig. 2b). These three natriuretic peptide binding proteins contain a
relatively large extracellular ligand binding domain and a single membrane-spanning region. The natriuretic peptide receptors A and B contain an equally large intracellular domain consisting of a kinase homology domain, dimerization domain,
and carboxyl-terminal guanylyl cyclase domain [50, 64]. Thus, NPR-A and NPR-
B signal by catalyzing the synthesis of the intracellular signaling molecule cyclic

3'-5'-guanosine monophosphate (cGMP) and activates cGMP-dependent protein kinase (PKG) and resulting in GTP-dependent phosphorylation. This phosphorylation initiates an intracellular signaling cascade that catalyzes a range of cardioprotective effects like vascular smooth muscle relaxation and increased blood flow [50, 64]. Unlike other NPRs, NPR-C contains an intracellular domain that lacks the guanylyl cyclase activity [51].

3.4.1 Natriuretic Peptide Receptor-A (NPR-A)

NPR-A is the principal receptor of ANP and BNP. The receptor has extracellular domain that contains three intracellular disulfide bonds, five N-linked glycosylation sites and exists as a homodimer or homotetramer in its native state with ligand-independent oligomerization [50, 51, 64]. NPs binds with NPR-A in the order of affinities as follows: ANP > BNP > CNP. The human NPR-A gene is located on chromosome 1q21–22. NPR-A is expressed in kidney, lung, adipose, adrenal, brain, heart, testis, and vascular smooth muscle tissue. Binding of NPs to this receptor activates cGMP signaling that increases vasodilatation, diuresis and natriuresis and decrease levels of renin and aldosterone, cell proliferation as well as cardiac fibrosis [50, 51, 64]. At physiological levels of ANP, BNP and CNP, NPR-A is activated by ANP and BNP but not by CNP.

3.4.2 Natriuretic Peptide Receptor-B (NPR-B)

NPR-B is the principal receptor of CNP. The receptor has a similar structure to NPR-A, and location of glycosylation sites, and intramolecular disulfide bonds. In rats, the NPR-B's extracellular and intracellular regions are identical to NPR-A. NPR-B binds NPs in the following order: CNP>ANP>BNP [50, 51, 64]. Similar to NPR-A, upon activation NPR-B dimerizes into a homodimer complex. The human NPR-B gene is located on chromosome 9p12–21 and the murine version, *Npr2*, is located on chromosome 4. NPR-B is expressed in bone, brain, fibroblasts, heart, kidney, liver, lung, uterine, and vascular smooth muscle tissue [50, 64]. Similarly to NPR-A activation of NPR-B by CNP activates cGMP signaling that leads to increase in vasodilatation and long bone growth as well as decrease in cell proliferation [49].

3.4.3 Natriuretic Peptide Receptor-C (NPR-C)

NPR-C has a large extracellular ligand binding domain like of NPR-A and NPR-B, a single membrane-spanning region with 37 intracellular amino acids [50, 51, 53]. The main function of NPR-C is to clear out circulating natriuretic peptides by receptor-mediated internalization and degradation. The human NPR-C gene is located on chromosome 5p13–14 [50, 51]. NPR-C is the most abundantly expressed receptor, and is normally found within close proximity to NPR-A and NPR-B [59].

3.5 Role of NPs in Cardiac Remodeling

ANP, BNP, and CNP natriuretic peptides are important in the integrated control of renal and cardiovascular function. Initially, the mRNA transcripts were reported for three receptors in rat and human heart, indicating a potential role for the involvement cardiac function [69]. In general, plasma natriuretic peptides are increased in patients with cardiac hypertrophy and systolic and diastolic dysfunctions. Several signaling pathways serve as agonists or antagonists to hypertrophic growth of cardiac myocytes. ANP is known to inhibit a variety of hypertrophic cellular signaling through the stimulation of NPR-A and the downstream production of cGMP leading to activation of PKG (protein kinase type I). *In vitro* studies have shown that ANP exerts its direct inhibitory effect on growth, at least partly through the induction of MAPK phosphatase-1, a dual serine/threonine and tyrosine phosphatase that specifically inactivates MAPK family members such as ERKs, JNK, and p38MAPKs which are involved in cell proliferation and hypertrophy [52].

Accumulating observations suggest that BNP may play important roles in ventricular remodeling. However, the precise functional significance of BNP is not yet understood. Thus, to understand its role in ventricular remodeling, Tamura et al. (2000) generated $Nppb^{-/-}$ (natriuretic peptide precursor B) mice by gene targeting and analyzed their phenotypes [70]. No signs of systemic hypertension and ventricular hypertrophy were noted in $Nppb^{-/-}$ mice, in response to ventricular pressure overload. However, focal fibrotic lesions were increased in size and number in $Nppb^{-/-}$ mice, whereas no focal fibrotic changes were found in wild-type littermates ($Nppb^{+/+}$ mice). This study established the BNP as a cardiomyocyte-derived antifibrotic factor *in vivo* and provided an evidence for its role as a local regulator of ventricular remodeling [70]. On the other hand, the NPR-A system plays a primary role in moderating cardiac hypertrophy *in vivo* independently of its effects on blood pressure regulation. Knowles et al. (2001) compared the response of $Npr1^{+/+}$ and $Npr1^{-/-}$ mice to a pressure overload induced by transverse aortic constriction (TAC). The result showed that $Npr1^{-/-}$ mice resulted in a 15-fold increase in atrial natriuretic peptide (ANP) expression, a 55 % increase in left ventricular weight/body weight (LV/BW), dilatation of the LV, and significant decline in cardiac function [71]. Finally, these results suggest that the NPRA system has direct antihypertrophic actions in the heart independent of its role in blood pressure control. TGF-β is a well-known factor which is associated with profibrotic processes in heart failure. Thus, an association study was conducted between TGF-β and BNP to understand whether BNP could inhibit TGF-β–induced fibrotic effects on primary human cardiac fibroblasts. BNP treatment resulted in a remarkable reduction in TGF-β effects. These findings demonstrated that BNP has a direct effect on cardiac fibroblasts to inhibit fibrotic responses via extracellular signal-related kinase signaling [71].

Nesiritide is a recombinant form of human BNP being used in the acute treatment of congestive heart failure of systolic dysfunction. The St Vincent's Screening to Prevent Heart Failure (STOP-HF) randomized trial showed that BNP-based screening and collaborative care reduced the combined rates of left ventricular systolic dysfunction, diastolic dysfunction, and HF as well as emergency hospitalizations for major adverse cardiovascular events among patients at risk of HF [72].

However, the Acute Study of Clinical Effectiveness of nesiritide and Decompensated Heart Failure (ASCEND-HF) trial showed nesiritide did not reduce the rate of recurrent heart failure hospitalization and did not affect 30-day all-cause mortality or worsening renal function [73].

BNP functions as an antifibrotic factor in the heart to prevent cardiac remodeling in pathological conditions. BNP plays a key role in the processes of extracellular matrix remodeling and wound-healing during the early phase after acute myocardial infarction [68]. The beneficial action of endogenous cardiac natriuretic peptides was also suggested against acute heart failure and attenuation of chronic cardiac remodeling after MI, by the activation of guanylyl cyclase-A. It was also reported that natriuretic peptides exerted beneficial effects by inhibition of the RAS. Knockout (KO) and wild-type (WT) mice lacking the natriuretic peptide receptor guanylyl cyclase-A were subjected to left coronary artery ligation. The enhanced myocardial fibrosis was observed in the KO mice and was virtually absent in infarcted double-KO mice which lack guanylyl cyclase-A and angiotensin II type 1a receptors. The higher levels of cardiac expression of ANP and BNP mRNA observed in KO mice early after MI. Taken together, these results indicate that KO mice have a diminished capacity to compensate for acute heart failure after MI that proves the critical role of natriuretic peptides in cardiac remodeling [62].

4   Sympathetic Nervous System

The Sympathetic Nervous System (SNS) evolved to enable fight-or-flight organismal response to a challenge. Such response requires enhanced cardiac performance. This enhancement is achieved through (i) increase in heart rate (positive chronotropy), (ii) increase in myocardial contractility (positive inotropy), (iii) acceleration of myocardial relaxation (positive lusitropy), and (iv) selective vasoconstriction (decrease in vessel diameter); e.g., decrease in venous capacitance and constriction of cutaneous vessels. Unlike many other organs, sympathetic and parasympathetic innervations of cardiovascular system are uneven. In the heart, atria receive both parasympathetic and sympathetic input whereas ventricles have predominantly sympathetic innervation. Similarly to ventricles, vasculature receives predominantly sympathetic innervation. Thus, the cardiovascular system is primarily regulated by the degree of sympathetic tone rather than by the balance between parasympathetic and sympathetic tones [74, 75].

Activation of SNS in the cardiovascular system results in the release of catecholamine norepinephrine (noradrenaline), which acts on adrenergic receptors (αAR and βAR) located on cell membrane. These receptors bind not only norepinephrine (noradrenalin), but also its analog, epinephrine (adrenaline), which is released from adrenal gland and represents about 80 % of total catecholamine release under normal conditions [76]. Therefore, it is justifiable to view adrenal gland and classical sympathetic nervous system as a combined adrenergic nervous system or ANS (Fig. 3a) [77].

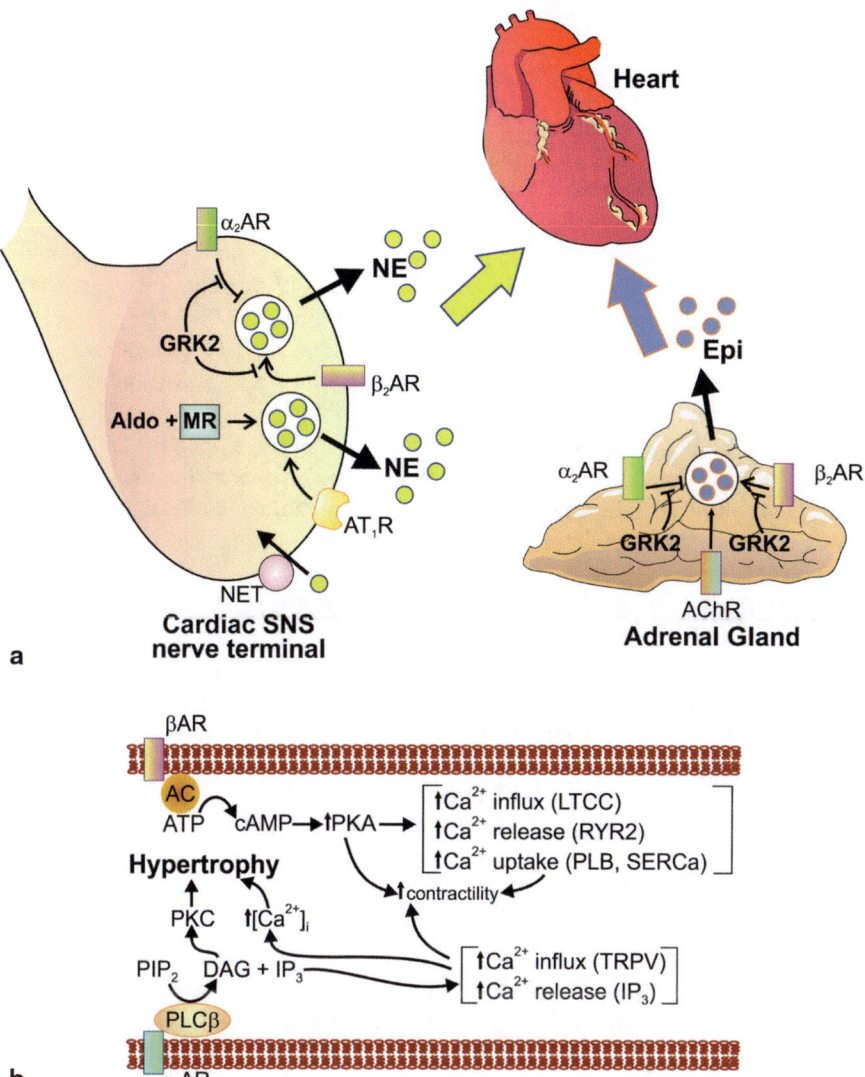

Fig. 3 Catecholamines and their interaction with alpha and beta-adrenergic receptors. a Combined action of sympathetic nervous system (*SNS*) and adrenal gland on the heart (see text for details). AChR, acetylcholine receptor; Aldo, aldosterone; AR, adrenergic receptor; AT_1R, angiotensin receptor 1; $G_{i/o}$, inhibitory or other G protein; G_s, stimulatory G protein; GRK2, G-protein-coupled receptor kinase type 2; MR, mineralocorticoid receptor; NE, norepinephrine; Epi, epinephrine; NET, norepinephrine transporter. **b** Adrenergic receptor signaling in myocytes (see text for details). AR, adrenergic receptor; DAG, 2-Diacylglycerol; IP_3, inositol-[1, 4, 5]-trisphosphate; LTCC, L-type calcium channel; PIP_2, phosphatidylinositol (4,5)-bisphosphate; PKA, protein kinase A; PKC, protein kinase C; PLB, phospholamban; PLC, phospholipase C; RyR2, ryanodine receptor type 2; SERCA, sarcoplasmic/endoplasmic reticulum Ca^{2+}-ATPase; TRPV, transient receptor potential vanilloid channel.

4.1 Adrenergic Receptors and Physiological Signaling.

The adrenergic receptors are G-protein coupled receptors and divided in three types with nine sub-types, as follows: α_1AR (α_{1A}, α_{1B}, α_{1D}), α_2AR (α_{2A}, α_{2B}, α_{2C}), and βAR (β_1, β_2, β_3) [77, 78]. In the human heart, βARs are the most numerous and comprise around 80% of all ARs. Expression of β subtypes is also uneven; e.g., β_1 accounts for 75–80% of βARs, β_2 accounts for 15–20%, and β_3 is only about 2–3% [79, 80]. β_1AR and β_2AR associate primarily with G_s and upon agonist stimulation activate adenylate cyclase (AC), which produces $3',5'$-monophosphate (cAMP), which, in turn, activates cAMP-dependent protein kinase A (PKA) (Fig. 3b). PKA phosphorylates multiple targets: (i) L-type Ca^{2+} channels (LTCC) promoting Ca^{2+} influx, (ii) ryanodine receptors (RYR2) enhancing Ca^{2+} release, (iii) phospholamban (PLB) which dis-inhibits sarcoplasmic reticulum Ca^{2+}-ATPase (SERCA) enhancing Ca^+ reuptake into sarcoplasmic reticulum and (iv) troponin I and myosin-binding protein C reducing Ca^{2+} sensitivity of myofilaments facilitating relaxation. Altogether, these changes result in enhanced contractility and relaxation (Fig. 3b). β_2AR can also couple to G_i in which case activation of β_2AR will inhibit AC and lower cAMP levels [77, 81].

α_1ARs are located predominantly in smooth muscle cells of vessels and their cardiac role is disputed. In smooth muscle cells [82], the α_1ARs couple to the $G_{q/11}$ protein and activate phospholipase C-β (PLC-β) (Fig. 3b). PLC-β produces inositol-[1, 4, 5]-trisphosphate (IP_3) and 2-diacylglycerol (DAG) from the cell membrane component, phospholipid phosphatidylinositol (4,5)-bisphosphate (PIP_2). IP_3 releases Ca^{2+} from intracellular stores (sarcoplasmic or endoplamic reticulum) by binding to IP_3 receptor channels in the membrane of the intracellular stores. At the same time, DAG activates PKC and transient receptor potential channels (TRPV), which promote Ca^{2+} influx (Fig. 3b). Increased intracellular Ca^{2+} concentration ($[Ca^{2+}]_i$) due to increased Ca^{2+} release and Ca^{2+} influx leads to increased contraction (vasoconstriction). Increased PKC activity coupled with elevated intracellular Ca^{2+} results in hypertrophy.

α_2ARs exhibit mostly vasodilatory effect. α_{2A}ARs and α_{2C}ARs are presynaptic inhibitory autoreceptors that inhibit norepinephrine release from sympathetic nerve terminals. These receptors also expressed in adrenal gland. Similarly to nerve terminals, activation of these receptors in the adrenal gland inhibits catecholamine release [83]. Deletion of both of these α_2ARs results in cardiac hypertrophy and HF due to chronically enhanced norepinephrine and epinephrine release from nerve terminals and adrenal gland [84, 85]. α_{2B}ARs are present in some vascular smooth muscle cells, and their activation can cause vasoconstriction of some vascular beds. However, overall action of α_2ARs is vasodilation and lowering of blood pressure [86, 87].

The effect of AR on epinephrine secretion and norepinephrine release is mediated by (i) G-protein-coupled receptor kinases (GRKs), (ii) angiotensin II (AngII), (iii) aldosterone, and (iv) acetylcholine (Fig. 3a). GRKs inhibit AR signaling by posphorylating ARs. GRK-phosphorylated ARs bind β-arrestins which uncouple G-proteins from AR receptors and silence its signaling [88, 89]. There are seven GRKs (GRK1-7), but GRK2 and GRK5 are of particular physiological importance since they are ubiquitous, regulate majority of GPCRs, and have high expression levels in cardiac and neuronal tissue [90, 91]. AngII stimulates the release and inhibits the reuptake of norepinephrine at SNS nerve terminals (Fig. 3a) [92]. Similarly to AngII, aldosterone (Aldo) reduces norepinephrine reuptake at nerve terminals contributing to elevated levels of norepineprine in HF [93, 94]. Finally, activation of the nicotinic cholinergic receptors (ACh) on chromaffin cells of the adrenal gland promotes epinephrine release [95].

4.2   AR Polymorphism

Polymorphism of β_1ARs is the-most-studied of all AR polymorphisms. In humans, β_1ARs exhibit polymorphism at two positions (Ser49Gly and Arg389Gly). Arg389 β_1AR hearts have significantly enhanced AC/PKA activity in comparison to Gly389 β_1AR hearts [95]. Similarly, mouse hearts with Arg389 exhibit higher contractility than Gly389 hearts, but higher deterioration and more prominent fibrosis by 9 month age [96]. This polymorphism in combination with 3 genetic polymorphisms of the β-subunit of G-protein (GNB3) is a predictor of the success of implantable cardioverter-defibrillator shock therapies in HF patients [97]. Gly49 β_1AR variant has increased agonist-dependent downregulation of receptor compare to Ser49 β_1AR [98] and confers improved survival in the absence of β-blocker and tends to improve response to β-blocker treatment [98–100]. However, some studies have found no difference between polymorphisms in cardiac outcomes [101–103].

β_2ARs have three non-synonymous polymorphisms: Gly16Arg, Gln27Glu, and Thr164Ile. Gly16Arg and Gln27Glu have enhanced agonist-promoted downregulation of the receptor, and Thr164Ile produces impaired receptor-G-protein coupling leading to reduced adenylate-cyclase-mediated signaling [81, 104]. α_{2C}AR human gene contains another important polymorphism, 4-amino-acid deletion (Δ322–325) that produces increased norepinephrine release from cardiac sympathetic nerve terminals [105]. This deletion in conjunction with the Arg389Gly (β_1AR) polymorphism was used to stratify patients by the clinical response to the β-blocker bucindolol into very favorable, favorable, and unfavorable response genotypes [106].

4.3   Pathophysiological Signaling of AR

In heart failure, norepinephrine and epinephrine levels as well as their turnovers are significantly higher than in healthy subjects [107]. For example, in untreated HF patients under maximal exercise conditions, cardiac norepinephrine spillover can be up to 50-fold higher than in healthy individuals [108]. These changes may be also accompanied by higher Ang II levels that lead to Ang II-dependent release of catecholamines. This additional release may further aggravate hemodynamic and left ventricular remodeling [109, 110]. Elevated levels of catecholamines lead to desensitization of heart to their stimulatory effects and chronically elevated norepinephrine release from the heart (increased norepinephrine spillover) due to stimulation of presynaptic $\beta_2 AR$ (Fig. 3a). These changes are accompanied by changes in receptor stoichiometry and altered signaling. The number of functional $\beta_1 AR$ is reduced resulting in a reduction of $\beta 1:\beta 2$ ratio from 75:20% to 50:50% in the failing heart [111, 112]. $\beta_2 AR$ signaling loses compartmentalization and resembles cAMP-dependent signaling similar to $\beta_1 AR$ [113]. Moreover, the expected inhibitory action of $\alpha_2 ARs$ is markedly blunted, increasing norepinephrine spillover observed in chronic HF [114] (Fig. 3a). Dysfunction of $\alpha_2 AR$ during HF may arise from increases in expression and activity of inhibitory GRK2 in peripheral sympathetic nerve terminals (Fig. 3a) contributing to the enhanced norepinephrine release and spillover [115, 116].

Another source of elevated levels of catecholamines is dysregulation of their release from adrenal gland. Many of the receptors in chromaffin cells are similar to cardiac AR in nerve endings. $\alpha_{2A} Rs$ inhibit secretion and $\beta_2 ARs$ enhance it (Fig. 3a) [84, 85, 87, 95]. Elevated GRK2 expression and activity has been shown in the heart during HF [117, 118], in some vascular beds in hypertension, [119] and in the adrenal gland during HF [120]. Adrenal GRK2 upregulation can be responsible for adrenal $\alpha_{2A} R$ dysfunction in chronic HF. GRK2 upregulation results in a loss of the inhibitory by these receptors, and subsequent chronically elevated catecholamine secretion (Fig. 3a) [120–124]. Moreover, specific inhibition of GRK2 *via* adenoviral-mediated adrenal gene delivery produces a significant reduction in levels of circulating catecholamine restoring both adrenal and cardiac function in HF [120]. Additionally, GRK2 knockout mice, which lack adrenal GRK2 expression from birth, display reduced circulating catecholamines in response to myocardial infarction and have preserved cardiac function and morphology [121].

4.4   Therapeutics in Heart Failure

Currently, pharmacological approach is a prevalent treatment of HF. Below we briefly discuss AR blockers and agonist as HF therapeutic agents as well as non-pharmacological approach (exercise training).

4.4.1   β-Blockers

B-blockers can be classified into four types: (i) nonsubtype-selective competitive blockers (propranolol, nadolol, timolol); (ii) higher affinity for the $\beta_1 AR$ than for the $\beta_2 AR$ (atenolol, metoprolol, bisoprolol); (iii) subtype-selective (celiprolol, nebivolol) and (iv) subtype-nonselective (bucindolol, carvedilol, labetalol) that can also block $\alpha_1 ARs$ causing peripheral vasodilation. Both subtype-selective and subtype-nonselective blockers have negative chronotropic and inotropic effects. $\beta_1 AR$-selective blockers have a lesser effect on the $\beta_2 AR$ and thus have less risk of peripheral vasoconstriction [125]. These blockers may also have less impaired exercise performance because of lesser inhibition of β_2AR, which is responsible for increased skeletal muscle blood flow during exercise. Finally, some β-blockers (pindolol, alprenolol, and oxprenolol) have intrinsic sympathomimetic activity, have a high propensity for arrhythmias, and thus should not be used for chronic HF treatment [126]. Chronic β-blocker therapy reverses left ventricular remodeling, reduces risk of hospitalization, improves survival, reduces risk of arrhythmias, improves coronary blood flow, and protects the heart against cardiotoxic overstimulation by the catecholamines. All of these effects decrease oxygen and energy demands of the heart and increase its oxygen and energy supply, thereby improving cardiac function and performance [77]. However, β-blockers approved for chronic HF should not be used for acute HF because they cause the acute decrease in cardiac output [127].

4.4.2 Alpha-Adrenergic Blockers

Alpha$_1$-blockers have produced poor results and are considered to be inappropriate for treating HF. Prazosin had worse outcomes than the combined vasodilator therapy of hydralazine and isosorbide dinitrate (BiDil)[128]. Possibly, because prazosin can increase catecholamine levels in a feedback manner, it overrides any potential benefit from α_1AR inhibition-induced vasodilation [129]. In addition to that, the doxazosin arm in the antihypertensive and lipid-lowering treatment to prevent heart attack trial (ALLHAT) was terminated prematurely because of higher HF incidence [130].

4.4.3 Alpha$_2$-Adrenergic Agonists

Activation of α_2ARs inhibits norepinephine release. Clonidine (a centrally acting α_2AR agonist) significantly reduces cardiac and renal sympathetic tones in HF patients without clinical deterioration [131]. However, the drug has not been evaluated by large clinical trials. Moxonidine (an α2AR and imidazoline receptors agonist [132] produces marked decrease in plasma norepinephrine [133]; however, it failed in clinical trials because it increased HF related mortality [134]. One of the explanations might be α_2AR desensitization and downregulation that accompanies HF [114], which limits efficacy of α_2AR agonists.

4.4.4 Exercise Training

Although exercise intolerance is a major symptom of chronic HF, exercise training has been shown to improve hemodynamics and muscle function as well as reduce sympathetic tone resulting in lower all-cause and cardiovascular mortalities [135, 136] The postulated explanations are improvements in arterial and chemoreflex controls, significant decrease in SNS outflow, increase in peripheral blood flow, and decrease in circulating proinflammatory cytokines [137]. Experiments in HF models demonstrated that exercise training improves cardiac βAR signaling and function, improves cardiac contractility and function, and restores normal SNS activity and circulating catecholamine levels [120, 138, 139].

5 Arginine Vasopressin (ADH)

Heart failure is commonly manifested as a syndrome of salt and water retention. One of the very common complications associated with HF, as well as with its treatment, is the development of hyponatremia. Hyponatremia, which is characterized by serum sodium concentration $[Na^+] < 134$ mmol/L, occurs in more than 20% of patients with HF and is an independent predictor of poor outcomes [140]. Hyponatremia in patients with HF is intimately associated with injudicious use of diuretic agents and can limit the use of diuretic therapy [141]. The development of hyponatremia in patients with HF is secondary to both neurohormonal and renal mechanisms that are operative in patients with impaired cardiac function [140, 142]. Essential to this process is the activation of the sympathetic nervous system, the renin–angiotensin–aldosterone axis, and arginine vasopressin (AVP) [143]. The use of RAS antagonists/inhibitors and SNS inhibitor has significantly improved clinical outcomes in HF. However, though excessive secretion of AVP has the potential for deleterious effects on various physiologic processes in HF, the possible benefits of blocking vasopressin in patients with heart failure remains unclear. Recently, the development of specific vasopressin receptor antagonists (aquaretic agents) has led to the potential therapeutic strategy, which is to reduce the volume overload that accompanies HF by directly antagonizing excessive activation of antidiuretic action of AVP. Concomitant with the use of these agents is the ability to treat hyponatremia associated with HF through the selective increase in renal free water excretion [144]. This section reviews the role of AVP as it relates to CHF and we will focus on the role of AVP in the HF and potential benefits of AVP antagonists as new therapeutic targets.

5.1 Physiology of Arginine Vasopressin

AVP, also known as antidiuretic hormone, is a nonapeptide secreted from the posterior pituitary gland. AVP is a nine amino acid peptide synthesized by neurosecretory cells located predominantly in the supraoptic and paraventricular hypothalamic nuclei. These neurons have axons terminating in the neural lobe of the posterior pituitary (neurohypophysis) that release vasopressin and oxytocin [145–147]. Vasopressin is produced in response to two broad classes of stimuli, namely, osmotic and non-osmotic. Osmoreceptors, also known as osmostats, are located in a small, discreet area of the hypothalamus just anterior to the third ventricle. Osmoreceptors are very sensitive to changes in plasma osmolality (± 1–2%). There is a close correlation between plasma osmolality and plasma AVP levels, osmoreceptors can either stimulate or inhibit vasopressin release from the hypothalamus via sensing very small changes in serum osmolality [148]. The osmotic threshold is also modulated by nonosmotic stimuli such as changes in the blood volume and/or pressure. The effects of changes in blood volume, cardiac output, and blood pressure on AVP secretion are mediated through the high- (located in aortic arch, carotid sinus) and low-pressure (located in left atrial) baroreceptors. Baroreceptors sense falls in blood pressure and directly stimulate neurones located in the supraoptic and paraventricular nuclei of the hypothalamus and that leads to the production of vasopressin [148]. Once the baroreceptors are stimulated, AVP production increases in logarithmic fashion and therefore, levels of AVP released in response to hypovolemia (non-osmotic release of AVP) are markedly higher than those achieved by osmotic stimulation.[149] Importantly, in physiologic conditions, the effects of changes in plasma osmolality generally take priority over volume in controlling AVP release, while in pathophysiologic conditions, the non-osmotic release of AVP might supersede the effect of osmolality on the AVP release [150]. Therefore, in patients with HF, the non-osmotic release of AVP is common [151].

Arginine vasopressin has three distinct receptor subtypes: V_{1a}, V_2, V_3 (previously known as V_{1b}). From cardiovascular perspective, the most important receptors are V_{1a} and V_2. V_{1a} receptors are located on the cardiac myocytes [152, 153] as well as vascular smooth muscle cells [154, 155], and hence can be considered as the cardiovascular AVP receptors, with effects on the maintenance and regulation of vascular tone and possibly myocardial function (Fig. 4). These are G_q-protein coupled receptors, which increases intracellular Ca^{2+} levels via the inositol triphosphate pathway (Fig. 4). The V_{1a} receptor has been shown to mediate increased protein synthesis in cardiomyocytes, including the contractile proteins, suggesting a possible role in cardiac hypertrophy and remodeling leading to the HF [152, 156, 157]. Stimulation of V_{1a} receptors in the smooth muscle cells leads in increased vascular smooth muscle cell contraction resulting in increased systemic vascular resistance, increased impedance to ventricular emptying (increased afterload) and thereby adversely affects the ventricular function which contributes to the progression of HF [158, 159]. V_2 receptors are located on the basolateral cells of the renal collecting duct. V_2 receptors are G_s-protein coupled receptors and the intracellular effects of this receptor subtype are mediated by the adenylate cyclase signaling pathway (Fig. 4). Binding of AVP to the V_2 receptors stimulates *de novo* synthesis and "shuttling" of aquaporin 2 water channels (AQP-2) from cytoplasmic vesicles to the luminal surface of the renal collecting duct cells, where they are inserted into the cell membrane and facilitate water transport across the collecting duct cells [160]. The net result is return of water to the circulation, dilution of the plasma volume, and production of concentrated, low-volume urine [161]. This effect may contribute to the volume expansion that exacerbates diastolic wall stress in HF, another mechanism that may contribute to ventricular remodeling and dysfunction [158, 161]. However, in the absence/deficiency of AVP mediated V_2 receptor activation, the collecting duct remains impermeable to water leading to production of the large-volume urine.

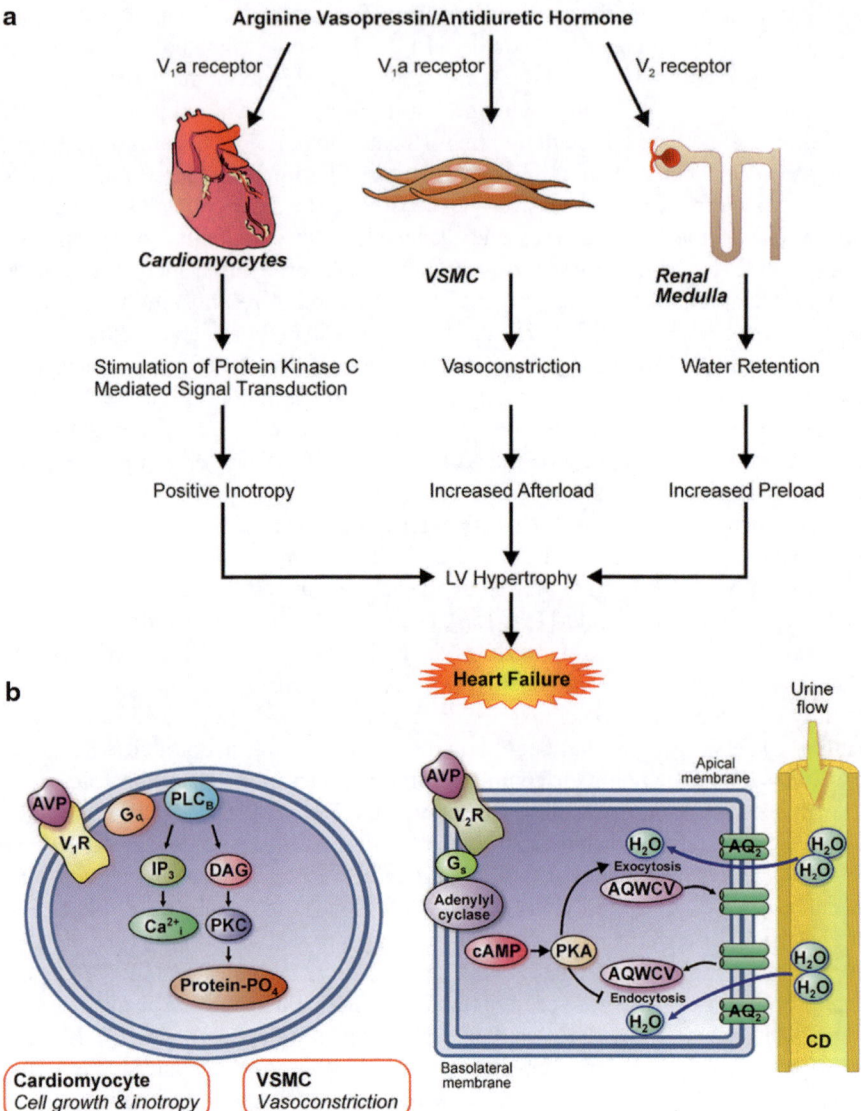

Fig. 4 Physiology of arginine vasopressin and its role in heart failure. Physiological actions of arginine vasopressin (AVP) are mediated via the activation of V_{1a} and V_2 receptors, leading to increased inotropic effects, increased systemic resistance-induced afterload and increased free water retention-induced preload. Elevated circulating AVP levels sustain the ventricular dysfunction via these effects. **a** Arginine vasopressin effects in the myocardial cells and vascular smooth muscle cells are mediated via V_{1a} receptor activation, which act through the G_q-protein coupled receptors activation-induced increase in intracellular Ca^{2+} levels. **b** A separate phosphorylation cascade occurs via diacylglycerol (DAG) and protein kinase C (PKC), which effects into vascular smooth muscle (VSMC) contraction and myocardial cell hypertrophy. Effects of AVP in the renal tubule are mediated via V_2 receptor activation, which act thorugh G_s protein-coupled receptors resulting in the cAMP formation induced-protein kinase A activation. This pathway increases the exocytosis, and inhibits endocytosis, of aquaporin water channel-containing vesicles (AQMCV), resulting in increases in aquaporin 2 (AQ2) channel formation and apical membrane insertion, which leads to increased free water retention

5.2 Role of AVP in Heart Failure

Blood pressures, intracardiac pressures, adrenergic central nervous stimuli and angiotensin II are the dominant non-osmotic factors which regulate AVP release. As discussed before, under normal physiological conditions these non-osmotic factors do not play a dominant role in the regulation of AVP release. However, in the pathophysiological condition, there appears to be a shift in regulation toward relatively greater influence of the non-osmotic mechanisms [162]. Assessment of AVP release after osmotic load with mannitol in the healthy controls and HF patients showed a greater release of AVP in the patients with HF. This implies that in the edematous state, the response to osmotic stimuli occurs at the lower plasma osmolality levels and are more pronounced [162].

Various preclinical and clinical studies have found elevated AVP levels in the plasma with HF. Like other cardiac neurohormones, elevated AVP levels have diagnostic and prognostic value in HF. In the Studies of Left Ventricular Dysfunction (SOLVD), plasma AVP levels were found to be significantly higher in patients with asymptomatic LV dysfunction in comparison to age-matched non-failing controls [163]. Plasma AVP levels were also significantly higher in patients with symptomatic HF compared with patients having asymptomatic LV dysfunction, suggesting the correlation of AVP levels with the progression of HF. The Survival and Ventricular Enlargement (SAVE) trial also showed high vasopressin levels were associated with worsened 1-year cardiovascular mortality [164]. The increased plasma AVP levels observed in HF seems paradoxical. In HF, due to the low cardiac output, baroreceptors perceive the body as being volume depleted. This results in the non-osmotic factor induced AVP release, allowing free water reabsorption by the kidney, via V_2 receptor activation, despite an already edematous state [143]. Thus, these baroreceptor responses on AVP release supersede both the left atrial stretch receptors as well as osmoreceptors. The reason for this shifting is not completely understood [148]. Additionally, recent studies have also demonstrated an increase in the relative density of the AVP-producing neurons in the supraoptic nuclei in patients with HF. This likely explains the chronic stimulation of AVP production in the patients with HF. As discussed above, a number of mechanisms related directly to the physiologic effects of AVP could underlie pathophysiologic contributions to the progression of HF. This chronic elevation of circulating AVP levels in HF patients may have detrimental myocardial effects. Intravenous infusions of AVP hormone in HF patients have shown to increase systemic vascular resistance leading to increased afterload on the ventricles [165], which along with the AVP-induced myocardial hypertrophic effects shown in the preclinical models could be deleterious to the disease progression in the HF patients [152, 157]. Elevated circulating AVP levels might also lead to severe hyponatremia, which is associated with poor outcome in HF, or aggravate the hyponatremia associated with the use of diuretics.

5.3   AVP Antagonists in Heart Failure

Based on the primary role AVP in free water retention and hyponatremia, AVP receptor antagonism would seem to be a rational therapeutic approach for HF. This
therapy would lead to "aquaresis", a selective increase in free water excretion by
the kidneys. AVP receptor antagonism is a novel therapeutic approach aimed at
interfering with the unfavorable actions of AVP in HF. Early investigations of AVP
antagonists involved the peptide analogues and were found effective in increasing
water excretion in the preclinical models. However, in humans, they exhibited poor
bioavailability, lacked a persistent effect, and also had partial agonistic effects, in
addition to the antagonistic effects [166]. Currently, several AVP antagonists are at
various stages of clinical trials for treating hyponatremia and/or HF. These agents
differ on the basis of their degree of specificity for the V_{1a} and V_2 receptors. However, the principal question in selecting the agent of choice for therapeutics of HF
is the selectivity of the AVP antagonist, whether it should be V_{1a} and V_2 receptors
blocker or selective V_2 receptor blocker. Nonpeptide AVP antagonists that target
either the V_2 receptor or a combination of the V_2 and V_{1a} receptors resulting in
aquaresis are now available.

5.3.1   V2 Receptor Antagonists

Tolvaptan, lixivaptan and satavaptan are non-peptide highly selective V_2 receptor
antagonists with the 29:1, 100:1 and 112:1 higher affinity for the binding to V_2 receptor compared to V_{1a} receptor, respectively [145]. They have shown potent aquauretic
properties in preclinical models. These antagonists have shown dose-dependent responses with increased free water clearance, less urinary loss of sodium than furosemide with no effect on serum creatinine [167–169]. Unlike the administration of loop
diuretics, antagonism of V_2 receptors appeared not to increase activation of the RAS.
Tolvaptan has been well studied in the preclinical models as well as clinical trials.
Clinical trial for tolvaptan in the patient with HF, utilizing different doses ranging
from 30 to 90 mg/d, at both the short-term and long-term effects have shown decreased body weight, mainly due to the loss of fluid and retention of Na^+ [168, 170].
The decrease in body weight was not associated with changes in heart rate, blood
pressure, renal function, or development of hypokalemia. However, in the long-term
follow up trial, 60 days in ACTIV in CHF to 9.9 months in EVEREST trial, there
was no difference in the end points of all-cause mortality, cardiovascular death, or
HF hospitalization between the groups [171]. Due to the lack of long-term mortality
assessments, the exact role of these agents in the treatment of HF and hyponatremia
is unclear. However, these agents have been shown to improve symptoms without an
adverse effect profile and thus may have a role in improving quality of life.

5.3.2 V_{1a} and V_2 Receptor Antagonists

Conivaptan and mozavaptan are the non-peptide AVP antagonist with high affinity to both, the V_{1a} and V_2 receptors. Conivaptan has shown to increase both urine volume and sodium concentration compared with placebo in the preclinical models [172]. However, both these agents can be given by oral as well as intravenous administrations; they are only approved for intravenous injections. Conivaptan has been studied for short-term intravenous treatment of euvolemic and hypervolemic hyponatremia and has recently approved by US Food and Drug Administration for this indication. As these agents antagonize the myocardial as well as vascular effects of AVP, it was hypothesized that this agent would reduce afterload and further improve HF symptoms. The effects of conivaptan were assessed in double-blind, placebo-controlled, randomized, multicenter study that enrolled 84 hospitalized patients with euvolemic or hypervolemic hyponatremia. Trial data showed increased Na^+ levels with conivaptan (20 mg-loading dose; 40 or 80 mg/d infusion for 4 days) along with normalization in about 69% patients post 4-days treatment. There was no difference in the incidence of death, side effects, and discontinuations of treatment compared with placebo, however, an increased incidence of adverse effects in dose related infusion-site reactions was present among patients receiving conivaptan versus those receiving placebo [173]. Clinical studies utilizing the V_{1a} and V_2 receptor antagonists in HF are very limited. In an early study focusing on the role of conivaptan in the HF patients, conivaptan showed dose-dependent fall in pulmonary capillary wedge pressure and right atrial pressure, along with a significant increase in urine output. There was no change in cardiac index, systemic/pulmonary vascular resistance, blood pressure, or heart rate in response to the conivaptan [174]. In a pilot, double-blind, multicenter trial with decompensated HF, conivaptan significantly increased urine output with no change in the respiratory symptoms [175, 176]. At this time, the role of combined V_{1a} and V_2 receptor antagonists in the therapy of HF remains undefined.

Currently, the therapeutic use of vasopressin antagonists has shown to safely and effectively reduce body weight in decompensated HF and to normalize sodium levels in hyponatremic patients and needs further investigations including multicenter, long-term clinical trials to demonstrate their effects on outcomes such as mortality and hospitalizations due to HF. These clinical trials will also help answer the important clinical questions such as the indications for AVP antagonists, duration of therapy, and combination therapy with other drugs for HF.

6 Conclusions

Neurohumoral activation plays a central pathogenic event in heart failure and is associated with adverse remodeling of the heart. Antagonism of various neurohumoral pathways has led to the development of successful therapies for heart failure such as inhibitors of the RAS and beta-blockers. However, targeting other pathways such as the natriuretic peptide and arginine vasopressin systems have failed to produce therapies which have improved the morbidity and mortality in heart failure patients. Clearly, more research is needed to enable the discovery of new therapies for heart failure and the use of a personalized and tailored approach can represent an ideal approach as the field moves forward.

Acknowledgements Gavin Y. Oudit is a Clinician-Investigator of the Alberta Innovates-Health Solutions and the Distinguish Clinician Scientist of the Heart and Stroke Foundation of Canada and Canadian Institutes of Health Research. Vaibhav B. Patel is supported by AI-HS Post-Doctoral Fellowship and Vaibhav B. Patel and Nirmal Parajuli are supported by Heart and Stroke Foundation of Canada Fellowships. Our research activities have been supported by operating grants from the Heart and Stroke Foundation of Canada and Canadian Institutes of Health Research.

References

1. Yancy CW, Jessup M, Bozkurt B, Butler J, Casey DE, Jr., Drazner MH, Fonarow GC, Geraci SA, Horwich T, Januzzi JL, Johnson MR, Kasper EK, Levy WC, Masoudi FA, McBride PE, McMurray JJ, Mitchell JE, Peterson PN, Riegel B, Sam F, Stevenson LW, Tang WH, Tsai EJ, Wilkoff BL (2013) 2013 ACCF/AHA Guideline for the management of heart failure: a report of the American College of Cardiology Foundation/American Heart Association Task Force on practice guidelines. Circulation
2. Go AS, Mozaffarian D, Roger VL, Benjamin EJ, Berry JD, Borden WB, Bravata DM, Dai S, Ford ES, Fox CS, Franco S, Fullerton HJ, Gillespie C, Hailpern SM, Heit JA, Howard VJ, Huffman MD, Kissela BM, Kittner SJ, Lackland DT, Lichtman JH, Lisabeth LD, Magid D, Marcus GM, Marelli A, Matchar DB, McGuire DK, Mohler ER, Moy CS, Mussolino ME, Nichol G, Paynter NP, Schreiner PJ, Sorlie PD, Stein J, Turan TN, Virani SS, Wong ND, Woo D, Turner MB (2013) Heart disease and stroke statistics—2013 update: a report from the American Heart Association. Circulation 127(1):e6–e245.
3. Scannell JW, Blanckley A, Boldon H, Warrington B (2012) Diagnosing the decline in pharmaceutical R & D efficiency. Nat Rev Drug Discov 11(3):191–200.
4. Editorial (2012) A marriage of convenience. Nature Medicine 18(4):469–470
5. The CONSENSUS Trial Study Group (1987) Effects of enalapril on mortality in severe congestive heart failure. Results of the Cooperative North Scandinavian Enalapril Survival Study (CONSENSUS). N Engl J Med 316(23):1429–1435.
6. McMurray JJ (2011) CONSENSUS to EMPHASIS: the overwhelming evidence which makes blockade of the renin-angiotensin-aldosterone system the cornerstone of therapy for systolic heart failure. Eur J Heart Fail 13(9):929–936.
7. Lin RC, Weeks KL, Gao XM, Williams RB, Bernardo BC, Kiriazis H, Matthews VB, Woodcock EA, Bouwman RD, Mollica JP, Speirs HJ, Dawes IW, Daly RJ, Shioi T, Izumo S, Febbraio MA, Du XJ, McMullen JR (2010) PI3K(p110 alpha) protects against myocardial infarction-induced heart failure: identification of PI3K-regulated miRNA and mRNA. Arterioscler Thromb Vasc Biol 30(4):724–732.
8. Shah AM, Mann DL (2011) In search of new therapeutic targets and strategies for heart failure: recent advances in basic science. Lancet 378(9792):704–712.
9. Mewton N, Liu CY, Croisille P, Bluemke D, Lima JA (2011) Assessment of myocardial fibrosis with cardiovascular magnetic resonance. J Am Coll Cardiol 57(8):891–903.
10. Kuwahara K, Kinoshita H, Kuwabara Y, Nakagawa Y, Usami S, Minami T, Yamada Y, Fujiwara M, Nakao K (2010) Myocardin-related transcription factor A is a common mediator of mechanical stress- and neurohumoral stimulation-induced cardiac hypertrophic signaling leading to activation of brain natriuretic peptide gene expression. Mol Cell Biol 30(17):4134–4148.
11. Rohini A, Agrawal N, Koyani CN, Singh R (2010) Molecular targets and regulators of cardiac hypertrophy. Pharmacol Res 61(4):269–280.

12. McMurray JJ (2010) Clinical practice. Systolic heart failure. N Engl J Med 362(3):228–238.
13. Fujita K, Maeda N, Sonoda M, Ohashi K, Hibuse T, Nishizawa H, Nishida M, Hiuge A, Kurata A, Kihara S, Shimomura I, Funahashi T (2008) Adiponectin protects against angiotensin II-induced cardiac fibrosis through activation of PPAR-alpha. Arterioscler Thromb Vasc Biol 28 (5):863–870.
14. Prasad A, Quyyumi AA (2004) Renin-angiotensin system and angiotensin receptor blockers in the metabolic syndrome. Circulation 110(11):1507–1512.
15. Zhong J, Basu R, Guo D, Chow FL, Byrns S, Schuster M, Loibner H, Wang XH, Penninger JM, Kassiri Z, Oudit GY (2010) Angiotensin-converting enzyme 2 suppresses pathological hypertrophy, myocardial fibrosis, and cardiac dysfunction. Circulation 122(7):717–728. (718 p following 728).
16. Pfeufer A, Osterziel KJ, Urata H, Borck G, Schuster H, Wienker T, Dietz R, Luft FC (1996) Angiotensin-converting enzyme and heart chymase gene polymorphisms in hypertrophic cardiomyopathy. Am J Cardiol 78(3):362–364
17. Kaschina E, Unger T (2003) Angiotensin AT1/AT2 receptors: regulation, signalling and function. Blood press 12(2):70–88
18. George AJ, Thomas WG, Hannan RD (2010) The renin-angiotensin system and cancer: old dog, new tricks. Nat Rev Cancer 10(11):745–759.
19. Mehta PK, Griendling KK (2007) Angiotensin II cell signaling: physiological and pathological effects in the cardiovascular system. Am J Physiol Cell Physiol 292(1):C82–97.
20. Fang F, Liu GC, Zhou X, Yang S, Reich HN, Williams V, Hu A, Pan J, Konvalinka A, Oudit GY, Scholey JW, John R (2013) Loss of ACE2 exacerbates murine renal ischemia-reperfusion injury. PloS ONE 8(8):e71433
21. Wang W, Bodiga S, Das SK, Lo J, Patel V, Oudit GY (2012) Role of ACE2 in diastolic and systolic heart failure. Heart Fail Rev 17(4–5):683–691
22. Connell JM, Davies E (2005) The new biology of aldosterone. J Endocrinol 186(1):1–20
23. Neves MF, Amiri F, Virdis A, Diep QN, Schiffrin EL (2005) Role of aldosterone in angiotensin II-induced cardiac and aortic inflammation, fibrosis, and hypertrophy. Can J Physiol Pharmacol 83(11):999–1006
24. Berk BC, Fujiwara K, Lehoux S (2007) ECM remodeling in hypertensive heart disease. J Clin Invest 117(3):568–575
25. Brown NJ (2008) Aldosterone and vascular inflammation. Hypertension 51(2):161–167
26. Schiffrin EL (2006) Effects of aldosterone on the vasculature. Hypertension 47(3):312–318
27. Grobe JL, Mecca AP, Mao H, Katovich MJ (2006) Chronic angiotensin-(1-7) prevents cardiac fibrosis in DOCA-salt model of hypertension. AmJ Physiol Heart Circ Physiol 290(6):H2417–H2423
28. Rosenkranz S (2004) TGF-beta1 and angiotensin networking in cardiac remodeling. Cardiovasc Res 63(3):423–432
29. Ichihara S, Senbonmatsu T, Price E Jr, Ichiki T, Gaffney FA, Inagami T (2001) Angiotensin II type 2 receptor is essential for left ventricular hypertrophy and cardiac fibrosis in chronic angiotensin II-induced hypertension. Circulation 104(3):346–351
30. Briones AM, Rodriguez-Criado N, Hernanz R, Garcia-Redondo AB, Rodrigues-Diez RR, Alonso MJ, Egido J, Ruiz-Ortega M, Salaices M (2009) Atorvastatin prevents angiotensin II-induced vascular remodeling and oxidative stress. Hypertension 54(1):142–149
31. Chen K, Chen J, Li D, Zhang X, Mehta JL (2004) Angiotensin II regulation of collagen type I expression in cardiac fibroblasts: modulation by PPAR-gamma ligand pioglitazone. Hypertension 44(5):655–661
32. Takimoto E, Kass DA (2007) Role of oxidative stress in cardiac hypertrophy and remodeling. Hypertension 49(2):241–248
33. Ramprasath T, Selvam GS (2013) Potential impact of genetic variants in Nrf2 regulated antioxidant genes and risk prediction of diabetes and associated cardiac complications. Curr Med Chem 20(37):4680–4693
34. Ramprasath T, Senthamizharasi M, Vasudevan V, Sasikumar S, Yuvaraj S, Selvam GS (2014) Naringenin confers protection against oxidative stress through upregulation of Nrf2 target genes in cardiomyoblast cells. J Physiol Biochem 70(2):407–15

35. Virdis A, Neves MF, Amiri F, Touyz RM, Schiffrin EL (2004) Role of NAD(P)H oxidase on vascular alterations in angiotensin II-infused mice. J Hypertens 22(3):535–542
36. Rey FE, Cifuentes ME, Kiarash A, Quinn MT, Pagano PJ (2001) Novel competitive inhibitor of NAD(P)H oxidase assembly attenuates vascular O(2)(-) and systolic blood pressure in mice. Circ Res 89(5):408–414
37. Touyz RM, Mercure C, He Y, Javeshghani D, Yao G, Callera GE, Yogi A, Lochard N, Reudelhuber TL (2005) Angiotensin II-dependent chronic hypertension and cardiac hypertrophy are unaffected by gp91phox-containing NADPH oxidase. Hypertension 45(4):530–537
38. Chen K, Mehta JL, Li D, Joseph L, Joseph J (2004) Transforming growth factor beta receptor endoglin is expressed in cardiac fibroblasts and modulates profibrogenic actions of angiotensin II. Circ Res 95(12):1167–1173
39. Iwanciw D, Rehm M, Porst M, Goppelt-Struebe M (2003) Induction of connective tissue growth factor by angiotensin II: integration of signaling pathways. Arterioscler Thromb Vasc Biol 23(10):1782–1787
40. Gradman AH, Schmieder RE, Lins RL, Nussberger J, Chiang Y, Bedigian MP (2005) Aliskiren, a novel orally effective renin inhibitor, provides dose-dependent antihypertensive efficacy and placebo-like tolerability in hypertensive patients. Circulation 111(8):1012–1018
41. Teo K, Yusuf S, Sleight P, Anderson C, Mookadam F, Ramos B, Hilbrich L, Pogue J, Schumacher H (2004) Rationale, design, and baseline characteristics of 2 large, simple, randomized trials evaluating telmisartan, ramipril, and their combination in high-risk patients: the Ongoing Telmisartan Alone and in Combination with Ramipril Global Endpoint Trial/Telmisartan Randomized Assessment Study in ACE Intolerant Subjects with Cardiovascular Disease (ONTARGET/TRANSCEND) trials. Am Heart J 148(1):52–61
42. Yusuf S, Sleight P, Pogue J, Bosch J, Davies R, Dagenais G (2000) Effects of an angiotensin-converting-enzyme inhibitor, ramipril, on cardiovascular events in high-risk patients. The Heart Outcomes Prevention Evaluation Study Investigators. N Engl J Med 342(3):145–153
43. Fyhrquist F, Saijonmaa O (2008) Renin-angiotensin system revisited. J Intern Med 264(3):224–236
44. Krum H, Driscoll A (2013) Management of heart failure. Med J Aust 199(5):334–339
45. Yahiro E, Miura S, Imaizumi S, Uehara Y, Saku K (2013) Chymase inhibitors. Curr Pharm Des 19(17):3065–3071
46. Shearer F, Lang CC, Struthers AD (2013) Renin-angiotensin-aldosterone system inhibitors in heart failure. Clin Pharmacol Ther 94(4):459–467
47. Braunwald E, Domanski MJ, Fowler SE, Geller NL, Gersh BJ, Hsia J, Pfeffer MA, Rice MM, Rosenberg YD, Rouleau JL (2004) Angiotensin-converting-enzyme inhibition in stable coronary artery disease. N Engl J Med 351(20):2058–2068
48. Benge CD, Muldowney JA 3rd (2012) The pharmacokinetics and pharmacodynamics of valsartan in the post-myocardial infarction population. Expert Opin Drug Metab Toxicol 8(11):1469–1482
49. Suzuki T, Yamazaki T, Yazaki Y (2001) The role of the natriuretic peptides in the cardiovascular system. Cardiovasc Res 51(3):489–494
50. Potter LR, Yoder AR, Flora DR, Antos LK, Dickey DM (2009) Natriuretic peptides: their structures, receptors, physiologic functions and therapeutic applications. Handb Exp Pharmacol 191:341–366
51. D'Alessandro R, Masarone D, Buono A, Gravino R, Rea A, Salerno G, Golia E, Ammendola E, Del Giorno G, Santangelo L, Russo MG, Calabro R, Bossone E, Pacileo G, Limongelli G (2013) Natriuretic peptides: molecular biology, pathophysiology and clinical implications for the cardiologist. Future Cardiol 9(4):519–534
52. Rubattu S, Sciarretta S, Valenti V, Stanzione R, Volpe M (2008) Natriuretic peptides: an update on bioactivity, potential therapeutic use, and implication in cardiovascular diseases. Am J Hypertens 21(7):733–741
53. Gao P, Huang L (2009) New insights into the role of natriuretic peptides in the regulation of apoptosis in cardiovascular system. Saudi Med J 30(5):595–604

54. Kim HN, Januzzi JL Jr (2011) Natriuretic peptide testing in heart failure. Circulation 123(18):2015–2019
55. Wei CM, Heublein DM, Perrella MA, Lerman A, Rodeheffer RJ, McGregor CG, Edwards WD, Schaff HV, Burnett JC Jr (1993) Natriuretic peptide system in human heart failure. Circulation 88(3):1004–1009
56. Federico C (2010) Natriuretic Peptide system and cardiovascular disease. Heart Views 11(1):10–15
57. McFarlane SI, Winer N, Sowers JR (2003) Role of the natriuretic peptide system in cardiorenal protection. Arch Intern Med 163(22):2696–2704
58. Mair J (2002) Role of cardiac natriuretic peptide testing in heart failure. Clin Chem 48(7):977–978
59. Daniels LB, Maisel AS (2007) Natriuretic peptides. J Am Coll Cardiol 50(25):2357–2368
60. Kousholt BS (2012) Natriuretic peptides as therapy in cardiac ischaemia/reperfusion. Dan Med J 59(6):B4469
61. Calvieri C, Rubattu S, Volpe M (2012) Molecular mechanisms underlying cardiac antihypertrophic and antifibrotic effects of natriuretic peptides. J Mol Med 90(1):5–13
62. Nakanishi M, Saito Y, Kishimoto I, Harada M, Kuwahara K, Takahashi N, Kawakami R, Nakagawa Y, Tanimoto K, Yasuno S, Usami S, Li Y, Adachi Y, Fukamizu A, Garbers DL, Nakao K (2005) Role of natriuretic peptide receptor guanylyl cyclase-A in myocardial infarction evaluated using genetically engineered mice. Hypertension 46(2):441–447
63. Nakao K, Ogawa Y, Suga S, Imura H (1992) Molecular biology and biochemistry of the natriuretic peptide system. I: Natriuretic peptides. J Hypertens 10(9):907–912
64. Takei Y (2000) Structural and functional evolution of the natriuretic peptide system in vertebrates. Int Rev Cytol 194:1–66
65. Gardner DG, Chen S, Glenn DJ, Grigsby CL (2007) Molecular biology of the natriuretic peptide system: implications for physiology and hypertension. Hypertension 49(3):419–426
66. Levin ER (1991) Atrial natriuretic peptide and endothelin: Interactions in the central nervous system and the periphery. Mol Cell Neurosci 2(3):189–201
67. Azizov VA, Muradova SR (2001) Atrial natriuretic peptide and cardiovascular system. Anadolu Kardiyol Derg 1(4):297–300
68. Kapoun AM, Liang F, O'Young G, Damm DL, Quon D, White RT, Munson K, Lam A, Schreiner GF, Protter AA (2004) B-type natriuretic peptide exerts broad functional opposition to transforming growth factor-beta in primary human cardiac fibroblasts: fibrosis, myofibroblast conversion, proliferation, and inflammation. Circ Res 94(4):453–461
69. Nunez DJ, Dickson MC, Brown MJ (1992) Natriuretic peptide receptor mRNAs in the rat and human heart. J Clin Invest 90(5):1966–1971
70. Tamura N, Ogawa Y, Chusho H, Nakamura K, Nakao K, Suda M, Kasahara M, Hashimoto R, Katsuura G, Mukoyama M, Itoh H, Saito Y, Tanaka I, Otani H, Katsuki M (2000) Cardiac fibrosis in mice lacking brain natriuretic peptide. Proc Natl Acad Sci U S A 97(8):4239–4244
71. Knowles JW, Esposito G, Mao L, Hagaman JR, Fox JE, Smithies O, Rockman HA, Maeda N (2001) Pressure-independent enhancement of cardiac hypertrophy in natriuretic peptide receptor A-deficient mice. J Clin Invest 107(8):975–984
72. Ledwidge M, Gallagher J, Conlon C, Tallon E, O'Connell E, Dawkins I, Watson C, O'Hanlon R, Bermingham M, Patle A, Badabhagni MR, Murtagh G, Voon V, Tilson L, Barry M, McDonald L, Maurer B, McDonald K (2013) Natriuretic peptide-based screening and collaborative care for heart failure: the STOP-HF randomized trial. JAMA 310(1):66–74
73. Armstrong PW, Rouleau JL (2008) A Canadian context for the Acute Study of Clinical Effectiveness of Nesiritide and Decompensated Heart Failure (ASCEND-HF) trial. Can J Cardiol 24(Suppl B):30B–32B
74. Triposkiadis F, Karayannis G, Giamouzis G, Skoularigis J, Louridas G, Butler J (2009) The sympathetic nervous system in heart failure physiology, pathophysiology, and clinical implications. J Am Coll Cardiol 54(19):1747–1762
75. Zipes DP (2008) Heart-brain interactions in cardiac arrhythmias: role of the autonomic nervous system. Cleve Clin J Med 75(Suppl 2):S94–96

76. Eaton MJ, Duplan H (2004) Useful cell lines derived from the adrenal medulla. Mol Cell Endocrinol 228(1–2):39–52
77. Lymperopoulos A, Rengo G, Koch WJ (2013) Adrenergic nervous system in heart failure: pathophysiology and therapy. Circ Res 113(6):739–753
78. Bylund DB, Eikenberg DC, Hieble JP, Langer SZ, Lefkowitz RJ, Minneman KP, Molinoff PB, Ruffolo RR Jr, Trendelenburg U (1994) International Union of Pharmacology nomenclature of adrenoceptors. Pharmacol Rev 46(2):121–136
79. Lymperopoulos A, Rengo G, Koch WJ (2012) GRK2 inhibition in heart failure: something old, something new. Curr Pharm Des 18(2):186–191
80. Brodde OE (1993) Beta-adrenoceptors in cardiac disease. Pharmacol Ther 60(3):405–430
81. Johnson JA, Liggett SB (2011) Cardiovascular pharmacogenomics of adrenergic receptor signaling: clinical implications and future directions. Clin Pharmacol Ther 89(3):366–378
82. Shannon R, Chaudhry M (2006) Effect of alpha1-adrenergic receptors in cardiac pathophysiology. Am Heart J 152(5):842–850
83. Hein L, Altman JD, Kobilka BK (1999) Two functionally distinct alpha2-adrenergic receptors regulate sympathetic neurotransmission. Nature 402(6758):181–184
84. Brede M, Wiesmann F, Jahns R, Hadamek K, Arnolt C, Neubauer S, Lohse MJ, Hein L (2002) Feedback inhibition of catecholamine release by two different alpha2-adrenoceptor subtypes prevents progression of heart failure. Circulation 106(19):2491–2496
85. Brede M, Nagy G, Philipp M, Sorensen JB, Lohse MJ, Hein L (2003) Differential control of adrenal and sympathetic catecholamine release by alpha 2-adrenoceptor subtypes. Mol Endocrinol 17(8):1640–1646
86. Philipp M, Brede M, Hein L (2002) Physiological significance of alpha(2)-adrenergic receptor subtype diversity: one receptor is not enough. Am J Physiol Regul Integr Comp Physiol 283(2):R287–R295
87. Philipp M, Hein L (2004) Adrenergic receptor knockout mice: distinct functions of 9 receptor subtypes. Pharmacol Ther 101(1):65–74
88. Ferguson SS (2001) Evolving concepts in G protein-coupled receptor endocytosis: the role in receptor desensitization and signaling. Pharmacol Rev 53(1):1–24
89. Reiter E, Lefkowitz RJ (2006) GRKs and beta-arrestins: roles in receptor silencing, trafficking and signaling. Trends Endocrinol Metab: TEM 17(4):159–165
90. Arriza JL, Dawson TM, Simerly RB, Martin LJ, Caron MG, Snyder SH, Lefkowitz RJ (1992) The G-protein-coupled receptor kinases beta ARK1 and beta ARK2 are widely distributed at synapses in rat brain. J Neurosci 12(10):4045–4055
91. Rockman HA, Koch WJ, Lefkowitz RJ (2002) Seven-transmembrane-spanning receptors and heart function. Nature 415(6868):206–212
92. Sumners C, Raizada MK (1986) Angiotensin II stimulates norepinephrine uptake in hypothalamus-brain stem neuronal cultures. AmJ Physiol 250(2 Pt 1):C236–C244
93. Weber MA, Purdy RE (1982) Catecholamine-mediated constrictor effects of aldosterone on vascular smooth muscle. Life Sci 30(23):2009–2017
94. Weber KT (2001) Aldosterone in congestive heart failure. N Engl J Med 345(23):1689–1697
95. Lymperopoulos A, Rengo G, Koch WJ (2007) Adrenal adrenoceptors in heart failure: fine-tuning cardiac stimulation. Trends Mol Med 13(12):503–511
96. Mialet Perez J, Rathz DA, Petrashevskaya NN, Hahn HS, Wagoner LE, Schwartz A, Dorn GW, Liggett SB (2003) Beta 1-adrenergic receptor polymorphisms confer differential function and predisposition to heart failure. Nat Med 9(10):1300–1305
97. Chemello D, Rohde LE, Santos KG, Silvello D, Goldraich L, Pimentel M, Rosa PR, Zimerman L, Clausell N (2010) Genetic polymorphisms of the adrenergic system and implantable cardioverter-defibrillator therapies in patients with heart failure. Europace 12(5):686–691
98. Levin MC, Marullo S, Muntaner O, Andersson B, Magnusson Y (2002) The myocardium-protective Gly-49 variant of the beta 1-adrenergic receptor exhibits constitutive activity and increased desensitization and down-regulation. J Biol Chem 277(34):30429–30435

99. Borjesson M, Magnusson Y, Hjalmarson A, Andersson B (2000) A novel polymorphism in the gene coding for the beta(1)-adrenergic receptor associated with survival in patients with heart failure. Eur Heart J 21(22):1853–1858

100. Terra SG, Hamilton KK, Pauly DF, Lee CR, Patterson JH, Adams KF, Schofield RS, Belgado BS, Hill JA, Aranda JM, Yarandi HN, Johnson JA (2005) Beta1-adrenergic receptor polymorphisms and left ventricular remodeling changes in response to beta-blocker therapy. Pharmacogenet Genomics 15(4):227–234

101. Sehnert AJ, Daniels SE, Elashoff M, Wingrove JA, Burrow CR, Horne B, Muhlestein JB, Donahue M, Liggett SB, Anderson JL, Kraus WE (2008) Lack of association between adrenergic receptor genotypes and survival in heart failure patients treated with carvedilol or metoprolol. J Am Coll Cardiol 52(8):644–651

102. de Groote P, Helbecque N, Lamblin N, Hermant X, Fadden E M, Foucher-Hossein C, Amouyel P, Dallongeville J, Bauters C (2005) Association between beta-1 and beta-2 adrenergic receptor gene polymorphisms and the response to beta-blockade in patients with stable congestive heart failure. Pharmacogenet Genomics 15(3):137–142

103. White HL, de Boer RA, Maqbool A, Greenwood D, van Veldhuisen DJ, Cuthbert R, Ball SG, Hall AS, Balmforth AJ (2003) An evaluation of the beta-1 adrenergic receptor Arg389Gly polymorphism in individuals with heart failure: a MERIT-HF sub-study. Eur J Heart Fail 5(4):463–468

104. Small KM, McGraw DW, Liggett SB (2003) Pharmacology and physiology of human adrenergic receptor polymorphisms. Annu Rev Pharmacol Toxicol 43:381–411

105. Small KM, Wagoner LE, Levin AM, Kardia SL, Liggett SB (2002) Synergistic polymorphisms of beta1- and alpha2C-adrenergic receptors and the risk of congestive heart failure. N Engl J Med 347(15):1135–1142

106. O'Connor CM, Fiuzat M, Carson PE, Anand IS, Plehn JF, Gottlieb SS, Silver MA, Lindenfeld J, Miller AB, White M, Walsh R, Nelson P, Medway A, Davis G, Robertson AD, Port JD, Carr J, Murphy GA, Lazzeroni LC, Abraham WT, Liggett SB, Bristow MR (2012) Combinatorial pharmacogenetic interactions of bucindolol and beta1, alpha2C adrenergic receptor polymorphisms. PloS ONE 7(10):e44324

107. Aggarwal A, Esler MD, Lambert GW, Hastings J, Johnston L, Kaye DM (2002) Norepinephrine turnover is increased in suprabulbar subcortical brain regions and is related to whole-body sympathetic activity in human heart failure. Circulation 105(9):1031–1033

108. Morris MJ, Cox HS, Lambert GW, Kaye DM, Jennings GL, Meredith IT, Esler MD (1997) Region-specific neuropeptide Y overflows at rest and during sympathetic activation in humans. Hypertension 29(1 Pt 1):137–143

109. Wang H, Huang BS, Ganten D, Leenen FH (2004) Prevention of sympathetic and cardiac dysfunction after myocardial infarction in transgenic rats deficient in brain angiotensinogen. Circ Res 94(6):843

110. Lindley TE, Doobay MF, Sharma RV, Davisson RL (2004) Superoxide is involved in the central nervous system activation and sympathoexcitation of myocardial infarction-induced heart failure. Circ Res 94(3):402–409

111. Bristow MR, Ginsburg R, Minobe W, Cubicciotti RS, Sageman WS, Lurie K, Billingham ME, Harrison DC, Stinson EB (1982) Decreased catecholamine sensitivity and beta-adrenergic-receptor density in failing human hearts. N Engl J Med 307(4):205–211

112. Bristow MR, Ginsburg R, Umans V, Fowler M, Minobe W, Rasmussen R, Zera P, Menlove R, Shah P, Jamieson S et al (1986) Beta 1- and beta 2-adrenergic-receptor subpopulations in nonfailing and failing human ventricular myocardium: coupling of both receptor subtypes to muscle contraction and selective beta 1-receptor down-regulation in heart failure. Circ Res 59(3):297–309

113. Nikolaev VO, Moshkov A, Lyon AR, Miragoli M, Novak P, Paur H, Lohse MJ, Korchev YE, Harding SE, Gorelik J (2010) Beta2-adrenergic receptor redistribution in heart failure changes cAMP compartmentation. Science 327(5973):1653–1657

114. Aggarwal A, Esler MD, Socratous F, Kaye DM (2001) Evidence for functional presynaptic alpha-2 adrenoceptors and their down-regulation in human heart failure. J Am Coll Cardiol 37(5):1246–1251

115. Neumeister A, Charney DS, Belfer I, Geraci M, Holmes C, Sharabi Y, Alim T, Bonne O, Luckenbaugh DA, Manji H, Goldman D, Goldstein DS (2005) Sympathoneural and adrenomedullary functional effects of alpha2C-adrenoreceptor gene polymorphism in healthy humans. Pharmacogenet Genomics 15(3):143–149

116. Lang CC, Stein CM, Nelson RA, He HB, Belas FJ, Blair IA, Wood M, Wood AJ (1997) Sympathoinhibitory response to clonidine is blunted in patients with heart failure. Hypertension 30(3 Pt 1):392–397

117. Rengo G, Lymperopoulos A, Leosco D, Koch WJ (2011) GRK2 as a novel gene therapy target in heart failure. J Mol Cell Cardiol 50(5):785–792

118. Rengo G, Perrone-Filardi P, Femminella GD, Liccardo D, Zincarelli C, de Lucia C, Pagano G, Marsico F, Lymperopoulos A, Leosco D (2012) Targeting the beta-adrenergic receptor system through G-protein-coupled receptor kinase 2: a new paradigm for therapy and prognostic evaluation in heart failure: from bench to bedside. Circ Heart Fail 5(3):385–391

119. Penn RB, Pronin AN, Benovic JL (2000) Regulation of G protein-coupled receptor kinases. Trends Cardiovasc Med 10(2):81–89

120. Lymperopoulos A, Rengo G, Funakoshi H, Eckhart AD, Koch WJ (2007) Adrenal GRK2 upregulation mediates sympathetic overdrive in heart failure. Nat Med 13(3):315–323

121. Lymperopoulos A, Rengo G, Gao E, Ebert SN, Dorn GW 2nd, Koch WJ (2010) Reduction of sympathetic activity via adrenal-targeted GRK2 gene deletion attenuates heart failure progression and improves cardiac function after myocardial infarction. J Biol Chem 285(21):16378–16386

122. Lymperopoulos A, Rengo G, Zincarelli C, Soltys S, Koch WJ (2008) Modulation of adrenal catecholamine secretion by in vivo gene transfer and manipulation of G protein-coupled receptor kinase-2 activity. Mol Ther 16(2):302–307

123. Rengo G, Leosco D, Zincarelli C, Marchese M, Corbi G, Liccardo D, Filippelli A, Ferrara N, Lisanti MP, Koch WJ, Lymperopoulos A (2010) Adrenal GRK2 lowering is an underlying mechanism for the beneficial sympathetic effects of exercise training in heart failure. AmJ Physiol Heart Circ Physiol 298(6):H2032–H2038

124. Rengo G, Lymperopoulos A, Zincarelli C, Femminella G, Liccardo D, Pagano G, de Lucia C, Cannavo A, Gargiulo P, Ferrara N, Perrone Filardi P, Koch W, Leosco D (2012) Blockade of beta-adrenoceptors restores the GRK2-mediated adrenal alpha(2) -adrenoceptor-catecholamine production axis in heart failure. Br J Pharmacol 166(8):2430–2440

125. Lopez-Sendon J, Swedberg K, McMurray J, Tamargo J, Maggioni AP, Dargie H, Tendera M, Waagstein F, Kjekshus J, Lechat P, Torp-Pedersen C (2004) Expert consensus document on beta-adrenergic receptor blockers. Eur Heart J 25(15):1341–1362

126. Kaumann AJ, Molenaar P (2008) The low-affinity site of the beta1-adrenoceptor and its relevance to cardiovascular pharmacology. Pharmacol Ther 118(3):303–336

127. Bristow M (2003) Antiadrenergic therapy of chronic heart failure: surprises and new opportunities. Circulation 107(8):1100–1102

128. Cohn JN, Archibald DG, Ziesche S, Franciosa JA, Harston WE, Tristani FE, Dunkman WB, Jacobs W, Francis GS, Flohr KH et al (1986) Effect of vasodilator therapy on mortality in chronic congestive heart failure. Results of a Veterans Administration Cooperative Study. N Engl J Med 314(24):1547–1552

129. Colucci WS, Williams GH, Braunwald E (1980) Increased plasma norepinephrine levels during prazosin therapy for severe congestive heart failure. Ann Intern Med 93(3):452–453

130. ALLHAT Collaborative Research Group (2000) Major cardiovascular events in hypertensive patients randomized to doxazosin vs chlorthalidone: the antihypertensive and lipid-lowering treatment to prevent heart attack trial (ALLHAT). JAMA 283(15):1967–1975

131. Giles TD, Thomas MG, Quiroz A, Rice JC, Plauche W, Sander GE (1987) Acute and short-term effects of clonidine in heart failure. Angiology 38(7):537–548

132. Ernsberger P, Meeley MP, Reis DJ (1988) An endogenous substance with clonidine-like properties: selective binding to imidazole sites in the ventrolateral medulla. Brain Res 441(1–2):309–318

133. Swedberg K, Bergh CH, Dickstein K, McNay J, Steinberg M (2000) The effects of moxonidine, a novel imidazoline, on plasma norepinephrine in patients with congestive heart failure. Moxonidine Investigators. J Am Coll Cardiol 35(2):398–404

134. Cohn JN, Pfeffer MA, Rouleau J, Sharpe N, Swedberg K, Straub M, Wiltse C, Wright TJ (2003) Adverse mortality effect of central sympathetic inhibition with sustained-release moxonidine in patients with heart failure (MOXCON). Eur J Heart Fail 5(5):659–667

135. Fraga R, Franco FG, Roveda F, de Matos LN, Braga AM, Rondon MU, Rotta DR, Brum PC, Barretto AC, Middlekauff HR, Negrao CE (2007) Exercise training reduces sympathetic nerve activity in heart failure patients treated with carvedilol. Eur J Heart Fail 9(6–7):630–636

136. O'Connor CM, Whellan DJ, Lee KL, Keteyian SJ, Cooper LS, Ellis SJ, Leifer ES, Kraus WE, Kitzman DW, Blumenthal JA, Rendall DS, Miller NH, Fleg JL, Schulman KA, McKelvie RS, Zannad F, Pina IL (2009) Efficacy and safety of exercise training in patients with chronic heart failure: HF-ACTION randomized controlled trial. JAMA 301(14):1439–1450

137. Adams V, Doring C, Schuler G (2008) Impact of physical exercise on alterations in the skeletal muscle in patients with chronic heart failure. Front Biosci 13:302–311

138. Leosco D, Rengo G, Iaccarino G, Filippelli A, Lymperopoulos A, Zincarelli C, Fortunato F, Golino L, Marchese M, Esposito G, Rapacciuolo A, Rinaldi B, Ferrara N, Koch WJ, Rengo F (2007) Exercise training and beta-blocker treatment ameliorate age-dependent impairment of beta-adrenergic receptor signaling and enhance cardiac responsiveness to adrenergic stimulation. Am J Physiol Heart Circ Physiol 293(3):H1596–H1603

139. Leosco D, Rengo G, Iaccarino G, Golino L, Marchese M, Fortunato F, Zincarelli C, Sanzari E, Ciccarelli M, Galasso G, Altobelli GG, Conti V, Matrone G, Cimini V, Ferrara N, Filippelli A, Koch WJ, Rengo F (2008) Exercise promotes angiogenesis and improves beta-adrenergic receptor signalling in the post-ischaemic failing rat heart. Cardiovasc Res 78(2):385–394

140. De Luca L, Klein L, Udelson JE, Orlandi C, Sardella G, Fedele F, Gheorghiade M (2005) Hyponatremia in patients with heart failure. Am J Cardiol 96(12 A):19L–23L

141. Sica DA (2006) Sodium and water retention in heart failure and diuretic therapy: basic mechanisms. Cleve Clin J Med 73(Suppl 2):S2–7. Discussion S30–S33

142. Goldberg A, Hammerman H, Petcherski S, Nassar M, Zdorovyak A, Yalonetsky S, Kapeliovich M, Agmon Y, Beyar R, Markiewicz W, Aronson D (2006) Hyponatremia and long-term mortality in survivors of acute ST-elevation myocardial infarction. Arch Intern Med 166(7):781–786

143. Schrier RW, Abraham WT (1999) Hormones and hemodynamics in heart failure. N Engl J Med 341(8):577–585

144. Lee CR, Watkins ML, Patterson JH, Gattis W, O'Connor CM, Gheorghiade M, Adams KF Jr (2003) Vasopressin: a new target for the treatment of heart failure. Am Heart J 146(1):9–18

145. Finley JJt, Konstam MA, Udelson JE (2008) Arginine vasopressin antagonists for the treatment of heart failure and hyponatremia. Circulation 118(4):410–421

146. Bankir L (2001) Antidiuretic action of vasopressin: quantitative aspects and interaction between V1a and V2 receptor-mediated effects. Cardiovasc Res 51(3):372–390

147. Thibonnier M (2003) Vasopressin receptor antagonists in heart failure. Curr Opin Pharmacol 3(6):683–687

148. Sanghi P, Uretsky BF, Schwarz ER (2005) Vasopressin antagonism: a future treatment option in heart failure. Eur Heart J 26(6):538–543

149. Kamath SA, Laskar SR, Yancy CW (2005) Novel therapies for heart failure: vasopressin and selective aldosterone antagonists. Congest Heart Fail 11(1):21–29

150. Ishikawa SE, Schrier RW (2003) Pathophysiological roles of arginine vasopressin and aquaporin-2 in impaired water excretion. Clin Endocrinol (Oxf) 58(1):1–17

151. Rosner MH, Ronco C (2010) Hyponatremia in heart failure: the role of arginine vasopressin and its antagonism. Congest Heart Fail 16(Suppl 1):S7–S14

152. Tahara A, Tomura Y, Wada K, Kusayama T, Tsukada J, Ishii N, Yatsu T, Uchida W, Tanaka A (1998) Effect of YM087, a potent nonpeptide vasopressin antagonist, on vasopressin-induced protein synthesis in neonatal rat cardiomyocyte. Cardiovasc Res 38(1):198–205

153. Xu YJ, Gopalakrishnan V (1991) Vasopressin increases cytosolic free [Ca2 +] in the neonatal rat cardiomyocyte. Evidence for V1 subtype receptors. Circ Res 69(1):239–245

154. Grillone LR, Clark MA, Godfrey RW, Stassen F, Crooke ST (1988) Vasopressin induces V1 receptors to activate phosphatidylinositol- and phosphatidylcholine-specific phospholipase C and stimulates the release of arachidonic acid by at least two pathways in the smooth muscle cell line, A-10. J Biol Chem 263(6):2658–2663

155. Stassen FL, Heckman G, Schmidt D, Aiyar N, Nambi P, Crooke ST (1987) Identification and characterization of vascular (V1) vasopressin receptors of an established smooth muscle cell line. Mol Pharmacol 31(3):259–266

156. Fukuzawa J, Haneda T, Kikuchi K (1999) Arginine vasopressin increases the rate of protein synthesis in isolated perfused adult rat heart via the V1 receptor. Mol Cell Biochem 195(1–2):93–98

157. Nakamura Y, Haneda T, Osaki J, Miyata S, Kikuchi K (2000) Hypertrophic growth of cultured neonatal rat heart cells mediated by vasopressin V(1 A) receptor. Eur J Pharmacol 391(1–2):39–48

158. Goldsmith SR (2006) The role of vasopressin in congestive heart failure. Clevel Clin J Med 73(Suppl 3):S19–S23

159. Nakamura T, Funayama H, Yoshimura A, Tsuruya Y, Saito M, Kawakami M, Ishikawa SE (2006) Possible vascular role of increased plasma arginine vasopressin in congestive heart failure. Int J Cardiol 106(2):191–195

160. Nielsen S, Chou CL, Marples D, Christensen EI, Kishore BK, Knepper MA (1995) Vasopressin increases water permeability of kidney collecting duct by inducing translocation of aquaporin-CD water channels to plasma membrane. Proc Natl Acad Sci U S A 92(4):1013–1017

161. Kalra PR, Anker SD, Coats AJ (2001) Water and sodium regulation in chronic heart failure: the role of natriuretic peptides and vasopressin. Cardiovasc Res 51(3):495–509

162. Uretsky BF, Verbalis JG, Generalovich T, Valdes A, Reddy PS (1985) Plasma vasopressin response to osmotic and hemodynamic stimuli in heart failure. Am J Physiol 248(3 Pt 2):H396–H402

163. Francis GS, Benedict C, Johnstone DE, Kirlin PC, Nicklas J, Liang CS, Kubo SH, Rudin-Toretsky E, Yusuf S (1990) Comparison of neuroendocrine activation in patients with left ventricular dysfunction with and without congestive heart failure. A substudy of the Studies of Left Ventricular Dysfunction (SOLVD). Circulation 82(5):1724–1729

164. Rouleau JL, Packer M, Moye L, de Champlain J, Bichet D, Klein M, Rouleau JR, Sussex B, Arnold JM, Sestier F et al (1994) Prognostic value of neurohumoral activation in patients with an acute myocardial infarction: effect of captopril. J Am Coll Cardiol 24(3):583–591

165. Goldsmith SR, Francis GS, Cowley AW Jr, Goldenberg IF, Cohn JN (1986) Hemodynamic effects of infused arginine vasopressin in congestive heart failure. J Am Coll Cardiol 8(4):779–783

166. Nicod P, Waeber B, Bussien JP, Goy JJ, Turini G, Nussberger J, Hofbauer KG, Brunner HR (1985) Acute hemodynamic effect of a vascular antagonist of vasopressin in patients with congestive heart failure. Am J Cardiol 55(8):1043–1047

167. Hirano T, Yamamura Y, Nakamura S, Onogawa T, Mori T (2000) Effects of the V(2)-receptor antagonist OPC-41061 and the loop diuretic furosemide alone and in combination in rats. J Pharmacol Exp Ther 292(1):288–294

168. Gheorghiade M, Konstam MA, Burnett JC Jr, Grinfeld L, Maggioni AP, Swedberg K, Udelson JE, Zannad F, Cook T, Ouyang J, Zimmer C, Orlandi C, Efficacy of Vasopressin Antagonism in Heart Failure Outcome Study With Tolvaptan I (2007) Short-term clinical effects of tolvaptan, an oral vasopressin antagonist, in patients hospitalized for heart failure: the EVEREST clinical status trials. JAMA 297 (12):1332–1343

169. Abraham WT, Shamshirsaz AA, McFann K, Oren RM, Schrier RW (2006) Aquaretic effect of lixivaptan, an oral, non-peptide, selective V2 receptor vasopressin antagonist, in New York Heart Association functional class II and III chronic heart failure patients. J Am Coll Cardiol 47(8):1615–1621

170. Costello-Boerrigter LC, Smith WB, Boerrigter G, Ouyang J, Zimmer CA, Orlandi C, Burnett JC Jr (2006) Vasopressin-2-receptor antagonism augments water excretion without changes in renal hemodynamics or sodium and potassium excretion in human heart failure. Am J Physiol Renal Physiol 290(2):F273–278

171. Gheorghiade M, Niazi I, Ouyang J, Czerwiec F, Kambayashi J, Zampino M, Orlandi C, Tolvaptan I (2003) Vasopressin V2-receptor blockade with tolvaptan in patients with chronic heart failure: results from a double-blind, randomized trial. Circulation 107(21):2690–2696

172. Naitoh M, Risvanis J, Balding LC, Johnston CI, Burrell LM (2002) Neurohormonal antagonism in heart failure; beneficial effects of vasopressin V(1a) and V(2) receptor blockade and ACE inhibition. Cardiovasc Res 54(1):51–57

173. Zeltser D, Rosansky S, van Rensburg H, Verbalis JG, Smith N, Conivaptan Study G (2007) Assessment of the efficacy and safety of intravenous conivaptan in euvolemic and hypervolemic hyponatremia. Am J Nephrol 27(5):447–457

174. Yatsu T, Tomura Y, Tahara A, Wada K, Tsukada J, Uchida W, Tanaka A, Takenaka T (1997) Pharmacological profile of YM087, a novel nonpeptide dual vasopressin V1A and V2 receptor antagonist, in dogs. Eur J Pharmacol 321(2):225–230

175. Udelson JE, Smith WB, Hendrix GH, Painchaud CA, Ghazzi M, Thomas I, Ghali JK, Selaru P, Chanoine F, Pressler ML, Konstam MA (2001) Acute hemodynamic effects of conivaptan, a dual V(1 A) and V(2) vasopressin receptor antagonist, in patients with advanced heart failure. Circulation 104(20):2417–2423

176. Goldsmith SR, Elkayam U, Haught WH, Barve A, He W (2008) Efficacy and safety of the vasopressin V1A/V2-receptor antagonist conivaptan in acute decompensated heart failure: a dose-ranging pilot study. J Card Fail 14(8):641–647

Natriuretic Peptides: Critical Regulators of Cardiac Fibroblasts and the Extracellular Matrix in the Heart

Hailey J. Jansen and Robert A. Rose

Abstract The mammalian heart contains numerous cell types with cardiac fibroblasts accounting for the majority of cells. These fibroblasts play essential roles in the heart including the synthesis and remodeling of the extracellular matrix (ECM), which is the component of the heart that includes interstitial collagens. In the setting of heart disease, including heart failure (HF), abnormal fibroblast proliferation and deposition of collagens leads to adverse structural remodeling, which is a major contributing factor to the progression of heart disease. Structural remodeling of the ECM in HF can increase stiffness of the myocardium leading to impaired cardiac performance and also increase the occurrence of cardiac arrhythmias due to impaired electrical conduction. Natriuretic peptides (NPs) are a family of cardioprotective hormones with numerous effects in the cardiovascular system. Included among these is the ability to prevent fibroblast proliferation and abnormal collagen deposition in the ECM. NPs elicit their effects by binding to three NP receptors denoted NPR-A, NPR-B and NPR-C. NPR-A and NPR-B are guanylyl cyclase-linked NPRs that elicit their effects by increasing cGMP levels. NPR-C is linked to the activation of inhibitory G-proteins (G_i). All three NPRs are expressed in cardiac fibroblasts and each has been shown to play a role in the ability of NPs to protect against adverse structural remodeling in the heart. The purpose of this chapter is to provide an overview of NPs and how they affect remodeling of the ECM in HF.

Keywords Natriuretic peptides · Natriuretic peptide receptors · Guanylyl cyclase · Adenylyl cyclase · Cyclic GMP · Cyclic AMP · Cardiac fibroblasts · Fibrosis

R. A. Rose (✉) · H. J. Jansen
Department of Physiology and Biophysics and School of Biomedical Engineering, Faculty of Medicine, Dalhousie University, Sir Charles Tupper Medical Building—Room 4J, 5850 College Street, PO Box 15000, Halifax, NS B3H 4R2, Canada
e-mail: robert.rose@dal.ca

© Springer International Publishing Switzerland 2015
I.M.C. Dixon, J. T. Wigle (eds.), *Cardiac Fibrosis and Heart Failure: Cause or Effect?*,
Advances in Biochemistry in Health and Disease 13, DOI 10.1007/978-3-319-17437-2_19

1   Introduction

The mammalian heart contains numerous cells types, including cardiac myocytes, cardiac fibroblasts, vascular smooth muscle cells, endothelial cells and others. Although cardiac myocytes account for the majority of the myocardial volume, cardiac fibroblasts are the most abundant cell type in the heart [1, 2]. These fibroblasts play essential roles in myocardial function in the normal heart and in the setting of heart disease. One critical function of the cardiac fibroblast is the synthesis and remodeling of the extracellular matrix (ECM), which is the component of the heart that includes interstitial collagens, proteoglycans, and glycoproteins. These components form a complex three dimensional network that is intricately involved in cardiac function. Some of the essential roles of the ECM include the formation of an organizational network that surrounds cellular structures, the creation of a scaffold for the myocyte and nonmyocyte cell populations in the heart, distribution of mechanical forces through the myocardium, mechanotransduction and fluid movement in the extracellular spaces.

The organization, composition and density of the ECM are highly dynamic and modulated under different physiological and pathophysiological conditions and this profoundly impacts cardiac function [1]. Collagen expression and accumulation are increased in the setting of heart failure (HF) in a process referred to as structural remodeling. As the density of the ECM affects compliance, the process of remodeling can be a pathological condition that leads to inappropriately enhanced fibrosis in the setting of HF. Specifically, this enhanced fibrosis in the diseased heart results in myocardial stiffness and diastolic dysfunction [3]. Enhanced fibrosis is also thought to increase the susceptibility to cardiac arrhythmias by slowing conduction and interfering with normal electrical propagation, which can lead to electrical reentry [4, 5]. Although remodeling of the ECM and enhanced fibrosis are hallmarks of HF there is still much that is unknown in terms of how the process is initiated and regulated.

Natriuretic peptides (NPs) are a family of cardioprotective hormones with numerous beneficial effects in the cardiovascular system [6, 7]. Although best known for their ability to regulate blood volume and blood pressure through effects in the kidneys and the vasculature, it is now known that NPs also have numerous additional effects. Included amongst these are potent effects on cardiac fibroblast function, ECM deposition and fibrosis. The purpose of this chapter is to provide an overview of natriuretic peptides and their role in the remodeling of the ECM that occurs in HF.

2   Natriuretic Peptides

In 1981, de Bold *et al.* infused atrial homogenates into rats and observed rapid and potent diuretic and natriuretic effects [8]. This landmark study ultimately led to the isolation and discovery of the first NP, atrial natriuretic peptide (ANP). B-type natriuretic peptide (BNP) and C-type natriuretic peptide (CNP) were subsequently identified and isolated from porcine brain extracts [9, 10]. BNP and CNP each exhibit profound relaxant effects on smooth muscle and following their discovery their presence in the heart was confirmed [7, 11]. *Dendroaspis* natriuretic peptide (DNP), a fourth member of the NP family, was initially identified in the venom of the Green Mamba snake [12]. There is evidence that DNP may also be present in human plasma and this NP has been shown to elicit relaxant responses in contracted aortic strips [13].

All NPs are synthesized as pre-pro-hormones that undergo posttranslational processing to form smaller, cyclical, biologically functional peptides [7]. NPs are structurally related and homology is observed in conserved residues within a 17 amino acid sequence flanked by cysteine residues (Fig. 1). A disulphide bridge is formed between these cysteine residues, creating a peptide ring. Structural variation occurs both within the cyclical structure and the amino- and carboxy-terminal tails of the NPs [7].

Pro-ANP and low levels of pro-BNP are stored within granules located in atrial myocytes [14, 15]. The dominant stimulus for release of these NPs from granules is atrial stretch in association with increased intravascular volume [16]. During exocytosis, pro-ANP is cleaved into the biologically active ANP by the transmembrane cardiac serine protease corin [17, 18]. ANP expression in the heart undergoes changes throughout development and in cardiac disease. For example, in addition to being expressed in the atria, ANP is expressed in fetal and neonate ventricles as well as hypertrophied ventricles in adults [19, 20]. ANP expression in normal adult ventricular tissue is very low. Circulating ANP levels increase by 10–30 fold in patients with congestive HF [21–23].

Within the ventricular myocardium BNP is constitutively expressed and released into the circulatory system. Ventricular BNP secretion is transcriptionally regulated and expression significantly increases in response to load induced ventricular wall stretch [24–26]. Plasma levels of BNP are 200–300 fold higher in patients with ventricular hypertrophy or those with congestive HF and, in some cases, circulating BNP levels exceed ANP levels [23].

Multiple CNP molecules have been identified. Pro-CNP is cleaved by the intracellular endoprotease furin to form a CNP molecule that is 53 amino acids in length (CNP-53) [27]. CNP-53 is located in cardiac tissue whereas a smaller 22 amino acid form of CNP (CNP-22) is detected in plasma [28, 29]. The enzyme responsible for the conversion between CNP-53 and CNP-22 remains unknown. Circulating levels of CNP are extremely low, approximately 1 fmol/l, and it is thought that CNP acts primarily as a paracrine molecule [30, 31]. As with ANP and BNP, circulating CNP levels are elevated in patients with congestive HF [7, 32].

NPs are rapidly cleared from the circulation via two mechanisms. First, NPs can be degraded by a membrane neutral endopeptidase called neprilysin, which cleaves peptides on the amino side of hydrophobic residues [33]. Interestingly, human BNP is more resistant to neprilysin hydrolysis compared to ANP [34]. The second component of NP degradation is coupled with the termination of surface receptor-mediated signaling though the internalization of the peptide-receptor complex. This is followed by hydrolytic degradation by lysosomes and recycling of a small pool of receptors back to the cell membrane [7].

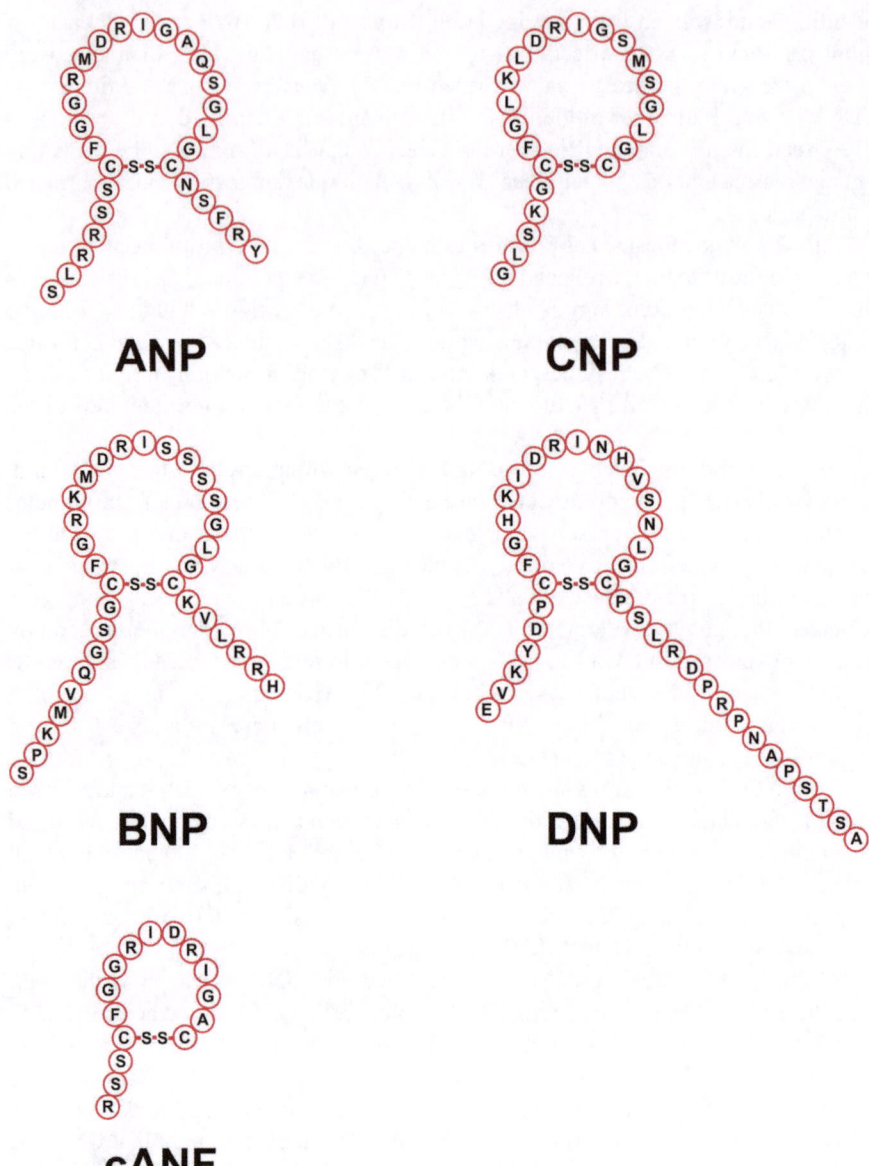

ANP

CNP

BNP

DNP

cANF

Fig. 1 Structure and amino acid sequence of natriuretic peptides. *ANP* atrial natriuretic peptide; *BNP* B-type natriuretic peptide; *CNP* C-type natriuretic peptide; *DNP Dendroaspis* natriuretic peptide; *cANF* synthetic natriuretic peptide receptor C (NPR-C) agonist

3   Natriuretic Peptide Receptors

NPs elicit their effects by binding to specific NP receptors (NPRs). There are currently three known NPRs denoted NPR-A, NPR-B and NPR-C (Fig. 2). NPR-A has binding affinity for ANP and BNP, while NPR-B preferentially binds CNP [7, 35].

NPR-A and NPR-B are coupled to intracellular particulate guanylyl cyclase (GC) enzymes. Following activation of these NPRs, GTP is converted into the second messenger cyclic guanosine monophosphate (cGMP); thus, NPR-A and NPR-B elicit their effects via changes in cGMP levels. Several downstream signaling molecules may be modulated by cGMP signaling including a cGMP-dependent protein kinase (PKG), cGMP regulated phosphodiesterases (PDEs), and cyclic nucleotide-gated ion channels [7].

NPR-C is the most abundantly expressed NPR and demonstrates similar binding affinity for all NPs [36, 37] (Fig. 2). In contrast to NPR-A and NPR-B, NPR-C is not directly coupled to changes in guanylyl cyclase signaling. Instead, NPR-C is coupled to the activation of inhibitory G-proteins (G_i) via specific 'G$_i$-activator domains' located within the 17 amino acid intracellular domain of the receptor [38, 39]. Following G_i activation, adenylyl cyclase (AC) activity is inhibited in a GTP-dependent fashion, which results in reductions in cAMP. Activation of NPR-C also results in the activation of the β isoform of phospholipase C (PLCβ), which converts phosphatidyl inositol bisphosphate (PIP_2) into inositol triphosphate (IP_3) and

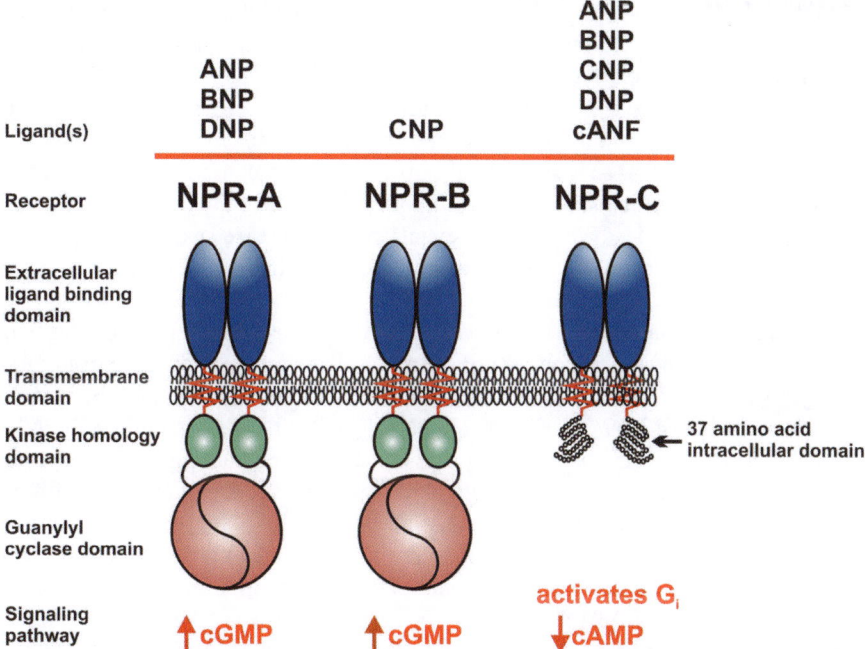

Fig. 2 Natriuretic peptide receptors and their ligand binding patterns. *NPR-A* natriuretic peptide receptor A; *NPR-B* natriuretic peptide receptor B; *NPR-C* natriuretic peptide receptor C. Note that *NPR-A* and *NPR-B* are guanylyl cyclase-linked receptors that mediate increases in cyclic guanosine monophosphate (*cGMP*). *NPR-C* has a short 37 amino acid intracellular domain that contains inhibitory G protein (G_i) activator sequences. As such NPR-C mediates a reduction in cyclic adenosine monophosphate (*cAMP*) levels

diacylglycerol (DAG). This leads to Ca^{2+} mobilization and protein kinase C (PKC) activation [36].

4   Natriuretic Peptides and Cardiac Fibrosis

NPs have been implicated in ECM remodeling and fibrosis in the heart. Some of this insight has been obtained from studies of genetically altered mice in which the NP system has been targeted. Specifically, enhanced cardiac fibrosis can be observed in NPR-A, BNP and ANP knockout animals depending on experimental conditions. Mice with global deletion of BNP ($Nppb^{-/-}$; $BNP^{-/-}$) display multifocal fibrotic lesions in the ventricles that are not observed in age matched wild type mice (Fig. 3a) [40]. When $BNP^{-/-}$ mice were subjected to aortic constriction, the level of ventricular fibrosis tripled compared to $BNP^{-/-}$ mice that received sham operations [40]. $BNP^{-/-}$ mice also exhibit increased transforming growth factor β3 (TGFβ3) and angiotensin converting enzyme (ACE) expression, suggesting that these pathways may be involved in the enhanced fibrosis characteristic of these mice.

ANP knockout mice ($Nppa^{-/-}$; $ANP^{-/-}$) are both hypertensive and hypertrophic at baseline [41, 42]. These mice display modest increases in collagen expression and collagen volume relative to wildtype controls; however, $ANP^{-/-}$ mice do not appear to exhibit the same degree of fibrosis as $BNP^{-/-}$ mice at baseline [43–45]. Nevertheless, $ANP^{-/-}$ mice subjected to pressure overload following transverse aortic constriction display profoundly worse fibrosis compared to sham operated $ANP^{-/-}$ mice, indicating that ANP is importantly involved in structural remodeling of the ECM in the heart, particularly in the setting of cardiac stress. This enhanced fibrotic response in $ANP^{-/-}$ mice occurred in association with increased expression of ECM proteins such as collagen I and III, matrix metalloproteinase 2 and tissue inhibitor of metalloproteinase 3 [44, 46].

Hearts from global NPR-A knockout ($NPR-A^{-/-}$) mice are both fibrotic (Fig. 3b) and hypertrophic in association with increased collagen deposition, increased procollagen I mRNA expression, and increased ANP and BNP mRNA expression in the ventricles [42, 47–49]. A cDNA microarray study in wild type and $NPR-A^{-/-}$ animals revealed significant alterations in gene expression patterns for genes from cell signaling pathways known to be involved in the development of cardiac fibrosis. These include, for example, fibroblast growth factor (FGF), collagens, matrix metalloproteinases, and multiple transcription factors including the histone deacetyltransferase 7a, myocyte-specific enhancer factor 2, calcineurin-nuclear factor of activated T cells, and GATA families [47, 50]. These observations in $NPR-A^{-/-}$ mice clearly suggest that the fibrotic phonotypes present in ANP and BNP knockout mice are at least partially due to a loss of NPR-A-dependent signaling. NPR-C has also been demonstrated to play an integral role in structural remodeling in the heart based on evidence that $NPR-C^{-/-}$ mice display enhanced fibrosis leading to atrial arrhythmias. Interestingly, fibrosis was restricted to the atrial myocardium in $NPR-C^{-/-}$ mice, while the ventricular myocardium was unaffected [51]. Collectively,

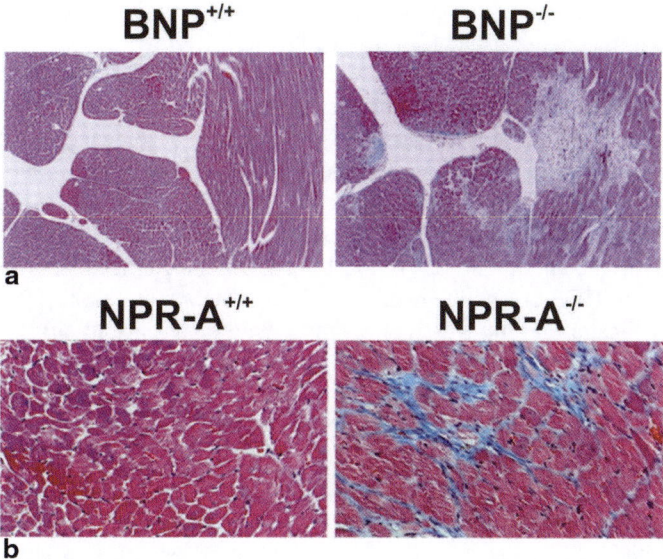

Fig. 3 Histological images of the ventricular myocardium in BNP and NPR-A knockout mice. Masson's trichrome stains from *BNP*$^{+/+}$ and *BNP*$^{-/-}$ mice (*panel A*) or *NPR-A*$^{+/+}$ and *NPR-A*$^{-/-}$ mice (*panel B*). Both *BNP*$^{-/-}$ and *NPR-A*$^{-/-}$ mice are characterized by ventricular fibrosis (*blue* color), which is not seen in wildtype mice. Data in *panel A* reproduced with permission from Tamura et al. (2000). Data in *panel B* reproduced with permission from Oliver et al. (1997)

these studies in genetically altered mice indicate that NPs play an essential protective role against adverse structural remodeling in the heart.

5   Effects of Natriuretic Peptides on Cardiac Fibroblasts

Although NPs are well known to be secreted from atrial granules located within atrial myocytes, it is now known that NPs are also made in, and secreted from cardiac fibroblasts [52, 53]. ANP and BNP mRNA can be detected in cultured fibroblasts from rats as young as 1 day old [54]. Furthermore, ANP and BNP proteins are readily detected by radioimmunoassay in the media from cultured fibroblasts. Similarly, CNP mRNA was detected in cultured ventricular fibroblasts isolated from 7-week-old rats and immunoreactive CNP was detected in the culture media [53]. Thus, NPs are synthesized in and secreted by cardiac fibroblasts.

In cultured neonatal rat ventricular fibroblasts, NPR-A, NPR-B, and NPR-C mRNAs are all expressed [55]. To determine the relative abundance of these NPRs, a Scatchard analysis was performed using cANF (Fig. 1), which is a selective agonist for NPR-C (Fig. 2) [56]. Using this approach, it has been estimated that 80 % of the total NPR population is NPR-C in cultured rat and human cardiac fibroblasts [35, 55, 57]. Interestingly, NPR-B may be more highly expressed in ventricular fibroblasts compared to cardiomyocytes [58]. All three NPRs have also been shown to be present in human cardiac fibroblasts [59].

NPs have potent antiproliferative and antimitogenic effects on cardiac fibroblasts. Using assays of radioactive thymidine incorporation into newly synthesized DNA, the potent effects of NPs on DNA synthesis have been quantified. In primary cultures of neonatal rat cardiac fibroblasts, ANP decreased the rate of DNA synthesis by approximately 40% under basal conditions [55, 60]. Cellular proliferation can also be induced by a number of hormones and growth factors including angiotensin II (Ang II), endothelin (ET), fibroblast growth factor (FGF), or insulin-like growth factor I (IGF-I), and the induction of proliferation by these compounds can be strongly antagonized by NPs. For example, in the presence of any of the above mentioned compounds, co-treatment with ANP inhibited agonist induced DNA synthesis [55]. In Ang II stimulated cultured adult rat cardiac fibroblasts, co-treatment with ANP (10^{-8} M) for 24 h resulted in 90% inhibition of cellular proliferation [61]. Similar antimitogenic effects were also observed in cells supplemented with BNP or CNP where FGF stimulated DNA synthesis rates were reduced by 25 and 21% respectively [55]. In primary human cardiac fibroblast cultures, co-treatment with BNP inhibited TGFβ induced cell proliferation by 65% [62]. In a separate study, application of BNP prevented 5-bromo-2′deoxyuridine (BrdU) incorporation into a human cardiac fibroblast cell line in which cellular proliferation was first stimulated with cardiotrophin-1 (CT-1) [59].

The vasoactive peptides Ang II and ET facilitate the enhanced cardiac fibroblast proliferation associated with cardiac fibrosis. Ang II, which is a peptide hormone, can elicit effects via the Ang II type 1 (AT_1) and type 2 (AT_2) receptors [63]. Cardiac fibroblasts express both AT_1 and AT_2 [64, 65] and it is thought that AT_1 mediates a number of the physiological and pathological effects of Ang II including fibroblast proliferation, collagen secretion, decreased collagenase activity, PLC activation, increased cytosolic calcium, and increased PKC activity [64–67].

ET-1 is a peptide growth factor initially described as a potent vasoconstrictor synthesized by cardiomyocytes and cardiac fibroblasts in the heart [68]. In cardiac fibroblasts, ET-1 levels are increased following activation of AT_1 [69, 70]. Interestingly, ET-1 levels are also increased in patients with HF. ET-1 promotes DNA synthesis following binding to the endothelin receptor ET_A, which stimulates cellular proliferation through the activation of PKC [71, 72].

Ang II and ET-1 stimulated DNA synthesis and cellular proliferation are inhibited in the presence of ANP, BNP, as well as 8-bromo-cGMP (a hydrolysis-resistant cGMP analogue) in culture media. The ET-1 promoter contains two regulatory elements responsible for basal transcriptional activity, a GATA element and activating protein-1 (AP-1) [73]. Mutation of specific sites in the proximal GATA element prevents the inhibitory effects of ANP on ET-1 induced DNA synthesis and cellular proliferation in cultured cardiac fibroblasts [60]. In this context, ANP is thought to function by inhibiting the ERK-dependent GATA4 phosphorylation required for binding to the ET-1 promoter. This in turn prevents ET-1 expression and subsequent

DNA synthesis and cellular proliferation. The fact that 8-bromo-cGMP elicits similar effects as ANP or BNP suggest that these NP effects are mediated by NPR-A.

6 Effects of Natriuretic Peptides on Collagen Synthesis

The interstitial collagens making up the ECM in the heart consist primarily of fibrillar collagen type I and collagen type III. The balance between the types of collagen present and the overall organization of these molecules within the heart play an important role in the mechanics of cardiac function. Increased levels of collagen type I is associated with myocardial stiffness whereas increased collagen type III is associated with compliance [74]. Collagen type I is several orders of magnitude stronger and stiffer than muscle [75].

NPs are very effective inhibitors of collagen synthesis in cardiac fibroblasts. In rat ventricular fibroblasts, the effects of ANP on collagen synthesis have been determined by quantifying hydroxyproline levels [57]. Treatment with TGFβ, Ang II, or serum results in a 1.3–3 fold increase in procollagen synthesis in cultured fibroblasts. The addition of ANP and zaprinast (a PDE5 inhibitor) to the culture media inhibited this increase [57]. In cultured canine ventricular fibroblasts, changes in *de novo* collagen synthesis were measured using [^{3}H]proline incorporation assays. In these experiments collagen synthesis was reduced by BNP in a concentration dependent manner. The maximum response was observed in the presence of 10^{-6} M BNP whereby [^{3}H]proline incorporation was inhibited by 29% [52]. Furthermore, RT-PCR experiments performed on TFGβ-stimulated primary human cardiac fibroblasts demonstrate increases in collagen I mRNA levels after 6, 24, and 48 h of exposure [62]; however, when cells were co-treated with BNP, this increase in collagen I expression was abolished. Western blots using collagen I antibodies further confirm these findings, whereby collagen levels increased by 3 fold in the presence of TFGβ and this effect was inhibited by 75% in the presence of BNP [62].

NPs also inhibit the effects of Ang II on collagen production by cardiac fibroblasts. For example, in Ang II stimulated rat cardiac fibroblasts, an 80% decrease in collagen synthesis was observed when cells were co-treated with ANP (10^{-8} M) for 24 h [61]. Similarly, in cultured neonatal rat cardiac fibroblasts, treatment with CNP (10^{-6} M) for 24 h caused a significant decrease in Ang II stimulated [^{3}H]proline incorporation [76]. This effect of CNP was blocked in the presence of Rp-8-pCPT-cGMP, a PKG inhibitor. Together, these experiments show that NPs have important inhibitory effects on collagen synthesis in cardiac fibroblasts.

Most of the effects of NPs on cardiac fibroblasts have been attributed to NPR-A and NPR-B activation. Consistent with this, intracellular levels of cGMP are dose dependently increased following exposure to NPs in cultured ventricular fibroblasts [52, 53, 58]. These increases in cGMP levels are correlated with decreases in collagen synthesis and DNA synthesis as determined by radioactive proline or thymidine incorporation assays, respectively. The addition of 8-bromo-cGMP to culture media mimics the effects of NPs on both DNA and collagen synthesis, thus suggesting that inhibition of DNA and collagen synthesis occurs in a GC dependent fashion [52, 53, 55, 57, 61, 76]. Furthermore, NPR-A knockdown in cultured adult rat cardiac fibroblasts treated with Ang II results in a twofold increase in collagen I expression and a threefold increase in collagen III expression [61]. Addition of a synthetic cGMP analog to the media of NPR-A knockdown fibroblast cultures prevented these changes in collagen expression further confirming that the NPs can affect remodeling of the ECM via an NPR-A/cGMP pathway.

Although most studies have focused on NPR-A and NPR-B mediated effects on NPs, emerging evidence suggests that NPR-C also plays an important role in cardiac fibroblast function. As mentioned above, NPR-C is clearly expressed in cardiac fibroblasts [55, 59]. In cultured human cardiac fibroblasts, CT-1 increases BrdU incorporation which is decreased in the presence of BNP [59]. Thus, BNP (which binds NPR-A and NPR-C) inhibits CT-1 stimulated DNA synthesis. To determine the contribution of NPR-A to this effect of BNP the NPR-A antagonist HS-142-1 was added to CT-1 and BNP co-treated fibroblasts. HS-142-1 had no effect on the BNP mediated inhibition of fibroblast proliferation. In contrast, the NPR-C agonist cANF inhibited the effects of BNP on proliferation indicating a role for NPR-C in the modulation of cardiac fibroblast proliferation.

7   Transforming Growth Factor β

TGFβ is critically involved in the regulation of cellular differentiation, proliferation, as well as extracellular matrix deposition and composition [77]. The TGFβ pathway affects fibrotic remodeling within the heart as it potently modulates cardiac fibroblast proliferation and production of ECM proteins including collagens and fibronectin. TGFβ1 expression and activity is increased as a result of AT_1 activation by Ang II in cultured rat cardiac fibroblasts [67]. TGFβ1 functions by binding to two cell membrane receptor kinases, TGFβRI and TGFβRII. Once activated, these kinases facilitate the phosphorylation of two downstream proteins, Smad2 and Smad3 [78], which can then form a complex with Smad4. This Smad complex translocates to the nucleus where it activates profibrotic gene programs [77, 79, 80]. There is a positive correlation between TGFβ1 levels and collagen content in the heart. For example, TGFβ1 deficient mice have decreased levels of fibrosis whereas TGFβ1 levels are elevated in patients with HF exhibiting enhanced ECM remodeling and fibrosis [74, 81].

In cultured human cardiac fibroblasts changes in gene expression patterns in TGFβ stimulated cells were determined using microarray analysis. In this study, it was found that TGFβ stimulation induced 394 and 501 gene expression changes at 24 and 48 h of treatment, respectively. When co-treated with BNP, 88 and 85 % of the TGFβ induced gene expression changes were abolished, including those involved in fibrosis and ECM production [62].

As discussed above, ANP acts as a negative modulator of cardiac remodeling and it appears that ANP functions, at least in part, by inhibiting the effects of TGFβ signaling in cardiac fibroblasts. This has been shown in cultured mouse cardiac fibroblasts pretreated with either ANP or cGMP prior to exposure to TGFβ1 for 24 h [82]. Pretreatment with ANP or cGMP resulted in a significant decrease in TGFβ induced collagen synthesis, fibroblast proliferation and pSmad3 translocation. Pretreatment with the PKG inhibitor KT5823 antagonized these inhibitory effects of ANP and cGMP. These findings indicate that ANP mediates its effects on TGFβ signaling and ECM remodeling via a cGMP-dependent mechanism. This study also shows that ANP inhibited the effect of TGFβ on collagen synthesis and fibroblast proliferation though the prevention of pSmad3 translocation into the nucleus [82].

8 Matrix Metalloproteinases and Tissue Inhibitors of Metalloproteinases

The structure of the fibrillar collagen scaffold within the heart results from an interplay between collagen synthesis and degradation. Matrix metalloproteinases (MMPs) are a family of proteins that play an essential role in matrix degradation [74]. In the diseased heart, increased MMP activity results in degradation of normal collagen and the development of interstitial deposits of poorly cross-linked collagens characteristic of those present in the fibrotic heart [83]. In NPR-A deficient animals, MMP2 and MMP9 protein levels are increased by 3 and 4 fold respectively in 4 week old animals and further increased by 22 weeks of age [49]. Furthermore, as discussed above, collagen levels are doubled in adult NPR-A$^{-/-}$ mice compared to their wild type littermates. Stimulation of cultured rat cardiac fibroblasts with Ang II results in increases in MMP2 and MMP9 mRNA expression as well as activity [61]. These alterations are also observed in fibroblasts in which NPR-A is knocked down. Conversely, when fibroblasts isolated from wildtype mice are co-treated with Ang II and ANP, the increase in MMP2 and MMP9 activity and expression is abolished [61]. Together, these findings suggest that ANP and NPR-A oppose Ang II induced MMP2 and MMP9 synthesis.

To unravel the underlying mechanism for these observations, the effects of ANP on second messenger levels were evaluated in Ang II-stimulated fibroblasts. Ang II stimulation activates nicotinamide adenine dinucleotide phosphate (NADPH) oxidase, resulting in the generation of reactive oxygen species (ROS) [84], which has been shown to induce cardiac fibroblast proliferation and increased fibrosis leading to the progression of end stage HF. Treatment of Ang II stimulated rat cardiac fibroblasts with ANP resulted in significantly decreased ROS levels as assessed by spectroflourometric analysis [61]. Conversely, in NPR-A knockdown experiments, ROS levels were further increased in the presence of Ang II relative to wildtype fibroblasts, but this could be abolished in the presence of 8-bromo-cGMP. Following ROS stimulation, nuclear factor-kappa-B (NF-κB) is translocated to the nucleus where binding of NF-κB to DNA results in the increased expression of ECM remodelers including collagens and MMP1, 3, and 9 [85]. In cultured rat cardiac fibroblasts treated with Ang II, nuclear translocation of NF-κβ was examined using confocal microscopy and an NF-κB antibody. In these studies, addition of ANP to the cultures inhibited NF-κB nuclear translocation and DNA binding [61]. This in turn would result in decreased expression of collagen type I, collagen type III, MMP2, and MMP9 transcripts.

Tissue inhibitors of metalloproteinases (TIMPs) are potent endogenous inhibitors of MMP activity. There are four TIMP isoforms detected in the heart including TIMP1, TIMP2, TIMP3, and TIMP4. X-ray crystallography studies have shown that TIMPs bind to the active site of MMPs, thereby preventing ECM substrate binding and inhibition of MMP activity [86, 87]. BNP appears to exert its effects on ECM remodeling in part through its effects on TIMP expression levels. In left ventricular tissues isolated from NPR-A$^{-/-}$ mice, TIMP1 and TIMP2 protein levels were significantly lower compared to wildtype mice [49]. In a different model, TIMP2 protein expression displayed a 12 % increase following 24 h of BNP treatment in cultured canine ventricular cardiac fibroblasts although TIMP1 levels remained unchanged [52]. Furthermore, in primary human cardiac fibroblast cultures, microarray analysis and RT-PCR experiments indicate that TIMP3 expression is increased in the presence of TGFβ [62]. The addition of BNP to these cells results in a downregulation of TIMP3 expression. Together, these studies suggest that NPs affect TIMP expression in cardiac fibroblasts; however, the mechanism by which NPs alter TIMP expression profiles or function remains largely unknown.

9   Chronic Natriuretic Peptide Treatment in the Diseased Heart

Myocardial infarction (MI) can be surgically induced in rodents by ligating the coronary artery, resulting in significant ventricular remodeling and a decline in cardiac function leading to HF. To study the effects of NPs in this disease model rats were subjected to MI and treated with a low dose (5 μg/kg/day) or a high dose (15 μg/kg/day) of BNP for 8 weeks beginning the day after the surgeries occurred [88]. Echocardiographic and hemodynamic measurements indicate that BNP treatment improved cardiac function compared to the untreated animals. In animals treated with BNP, histological analysis of excised hearts showed a significant decrease in the amount of collagen deposited within the ventricles. Both plasma and myocardium Ang II levels were significantly higher in vehicle-treated animals compared to those receiving BNP treatment. Furthermore, in animals treated with BNP there was a significant decrease in TGFβ1 and Smad2 mRNA and protein expression despite an increase in Ang II expression. This suggests that BNP both counteracts the harmful effects of increased Ang II levels and inhibits TGFβ1/Smad2 signaling resulting in less detrimental ECM remodeling following MI. These beneficial effects of BNP were more pronounced in animals treated with 15 μg/kg/day compared to the lower dose of 5 μg/kg/day.

A separate study infused CNP for 2 weeks in rats subjected to experimental MI. CNP (0.1 µg/kg/day) was delivered intravenously using osmotic mini-pumps starting 4 days following surgery and continuing for 2 weeks [76]. In this study, CNP infusion significantly prevented left ventricular enlargement and reduction in cardiac function caused by MI. Autoradiograms and qPCR experiments showed a significant decrease in the amount of collagen I and collagen III protein and mRNA expression in the ventricles of CNP treated animals. Interestingly, endogenous expression of CNP mRNA initially increased four-fold on day 3 in the infarcted left ventricle and gradually decreased to the end of the treatment period at day 18. Histological analysis revealed that CNP was concentrated at the infarct and border zone on day 7 following MI. Thus, CNP also acts as a cardioprotective agent following MI.

The cardioprotective effects of CNP have also been investigated in mice chronically treated with Ang II, which is a well-established model of cardiac hypertrophy and fibrosis [89, 90]. In a recent study using this model, mice were treated with Ang II (3.2 mg/kg/day) for 2 weeks and a subset of animals were co-treated with CNP (0.05 µg/kg/min) also for 2 weeks [91]. As expected, Ang II treated mice showed clear signs of cardiac dysfunction and had increased levels fibrosis, collagen expression, and ROS production. Co-treatment with CNP resulted in a significant decrease in the level of interstitial fibrosis and collagen type I and III mRNA expression compared to vehicle-treated animals (Fig. 4). CNP infusion completely prevented Ang

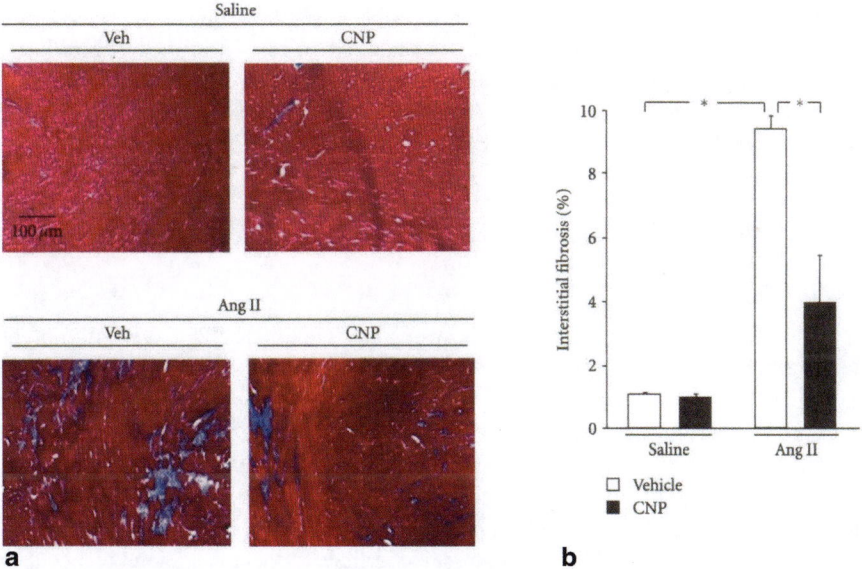

Fig. 4 Effects of CNP on Ang II induced ventricular fibrosis in mice. **a** Masson's trichrome stains of myocardium in saline and Ang II treated mice cotreated with CNP or vehicle. **b** quantification of interstitial fibrosis in Ang II and/or CNP treated mice. Ang II induces ventricular fibrosis, which is significantly attenuated by cotreatment with CNP. Data reproduced with permission from Izumiya et al. (2012)

II-induced cardiac superoxide production and significantly reduced the expression of the NADPH oxidase subunit NOX4. Interestingly, CNP infusion also prevented the upregulation of ANP and BNP mRNA expression seen in the vehicle treated control animals. Combined, these data further support the notion that CNP acts as a protective agent within the heart preventing Ang II-induced cardiac remodeling and superoxide production.

Most recently, attention has been given to the development of designer synthetic NPs that may be particularly effective in the treatment of HF and its associated complications. One example of this is the peptide CD-NP (also known as cenderitide), which is a chimeric peptide that combines CNP with the C-terminal tail of DNP [92, 93] (Fig. 5a). CD-NP is able to bind and activate all three NPRs [94]. The effects of CD-NP on ECM remodeling have been tested in an experimental model of cardiac fibrosis induced by unilateral nephrectomy in rats [95]. In this study, a 2 week subcutaneous infusion of CD-NP significantly suppressed left ventricular fibrosis (Fig. 5b) and preserved systolic and diastolic function compared to vehicle treated rats with unilateral nephrectomy. This same report also demonstrated that CD-NP could increase cGMP production in cells heterologously expressing NPR-A and NPR-B; however, this was not confirmed specifically in cardiac fibroblasts.

A separate investigation assessed the ability to slowly release CD-NP from bio-degradable polymeric films [96], which could have important implications for the therapeutic use of CD-NP in conjunction with cardiac patches. Importantly, the bio-activity of CD-NP released from these patches was assessed by measuring the effects of released peptide on human cardiac fibroblasts. These studies demonstrate that the released CD-NP is able to inhibit fibroblast proliferation and suppress DNA synthesis in association with increased production of cGMP in these fibroblasts. This suggests that CD-NP may have beneficial therapeutic effects, which involve the prevention of ECM remodeling. Furthermore, these effects at least partially involve the NPR-A and NPR-B receptors.

10   Summary and Conclusions

When considered collectively, there is strong evidence that NPs have both potent antiproliferative and antifibrotic effects on cardiac fibroblasts. As such, NPs play an important protective role against adverse structural remodeling of the ECM in the normal heart and in the setting of cardiovascular disease. Despite these clear beneficial effects, there are several areas of ongoing investigation that will improve our understanding of how NPs protect against remodeling of the ECM. For example, most of the effects of NPs on cardiac fibroblasts have been attributed to the GC-linked NPR-A and NPR-B receptors. Nevertheless, there is some evidence that NPR-C, which is highly expressed in cardiac fibroblasts, may also be involved. As most naturally occurring and synthetic NPs are able to bind multiple NPRs, it seems critical that ongoing studies consider how simultaneous activation of the GC-linked NPRs and NPR-C results in the overall effects of NPs on the ECM.

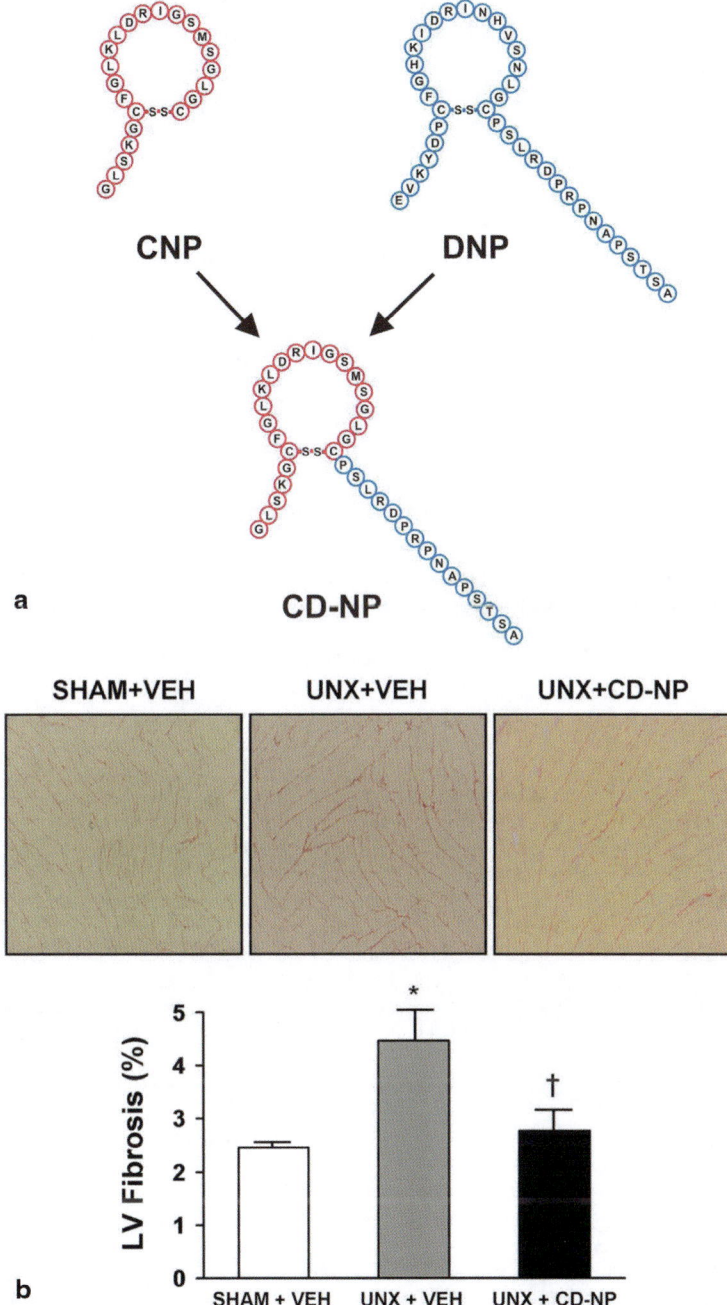

Fig. 5 Structure and antifibrotic effects of the chimeric natriuretic peptide CD-NP. **a** Structure and amino acid sequence of CD-NP, which is formed by combining CNP with the C-terminal tail of DNP. **b** Picrosirius red histology images and quantification of ventricular fibrosis from control rats (sham + vehicle), rats subjected to unilateral nephrectomy (*UNX*) to induce cardiac fibrosis, and rats subjected to UNX treated with CD-NP for 2 weeks. Data reproduced with permission from Martin et al. (2012)

Another emerging area of investigation is related to NP effects on ion channels in cardiac fibroblasts. These fibroblasts express a number of potassium and transient receptor potential (TRP) channels [97, 98]. NPs have been shown to activate non-selective cation currents that are likely carried by members of the TRP-C family of ion channels [98]. Furthermore, these same TRP-C channels have been shown to mediate an influx of Ca^{2+} into cardiac fibroblasts, which has implications for arrhythmogenesis in the heart [99, 100]. It is presently unknown whether the effects of NPs on fibroblast ion channels and Ca^{2+} homeostasis are directly linked to the protective effects of NPs against structural remodeling; thus, this will require ongoing investigation.

Finally, the recent development of chimeric NPs, such as CD-NP, and the possibility of delivering NPs via synthetic patches, highlights the exciting potential for the therapeutic use of NPs for the prevention of adverse structural remodeling and fibrosis in the heart. Continued investigation into the design of these synthetic NPs and the methods for their chronic delivery to patients is needed to bring this to fruition.

Progress in each of the above mentioned areas, in combination with the information already known regarding the effects of NPs on remodeling of the ECM, will greatly impact the strong potential for the use of NPs for the prevention of adverse structural remodeling and fibrosis in human HF patients.

References

1. Baudino TA, Carver W, Giles W, Borg TK (2006) Cardiac fibroblasts: friend or foe? Am J Physiol Heart Circ Physiol 291(3):H1015–1026. doi:10.1152/ajpheart.00023.2006
2. Souders CA, Bowers SL, Baudino TA (2009) Cardiac fibroblast: the renaissance cell. Circ Res 105(12):1164–1176. doi:10.1161/CIRCRESAHA.109.209809
3. Camelliti P, Borg TK, Kohl P (2005) Structural and functional characterisation of cardiac fibroblasts. Cardiovasc Res 65(1):40–51. doi:10.1016/j.cardiores.2004.08.020
4. Wolf RM, Glynn P, Hashemi S, Zarei K, Mitchell CC, Anderson ME, Mohler PJ, Hund TJ (2013) Atrial fibrillation and sinus node dysfunction in human ankyrin-B syndrome: a computational analysis. Am J Physiol Heart Circ Physiol 304(9):H1253–1266. doi:10.1152/ajpheart.00734.2012
5. Yue L, Xie J, Nattel S (2011) Molecular determinants of cardiac fibroblast electrical function and therapeutic implications for atrial fibrillation. Cardiovasc Res 89(4):744–753. doi:10.1093/cvr/cvq329
6. Levin ER, Gardner DG, Samson WK (1998) Natriuretic peptides. N Engl J Med 339(5):321–328. doi:10.1056/NEJM199807303390507
7. Potter LR, Abbey-Hosch S, Dickey DM (2006) Natriuretic peptides, their receptors, and cyclic guanosine monophosphate-dependent signaling functions. Endocr Rev 27(1):47–72. doi:10.1210/er.2005-0014
8. de Bold AJ, Borenstein HB, Veress AT, Sonnenberg H (1981) A rapid and potent natriuretic response to intravenous injection of atrial myocardial extract in rats. Life Sci 28(1):89–94
9. Sudoh T, Kangawa K, Minamino N, Matsuo H (1988) A new natriuretic peptide in porcine brain. Nature 332(6159):78–81. doi:10.1038/332078a0

10. Sudoh T, Minamino N, Kangawa K, Matsuo H (1990) C-type natriuretic peptide (CNP): a new member of natriuretic peptide family identified in porcine brain. Biochem Biophys Res Commun 168(2):863–870
11. Minamino N, Aburaya M, Ueda S, Kangawa K, Matsuo H (1988) The presence of brain natriuretic peptide of 12,000 daltons in porcine heart. Biochem Biophys Res Commun 155(2):740–746
12. Schweitz H, Vigne P, Moinier D, Frelin C, Lazdunski M (1992) A new member of the natriuretic peptide family is present in the venom of the green mamba (Dendroaspis angusticeps). J Biol Chem 267(20):13928–13932
13. Schirger JA, Heublein DM, Chen HH, Lisy O, Jougasaki M, Wennberg PW, Burnett JC Jr. (1999) Presence of Dendroaspis natriuretic peptide-like immunoreactivity in human plasma and its increase during human heart failure. Mayo Clin Proc Mayo Clin 74(2):126–130. doi:10.4065/74.2.126
14. Thibault G, Charbonneau C, Bilodeau J, Schiffrin EL, Garcia R (1992) Rat brain natriuretic peptide is localized in atrial granules and released into the circulation. Am J Physiol 263(2 Pt 2):R301–309
15. de Bold AJ, Bruneau BG, Kuroski de Bold ML (1996) Mechanical and neuroendocrine regulation of the endocrine heart. Cardiovasc Res 31(1):7–18
16. Edwards BS, Zimmerman RS, Schwab TR, Heublein DM, Burnett JC Jr. (1988) Atrial stretch, not pressure, is the principal determinant controlling the acute release of atrial natriuretic factor. Circ Res 62(2):191–195
17. Inagami T (1989) Atrial natriuretic factor. J Biol Chem 264(6):3043–3046
18. Yan W, Wu F, Morser J, Wu Q (2000) Corin, a transmembrane cardiac serine protease, acts as a pro-atrial natriuretic peptide-converting enzyme. Proc Natl Acad Sci U S A 97(15):8525–8529. doi:10.1073/pnas.150149097
19. Gu J, D'Andrea M, Seethapathy M (1989) Atrial natriuretic peptide and its messenger ribonucleic acid in overloaded and overload-released ventricles of rat. Endocrinology 125(4):2066–2074. doi:10.1210/endo-125-4-2066
20. Saito Y, Nakao K, Arai H, Nishimura K, Okumura K, Obata K, Takemura G, Fujiwara H, Sugawara A, Yamada T et al (1989) Augmented expression of atrial natriuretic polypeptide gene in ventricle of human failing heart. J Clin Invest 83(1):298–305. doi:10.1172/JCI113872
21. Cody RJ, Atlas SA, Laragh JH, Kubo SH, Covit AB, Ryman KS, Shaknovich A, Pondolfino K, Clark M, Camargo MJ et al (1986) Atrial natriuretic factor in normal subjects and heart failure patients. Plasma levels and renal, hormonal, and hemodynamic responses to peptide infusion. J Clin Invest 78(5):1362–1374. doi:10.1172/JCI112723
22. Mukoyama M, Nakao K, Hosoda K, Suga S, Saito Y, Ogawa Y, Shirakami G, Jougasaki M, Obata K, Yasue H et al (1991) Brain natriuretic peptide as a novel cardiac hormone in humans. Evidence for an exquisite dual natriuretic peptide system, atrial natriuretic peptide and brain natriuretic peptide. J Clin Invest 87(4):1402–1412. doi:10.1172/JCI115146
23. Wei CM, Heublein DM, Perrella MA, Lerman A, Rodeheffer RJ, McGregor CG, Edwards WD, Schaff HV, Burnett JC Jr. (1993) Natriuretic peptide system in human heart failure. Circulation 88(3):1004–1009
24. Grepin C, Dagnino L, Robitaille L, Haberstroh L, Antakly T, Nemer M (1994) A hormone-encoding gene identifies a pathway for cardiac but not skeletal muscle gene transcription. Mol Cell Biol 14(5):3115–3129
25. Thuerauf DJ, Hanford DS, Glembotski CC (1994) Regulation of rat brain natriuretic peptide transcription. A potential role for GATA-related transcription factors in myocardial cell gene expression. J Biol Chem 269(27):17772–17775
26. Nakagawa O, Ogawa Y, Itoh H, Suga S, Komatsu Y, Kishimoto I, Nishino K, Yoshimasa T, Nakao K (1995) Rapid transcriptional activation and early mRNA turnover of brain natriuretic peptide in cardiocyte hypertrophy. Evidence for brain natriuretic peptide as an "emergency" cardiac hormone against ventricular overload. J Clin Invest 96(3):1280–1287. doi:10.1172/JCI118162
27. Wu C, Wu F, Pan J, Morser J, Wu Q (2003) Furin-mediated processing of Pro-C-type natriuretic peptide. J Biol Chem 278(28):25847–25852. doi:10.1074/jbc.M301223200

28. Minamino N, Aburaya M, Kojima M, Miyamoto K, Kangawa K, Matsuo H (1993) Distribution of C-type natriuretic peptide and its messenger RNA in rat central nervous system and peripheral tissue. Biochem Biophys Res Commun 197:326–335

29. Stingo AJ, Clavell AL, Heublein DM, Wei CM, Pittelkow MR, Burnett JC Jr. (1992) Presence of C-type natriuretic peptide in cultured human endothelial cells and plasma. Am J Physiol 263(4 Pt 2):H1318–1321

30. Igaki T, Itoh H, Suga S, Komatsu Y, Ogawa Y, Doi K, Yoshimasa T, Nakao K (1996) Insulin suppresses endothelial secretion of C-type natriuretic peptide, a novel endothelium-derived relaxing peptide. Diabetes 45(Suppl 3):S62–64

31. Pandey KN (2005) Biology of natriuretic peptides and their receptors. Peptides 26(6):901–932. doi:10.1016/j.peptides.2004.09.024

32. Del Ry S, Passino C, Maltinti M, Emdin M, Giannessi D (2005) C-type natriuretic peptide plasma levels increase in patients with chronic heart failure as a function of clinical severity. Eur J Heart Fail 7(7):1145–1148. doi:10.1016/j.ejheart.2004.12.009

33. Kerr MA, Kenny AJ (1974) The purification and specificity of a neutral endopeptidase from rabbit kidney brush border. Biochem J 137(3):477–488

34. Smith MW, Espiner EA, Yandle TG, Charles CJ, Richards AM (2000) Delayed metabolism of human brain natriuretic peptide reflects resistance to neutral endopeptidase. J Endocrinol 167(2):239–246

35. Anand-Srivastava MB, Trachte GJ (1993) Atrial natriuretic factor receptors and signal transduction mechanisms. Pharmacol Rev 45(4):455–497

36. Anand-Srivastava MB (2005) Natriuretic peptide receptor-C signaling and regulation. Peptides 26(6):1044–1059. doi:10.1016/j.peptides.2004.09.023

37. Rose RA, Giles WR (2008) Natriuretic peptide C receptor signalling in the heart and vasculature. J Physiol 586(2):353–366. doi:10.1113/jphysiol.2007.144253

38. Pagano M, Anand-Srivastava MB (2001) Cytoplasmic domain of natriuretic peptide receptor C constitutes Gi activator sequences that inhibit adenylyl cyclase activity. J Biol Chem 276(25):22064–22070. doi:10.1074/jbc.M101587200

39. Zhou H, Murthy KS (2003) Identification of the G protein-activating sequence of the single-transmembrane natriuretic peptide receptor C (NPR-C). Am J Physiol Cell Physiol 284(5):C1255–1261. doi:10.1152/ajpcell.00520.2002

40. Tamura N, Ogawa Y, Chusho H, Nakamura K, Nakao K, Suda M, Kasahara M, Hashimoto R, Katsuura G, Mukoyama M, Itoh H, Saito Y, Tanaka I, Otani H, Katsuki M (2000) Cardiac fibrosis in mice lacking brain natriuretic peptide. Proc Natl Acad Sci U S A 97(8):4239–4244. doi:10.1073/pnas.070371497

41. Lopez MJ, Wong SK, Kishimoto I, Dubois S, Mach V, Friesen J, Garbers DL, Beuve A (1995) Salt-resistant hypertension in mice lacking the guanylyl cyclase-A receptor for atrial natriuretic peptide. Nature 378(6552):65–68. doi:10.1038/378065a0

42. Oliver PM, Fox JE, Kim R, Rockman HA, Kim HS, Reddick RL, Pandey KN, Milgram SL, Smithies O, Maeda N (1997) Hypertension, cardiac hypertrophy, and sudden death in mice lacking natriuretic peptide receptor A. Proc Natl Acad Sci U S A 94(26):14730–14735

43. Franco V, Chen YF, Feng JA, Li P, Wang D, Hasan E, Oparil S, Perry GJ (2006) Eplerenone prevents adverse cardiac remodelling induced by pressure overload in atrial natriuretic peptide-null mice. Clin Exp Pharmacol Physiol 33(9):773–779. doi:10.1111/j.1440-1681.2006.04434.x

44. Franco V, Chen YF, Oparil S, Feng JA, Wang D, Hage F, Perry G (2004) Atrial natriuretic peptide dose-dependently inhibits pressure overload-induced cardiac remodeling. Hypertension 44(5):746–750. doi:10.1161/01.HYP.0000144801.09557.4c

45. Wang D, Gladysheva IP, Fan TH, Sullivan R, Houng AK, Reed GL (2014) Atrial natriuretic Peptide affects cardiac remodeling, function, heart failure, and survival in a mouse model of dilated cardiomyopathy. Hypertension 63(3):514–519. doi:10.1161/HYPERTENSIONAHA.113.02164

46. Wang D, Oparil S, Feng JA, Li P, Perry G, Chen LB, Dai M, John SW, Chen YF (2003) Effects of pressure overload on extracellular matrix expression in the heart of the atrial natriuretic peptide-null mouse. Hypertension 42(1):88–95. doi:10.1161/01.HYP.0000074905.22908.A6

47. Ellmers LJ, Knowles JW, Kim HS, Smithies O, Maeda N, Cameron VA (2002) Ventricular expression of natriuretic peptides in Npr1(-/-) mice with cardiac hypertrophy and fibrosis. Am J Physiol Heart Circ Physiol 283(2):H707–714. doi:10.1152/ajpheart.00677.2001

48. Kuhn M, Holtwick R, Baba HA, Perriard JC, Schmitz W, Ehler E (2002) Progressive cardiac hypertrophy and dysfunction in atrial natriuretic peptide receptor (GC-A) deficient mice. Heart 87(4):368–374

49. Vellaichamy E, Khurana ML, Fink J, Pandey KN (2005) Involvement of the NF-kappa B/ matrix metalloproteinase pathway in cardiac fibrosis of mice lacking guanylyl cyclase/natriuretic peptide receptor A. J Biol Chem 280(19):19230–19242. doi:10.1074/jbc.M411373200

50. Ellmers LJ, Scott NJ, Piuhola J, Maeda N, Smithies O, Frampton CM, Richards AM, Cameron VA (2007) Npr1-regulated gene pathways contributing to cardiac hypertrophy and fibrosis. J Mol Endocrinol 38(1–2):245–257. doi:10.1677/jme.1.02138

51. Egom EE, Vella K, Hua R, Jansen HJ, Moghtadaei M, Polina I, Bogachev O, Hurnik R, Mackasey M, Rafferty S, Ray, G, Rose RA (2015) Impaired sinoatrial node function and increased susceptibility to atrial fibrillation in mice lacking natriuretic peptide receptor C. J Physiol 593:1127–1146.

52. Tsuruda T, Boerrigter G, Huntley BK, Noser JA, Cataliotti A, Costello-Boerrigter LC, Chen HH, Burnett JC Jr. (2002) Brain natriuretic Peptide is produced in cardiac fibroblasts and induces matrix metalloproteinases. Circ Res 91(12):1127–1134

53. Horio T, Tokudome T, Maki T, Yoshihara F, Suga S, Nishikimi T, Kojima M, Kawano Y, Kangawa K (2003) Gene expression, secretion, and autocrine action of C-type natriuretic peptide in cultured adult rat cardiac fibroblasts. Endocrinology 144(6):2279–2284

54. Harada E, Nakagawa O, Yoshimura M, Harada M, Nakagawa M, Mizuno Y, Shimasaki Y, Nakayama M, Yasue H, Kuwahara K, Saito Y, Nakao K (1999) Effect of interleukin-1 beta on cardiac hypertrophy and production of natriuretic peptides in rat cardiocyte culture. J Mol Cell Cardiol 31(11):1997–2006. doi:10.1006/jmcc.1999.1030

55. Cao L, Gardner DG (1995) Natriuretic peptides inhibit DNA synthesis in cardiac fibroblasts. Hypertension 25(2):227–234

56. Anand-Srivastava MB, Sairam MR, Cantin M (1990) Ring-deleted analogs of atrial natriuretic factor inhibit adenylate cyclase/cAMP system. Possible coupling of clearance atrial natriuretic factor receptors to adenylate cyclase/cAMP signal transduction system. J Biol Chem 265(15):8566–8572

57. Redondo J, Bishop JE, Wilkins MR (1998) Effect of atrial natriuretic peptide and cyclic GMP phosphodiesterase inhibition on collagen synthesis by adult cardiac fibroblasts. Br J Pharmacol 124(7):1455–1462. doi:10.1038/sj.bjp.0701994

58. Doyle DD, Upshaw-Earley J, Bell EL, Palfrey HC (2002) Natriuretic peptide receptor-B in adult rat ventricle is predominantly confined to the nonmyocyte population. Am J Physiol Heart Circ Physiol 282(6):H2117–2123. doi:10.1152/ajpheart.00988.2001

59. Huntley BK, Sandberg SM, Noser JA, Cataliotti A, Redfield MM, Matsuda Y, Burnett JC Jr. (2006) BNP-induced activation of cGMP in human cardiac fibroblasts: interactions with fibronectin and natriuretic peptide receptors. J Cell Physiol 209(3):943–949. doi:10.1002/jcp.20793

60. Glenn DJ, Rahmutula D, Nishimoto M, Liang F, Gardner DG (2009) Atrial natriuretic peptide suppresses endothelin gene expression and proliferation in cardiac fibroblasts through a GATA4-dependent mechanism. Cardiovasc Res 84(2):209–217. doi:10.1093/cvr/cvp208

61. Parthasarathy A, Gopi V, Umadevi S, Simna A, Sheik MJ, Divya H, Vellaichamy E (2013) Suppression of atrial natriuretic peptide/natriuretic peptide receptor-A-mediated signaling upregulates angiotensin-II-induced collagen synthesis in adult cardiac fibroblasts. Mol Cell Biochem 378(1–2):217–228. doi:10.1007/s11010-013-1612-z

62. Kapoun AM, Liang F, O'Young G, Damm DL, Quon D, White RT, Munson K, Lam A, Schreiner GF, Protter AA (2004) B-type natriuretic peptide exerts broad functional opposition to

transforming growth factor-beta in primary human cardiac fibroblasts: fibrosis, myofibro-blast conversion, proliferation, and inflammation. Circ Res 94(4):453–461. doi:10.1161/01. RES.0000117070.86556.9F

63. Sechi LA, Griffin CA, Grady EF, Kalinyak JE, Schambelan M (1992) Characterization of angiotensin II receptor subtypes in rat heart. Circ Res 71(6):1482–1489

64. Villarreal FJ, Kim NN, Ungab GD, Printz MP, Dillmann WH (1993) Identification of func-tional angiotensin II receptors on rat cardiac fibroblasts. Circulation 88(6):2849–2861

65. Crabos M, Roth M, Hahn AW, Erne P (1994) Characterization of angiotensin II receptors in cultured adult rat cardiac fibroblasts. Coupling to signaling systems and gene expression. J Clin Invest 93(6):2372–2378. doi:10.1172/JCI117243

66. Iwami K, Ashizawa N, Do YS, Graf K, Hsueh WA (1996) Comparison of ANG II with other growth factors on Egr-1 and matrix gene expression in cardiac fibroblasts. Am J Physiol 270(6 Pt 2):H2100–2107

67. Lijnen PJ, Petrov VV, Fagard RH (2001) Angiotensin II-induced stimulation of collagen se-cretion and production in cardiac fibroblasts is mediated via angiotensin II subtype 1 recep-tors. J Renin-Angiotensin-Aldosterone Syst 2(2):117–122. doi:10.3317/jraas.2001.012

68. Yanagisawa M, Kurihara H, Kimura S, Tomobe Y, Kobayashi M, Mitsui Y, Yazaki Y, Goto K, Masaki T (1988) A novel potent vasoconstrictor peptide produced by vascular endothelial cells. Nature 332(6163):411–415. doi:10.1038/332411a0

69. Gray MO, Long CS, Kalinyak JE, Li HT, Karliner JS (1998) Angiotensin II stimulates car-diac myocyte hypertrophy via paracrine release of TGF-beta 1 and endothelin-1 from fibro-blasts. Cardiovasc Res 40(2):352–363

70. Porter KE, Turner NA (2009) Cardiac fibroblasts: at the heart of myocardial remodeling. Pharmacol Ther 123(2):255–278. doi:10.1016/j.pharmthera.2009.05.002

71. Fujisaki H, Ito H, Hirata Y, Tanaka M, Hata M, Lin M, Adachi S, Akimoto H, Marumo F, Hiroe M (1995) Natriuretic peptides inhibit angiotensin II-induced proliferation of rat car-diac fibroblasts by blocking endothelin-1 gene expression. J Clin Invest 96(2):1059–1065. doi:10.1172/JCI118092

72. Piacentini L, Gray M, Honbo NY, Chentoufi J, Bergman M, Karliner JS (2000) Endothelin-1 stimulates cardiac fibroblast proliferation through activation of protein kinase C. J Mol Cell Cardiol 32(4):565–576. doi:10.1006/jmcc.2000.1109

73. Kawana M, Lee ME, Quertermous EE, Quertermous T (1995) Cooperative interaction of GATA-2 and AP1 regulates transcription of the endothelin-1 gene. Mol Cell Biol 15(8):4225–4231

74. Li YY, McTiernan CF, Feldman AM (2000) Interplay of matrix metalloproteinases, tissue in-hibitors of metalloproteinases and their regulators in cardiac matrix remodeling. Cardiovasc Res 46(2):214–224

75. Whittaker P (1995) Unravelling the mysteries of collagen and cicatrix after myocardial in-farction. Cardiovasc Res 29(6):758–762

76. Soeki T, Kishimoto I, Okumura H, Tokudome T, Horio T, Mori K, Kangawa K (2005) C-type natriuretic peptide, a novel antifibrotic and antihypertrophic agent, prevents cardiac remodeling after myocardial infarction. J Am Coll Cardiol 45(4):608–616. doi:10.1016/j. jacc.2004.10.067

77. Leask A, Abraham DJ (2004) TGF-beta signaling and the fibrotic response. FASEB J 18(7):816–827. doi:10.1096/fj.03-1273rev

78. Calvieri C, Rubattu S, Volpe M (2012) Molecular mechanisms underlying cardiac antihyper-trophic and antifibrotic effects of natriuretic peptides. J Mol Med 90(1):5–13. doi:10.1007/ s00109-011-0801-z

79. Hao J, Ju H, Zhao S, Junaid A, Scammell-La Fleur T, Dixon IM (1999) Elevation of expres-sion of Smads 2, 3, and 4, decorin and TGF-beta in the chronic phase of myocardial infarct scar healing. J Mol Cell Cardiol 31(3):667–678. doi:10.1006/jmcc.1998.0902

80. Hao J, Wang B, Jones SC, Jassal DS, Dixon IM (2000) Interaction between angiotensin II and Smad proteins in fibroblasts in failing heart and in vitro. Am J Physiol Heart Circ Physiol 279(6):H3020–3030

81. Brooks WW, Conrad CH (2000) Myocardial fibrosis in transforming growth factor beta(1) heterozygous mice. J Mol Cell Cardiol 32(2):187–195. doi:10.1006/jmcc.1999.1065

82. Li P, Wang D, Lucas J, Oparil S, Xing D, Cao X, Novak L, Renfrow MB, Chen YF (2008) Atrial natriuretic peptide inhibits transforming growth factor beta-induced Smad signaling and myofibroblast transformation in mouse cardiac fibroblasts. Circ Res 102(2):185–192. doi:10.1161/CIRCRESAHA.107.157677

83. Gunja-Smith Z, Morales AR, Romanelli R, Woessner JF Jr. (1996) Remodeling of human myocardial collagen in idiopathic dilated cardiomyopathy. Role of metalloproteinases and pyridinoline cross-links. Am J Pathol 148(5):1639–1648

84. Sorescu D, Griendling KK (2002) Reactive oxygen species, mitochondria, and NAD(P)H oxidases in the development and progression of heart failure. Congest Heart Fail 8(3):132–140

85. Bond M, Chase AJ, Baker AH, Newby AC (2001) Inhibition of transcription factor NF-kappaB reduces matrix metalloproteinase-1, -3 and -9 production by vascular smooth muscle cells. Cardiovasc Res 50(3):556–565

86. Fernandez-Catalan C, Bode W, Huber R, Turk D, Calvete JJ, Lichte A, Tschesche H, Maskos K (1998) Crystal structure of the complex formed by the membrane type 1-matrix metalloproteinase with the tissue inhibitor of metalloproteinases-2, the soluble progelatinase A receptor. EMBO J 17(17):5238–5248. doi:10.1093/emboj/17.17.5238

87. Moore L, Fan D, Basu R, Kandalam V, Kassiri Z (2012) Tissue inhibitor of metalloproteinases (TIMPs) in heart failure. Heart Fail Rev 17(4–5):693–706. doi:10.1007/s10741-011-9266-y

88. He J, Chen Y, Huang Y, Yao F, Wu Z, Chen S, Wang L, Xiao P, Dai G, Meng R, Zhang C, Tang L, Huang Y, Li Z (2009) Effect of long-term B-type natriuretic peptide treatment on left ventricular remodeling and function after myocardial infarction in rats. Eur J Pharmacol 602(1):132–137. doi:10.1016/j.ejphar.2008.10.064

89. Sun Y, Cleutjens JP, Diaz-Arias AA, Weber KT (1994) Cardiac angiotensin converting enzyme and myocardial fibrosis in the rat. Cardiovasc Res 28(9):1423–1432

90. Kim S, Ohta K, Hamaguchi A, Yukimura T, Miura K, Iwao H (1995) Angiotensin II induces cardiac phenotypic modulation and remodeling in vivo in rats. Hypertension 25(6):1252–1259

91. Izumiya Y, Araki S, Usuku H, Rokutanda T, Hanatani S, Ogawa H (2012) Chronic C-type natriuretic peptide infusion attenuates angiotensin II-induced myocardial superoxide production and cardiac remodeling. Int J Vasc Med 2012:246058. doi:10.1155/2012/246058

92. Rose RA (2010) CD-NP, a chimeric natriuretic peptide for the treatment of heart failure. Curr Opin Investig Drugs 11(3):349–356

93. Lee CY, Lieu H, Burnett JC Jr. (2009) Designer natriuretic peptides. J Investig Med 57(1):18–21. doi:10.231/JIM.0b013e3181946fb2

94. Dickey DM, Burnett JC Jr., Potter LR (2008) Novel bifunctional natriuretic peptides as potential therapeutics. J Biol Chem 283(50):35003–35009. doi:10.1074/jbc.M804538200

95. Martin FL, Sangaralingham SJ, Huntley BK, McKie PM, Ichiki T, Chen HH, Korinek J, Harders GE, Burnett JC Jr. (2012) CD-NP: a novel engineered dual guanylyl cyclase activator with anti-fibrotic actions in the heart. PloS One 7(12):e52422. doi:10.1371/journal.pone.0052422

96. Ng XW, Huang Y, Chen HH, Burnett JC Jr., Boey FY, Venkatraman SS (2013) Cenderitide-eluting film for potential cardiac patch applications. PloS One 8(7):e68346. doi:10.1371/journal.pone.0068346

97. Chilton L, Ohya S, Freed D, George E, Drobic V, Shibukawa Y, Maccannell KA, Imaizumi Y, Clark RB, Dixon IM, Giles WR (2005) K+ currents regulate the resting membrane potential, proliferation, and contractile responses in ventricular fibroblasts and myofibroblasts. Am J Physiol Heart Circ Physiol 288(6):H2931–2939. doi:10.1152/ajpheart.01220.2004

98. Rose RA, Hatano N, Ohya S, Imaizumi Y, Giles WR (2007) C-type natriuretic peptide activates a non-selective cation current in acutely isolated rat cardiac fibroblasts via natriuretic peptide C receptor-mediated signalling. J Physiol 580(Pt 1):255–274. doi:10.1113/jphysiol.2006.120832

99. Harada M, Luo X, Qi XY, Tadevosyan A, Maguy A, Ordog B, Ledoux J, Kato T, Naud P, Voigt N, Shi Y, Kamiya K, Murohara T, Kodama I, Tardif JC, Schotten U, Van Wagoner DR, Dobrev D, Nattel S (2012) Transient receptor potential canonical-3 channel-dependent fibroblast regulation in atrial fibrillation. Circulation 126(17):2051–2064. doi:10.1161/CIR-CULATIONAHA.112.121830
100. Rose RA, Belke DD, Maleckar MM, Giles WR (2012) Ca2+ entry through TRP-C channels regulates fibroblast biology in chronic atrial fibrillation. Circulation 126(17):2039–2041. doi:10.1161/CIRCULATIONAHA.112.138065

Cardiac Tissue Engineering for the Treatment of Heart Failure Post-Infarction

Jacqueline S. Wendel and Dr. Robert. T. Tranquillo

Abstract Cell-based therapy has become an attractive solution to the high incidence of heart failure post-infarction. Many current approaches to cell delivery
post-infarction result in poor cell engraftment, resulting in limited functional benefits. Thus, the use of engineered tissues to deliver cells to the injured myocardium
or replace myocardium post infarction has been a topic of increasing interest. Tissue
engineering provides a platform for the delivery of a large number of cells to the
injured myocardium with high retention, allowing for *in vitro* development of cellular organization, intracellular communication and ECM deposition. This chapter
will discuss the currently used methods to create engineered cardiac tissues, including scaffolds, cells, and cellular conditioning. This chapter will also review the efficacy of these patches in limiting left ventricular remodeling post-infarction *in vivo*.

Keywords Tissue engineering · Myocardial infarction · Embryonic stem cells ·
Induced pluripotent stem cells · Endothelial cells · Bioreactor conditioning · Cyclic
stretch · Electrical stimulation · Perfusion · Biomaterials

1 Introduction

Heart failure has many causes, but the most prevalent initiator is a myocardial infarction, in which myocardial tissue is deprived of oxygen for an extended period
of time, usually through a blockage of one of the coronary arteries. In the ischemia
induced by an infarction, one-quarter of the 4 billion cells in the left ventricle can
be lost [1]. With this level of cell death, heart failure can develop as a result of the
limited capacity of the injured myocardial tissue to recover or regenerate, leading
to fibrosis and scar formation of the damaged myocardium, left ventricular dilation
and thinning due to a resulting pressure and volume overload, and the inhibition of
proper action potential propagation. Current treatments, whether they are pharma-

R. T. Tranquillo (✉) · J. S. Wendel
Department of Biomedical Engineering, , University of Minnesota, 7-114 Hasselmo Hall,
Minneapolis, MN 55455, USA
e-mail: wende045@umn.edu

© Springer International Publishing Switzerland 2015   405
I.M.C. Dixon, J. T. Wigle (eds.), *Cardiac Fibrosis and Heart Failure: Cause or Effect?,*
Advances in Biochemistry in Health and Disease 13, DOI 10.1007/978-3-319-17437-2_20

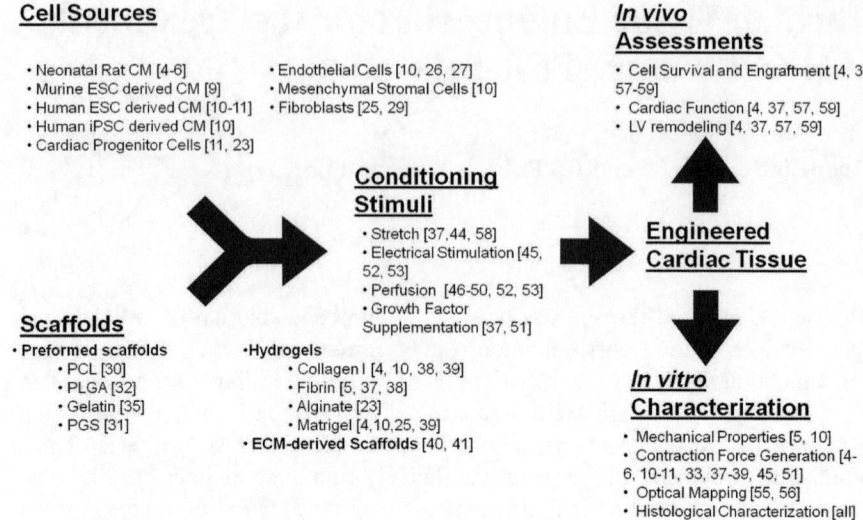

Cell Sources

- Neonatal Rat CM [4-6]
- Murine ESC derived CM [9]
- Human ESC derived CM [10-11]
- Human iPSC derived CM [10]
- Cardiac Progenitor Cells [11, 23]

- Endothelial Cells [10, 26, 27]
- Mesenchymal Stromal Cells [10]
- Fibroblasts [25, 29]

In vivo Assessments

- Cell Survival and Engraftment [4, 37, 57-59]
- Cardiac Function [4, 37, 57, 59]
- LV remodeling [4, 37, 57, 59]

Conditioning Stimuli

- Stretch [37,44, 58]
- Electrical Stimulation [45, 52, 53]
- Perfusion [46-50, 52, 53]
- Growth Factor Supplementation [37, 51]

Engineered Cardiac Tissue

Scaffolds

- Preformed scaffolds
  - PCL [30]
  - PLGA [32]
  - Gelatin [35]
  - PGS [31]

- Hydrogels
  - Collagen I [4, 10, 38, 39]
  - Fibrin [5, 37, 38]
  - Alginate [23]
  - Matrigel [4,10,25, 39]
- ECM-derived Scaffolds [40, 41]

In vitro Characterization

- Mechanical Properties [5, 10]
- Contraction Force Generation [4-6, 10-11, 33, 37-39, 45, 51]
- Optical Mapping [55, 56]
- Histological Characterization [all]

Fig. 1 Construction and characterization of cardiac patches

ceutical or medical device-based, act merely as palliative measures and the only effective long term treatment to date for heart failure is to replace the damaged myocardium via total heart transplantation [2].

With the limited number of donors and the inherent risks of surgery and immunogenicity, cellular therapy has become an attractive solution to the high incidence of heart failure post-infarction. However, direct injection of cells into the myocardium has shown limited efficacy due to poor grafting efficiency [3]. Low rates of retention may be a product of cell loss due to inability to create focal adhesions with neighboring cells, the inflammatory response to myocardial injury, or the hypoxic environment of the infarct zone. Thus, the use of engineered cardiac tissues has been a topic of increasing interest. Tissue engineering not only provides a platform for the delivery of a large number of cells to the injured myocardium, it provides a means to replace damaged myocardium. Tissue engineering allows for *in vitro* development of cellular organization, intracellular communication and ECM deposition while also isolating the cells from the inflammatory infarct environment when implanted *in vivo* (see Fig. 1).

In this chapter, we will review the currently used methods used to create engineered cardiac tissues, or "cardiac patches", and their efficacy when delivered *in vivo*. We will begin by surveying the cells, scaffold and *in vitro* conditioning used to create these patches, and then move on to how they are characterized, and finally their efficacy to limit left ventricular remodeling post-infarction.

2 Cells for Cardiac Tissue Engineering

2.1 Neonatal Rat Cardiomyocytes

Until functional cardiomyocytes were able to be differentiated from human pluripotent stem cell sources, primary isolates of neonatal rat cardiomyocytes were the standard cell source for the development of cardiac patches [4–6]. These cells have been well characterized *in vitro* and can be obtained in large numbers, making them a useful tool for developing cardiac patches and interrogating the effects of *in vitro* conditioning on these tissues.

2.2 Pluripotent Stem Cell Derived Cardiomyocytes

Stem cells, whether taken from an embryo or reprogrammed from an adult cell, have enormous therapeutic potential due to their ability to be expanded in an undifferentiated state, and differentiated into cells of any lineage found in the body. In their undifferentiated state, pluripotent stem cells cannot be transplanted into patients as they can form teratomas. However, as the ability to not only differentiate pluripotent stem cells into functional cardiomyocytes but also in numbers large enough to be used therapeutically has become available, these cells have become the new standard cell source for creating cardiac patches.

2.2.1 Embryonic Stem Cell Derived Cardiomyocytes

Spontaneous differentiation of embryonic stem cells (ESCs) to cardiogenic cells was first observed in mouse ESCs in 1981 when these cells were cultured in 3D aggregates, called embryoid bodies [7]. Human ESCs were first obtained from a blastocyst in 1998 [8], but it wasn't shown until 2001 that they could successfully be differentiated into functional cardiomyocytes that exhibited contractile force generation and recordable action potentials [9]. Human ESC-CMs are now commonly used as a cell source for cardiac tissue engineering [10, 11].

2.2.2 Induced Pluripotent Stem Cell Derived Cardiomyocytes

In 2006 it was discovered that the introduction of four transcription factors (Oct3/4, Sox2, c-Myc, and Klf4) into adult murine fibroblasts reprogrammed these cells into an embryonic state, called induced pluripotent stem cells (iPSCs) [12]. This was a landmark discovery, as the reprogramming of adult cells into an embryonic state not only alleviates the ethical concerns of using embryonic cardiomyocytes but also allows for potentially autologous cell transplantation, eliminating the immunogenicity obstacle of allogeneic transplantation. Human iPSCs were reported in 2007 using the same factors or a new combination of Oct4, NANOG, LIN28, and SOX2 [13, 14], and were differentiated into functional cardiomyocytes in 2009 [15]. They have been used in parallel with hESC-CMs for tissue engineering applications, and will likely become the preferred cell source for cardiac patches for the foreseeable future [10].

2.2.3 Differentiation Methods

Initially, pluripotent stem cell derived cardiomyocytes were differentiated via the embryoid body method, where undifferentiated pluripotent stem cells are cultured in 3D aggregates on top of irradiated mouse embryonic fibroblast feeder cells and allowed to spontaneously differentiate into cells from all three embryonic germ layers, among them being cardiomyocytes. However, as can be expected, this method results in low percentages and yields of cardiomyocytes and requires purification steps in order to obtain high percentages of cardiomyocytes. In efforts to efficiently obtain high yields of cardiomyocytes from stem cell differentiation, Zhang et al developed a new method in which cells are cultured between two layers of matrigel [16], resulting in a significant increase in differentiation efficiency to cardiomyocytes. Recently, small molecule methods of reprogramming and differentiation under defined conditions have been developed to eliminate the need for viral vectors and to increase reprogramming efficiency [17].

2.3 Cardiac Progenitor Cells

The heart has a limited capacity to regenerate via the presence of sparsely distributed resident cardiac progenitor cells (CPCs) and cardiac side population (SP) cells within the myocardium, approximately one CPC for every 30,000–40,000 cells in the myocardium [18]. These cells have been isolated based on their expression of a number of cell surface markers and gene expression, including c-kit [19], Sca-1 [20], and islet-1 [21] as well as from their migration of progenitor cells out of excised cardiac tissue when cultured *in vitro* [22]. Cells selected through these methods display the capacity to self-renew and differentiate into cardiomyocyte as well as endothelial phenotypes and can be used to create cardiac patches [11, 23].

2.4 Non-Cardiomyocyte Cells and Co-Culture of Cells for Cardiac Patches

Cardiomyocytes are not the only cell type used to create engineered tissues for cardiac repair. Non-cardiomyocyte cells are required for the creation of cardiac patches that utilize compacting biopolymer hydrogels, as cardiomyocytes do not contract collagen and fibrin hydrogels. Despite improvements in differentiation techniques, pluripotent stem cell-derived CM populations are only 60–90% pure CMs, with the remaining 10–40% of cells primarily being fibroblast-like cells, though those cells remain poorly characterized. Neonatal rat cell isolates also contain between 50–60% of a heterogeneous population of non-CM cells, consisting primarily of fibroblasts, but also containing smooth muscle cells and endothelial cells from the vasculature and other interstitial cells from the myocardium [24]. The addition or inclusion of non-CM cells into cardiac patches has proven to be beneficial. Co-culture of hESC and hiPSc-derived cardiomyocytes with endothelial cells with or without mesenchymal stromal cells has resulted in an increase in contractile force generation by the resulting cardiac patches [10], and the force generated per cardiomyocyte has shown to be higher with less pure populations of hESC-CMs [25]. Other non-CM cells used in cardiac tissue engineering include blood outgrowth endothelial cells (BOECs) [26], human umbilical endothelial cells (HUVECs) [27], pericytes (PCs) [28], and human dermal fibroblasts (HDFs) [29]. These cells are used either in combination with cardiomyocytes or by themselves in engineered tissues, all with the goal of rescuing the injured myocardium post-infarction.

3 Scaffolds for Cardiac Tissue Engineering

Cardiac patches are typically formed by the *in vitro* seeding and culturing cells in 3-dimensional scaffolds. The choice of scaffold influences cell survival, phenotype, and function through its physical and chemical properties, including stiffness, microstructure, and surface chemistry. Two general classes of scaffolds are utilized to create cardiac patches: prefabricated scaffolds, in which cells are seeded onto a pre-existing scaffold structure; and biopolymer scaffolds, in which cells are present during scaffold polymerization and become entrapped in a fibrous network.

3.1 Prefabricated Scaffolds

Synthetic biodegradable, elastomeric polymers including polycaprolactone (PCL) [30], poly(glycerol-sebacate) (PGS) [31], and poly(lactide-co-glycolide) (PLGA) [32] copolymers and natural materials such as collagen I [33], alginate [34], and gelatin [35] are materials that have been used to create prefabricated foam or nanofiber scaffolds. Foam scaffolds allow for control of pore size and structure through the use of different material processing techniques. Highly porous scaffolds have proven successful in creating cardiac patches with high cell viability, but they are often isotropic and do not result in cellular alignment. Nanofiber scaffolds can be created through electrospinning [36], in which fiber diameter and orientation can be controlled, mimicking organization of collagen fibers in the native myocardium. These scaffolds can be fabricated in advance and can incorporate surface modifications to enhance cell adhesion and activity, such as the conjugation of RGD peptides [23].

3.2   Biopolymer Scaffolds

Biopolymers are another class of scaffolds used in cardiac tissue engineering. These are native protein fibril networks in hydrogel form made from either collagen I or fibrin [5, 37, 38]. Rather than being seeded onto pre-formed scaffolds, cells are suspended in a gel-forming solution and fibrillogenesis occurs around the cells, entrapping them in a hydrated, fibrillar network. Initially, collagen and fibrin gels are extremely soft anisotropic. Over time, entrapped non-CM cells compact the gel, resulting in a denser fibril network. Eventually, the entrapped cells not only compact the gel, but begin to degrade the fibrils (in the case of fibrin gels) and replace them with cell-produced matrix. By constraining the compaction of the gel, fibrillar and cellular alignment can be induced in these gels. Induction of alignment allows the native architecture of the myocardium to be mimicked and has been shown to result in an increase in both contractile force generation by entrapped cardiomyocytes and the presence of gap junctions between cardiomyocytes [5]. Collagen and fibrin biopolymer scaffolds can be altered by the addition of Matrigel [10, 25, 39]. Other biopolymer scaffold materials include alginate, which can be 3D printed into various geometries [23].

3.3   Tissue-Based Scaffolds

Decellularized heart tissue has been used as a scaffold to culture cardiac cells. Heart tissues are decellularized by perfusion with SDS followed by Triton-X to remove cells while leaving most of the heart ECM content and structure intact [40]. Cells can then be reperfused into the heart [40], or entrapped in an ECM-hydrogel matrix to create cardiac patches [41].

3.4   Scaffold-Free Tissues

Scaffold-free approaches have been used to create cardiac patches in the form of cell sheets [42] and cell aggregate patches [43]. Cell sheets are created by seeding cells onto temperature-responsive membranes, allowing them to deposit cell-produced ECM and form cellular connections, and releasing them from the culture surfaces through temperature reduction. Cell sheets can then be stacked to create thin 3D tissues or layered with cell sheets of endothelial cells to facilitate rapid vascularization upon implantation [42]. Cell aggregate patches are prepared by culturing cells on a low-attachment plate placed on an orbital shaker, allowing for large numbers of cells to aggregate in solution and form tissues.

4 *In vitro* Cellular Conditioning

In addition to seeding cells onto or into scaffolds, external stimulation of the cells is often required to induce them to convert the scaffold into a functional cardiac tissue. This may be achieved through mechanical, electrical, or pharmacological stimulation.

4.1 Mechanical Stimulation

Cardiomyocytes in the native myocardium experience cyclic mechanical stretch as part of the cardiac cycle. Taking inspiration from this, *in vitro* mechanical stimulation has been widely used to stimulate alignment, hypertrophy, and maturation of cardiomyocytes in engineered tissues. Stimulation through cyclic stretching has been achieved for ring-shaped cardiac patches using two methods: looping the patch around one fixed post and another coupled to a motorized stretching device [44], or through pneumatically-controlled distension of a latex mandrel on which the patch is mounted [37]. In both instances, cyclic stretching resulted in over a two-fold increase in contractile force generation.

4.2 Electrical Stimulation

Ventricular cardiomyocytes in the heart beat under the continuous regulation of their electrical activity through the cardiac conduction system originating at the sinoatrial node. *In vitro*, cardiomyocytes in cardiac patches beat spontaneously, but often at an irregular and variable rate, and not in synchrony with cardiomyocytes in other regions of the patch. Electrical stimulation of cardiac patches throughout culture has resulted in increases in the maximum frequency at which these patches can be stimulated and elicit a contractile response synchronous with pacing, a reduction in the voltage threshold at which cardiomyocytes contract, increased contractile force and also stimulated cardiomyocyte hypertrophy [45].

4.3 In vitro Perfusion

Due to the high metabolic demand of cardiomyocytes, diffusion limitations hamper the ability to make the thicker cardiac patches that are necessary to achieve clinical relevance. To increase oxygen and nutrient availability to cells, different methods of *in vitro* perfusion have been utilized in cardiac patches. Pulsatile perfusion of cell culture medium through porous pre-formed scaffolds has proven effective in increasing cell viability and myofibril assembly through media flow [6] and activates the ERK 1/2 signaling pathway [46]. In biopolymer hydrogels, microchannels have been included in the gel [47], and perfusable microvessels have been created within the gel to allow for medium flow through the gel during culture [48–50]. Though promising, this latter approach has yet to be combined with cardiomyocyte culture *in vitro*.

4.4   Medium Supplementation

Supplementation of culture medium with pharmacological agents can also be used to stimulate development of cardiac patches. The addition of platelet-derived growth factor BB (PDGF-BB) to cultures of collagen/matrigel scaffolds containing neonatal rat cardiomyocytes protected cells from apoptotic death, thus increasing the final contractile performance of the patch [51]. Insulin and ascorbic acid have been used to promote matrix deposition by entrapped non-CM cells [37].

4.5   Combination Treatments

In addition to exposing cardiac patches to one form of *in vitro* conditioning, combinations of the above mentioned techniques have been used in combination to further condition tissues. Electrical stimulation has been used in concert with perfusion and resulted in an increase in cell elongation, enhanced expression of gap junction protein connexin-43, increase in cell number, and an increase in contractile performance [52, 53]. However, the combination of stretch and β-adrenergic stimulation resulted in no additional benefit [53].

5   Characterization of Cardiac Patch Function

One additional benefit of using engineered tissues to treat heart failure post-infarction is that these engineered tissues and the cells entrapped in them can be functionally characterized *in vitro*, allowing for quality control and a more thorough understanding of the benefits of cell transplantation. Functional performance of patches is primarily assessed through contractile force measurements and optical mapping of electrical conductivity.

5.1   Contractile Force Measurements

Cardiac patches should, once developed, function as a piece of cardiac muscle tissue and generate contractile stresses close to physiological values. Quantification of this contractile activity is widely used to assess the quality of a cardiac patch. Contractile activity can be assessed through the magnitude of patch contraction both spontaneously and in response to pacing, the maximum frequency by which a patch can be paced and still elicit a contractile response in phase with the electrical stimulus, the force-frequency relationship of contractile amplitude with increasing pacing frequency, and the contractile response of the patch to exposure to pharmacological agents. Such agents include β-adgrenergic agonists, gap junction blocker 1-Heptanol [54], or cholinergic agonist carbacol [39].

5.2 Optical Mapping

Electrical activity of cardiac patches can be assessed through optical mapping. To do this, tissues are submerged in medium containing a voltage sensitive dye, such as Di-4ANEPPS or ANNINE-6. Tissues are then point-stimulated and the action potential propagation is recorded by a high speed camera. This data can be analyzed to assess both conduction velocity (the rate at which action potentials propagate across the tissue), and action potential duration (the time it takes for a cell to re polarize to either 50 or 90 % of its resting potential). This data can be used to obtain the conduction velocities of action potentials across the patch, as well as the action potential duration. These values can then be compared to physiological values to assess the functional quality of the cardiac patch [55]. Additionally, calcium transients can be similarly imaged with a calcium sensitive dye [56].

6 *In vivo* Assessments of Cardiac Patches

Although cardiac patches can be fully characterized *in vitro* and can be conditioned *in vitro* to obtain functional values close to that of the native myocardium, cardiac patches can only truly be effective if their implantation results in the restoration of cardiac function post-infarction. Some studies have made progress towards achieving this goal, utilizing both small and large animal models, and acute or delayed implantation. The timing of patch implantation is a crucial aspect of *in vivo* assessments as a way to help determine the mechanisms by which any benefits provided by the patch occur. Acute transplantation is transplantation of the patch immediately after infarction, delivering it during the initial inflammatory phase before fibrosis begins to occur. Chronic, or delayed transplantation, requires a second surgery and applies the patch after the inflammatory phase has passed and either before or after mature scar formation, depending on the time of implantation.

6.1 Rodent Models

Rodent models of myocardial infarction treatments are commonly used as the first initial model due to the smaller size of patch required for treatment. Acute placement of a cardiac patch using fibrin as the scaffold and syngeneic neonatal rat cardiac cells, with *in vitro* stretch conditioning of the patch prior to implantation, resulted in minimal scar formation and restoration of cardiac function 4 weeks post-infarct, with 36% cardiomyocyte retention [37]. A similar patch made with a collagen/Matrigel scaffold rather than fibrin and implanted 2 weeks post-infarction limited further scar formation and preserved cardiac function from the time of implantation [4]. Transplantation of a layered neonatal rat cell sheet patch into a nude rat infarct model 14 days post-infarction resulted in significant improvements in cardiac function, though cell retention was minimal after 4 weeks [57]. Few studies to date have been published assessing the functional consequences of a human stem cell derived cardiac patch in an infarcted rat model. Acute transplantation into non-infarcted rats of scaffold-free tissue patches utilizing hESC-CMs and HUVECs resulted in rapid vascularization of the patches *in vivo*. Patches containing only an enriched CM population resulted in poor CM retention [58].

6.2   Large Animal Models

The use of large animal models of myocardial infarction is a necessary step towards bringing cardiac patches to clinical relevance Few studies to date have been able to generate the large numbers of human stem cell derived cardiomyocytes required to create cardiac patches large enough to be implemented in large animal models. Human ESC-CMs in cell sheet form were transplanted into a porcine ischemia-reperfusion model 4 weeks after ligation in a feasibility and safety study, and it was found that the cell sheet treatment resulted in a reduction in LV dilation and an improvement in cardiac function both 4 and 8 weeks post-transplantation [59].

7   Current Deficiencies and Obstacles

Despite the many advances in cardiac tissue engineering, there remain deficiencies in the field and many obstacles still to be overcome. Contractile force generation and cellularity of cardiac patches average far below physiological values of 44 mN/mm^2 and 20–40 M cells/gram for human ventricular myocardium and 56.4 N/mm^2 rat ventricular myocardium [60]. Diffusional limitations may contribute to this, and may be overcome by the development of a microvascular network- withing the patch during its fabrication. As mentioned earlier, progress has been made, but microvascular network have yet to be created and perfused in the presence of a cardiomyocyte-containing patch of size relevant to even a rat infarct. Other issues that will be addressed in the coming years are maturation mismatch of donor cells to recipient hearts. Currently used stem cell-derived cardiomyocytes are a mixture of atrial, ventricular, and pacemaker phenotypes, and all exhibit a fetal or neonatal phenotype. Immature phenotypes also have different electrophysiological, metabolic, and contractile properties than adult CMs. It has yet to be determined if transplantation of immature cells will elicit arrhythmic events, or if cells need to be differentiated to maturity prior to transplantation. However, more mature cells may not survive transplantation with the same efficiency as less mature cells.

References

1. Caulfield JB, Leinbach R, Gold H (1976) The relationship of myocardial infarct size and prognosis. Circulation 53(3 Suppl):I141–144
2. Miniati DN, Robbins RC (2011) Heart transplantation: a thirty-year perspective. Annu Rev Med 53(1):189–205
3. Reinecke H, Murry CE (2002) Taking the death toll after cardiomyocyte grafting: a reminder of the importance of quantitative biology. J Mol Cell Cardiol 34(3):251–253
4. Zimmermann WH, Melnychenko I, Wasmeier G, Didie M, Naito H, Nixdorff U, Hess A, Budinsky L, Brune K, Michaelis B, Dhein S, Schwoerer A, Ehmke H, Eschenhagen T (2006) Engineered heart tissue grafts improve systolic and diastolic function in infarcted rat hearts. Nat Med 12(4):452–458
5. Black LD 3rd, Meyers JD, Weinbaum JS, Shvelidze YA, Tranquillo RT (2009) Cell-induced alignment augments twitch force in fibrin gel-based engineered myocardium via gap junction modification. Tissue Eng Part A 15(10):3099–3108
6. Brown MA, Iyer, Rohin K., Radisic, M. (2008) Pulsatile perfusion bioreactor for cardiac tissue engineering. Biotechnol Prog 2008:907–920
7. Evans MJ, Kaufman MH (1981) Establishment in culture of pluripotential cells from mouse embryos. Nature 292(5819):154–156
8. Thomson JA, Itskovitz-Eldor J, Shapiro SS, Waknitz MA, Swiergiel JJ, Marshall VS, Jones JM (1998) Embryonic stem cell lines derived from human blastocysts. Science 282(5391):1145–1147
9. Kehat I, Kenyagin-Karsenti D, Snir M, Segev H, Amit M, Gepstein A, Livne E, Binah O, Itskovitz-Eldor J, Gepstein L (2001) Human embryonic stem cells can differentiate into myocytes with structural and functional properties of cardiomyocytes. J Clin Invest 108(3):407–414. doi:10.1172/jci12131
10. Tulloch NL, Muskheli V, Razumova MV, Korte FS, Regnier M, Hauch KD, Pabon L, Reinecke H, Murry CE (2011) Growth of engineered human myocardium with mechanical loading and vascular coculture. Circ Res 109(1):47–59
11. Liau B, Christoforou N, Leong KW, Bursac N (2011) Pluripotent stem cell-derived cardiac tissue patch with advanced structure and function. Biomaterials 32(35):9180–9187
12. Takahashi K, Yamanaka S (2006) Induction of pluripotent stem cells from mouse embryonic and adult fibroblast cultures by defined factors. Cell 126(4):663–676
13. Yu J, Vodyanik MA, Smuga-Otto K, Antosiewicz-Bourget J, Frane JL, Tian S, Nie J, Jonsdottir GA, Ruotti V, Stewart R, Slukvin, II, Thomson JA (2007) Induced pluripotent stem cell lines derived from human somatic cells. Science 318(5858):1917–1920
14. Takahashi K, Tanabe K, Ohnuki M, Narita M, Ichisaka T, Tomoda K, Yamanaka S (2007) Induction of pluripotent stem cells from adult human fibroblasts by defined factors. Cell 131(5):861–872
15. Zhang J, Wilson GF, Soerens AG, Koonce CH, Yu J, Palecek SP, Thomson JA, Kamp TJ (2009) Functional cardiomyocytes derived from human induced pluripotent stem cells. Circ Res 104(4):e30–e41
16. Zhang J, Klos M, Wilson GF, Herman AM, Lian X, Raval KK, Barron MR, Hou L, Soerens AG, Yu J, Palecek SP, Lyons GE, Thomson JA, Herron TJ, Jalife J, Kamp TJ (2012) Extracellular matrix promotes highly efficient cardiac differentiation of human pluripotent stem cells: the matrix sandwich method. Circ Res 111(9):1125–1136

17. Lian X, Zhang J, Azarin SM, Zhu K, Hazeltine LB, Bao X, Hsiao C, Kamp TJ, Palecek SP (2013) Directed cardiomyocyte differentiation from human pluripotent stem cells by modulating Wnt/beta-catenin signaling under fully defined conditions. Nat Protoc 8(1):162–175

18. Leri A, Kajstura J, Anversa P (2005) Cardiac stem cells and mechanisms of myocardial regeneration. Physiol Rev 85:1373–1416

19. Beltrami AP, Barlucchi L, Torella D, Baker M, Limana F, Chimenti S, Kasahara H, Rota M, Musso E, Urbanek K, Leri A, Kajstura J, Nadal-Ginard B, Anversa P (2003) Adult cardiac stem cells are multipotent and support myocardial regeneration. Cell 114(6):763–776

20. Oh H, Bradfute SB, Gallardo TD, Nakamura T, Gaussin V, Mishina Y, Pocius J, Michael LH, Behringer RR, Garry DJ, Entman ML, Schneider MD (2003) Cardiac progenitor cells from adult myocardium: homing, differentiation, and fusion after infarction. Proc Natl Acad Sci U S A 100(21):12313–12318

21. Laugwitz K-L, Moretti A, Lam J, Gruber P, Chen Y, Woodard S, Lin L-Z, Cai C-L, Lu MM, Reth M, Platoshyn O, Yuan JX-J, Evans S, Chien KR (2005) Postnatal isl1+ cardioblasts enter fully differentiated cardiomyocyte. Nature 433(7026):647–653

22. Messina E, De Angelis L, Frati G, Morrone S, Chimenti S, Fiordaliso F, Salio M, Battaglia M, Latronico MVG, Coletta M, Vivarelli E, Frati L, Cossu G, Giacomello A (2004) Isolation and expansion of adult cardiac stem cells from human and murine heart. Circ Res 95(9):911–921

23. Gaetani R, Doevendans PA, Metz CH, Alblas J, Messina E, Giacomello A, Sluijter JP (2012) Cardiac tissue engineering using tissue printing technology and human cardiac progenitor cells. Biomaterials 33(6):1782–1790

24. Banerjee I, Fuseler JW, Price RL, Borg TK, Baudino TA (2007) Determination of cell types and numbers during cardiac development in the neonatal and adult rat and mouse. Am J Physiol Heart Circ Physiol 293:H1883–H1891

25. Zhang D, Shadrin IY, Lam J, Xian HQ, Snodgrass HR, Bursac N (2013) Tissue-engineered cardiac patch for advanced functional maturation of human ESC-derived cardiomyocytes. Biomaterials 34(23):5813–5820

26. Morin KT, Tranquillo RT (2011) Guided sprouting from endothelial spheroids in fibrin gels aligned by magnetic fields and cell-induced gel compaction. Biomaterials 32(26):6111–6118

27. Chouinard JA, Gagnon S, Couture MG, Levesque A, Vermette P (2009) Design and validation of a pulsatile perfusion bioreactor for 3D high cell density cultures. Biotechnol Bioeng 104(6):1215–1223

28. Stratman AN, Malotte KM, Mahan RD, Davis MJ, Davis GE (2009) Pericyte recruitment during vasculogenic tube assembly stimulates endothelial basement membrane matrix formation. Blood 114:5091–5101

29. Kellar RS, Landeen LK, Shepherd BR, Naughton GK, Ratcliffe A, Williams SK (2001) Scaffold-based three-dimensional human fibroblast culture provides a structural matrix that supports angiogenesis in infarcted heart tissue. Circulation 104(17):2063–2068

30. Yeong W, Sudarmadji N, Yu H, Chua C, Leong K, Venkatraman S, Boey Y, Tan L (2010) Porous polycaprolactone scaffold for cardiac tissue engineering fabricated by selective laser sintering. Acta Biomater 6(6):2028–2034

31. Engelmayr GC Jr, Cheng M, Bettinger CJ, Borenstein JT, Langer R, Freed LE (2008) Accordion-like honeycombs for tissue engineering of cardiac anisotropy. Nat Mater 7(12):1003–1010. 6

32. Caspi O, Lesman A, Basevitch Y, Gepstein A, Arbel G, Habib IH, Gepstein L, Levenberg S (2007) Tissue engineering of vascularized cardiac muscle from human embryonic stem cells. Circ Res 100(2):263–272

33. Radisic M, Euloth M, Yang L, Langer R, Freed LE, Vunjak-Novakovic G (2003) High-density seeding of myocyte cells for cardiac tissue engineering. Biotechnol Bioeng 82(4):403–414

34. Dar A, Shachar M, Leor J, Cohen S (2002) Optimization of cardiac cell seeding and distribution in 3D porous alginate scaffolds. Biotechnol Bioeng 80(3):305–312

35. Li R-K, Jia Z-Q, Weisel RD, Mickle DAG, Choi A, Yau TM (1999) Survival and function of bioengineered cardiac grafts. Circulation 100(suppl 2):II63–Ii69

36. Ifkovits JL, Devlin JJ, Eng G, Martens TP, Vunjak-Novakovic G, Burdick JA (2009) Bio-degradable fibrous scaffolds with tunable properties formed from photo-cross-linkable poly(glycerol sebacate). ACS Appl Mater Interfaces 1(9):1878–1886

37. Wendel JS, Ye L, Zhang P, Tranquillo RT, Zhang JJ (2014) Functional consequences of a tissue-engineered myocardial patch for cardiac repair in a rat infarct model. Tissue Eng Part A 20(7–8):1325–1335

38. Boudou T, Legant WR, Mu A, Borochin MA, Thavandiran N, Radisic M, Zandstra PW, Epstein JA, Margulies KB, Chen CS (2012) A microfabricated platform to measure and manipulate the mechanics of engineered cardiac microtissues. Tissue Eng Part A 18(9–10):910–919

39. Zimmermann WH (2001) Tissue engineering of a differentiated cardiac muscle construct. Circ Res 90 (2):223–230

40. Ott HC, Matthiesen TS, Goh SK, Black LD, Kren SM, Netoff TI, Taylor DA (2008) Perfusion-decellularized matrix: using nature's platform to engineer a bioartificial heart. Nat Med 14(2):213–221

41. Duan Y, Liu Z, O'Neill J, Wan LQ, Freytes DO, Vunjak-Novakovic G (2011) Hybrid gel composed of native heart matrix and collagen induces cardiac differentiation of human embryonic stem cells without supplemental growth factors. J Cardiovasc Transl Res 4(5):605–615

42. Sekine H, Shimizu T, Hobo K, Sekiya S, Yang J, Yamato M, Kurosawa H, Kobayashi E, Okano T (2008) Endothelial cell coculture within tissue-engineered cardiomyocyte sheets enhances neovascularization and improves cardiac function of ischemic hearts. Circulation 118(14 Suppl):S 145–S152

43. Stevens KR, Kreutziger KL, Dupras SK, Korte FS, Regnier M, Muskheli V, Nourse MB, Bendixen K, Reinecke H, Murry CE (2009) Physiological function and transplantation of scaffold-free and vascularized human cardiac muscle tissue. Proc Natl Acad Sci U S A 106(39):16568–16573

44. Fink C, Ergün S, Kralisch D, Remmers U, Weil J, Eschenhagen T (2000) Chronic stretch of engineered heart tissue induces hypertrophy and functional improvement. FASEB J 14 (5):669–679

45. Chiu LLY, Iyer RK, King J-P, Radisic M (2011) Biphasic electrical field stimulation aids in tissue engineering of multicell-type cardiac organoids. Tissue Eng Part A 17(11–12):1465–1477

46. Dvir T, Levy O, Shachar M, Granot Y, Cohen S (2007) Activation of the ERK1/2 cascade via pulsatile interstitial fluid flow promotes cardiac tissue assembly. Tissue Eng 13(9):2185–2193

47. Kolesky DB, Truby RL, Gladman AS, Busbee TA, Homan KA, Lewis JA (2014) 3D bioprinting of vascularized, heterogeneous cell-laden tissue constructs. Adv Mater 26(19):3124–3130

48. Morin KT, Dries-Devlin JL, Tranquillo RT (2014) Engineered microvessels with strong alignment and high lumen density via cell-induced fibrin gel compaction and interstitial flow. Tissue Eng Part A 20(3–4):553–565

49. Moya ML, Hsu YH, Lee AP, Hughes CC, George SC (2013) *In vitro* perfused human capillary networks. Tissue Eng Part C Methods 19(9):730–737

50. Raghavan S, Nelson CM, Baranski JD, Lim E, Chen CS (2010) Geometrically controlled endothelial tubulogenesis in micropatterned gels. Tissue Eng Part A 16(7):2255–2263

51. Vantler M, Karikkineth BC, Naito H, Tiburcy M, Didie M, Nose M, Rosenkranz S, Zimmermann WH (2010) PDGF-BB protects cardiomyocytes from apoptosis and improves contractile function of engineered heart tissue. J Mol Cell Cardiol 48(6):1316–1323

52. Morgan KY, Black LD (2014) Mimicking isovolumic contraction with combined electromechanical stimulation improves the development of engineered cardiac constructs. Tissue Eng Part A. 20(11–12):1654–1667. doi:10.1089/ten.TEA.2013.0355

53. Barash Y, Dvir T, Tandeitnik P, Ruvinov E, Guterman H, Cohen S (2011) Electric field stimulation integrated into perfusion bioreactor for cardiac tissue engineering. Tissue Eng Part C Methods 16(6):1417–1426

54. Shapira-Schweitzer K, Habib M, Gepstein L, Seliktar D (2009) A photopolymerizable hydrogel for 3-D culture of human embryonic stem cell-derived cardiomyocytes and rat neonatal cardiac cells. J Mol Cell Cardiol 46(2):213–224
55. Radisic M PD, Fast VG PhD, Sharifov OF PhD, Iyer RK BASc, Park H PhD, Vunjak-Novakovic G PhD (2009) Optical mapping of impulse propagation in engineered cardiac tissue. Tissue Eng Part A 15(4):851–860
56. Herron TJ, Lee P, Jalife J (2012) Optical imaging of voltage and calcium in cardiac cells & tissues. Circ Res 110(4):609–623
57. Sekine H, Shimizu T, Dobashi I, Matsuura K, Hagiwara N, Takahashi M, Kobayashi E, Yamato M, Okano T (2011) Cardiac cell sheet transplantation improves damaged heart function via superior cell survival in comparison with dissociated cell injection. Tissue Eng Part A 17(23–24):2973–2980
58. Mihic A, Li J, Miyagi Y, Gagliardi M, Li SH, Zu J, Weisel RD, Keller G, Li RK (2014) The effect of cyclic stretch on maturation and 3D tissue formation of human embryonic stem cell-derived cardiomyocytes. Biomaterials 35(9):2798–2808
59. Kawamura M, Miyagawa S, Miki K, Saito A, Fukushima S, Higuchi T, Kawamura T, Kuratani T, Daimon T, Shimizu T, Okano T, Sawa Y (2012) Feasibility, safety, and therapeutic efficacy of human induced pluripotent stem cell-derived cardiomyocyte sheets in a porcine ischemic cardiomyopathy model. Circulation 126:S29–S37
60. Hasenfuss G, Mulieri LA, Blanchard EM, Holubarsch C, Leavitt BJ, Ittleman F, Alpert NR (1991) Energetics of isometric force development in control and volume-overload human myocardium. Comparison with animal species. Circ Res 68(3):836–846

Mechanisms of Cardiac Valve Failure and the Development of Tissue Engineered Heart Valves

Meghana R. K. Helder and Robert D. Simari

Abstract Calcification of the aortic valve results in valvular dysfunction and is an important cause of morbidity and mortality. Our understanding of the process of aortic valve calcification has changed from a passive wear and tear process to that of an actively regulated process with known molecular mediators. Prior to calcification of the valve, activated valvular interstitial cells coordinate maladaptive extracellular matrix remodeling of the leaflets. Bioprosthetic heart valves used to surgically replace stenotic aortic valves have excellent hemodynamic profiles and are anti-thrombogenic. However, they fail due to similar mechanisms as native aortic valves and thus durability is a limiting factor. Mechanical prostheses necessitate anticoagulation. Ideal heart valve substitutes would be non-thrombogenic, maintain excellent hemodynamics, but would be durable and may hold the promise of growth. The basics of tissue engineering include fabricating a scaffold onto which autologous cells may be incorporated. The cells then transform the scaffold to autologous tissue with the ability to function in its desired location in the body. Popular scaffolds are decellularized allografts or xenogeneic aortic valves as they have the complex structure of the aortic valve still intact. Recellularization with valvular endothelial cells has been successful but avenues for recellularizing valvular interstitial cells are still being pursued. Alternative methods for generating scaffolds include three dimensional bioprinting and electrospinning.

Keywords Tissue engineering · Aortic valve stenosis · Decellularization

R. D. Simari (✉)
Executive Dean School of Medicine, University of Kansas School of Medicine,
3901 Rainbow Boulevard, Kansas City, KS 66160, USA
e-mail: rsimari@kumc.edu

M. R. K. Helder
Divisions of Cardiovascular Surgery and Diseases, Mayo Clinic, Rochester, MN, USA

© Springer International Publishing Switzerland 2015
I.M.C. Dixon, J. T. Wigle (eds.), *Cardiac Fibrosis and Heart Failure: Cause or Effect?*,
Advances in Biochemistry in Health and Disease 13, DOI 10.1007/978-3-319-17437-2_21

1 Native Aortic Valve Stenosis Mechanisms

Aortic valve sclerosis and stenosis affects up to 26 % of patients over the age of 65 in the United States [1]. This process is characterized by a fibrous thickening of the valve cusps followed by areas of calcification as the process becomes more severe [2]. Much research has focused on revising the theory that aortic valve stenosis (AS) is a degenerative wear-and-tear process that occurs with aging, but is in-fact is an actively regulated process of osteoblast and osteoclast-like cells [3]. The calcification process of AS has been described well, but extracellular matrix (ECM) remodeling likely occurs prior to and contributes to calcification of the valve [2].

The earliest initiating factor in AS is endothelial dysfunction and disruption that then triggers a cascade [4] attracting a multitude of inflammatory cells to the area. Monocytes are attracted and enter the sub endothelial layer through the disrupted endothelial layer with the aid of adhesion molecules [5]. Patients with AS show increased levels of soluble E-selectin and cellular adhesion molecules which are markers of endothelial activation and evidence of monocyte infiltration [6]. T lymphocytes are also attracted to the area of injury. They become activated within the subendothelium and release transforming growth factor $\beta 1$ (TGF $\beta 1$) and interleukin-1β [7]. TGF $\beta 1$ has a large role in activating valvular interstitial cells (VICs) [8].

TGF $\beta 1$ has been shown to regulate VICs, not only in terms of turning VICs into osteoblast-like cells but also in regards to ECM remodeling. When VICs are cultured in the presence of TGF $\beta 1$, α-smooth muscle actin (α-SMA) fibers align and pull away from the rigid plastic surface displaying activation and myofibroblast type of activity. TGF $\beta 1$ has also been shown to inhibit VIC proliferation and apoptosis, [8]. Thus, activated VICs are left to continue the pro-inflammatory milieu.

Once VICs are activated, they become myofibroblast-like and express α-SMA [9]. Activated VICs produce pro-inflammatory mediators and toll-like receptors, particularly 2 and 4, have been shown to regulate the production of these mediators *in vitro* [10]. Pro-inflammatory matrix metalloproteinases (MMPs) remodel the ECM of the valve, which results in a more fibrotic valve [11]. Matrix metalloproteinase-3 has been found to be significantly higher in patients in AS than in patients with normal aortic valves. Tissue inhibitor of metalloproteinases-1 activity was also seven times higher in patients with AS compared to patients with normal aortic valves. Both these findings imply increased remodeling of the ECM that is perhaps dysregulated [12]. Molecular studies show a complex interplay between MMPs, tumor necrosis factor-α, interleukin-1β, and RANKL in regulating the process of valve fibrosis and degeneration by promoting cell proliferation and remodeling [13], [14]. Once the aortic valve is calcified, patients typically experience symptoms of angina, fatigue, and shortness of breath. When symptoms are present, average survival is 2 to 3 years with a high risk of sudden death. Aortic valve replacement is indicated in patients with symptomatic aortic valve stenosis [15] and is the best chance for increased survival.

Currently available valve replacements can be classified as mechanical or bioprosthetic. Mechanical valves can last a lifetime, but currently require chronic anticoagulation with warfarin. Bioprosthetic heart valves (BHVs) have the advantage of excellent hemodynamics [16] and avoidance of anticoagulation and are preferable in patients over the age of 65, the majority of patients requiring an aortic valve replacement [17]. Bioprosthetic valves can be autografts (Ross Procedure), allografts, or xenografts. Allografts are cadaveric aortic valves that are cryopreserved. Xenografts are either porcine or bovine in nature and are glutaraldehyde fixed.

2   Bioprosthetic Valve Failure Mechanisms

Twenty to thirty percent of most BHVs fail within 12 to 15 years post-implantation [18]. The methods of failure are still not completely understood but degradation of the ECM and calcification are noted. Studies show degradation of glycosaminoglycans and collagen in failing BHVs [19]. However, there must be some external stimuli that cause BHVs to deteriorate when normal aortic valves not subjected to the above discussed mechanisms do not just "wear out." Glutaraldehyde maybe one of these inciting factors. Fixation with glutaraldehyde is conducted to protect the xenograft from the human immune system. However cells and ECM proteins are devitalized secondary to this fixation and become more phosphorus-rich structures that are sites for calcium infiltration and nodule formation [20]. Also the amount of cross-linking promoted by glutaraldehyde may determine tissue stability and calcification [21].

Even though glutaraldehyde fixation is meant to protect the BHV from immune detection, it may not be completely effective in accomplishing this task. The risk of tissue degeneration is higher in patients <35 years of age [22] and is attributed to an increased immune competence in younger patients. Humoral immune components trigger complement-mediated destruction of BHVs. Moczar et al. found evidence of IgG and complement proteins on BHVs in ventricular support systems implanted in younger patients [23] showing that antibodies are attracted to antigens on the BHV. There has also been evidence of increased T cells, macrophages, and antibody levels in animals receiving xenogeneic valves as compared to the animals receiving allogeneic valves [24].

The largest stimulus for immune activation may be xenoantibodies. Eighty percent of xenoantibodies are against the antigen, galactose-galactose-galactose-α-1,3-glactose β-1,4-N-acetylglucosamine (α-Gal). This antigen is added on to most xenogeneic proteins during post-translational modification and is prevalent throughout all tissues that are not from humans or non-human primates. Current BHVs express α-Gal antigens on the surface [25] even with glutaraldehyde fixation. Also patients who have had BHVs implanted show higher levels of antibodies to α-Gal after operation [26], again even with glutaraldehyde fixation. Thus, immune-mediate mechanisms definitely contribute to the failure of BHVs that includes a combination of calcification and ECM remodeling [27].

3   Novel Methods of Creating Bioprosthetic Heart Valves

Bioprosthetic heart valves are meant to mimic the native aortic valve. An ideal aortic valve substitute would mimic the native valve completely. Current BHVs maintain the anti-thrombogenicity and hemodynamic profile of native aortic valves and these characteristics should be incorporated. However, areas of improvement from current BHVs include complete prevention of immune system activation and the capacity to grow. A promising method of achieving this is through tissue engineering.

Tissue engineering in its most basic form is taking a scaffold made of a biocompatible material that can be populated with the patient's own cells and then is maintained or changed by those cells to achieve its intended structure and function. Each level of tissue engineering a valve has many options and an overview is presented here.

Decellularization of allogeneic and xenogeneic valves has been used to generate scaffolds for tissue engineering. Decellularization affords the use of a pre-existing complex scaffold that maintains native valve structure. The process of decellularization rids the valve of all DNA and cellular material. This theoretically abolishes the need for fixation and reduces immune response [28], [29]. Decellularization does preserve, grossly, the integrity of the ECM and thus the strength and elasticity needed for the valve to function and be durable [30]. However, microscopic examination of the ECM integrity does show some differences to the native valve [31], but this difference may not be functionally relevant.

While it is the intent of decellularization to remove all cellular and DNA material, ECM proteins are still left behind. The α-gal antigen, as described above, is a glycopeptide that is ubiquitous in porcine tissue as it is a common post-translational modification step in most porcine proteins. Decellularization has been found to remove α-gal from valves [32]. Addition of the enzyme alpha-galactosidase may also be used for targeted removal of the α-gal antigen [33]. Pigs in which α-gal has been genetically deleted have been developed and could provide tissue for bioprosthetic valves without this xenoantigen. These adjuncts to decellularization represent progress in removing all the immune stimuli associated with xenogeneic materials for BHVs and thus reduce the impetus for calcification and subsequent failure.

Several human trials of decellularized valves have already been conducted. The least risky avenue to introduce this approach was with the use allografts. Allografts are already accepted aortic valve substitutes and any decellularization process may improve the valve with minimal adverse effects.

The Synergraft™ by Cryolife was the first decellularized human aortic valve allograft available. O'Brien et al. performed the first experimental studies in sheep and showed promising results [34]. Elkins et al. showed decreased staining for MHA class I and II proteins along with decreased panel reactive antibodies in patients implanted with decellularized pulmonary allografts [35]. The Mayo Clinic published our cohort of 22 patients in whom the Synergraft™ was implanted in the aortic position and had good short-term results [36]. Mid-term survival is promising as well. At a mean follow-up of 19 months, there were no aortic valve re-interventions [37] and a follow-up of 52 months, there was still no need for reoperations in patients with Synergraft™ allografts [38].

Human pulmonary allografts reseeded with autologous peripheral mononuclear cells were implanted into two pediatric patients. There were no signs of valve degeneration during 3.5 years of follow up in either child [39].

The limited supply of allografts forces us to look to xenografts for possible aortic valve substitutes. Xenografts were treated with the same decellularization protocol and four xeno-Synergrafts™ were implanted in the right ventricular outflow tract of children requiring reconstruction and limited other options. The results were dismal. Three patients died between postoperative day 7 and 1 year. In the fourth patient, the graft was explanted on postoperative day 2 because of the known graft failure in the 3rd patient [40]. In examination of the failed grafts, incomplete decellularization of the xenogeneic valves was evident and immune-mediated failure was likely the culprit[41]. The consequences of incomplete decellularization are more dire in xenogeneic valves than in allogeneic valves because of the degree of immune reaction elicited.

Besides these initial clinical applications, decellularized xenograft valves have remained in the experimental realm. Multiple animal studies have been conducted that have taken decellularized aortic valves and implanted them into juvenile sheep either in the aortic or pulmonary position [42–46]. In all of these studies, a confluent endothelial cell layer has been noted at explantation with good hemodynamic function and no thromboembolic complications of note.

Even though decellularization of animal aortic valves has been the most popular route for creating a natural scaffold, other groups have tackled this task by creating novel tissue. Syedain et al. created a scaffold utilizing fibroblasts in a fibrin gel and then decellularizing the resulting tissue to create a decellularized tube. This tube is placed over a trident stent and the result is a functional valve with short term function in a sheep pulmonary artery [47]. A similar approach was taken by Dijkman et al., who utilized biodegradable polyglycolic mesh leaflets seeded with autologous bone marrow-derived mesenchymal stromal cells (MSCs) and then decellularized the valves. These valves avoid the risk of xenograft-specific immune responses because of the polymer based initial scaffolds [48]. Transcatheter implantation of these valves in the pulmonary position of sheep showed regurgitation and possible evidence of leaflet contraction [49].

Scaffolds, whether decellularized or created are meant to serve as sites for (re) cellularization of some variety, be it *in vitro* prior to implantion or *in-vivo*, relying on the patient to repopulate the scaffold after implantation. Along with discussing the options for recellularization, cell sources and cell types used for recellularization must also be discussed. There are two primary types of cells in a native human aortic valve; valvular endothelial cells (VECs) and VICs. Both serve unique functions that serve to extend valve durability.

Valvular endothelial cells have several physiologic roles both in regards to the circulation and valve tissue. A monolayer of endothelial cells lines all surfaces in the body that are in contact with blood. The monolayer serves a regulatory role in the coagulation cascade, allowing platelet adhesion only in times of injury [50]. In the vasculature, beyond coagulation and the resulting thrombosis, endothelial cells are well known to influence the size of a blood vessel by releasing proteins such as nitric oxide. However, not much is known about the valvular endothelial cells in this regard. *In-vitro* studies have shown that VECs respond to vasoactive agents and thus may have an effect on the morphology of the valve itself [51]. Beyond morphologic and coagulation regulation, VECs seem to modulate VIC phenotype. In co-culture, VECs decrease the α-SMA expression from VICs. Thus, valvular endothelial cells are important to the function of the valves and do not serve just a protective role [52].

In-vivo implantation of decellularized aortic valves show regrowth of a mono-layer of endothelial cells at 5-months [43, 45]. Thus, there is proof of *in-vivo* recellularization with endothelial cells. The source of these cells may include neighboring endothelium or circulating sources including endothelial progenitor cells (EPCs). The debate of the benefits of *in vitro* versus *in vivo* endothelial recel-lularization continues. No study to date has pinpointed the time it takes to repopulate a scaffold *in-vivo* or the origin of the endothelial cells that do repopulate the scaffold. Important questions remain however regarding the translation of what is seen in the animal model to the human patient. There is a large decrease in the number of EPCs in the circulation of patients with aortic stenosis [53]. Thus, the circulatory cells of a juvenile sheep are much different than those of an elderly patient with aortic stenosis.

Recellularization *in-vitro* allows for autologous seeding with minimally invasive techniques. Endothelial outgrowth cell (EOCs) have been extensively character-ized and have great proliferative potential can be derived from venous blood with a single needle stick [54]. However, no studies to date have sought to compare the benefits of *in-vitro* recellularization versus *in-vivo* recellularization directly.

Valvular interstitial cells are typically identified as the cells that maintain the ECM of the valve cusp, but have many functions and forms beyond this mainte-nance role. Most of the time, VICs in the valve cusp are quiescent and focus on maintaining valve structure and function. Progenitor VICs (pVICs) are found in the bone marrow and circulation but also some are likely residents in the valve itself [9]. These cells are precursors to activated VICs (aVICs), which as discussed above, are the progenitors of maladaptive remodeling (Fig. 1).

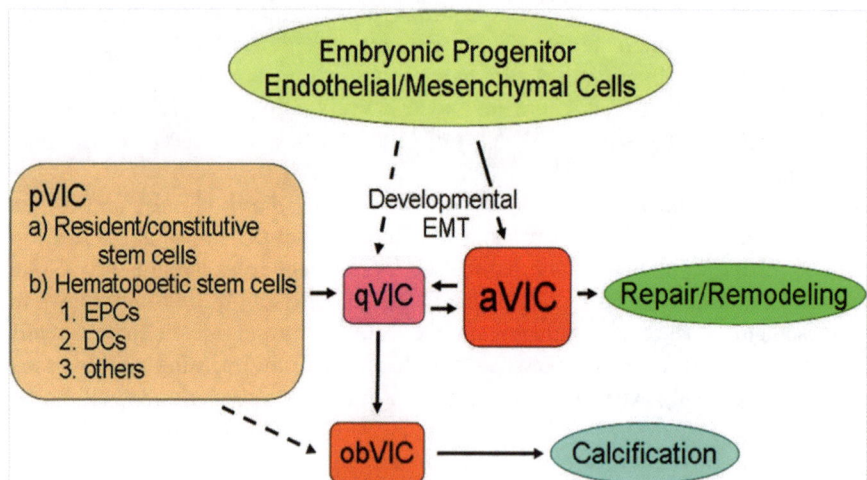

Fig. 1 VICs exist in different configurations within the valve. pVICs are the resident form that give rise to qVICs. qVICs can become activated and lead to repair and remodeling of the valve. qVICs can also be differentiated into obVICs which can lead to calcification of the valve. From Liu et al. [9], permission to reuse obtained

Decellularized porcine aortic valves have been recellularized with VICs *in-vitro* by seeding with the aid of fibronectin. In culture, VICs seem to differentiate into myofibroblasts, fibroblasts, smooth muscle cells, and even endothelial cells [55]. It will be important to characterize the cellular structure of recellularized valves prior to implantation for *in vivo* studies. No clinically viable sources of VICs that can be used for *in vitro* recellularization have been identified to date.

For this reason, bone marrow-derived mesenchymal stromal cells (MSCs) have been utilized for recellularization thus far in animal studies. Bone marrow derived mesenchymal stem cells were injected into the decellularized porcine pulmonary valve and then implanted into the pulmonary position in a lamb. In this study, they saw complete re-endothelialization at 4 months along with very few inflammatory cells and ECM remodeling [56]. *In vitro* recellularization with MSCs for 30 days in static culture resulted in an endothelial layer along with the presence of smooth muscle cells, osteogenic and adipogenic cells in the valve itself [57]. Even though smooth muscle and osteogenic cell types are present at times in the native leaflet, they are often present in pathologic conditions. Thus, the multipotent differentiation potential of MSCs might be detrimental when implanted into a pathologic milieu of the elderly patient that already produced stenosis of their native valve.

4   Alternative Methods of Creating Scaffolds

The natural scaffold of a decellularized xenograft or allograft still may carry the possibility of an inflammatory response. A synthetic biodegradable scaffold seeded with autologous cells may provide a platform to create autologous implantable tissue, without worry of an inflammatory response. Mayer et al. were the first to replace a valve leaflet in a lamb with a polymer scaffold seeded with autologous endothelial cells and fibroblasts [58]. The same group constructed a tri-leaflet heart valve in much the same way which demonstrated good hemodynamic function in the pulmonary position in a juvenile sheep at 120 days [59]. Tri-leaflet heart valves scaffolds were made utilizing biodegradable Fastacryl stent material and leaflets were cut out of polyglycolic acid meshes [60]. These synthetic heart valves were then seeded with cells from human saphenous vein and collagen and glycosaminoglycan content was seen to increase up to 3 weeks. The strength of the tissue was also shown to be increased in the synthetic heart valves seeded with cells compared to those that were not seeded with cells [61].

Other technologies have been implemented in developing a tissue-engineered aortic valve substitutes. These methods have yet to reach animal experimental stages but are promising venues. Duan et al. have been able to utilize a 3D bioplotter to print an aortic valve (Fig. 2). 3D bioplotters have the ability to print materials based on a computer aided drafting file much in the same way as an inkjet printer except by introducing another dimension. The difference between a 3D printer and a 3D bioplotter is that cells can be printed with materials in a 3D bioplotter so that the recellularization step is done in conjunction with creating the scaffold. Also, the

Fig. 2 Photograph of an
aortic valve printed on a
bioplotter by Duan et al. [62],
permission to reuse obtained

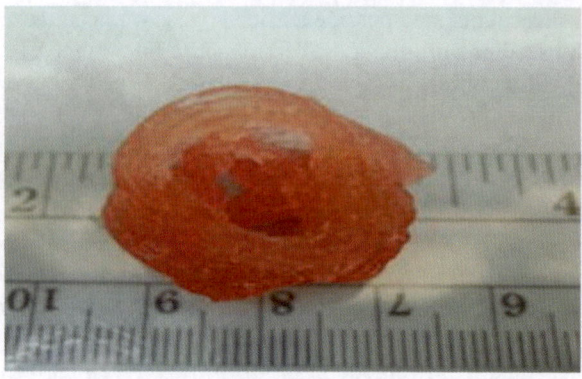

entire process is done in a sterile manner. Duan et al. used a combination of alginate
and hydrogel preparations to create the scaffold and utilized VICs for the cellular
material. The valves imprinted with VICs showed an increase in type I collagen
after 7 days of incubation, showing that the cells are laying down their own ECM.
These results show that 3D printing of aortic valves can be accomplished, but much
ground must be covered before *in-vivo* testing can be attained [62].

Another methodology of creating a tissue engineered valve is electrospinning.
Electrospinning is a process of using an electrical charge to draw nano or micro
scale fibers from a specified liquid. The fibers are collected on a charged scaffold
and the scaffold is created for the particular function. This process has been shown
to produce a functional valve [63]. The process was modified further to produce
a hybrid design. Two polymers were utilized to electrospin a heart valve scaffold
that was seeded with VICs and layered with VECs. This leaflet was tested in conjunction with fresh porcine leaflets in a bioreactor which provides a simulation of
human pressures and flows. In bioreactor stimulation and testing, the electrospun
leaflet opened and closed in step with native leaflets [64] (Fig. 3). Again, advances
are necessary before *in vivo* experimentation with electrospun valves, but this process has the unique advantage of being able to align cells in the ways that mimic
the native aortic valve and being able to duplicate the microstructure of the valve.

5 Conclusions

Many viable methods exist to create an ideal heart valve substitute that is not only
durable but may also not require anticoagulation. Some tissue-engineering methods such as decellularization of allografts have already been implemented in clinical practice and the clinical outcomes of these methodological changes are being
evaluated. Extensive animal testing has already been conducted on decellularized
xenogeneic scaffolds and results are encouraging. However with devastating first
clinical trials, it is appropriate and prudent to tread carefully before reintroducing
this technology into the clinical realms.

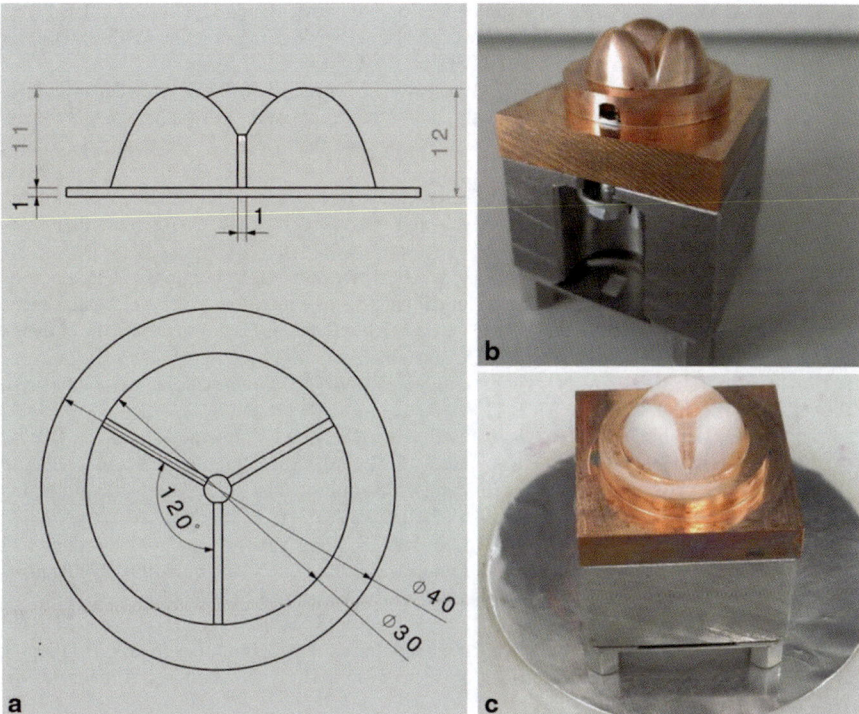

Fig. 3 Electrospinning of valve includes design of a mold using computer aided drafting (**a**) that is then utilized to capture the synthetic fibers (**b**). The mold with a partially electrospun valve is shown by Hinderer et al. [64], permission to reuse obtained

References

1. Otto CM, Lind BK, Kitzman DW, Gersh BJ, Siscovick DS (1999) Association of aortic-valve sclerosis with cardiovascular mortality and morbidity in the elderly. N Engl J Med 341(3):142–147. doi:10.1056/NEJM199907153410302
2. Kaden JJ, Dempfle CE, Grobholz R, Fischer CS, Vocke DC, Kilic R, Sarikoc A, Pinol R, Hagl S, Lang S, Brueckmann M, Borggrefe M (2005) Inflammatory regulation of extracellular matrix remodeling in calcific aortic valve stenosis. Cardiovasc Pathol 14(2):80–87. doi:10.1016/j.carpath.2005.01.002
3. Miller JD, Weiss RM, Heistad DD (2011) Calcific aortic valve stenosis: methods, models, and mechanisms. Circ Res 108(11):1392–1412. doi:10.1161/CIRCRESAHA.110.234138
4. Freeman RV, Otto CM (2005) Spectrum of calcific aortic valve disease: pathogenesis, disease progression, and treatment strategies. Circulation 111(24):3316–3326. doi:10.1161/CIRCULATIONAHA.104.486738
5. Jian B, Narula N, Li QY, Mohler ER, 3rd, Levy RJ (2003) Progression of aortic valve stenosis: TGF-beta1 is present in calcified aortic valve cusps and promotes aortic valve interstitial cell calcification via apoptosis. Ann Thorac Surg 75(2):457–465; (discussion 465–456)
6. Ghaisas NK, Foley JB, O'Briain DS, Crean P, Kelleher D, Walsh M (2000) Adhesion molecules in nonrheumatic aortic valve disease: endothelial expression, serum levels and effects of valve replacement. J Am Coll Cardiol 36(7):2257–2262

7. Olsson M, Dalsgaard CJ, Haegerstrand A, Rosenqvist M, Ryden L, Nilsson J (1994) Accumulation of T lymphocytes and expression of interleukin-2 receptors in nonrheumatic stenotic aortic valves. J Am Coll Cardiol 23(5):1162–1170
8. Walker GA, Masters KS, Shah DN, Anseth KS, Leinwand LA (2004) Valvular myofibroblast activation by transforming growth factor-beta: implications for pathological extracellular matrix remodeling in heart valve disease. Circ Res 95(3):253–260. doi:10.1161/01. RES.0000136520.07995.aa
9. Liu AC, Joag VR, Gotlieb AI (2007) The emerging role of valve interstitial cell phenotypes in regulating heart valve pathobiology. Am J Pathol 171(5):1407–1418. doi:10.2353/ajpath.2007.070251
10. Meng X, Ao L, Song Y, Babu A, Yang X, Wang M, Weyant MJ, Dinarello CA, Cleveland JC Jr, Fullerton DA (2008) Expression of functional Toll-like receptors 2 and 4 in human aortic valve interstitial cells: potential roles in aortic valve inflammation and stenosis. Am J Physiol Cell Physiol 294(1):C29–35. doi:10.1152/ajpcell.00137.2007
11. Taylor PM, Batten P, Brand NJ, Thomas PS, Yacoub MH (2003) The cardiac valve interstitial cell. Int J Biochem Cell Biol 35(2):113–118
12. Fondard O, Detaint D, Iung B, Choqueux C, Adle-Biassette H, Jarraya M, Hvass U, Couetil JP, Henin D, Michel JB, Vahanian A, Jacob MP (2005) Extracellular matrix remodelling in human aortic valve disease: the role of matrix metalloproteinases and their tissue inhibitors. Eur Heart J 26(13):1333–1341. doi:10.1093/eurheartj/ehi248
13. Kaden JJ, Dempfle CE, Kilic R, Sarikoc A, Hagl S, Lang S, Brueckmann M, Borggrefe M (2005) Influence of receptor activator of nuclear factor kappa B on human aortic valve myofibroblasts. Exp Mol Pathol 78(1):36–40. doi:10.1016/j.yexmp.2004.09.001
14. Kaden JJ, Dempfle CE, Grobholz R, Tran HT, Kilic R, Sarikoc A, Brueckmann M, Vahl C, Hagl S, Haase KK, Borggrefe M (2003) Interleukin-1 beta promotes matrix metalloproteinase expression and cell proliferation in calcific aortic valve stenosis. Atherosclerosis 170(2):205–211
15. Bonow RO, Carabello BA, Chatterjee K, de Leon AC Jr, Faxon DP, Freed MD, Gaasch WH, Lytle BW, Nishimura RA, O'Gara PT, O'Rourke RA, Otto CM, Shah PM, Shanewise JS (2008) 2008 focused update incorporated into the ACC/AHA 2006 guidelines for the management of patients with valvular heart disease: a report of the American College of Cardiology/ American Heart Association Task Force on Practice Guidelines (Writing Committee to revise the 1998 guidelines for the management of patients with valvular heart disease). Endorsed by the Society of Cardiovascular Anesthesiologists, Society for Cardiovascular Angiography and Interventions, and Society of Thoracic Surgeons. J Am Coll Cardiol 52(13):e1–142. doi:10.1016/j.jacc.2008.05.007
16. Nollert G, Miksch J, Kreuzer E, Reichart B (2003) Risk factors for atherosclerosis and the degeneration of pericardial valves after aortic valve replacement. J Thorac Cardiovasc Surg 126(4):965–968. doi:10.1016/S0022
17. Pibarot P, Dumesnil JG (2009) Prosthetic heart valves: selection of the optimal prosthesis and long-term management. Circulation 119(7):1034–1048. doi:10.1161/CIRCULATIONAHA.108.778886
18. Jamieson WR, Munro AI, Miyagishima RT, Allen P, Burr LH, Tyers GF (1995) Carpentier-Edwards standard porcine bioprosthesis: clinical performance to seventeen years. Ann Thorac Surg 60(4):999–1006; discussion 1007
19. Vyavahare N, Ogle M, Schoen FJ, Zand R, Gloeckner DC, Sacks M, Levy RJ (1999) Mechanisms of bioprosthetic heart valve failure: fatigue causes collagen denaturation and glycosaminoglycan loss. J Biomed Mater Res 46(1):44–50
20. Chen W, Schoen FJ, Levy RJ (1994) Mechanism of efficacy of 2-amino oleic acid for inhibition of calcification of glutaraldehyde-pretreated porcine bioprosthetic heart valves. Circulation 90(1):323–329
21. Golomb G, Schoen FJ, Smith MS, Linden J, Dixon M, Levy RJ (1987) The role of glutaraldehyde-induced cross-links in calcification of bovine pericardium used in cardiac valve bioprostheses. Am J Pathol 127(1):122–130

22. Cohn LH, Collins JJ Jr, DiSesa VJ, Couper GS, Peigh PS, Kowalker W, Allred E (1989) Fifteen-year experience with 1678 Hancock porcine bioprosthetic heart valve replacements. Ann Surg 210(4):435–442; (discussion 442–433)

23. Moczar M, Houel R, Ginat M, Clerin V, Wheeldon D, Loisance D (2000) Structural changes in porcine bioprosthetic valves of a left ventricular assist system in human patients. J Heart Valve Dis 9(1):88–95; (discussion 95–86)

24. Manji RA, Zhu LF, Nijjar NK, Rayner DC, Korbutt GS, Churchill TA, Rajotte RV, Koshal A, Ross DB (2006) Glutaraldehyde-fixed bioprosthetic heart valve conduits calcify and fail from xenograft rejection. Circulation 114(4):318–327. doi:10.1161/CIRCULATIONAHA.105.549311

25. McGregor CG, Carpentier A, Lila N, Logan JS, Byrne GW (2011) Cardiac xenotransplantation technology provides materials for improved bioprosthetic heart valves. J Thorac Cardiovasc Surg 141(1):269–275. doi:10.1016/j.jtcvs.2010.08.064

26. Konakci KZ, Bohle B, Blumer R, Hoetzenecker W, Roth G, Moser B, Boltz-Nitulescu G, Gorlitzer M, Klepetko W, Wolner E, Ankersmit HJ (2005) Alpha-Gal on bioprostheses: xenograft immune response in cardiac surgery. Eur J Clin Invest 35(1):17–23. doi:10.1111/j.1365-2362.2005.01441.x

27. Srivatsa SS, Harrity PJ, Maercklein PB, Kleppe L, Veinot J, Edwards WD, Johnson CM, Fitzpatrick LA (1997) Increased cellular expression of matrix proteins that regulate mineralization is associated with calcification of native human and porcine xenograft bioprosthetic heart valves. J Clin Invest 99(5):996–1009. doi:10.1172/JCI119265

28. Grauss RW, Hazekamp MG, van Vliet S, Gittenberger-de Groot AC, DeRuiter MC (2003) Decellularization of rat aortic valve allografts reduces leaflet destruction and extracellular matrix remodeling. J Thorac Cardiovasc Surg 126(6):2003–2010. doi:10.1016/S0022

29. Meyer SR, Nagendran J, Desai LS, Rayat GR, Churchill TA, Anderson CC, Rajotte RV, Lakey JR, Ross DB (2005) Decellularization reduces the immune response to aortic valve allografts in the rat. J Thorac Cardiovasc Surg 130(2):469–476. doi:10.1016/j.jtcvs.2005.03.021

30. Korossis SA, Wilcox HE, Watterson KG, Kearney JN, Ingham E, Fisher J (2005) In-vitro assessment of the functional performance of the decellularized intact porcine aortic root. J Heart Valve Dis 14(3):408–421; (discussion 422)

31. Liao J, Joyce EM, Sacks MS (2008) Effects of decellularization on the mechanical and structural properties of the porcine aortic valve leaflet. Biomaterials 29(8):1065–1074. doi:10.1016/j.biomaterials.2007.11.007

32. Naso F, Gandaglia A, Bottio T, Tarzia V, Nottle MB, d'Apice AJ, Cowan PJ, Cozzi E, Galli C, Lagutina I, Lazzari G, Iop L, Spina M, Gerosa G (2013) First quantification of alpha-Gal epitope in current glutaraldehyde-fixed heart valve bioprostheses. Xenotransplantation 20(4):252–261. doi:10.1111/xen.12044

33. Goncalves AC, Griffiths LG, Anthony RV, Orton EC (2005) Decellularization of bovine pericardium for tissue-engineering by targeted removal of xenoantigens. J Heart Valve Dis 14(2):212–217

34. O'Brien MF, Goldstein S, Walsh S, Black KS, Elkins R, Clarke D (1999) The SynerGraft valve: a new acellular (nonglutaraldehyde-fixed) tissue heart valve for autologous recellularization first experimental studies before clinical implantation. Semin Thorac Cardiovasc Surg 11(4 Suppl 1):194–200

35. Elkins RC, Dawson PE, Goldstein S, Walsh SP, Black KS (2001) Decellularized human valve allografts. Ann Thorac Surg 71(5 Suppl):S428–S432

36. Zehr KJ, Yagubyan M, Connolly HM, Nelson SM, Schaff HV (2005) Aortic root replacement with a novel decellularized cryopreserved aortic homograft: postoperative immunoreactivity and early results. J Thorac Cardiovasc Surg 130(4):1010–1015. doi:10.1016/j.jtcvs.2005.03.044

37. da Costa FD, Costa AC, Prestes R, Domanski AC, Balbi EM, Ferreira AD, Lopes SV (2010) The early and midterm function of decellularized aortic valve allografts. Ann Thorac Surg 90(6):1854–1860. doi:10.1016/j.athoracsur.2010.08.022

38. Bechtel JF, Stierle U, Sievers HH (2008) Fifty-two months' mean follow up of decellularized SynerGraft-treated pulmonary valve allografts. J Heart Valve Dis 17(1):98–104; (discussion 104)

39. Cebotari S, Lichtenberg A, Tudorache I, Hilfiker A, Mertsching H, Leyh R, Breymann T, Kallenbach K, Maniuc L, Batrinac A, Repin O, Maliga O, Ciubotaru A, Haverich A (2006) Clinical application of tissue engineered human heart valves using autologous progenitor cells. Circulation 114(1 Suppl):I132–137. doi:10.1161/CIRCULATIONAHA.105.001065

40. Simon P, Kasimir MT, Seebacher G, Weigel G, Ullrich R, Salzer-Muhar U, Rieder E, Wolner E (2003) Early failure of the tissue engineered porcine heart valve SYNERGRAFT in pediatric patients. Eur J Cardiothorac Surg 23(6):1002–1006; (discussion 1006)

41. Kasimir MT, Rieder E, Seebacher G, Nigisch A, Dekan B, Wolner E, Weigel G, Simon P (2006) Decellularization does not eliminate thrombogenicity and inflammatory stimulation in tissue-engineered porcine heart valves. J Heart Valve Dis 15(2):278–286

42. Steinhoff G, Stock U, Karim N, Mertsching H, Timke A, Meliss RR, Pethig K, Haverich A, Bader A (2000) Tissue engineering of pulmonary heart valves on allogenic acellular matrix conduits: in vivo restoration of valve tissue. Circulation 102(19 Suppl 3):III50–III55

43. Baraki H, Tudorache I, Braun M, Hoffler K, Gorler A, Lichtenberg A, Bara C, Calistru A, Brandes G, Hewicker-Trautwein M, Hilfiker A, Haverich A, Cebotari S (2009) Orthotopic replacement of the aortic valve with decellularized allograft in a sheep model. Biomaterials 30(31):6240–6246. doi:10.1016/j.biomaterials.2009.07.068

44. Dohmen PM, Costa F, Lopes SV, Yoshi S, Souza FP, Vilani R, Costa MB, Konertz W (2005) Results of a decellularized porcine heart valve implanted into the juvenile sheep model. Heart Surg Forum 8(2):E100–E104; discussion E104. doi:10.1532/HSF98.20041140

45. Akhyari P, Kamiya H, Gwanmesia P, Aubin H, Tschierschke R, Hoffmann S, Karck M, Lichtenberg A (2010) In vivo functional performance and structural maturation of decellularised allogenic aortic valves in the subcoronary position. Eur J Cardiothorac Surg 38(5):539–546. doi:10.1016/j.ejcts.2010.03.024

46. Tudorache I, Calistru A, Baraki H, Meyer T, Hoffler K, Sarikouch S, Bara C, Gorler A, Hartung D, Hilfiker A, Haverich A, Cebotari S (2013) Orthotopic replacement of aortic heart valves with tissue-engineered grafts. Tissue Eng Part A 19(15–16):1686–1694. doi:10.1089/ten.TEA.2012.0074

47. Syedain ZH, Meier LA, Reimer JM, Tranquillo RT (2013) Tubular heart valves from decellularized engineered tissue. Ann Biomed Eng 41(12):2645–2654. doi:10.1007/s10439-013-0872-9

48. Dijkman PE, Driessen-Mol A, Frese L, Hoerstrup SP, Baaijens FP (2012) Decellularized homologous tissue-engineered heart valves as off-the-shelf alternatives to xeno- and homografts. Biomaterials 33(18):4545–4554. doi:10.1016/j.biomaterials.2012.03.015

49. Driessen-Mol A, Emmert MY, Dijkman PE, Frese L, Sanders B, Weber B, Cesarovic N, Sidler M, Leenders J, Jenni R, Grunenfelder J, Falk V, Baaijens FP, Hoerstrup SP (2013) Transcatheter implantation of homologous "off-the-shelf" tissue engineered heart valves with self-repair capacity: long term functionality and rapid in vivo remodeling in sheep. J Am Coll Cardiol. doi:10.1016/j.jacc.2013.09.082

50. Lee DJ, Steen J, Jordan JE, Kincaid EH, Kon ND, Atala A, Berry J, Yoo JJ (2009) Endothelialization of heart valve matrix using a computer-assisted pulsatile bioreactor. Tissue Eng Part A 15(4):807–814. doi:10.1089/ten.tea.2008.0250

51. Pompilio G, Rossoni G, Sala A, Polvani GL, Berti F, Dainese L, Porqueddu M, Biglioli P (1998) Endothelial-dependent dynamic and antithrombotic properties of porcine aortic and pulmonary valves. Ann Thorac Surg 65(4):986–992

52. Butcher JT, Nerem RM (2007) Valvular endothelial cells and the mechanoregulation of valvular pathology. Philos Trans R Soc Lond B Biol Sci 362(1484):1445–1457. doi:10.1098/rstb.2007.2127

53. Matsumoto Y, Adams V, Walther C, Kleinecke C, Brugger P, Linke A, Walther T, Mohr FW, Schuler G (2009) Reduced number and function of endothelial progenitor cells in patients

with aortic valve stenosis: a novel concept for valvular endothelial cell repair. Eur Heart J 30(3):346–355. doi:10.1093/eurheartj/ehn501

54. Lin Y, Weisdorf DJ, Solovey A, Hebbel RP (2000) Origins of circulating endothelial cells and endothelial outgrowth from blood. J Clin Invest 105(1):71–77. doi:10.1172/JCI8071

55. Bertipaglia B, Ortolani F, Petrelli L, Gerosa G, Spina M, Pauletto P, Casarotto D, Marchini M, Sartore S (2003) Cell characterization of porcine aortic valve and decellularized leaflets repopulated with aortic valve interstitial cells: the VESALIO Project (Vitalitate Exornatum Succedaneum Aorticum Labore Ingenioso Obtenibitur). Ann Thorac Surg 75(4):1274–1282

56. Vincentelli A, Wautot F, Juthier F, Fouquet O, Corseaux D, Marechaux S, Le Tourneau T, Fabre O, Susen S, Van Belle E, Mouquet F, Decoene C, Prat A, Jude B (2007) *In vivo* autologous recellularization of a tissue-engineered heart valve: are bone marrow mesenchymal stem cells the best candidates? J Thorac Cardiovasc Surg 134(2):424–432. doi:10.1016/j.jtcvs.2007.05.005

57. Iop L, Renier V, Naso F, Piccoli M, Bonetti A, Gandaglia A, Pozzobon M, Paolin A, Ortolani F, Marchini M, Spina M, De Coppi P, Sartore S, Gerosa G (2009) The influence of heart valve leaflet matrix characteristics on the interaction between human mesenchymal stem cells and decellularized scaffolds. Biomaterials 30(25):4104–4116. doi:10.1016/j.biomaterials.2009.04.031

58. Shinoka T, Breuer CK, Tanel RE, Zund G, Miura T, Ma PX, Langer R, Vacanti JP, Mayer JE Jr (1995) Tissue engineering heart valves: valve leaflet replacement study in a lamb model. Ann Thorac Surg 60(6 Suppl):S513–S516

59. Sodian R, Hoerstrup SP, Sperling JS, Daebritz S, Martin DP, Moran AM, Kim BS, Schoen FJ, Vacanti JP, Mayer JE Jr (2000) Early *in vivo* experience with tissue-engineered trileaflet heart valves. Circulation 102(19 Suppl 3):III22–III29

60. Mol A, Driessen NJ, Rutten MC, Hoerstrup SP, Bouten CV, Baaijens FP (2005) Tissue engineering of human heart valve leaflets: a novel bioreactor for a strain-based conditioning approach. Ann Biomed Eng 33(12):1778–1788. doi:10.1007/s10439-005-8025-4

61. Mol A, Rutten MC, Driessen NJ, Bouten CV, Zund G, Baaijens FP, Hoerstrup SP (2006) Autologous human tissue-engineered heart valves: prospects for systemic application. Circulation 114(1 Suppl):I152–I158. doi:10.1161/CIRCULATIONAHA.105.001123

62. Duan B, Hockaday LA, Kang KH, Butcher JT (2013) 3D bioprinting of heterogeneous aortic valve conduits with alginate/gelatin hydrogels. J Biomed Mater Res A 101(5):1255–1264. doi:10.1002/jbm.a.34420

63. Del Gaudio CB, A; Grigioni, M (2007) Electrospun bioresorbable trileaflet heart valve prosthesis for tissue engineering: *in vitro* functional assessment of a pulmonary cardiac valve design. Annali dell'Istituto superiore di sanita 44(2):178–186

64. Hinderer S, Seifert J, Votteler M, Shen N, Rheinlaender J, Schaffer TE, Schenke-Layland K (2014) Engineering of a bio-functionalized hybrid off-the-shelf heart valve. Biomaterials 35(7):2130–2139. doi:10.1016/j.biomaterials.2013.10.080

Index

© Springer International Publishing Switzerland 2015

433

I.M.C. Dixon, J. T. Wigle (eds.), *Cardiac Fibrosis and Heart Failure: Cause or Effect?,*
Advances in Biochemistry in Health and Disease 13, DOI 10.1007/978-3-319-17437-2

Printed by Printforce, the Netherlands